# Handbook of
# ELECTRONIC MESSAGING

*Edited by*
Nancy A. Cox

AUERBACH

**Library of Congress Cataloging-in-Publication Data**

Handbook of electronic messaging / Nancy A. Cox, editor.
   p. cm.
   Includes index.
   ISBN 0-8493-9946-7 (alk. paper)
   1. Telecommunication—Handbooks, manuals, etc. 2. Electronic mail
systems—Handbooks, manuals, etc.  I. Cox, Nancy, 1950–    .
TK5101.H284   1997
004.692—dc21
                                                                97-30660
                                                                    CIP

   This book contains information obtained from authentic and highly regarded sources. Reprinted material is quoted with permission, and sources are indicated. A wide variety of references are listed. Reasonable efforts have been made to publish reliable data and information, but the author and the publisher cannot assume responsibility for the validity of all materials or for the consequences of their use.

   Neither this book nor any part may be reproduced or transmitted in any form or by any means, electronic or mechanical, including photocopying, microfilming, and recording, or by any information storage or retrieval system, without prior permission in writing from the publisher.

   All rights reserved. Authorization to photocopy items for internal or personal use, or the personal or internal use of specific clients, may be granted by CRC Press LLC, provided that $.50 per page photocopied is paid directly to Copyright Clearance Center, 27 Congress Street, Salem, MA 01970 USA. The fee code for users of the Transactional Reporting Service is ISBN 0-8493-9946-7/98/$0.00+$.50. The fee is subject to change without notice. For organizations that have been granted a photocopy license by the CCC, a separate system of payment has been arranged.

   The consent of CRC Press LLC does not extend to copying for general distribution, for promotion, for creating new works, or for resale. Specific permission must be obtained in writing from CRC Press LLC for such copying.

   Direct all inquiries to CRC Press LLC, 2000 Corporate Blvd., N.W., Boca Raton, Florida 33431.

© 1998 by CRC Press LLC

No claim to original U.S. Government works
International Standard Book Number 0-8493-9946-7
Library of Congress Card Number 97-30660
Printed in the United States of America  1 2 3 4 5 6 7 8 9 0
Printed on acid-free paper

# Dedication

To each and every individual contributor whose high level of experience, industry expertise, and insightful writing made this compendium of electronic messaging knowledge possible.

# Contributors

AUDREY AUGUN *Principal, Augun & Associates, Hollis NH, Aaugen7457@aol.com*
PETER BECK *Consulting Engineer, SSDS (a division of TCI), Fairfax VA, Peter.Beck@ssds.com*
LEE BENJAMIN *Managing Consultant, New Technology Partners (1997 Microsoft Solution Provider of the Year), Bedford NH, LBenjamin@NTP.com*
GARY CANNON *Senior Consultant, Control Data Systems, Arden Hills MN, gcannon@idt.net*
RUSSELL W. CHUNG *Senior Consultant, American Eagle Group, Sunland CA, russ.chung@ameagle.com*
DALE COHEN *Managing Consultant, Software Spectrum Technology Services Group, Chicago IL, Dale.Cohen@swspectrum.com*
RHONDA DELMATER *Program Manager for Healthcare Programs, Computer Science Innovations, Inc., Melbourne FL*
RIK DRUMMOND *CEO, Drummond Group, 3120 Clover Meadow Drive, Fort Worth TX; Chair for the IETF EDINT (EDI over the Internet) Workgroup; Executive Director for Commerce Net's EDI and Network Services Portfolio; drummond@onramp.NET*
ERIC E. FAUNCE *Director, Messaging Marketing, Lotus Development Corp., Cambridge MA, eric_faunce@lotus.com*
BRUCE GREENBLATT *Directory Architect, Novell, Inc., San Jose CA, bgg@novell.com*
CYDE F. HAGGARD *Computer Scientist, Computer Sciences Corp., Fort Worth TX, chaggard@csc.com*
BARBARA J. HALEY *University of Georgia, Athens GA*
GILBERT HELD *Director, 4-Degree Consulting, Macon GA, 2358068@mcimail.com*
SATHVIK KRISHNAMURTHY *Vice President and General Manager, Worldtalk Internet Security, Worldtalk Corp., Santa Clara CA, sathvik@worldtalk.com*
SUE K. LEBECK *Messaging Architect, Tandem Computers, Inc., Clipertino CA, sue_lebeck@tandem.com*
NAOMI S. LEVENTHAL *Project Manager, ANDRULIS Research Corp., Arlington VA*
DAVID LITWACK *President, dml Associates, Fairfax VA, 4623076@mcimail.com*

## Contributors

CHRISTOPHER J. "KIT" LUEDER *Member of the Technical Staff, The MITRE Corp., McLean VA, kit@mitre.org*

PHILIP Q. MAIER *Program Manager, Secure Network Initiative, Lockheed Martin Enterprise Information Systems, Sunnyvale CA, Phil.Maier@lmco.com*

LYNDA L. MCGHIE *Manager, Corporate Information Security, Lockheed Martin Corp., Bethesda MD, lmcghie@lmsc.lockheed.com*

STEWART S. MILLER *CEO, Executive Information Services, Carlsbad CA, 1-800-IT-MAVEN*

ROGER MIZUMORI *Principal, Rapport Communications, Bellevue WA, Roger-miz@aol.com*

BRETT MOLOTSKY *Lotus Notes Product Manager, Omicron Consulting, Philadelphia PA*

ALEX MORGAN *Program Manager, Hewlett Packard Co., Cuppertino CA, Alex_Morgan@hp.com*

ED OWENS *V.P. of Architecture, Enterprise Solutions, Ltd., Westlake Village CA, ed.owens@esltd.com*

CARROLL M. PEARSON *Lockheed Martin Corp., Sunnyvale CA, persncaa@alisal.lockheed.com*

GORDON PRESTON *Consulting Manager, Bell Atlantic Network Integration, gpreston@bani.com*

SEBASTIAN M. RAINONE, J.D., LL.M. *Business Law Department, College of Commerce and Finance, Villanova University, Villanova PA, rainone@cf_faculty.vill.edu*

MARTIN SCHLEIFF *Technical Lead, Boeing Corp., Seattle WA, Schleiff@boeing.com*

JANICE C. SIPIOR, PH.D. *Management Department, College of Commerce and Finance, Villanova University, Villanova PA, sipior@ucis.vill.edu*

BURKE J. WARD, J.D., LL.M. *Business Law Department, The Graduate Tax Program, College of Commerce and Finance, Villanova University, Villanova PA, ward@cf_faculty.vill.edu*

HUGH J. WATSON *Chair of Business Administration, C. Herman and Mary Virginia Terry College of Business, Athens GA*

DAVID A. ZIMMER *President, American Eagle Group, Warrington PA, dazimmer@ameagle.com*

# Contents

**INTRODUCTION** ................................................. xi

**SECTION I**
**THE ELECTRONIC MESSAGING ENVIRONMENT** .................... 1

1. Strategic Growth of Electronic Messaging .................. 3
   *Rhonda Delmater*
2. Local Area Network (LAN) Messaging ....................... 11
   *Russell W. Chung*
3. Popular E-Mail Systems .................................. 23
   *Gary Cannon*
4. An Introduction to Microsoft Exchange Server ............. 37
   *Lee Benjamin*
5. Novell Messaging Products ............................... 59
   *Bruce Greenblatt*
6. HP OpenMail ............................................. 83
   *Alex Morgan*
7. The Defense Message System ............................. 101
   *Christopher J. "Kit" Lueder*
8. Electronic Messaging in the Healthcare Industry ......... 125
   *Rhonda Delmater*
9. Enterprise Directory Services .......................... 135
   *Martin Schleiff*
10. Directory Synchronization ............................. 149
    *Sathvik Krishnamurthy*
11. X.500 Directory Services: A Business Process Enabler .. 159
    *Roger Mizumori*
12. X.500 Directory Services: Under the Covers ............ 185
    *Roger Mizumori*
13. Messaging Gateways .................................... 213
    *Peter M. Beck*
14. X.400 vs. SMTP ........................................ 243
    *Gordon L. Preston*
15. Value-Added Networks: Marketplace Trends .............. 255
    *David A. Zimmer*

16  Commercial Online Services in Review .................... 275
    *Stewart S. Miller*
17  Selecting Internet Service Providers ...................... 283
    *Stewart S. Miller*
18  Using Internet Resources................................ 293
    *Stewart S. Miller*
19  Netting Web Customers................................. 317
    *Stewart S. Miller*

**SECTION II**
**LEVERAGING THE ELECTRONIC MESSAGING**
**INFRASTRUCTURE** .............................................. 325

20  Exploiting and Extending E-Mail.......................... 327
    *Audrey Augun and Eric E. Faunce*
21  Electronic Commerce................................... 345
    *Rik Drummond*
22  Developing a Trusted Infrastructure for Electronic
    Commerce Services ...................................... 361
    *David Litwack*
23  The World of Electronic Commerce........................ 373
    *David A. Zimmer*
24  Intranets: Notes vs. the Internet.......................... 393
    *Brett Molotsky*
25  Using Groupware to Enhance Team Decision-Making ........ 401
    *Naomi S. Leventhal*
26  Using Lotus Notes in Executive Information Systems ........ 413
    *Barbara J. Haley and Hugh J. Watson*
27  Universal Message Services ............................. 423
    *David A. Zimmer*
28  Client/Server Messaging and the Mobile Worker............ 445
    *David A. Zimmer*
29  Telecommuting ........................................ 461
    *David A. Zimmer*

**SECTION III**
**MANAGING ELECTRONIC MESSAGING SYSTEMS** ................. 501

30  Introduction to Messaging Management.................... 503
    *Roger Mizumori, Sue Lebeck, and Ed Owens*
31  Implementing Electronic Messaging Systems and
    Infrastructures ......................................... 519
    *Dale Cohen*
32  Selecting Electronic Messaging Products ................... 537
    *Rhonda Delmater*

| | | |
|---|---|---|
| 33 | Performance Measurement............................... | 547 |
| | *Clyde F. Haggard* | |
| 34 | Using Statistical Process Control to Manage Message Delivery............................................... | 559 |
| | *Clyde F. Haggard* | |
| 35 | Information Security .................................... | 577 |
| | *Phillip Q. Maier and Lynda L. McGhie* | |
| 36 | Securing Electronic Messages .......................... | 597 |
| | *Gilbert Held* | |
| 37 | E-Mail Security and Privacy ............................. | 605 |
| | *Stewart S. Miller* | |
| 38 | Creating Policy Relative to E-Mail ...................... | 613 |
| | *Carroll M. Pearson* | |
| | Appendix to Chapter 38 Guidelines for Electronic Mail Privacy .................... | 625 |
| 39 | Ethical Management of Employee E-Mail Privacy .......... | 629 |
| | *Janice C. Sipior, Burke T. Ward, and Sebastian M. Rainone* | |

**ABOUT THE EDITOR** ......................................... 641

**GLOSSARY**................................................... 643

**INDEX** ...................................................... 663

# Introduction

*Global competition.* Never has a phrase sent a thrill of challenge and anticipation to so many. And never before has a distributed application — electronic messaging — had the potential to catapult the businesses and individuals fully embracing the technology successfully into the next century.

Just as messaging has been transformed in a few short years from an internal communications tool to one hosting millions of users worldwide, businesses will undergo a swift transition from mass domestic marketing of their goods and services to selling to the individual, anywhere, anytime, anyhow. Messaging is uniquely positioned to fulfill this promise. Within this handbook, we'll examine the current messaging environment, ways to leverage your investment in a messaging infrastructure, and how to manage it all.

## A BRIEF HISTORY OF MESSAGING

Things used to be simple in the days of mainframe-based E-mail systems such as IBM's Professional Office System. Everyone used the same terminal platform for the exchange of text-based messages and documents. Administration was performed at one central location and, if you needed to exchange mail with someone at another company, you just installed a circuit and hooked their PROFS system into yours. That was the beauty of having everyone on the same messaging system.

Widespread deployment of the personal computer changed messaging's landscape forever. With low-cost hardware and software and high returns on output, the swelling ranks of PC users soon demanded to be connected to everyone else in the realm. The local area network and the workgroup came of age (although not without growing pains). Little islands of people, technology, and processes were created and began to flourish.

Along with a distributed user base came decentralized administration and all the attendant problems of managing an increasingly complex environment. Plus, autonomous business groups suddenly had carte blanche to buy whatever products they needed to get the job done. Pockets of different PC-based mail systems popped up literally overnight across the en-

*Introduction*

terprise. These systems offered something the mainframe-based systems couldn't — an attractive user interface and the ability to exchange binary file attachments. All those lavish spreadsheets and word processing documents and images, so closely held and hand-carried between users, could now be sent electronically. And, of course, all the users on the various systems insisted on being able to exchange these attachments and messages with everyone else.

Initially, the demand for connectivity between disparate messaging platforms drove the development of a wide variety of gateways and E-mail bridges. Although these products provided a much-needed service, connecting two different messaging platforms such as PROFS and DEC's All-in-One or cc:Mail and Microsoft Mail, extensive use in an enterprise with more than two messaging systems proved a management nightmare.

Gateway growth became exponential, and messaging managers began looking in earnest for an "any-to-any" type of backbone or infrastructure-based solution. This need forced the development and deployment of international messaging standards such as Simple Mail Transport Protocol (SMTP) and X.400 for the exchange of messages, regardless of the user's computing platform or E-mail system. Backbones based on either of these two protocols, or offering both, gave users the ability to send their messages and attachments to anyone else on a compliant messaging system.

**THE RISE OF DIRECTORY SERVICES**

This level of connectivity rapidly brought about a critical need for efficient and accurate directories, both of internal users and those outside the enterprise. Keeping the directories updated, replicated, and synchronized became a full-time job. These needs compelled the industry to establish an international standard known as X.500.

This model for a distributed electronic directory had huge potential but was not widely deployed because of the resource-intensive nature of the programming code and the lack of commercial-grade products. Luckily, a streamlined version of the standard providing access to the directory, called Lightweight Directory Access Protocol (LDAP), was adopted by Netscape Communications, the largest Web browser vendor. Today the worldwide user base is reaping significant benefits in enhanced search and retrieval capabilities based on LDAP.

**A PORTAL TO THE GLOBAL ELECTRONIC MARKETPLACE**

Once messaging infrastructures were deployed within an enterprise, the environment could be leveraged to provide the underpinnings for mission-critical applications extending the business out to customers, suppliers, and other parties. Messaging enables such applications as electronic commerce, calendaring and scheduling, and workflow and document management. In addition, vendors are now providing universal messaging services to take all manner of messages — E-mail, voice, fax, and pager — and swiftly ship them to the recipient. As more employers provide telecommuting opportunities for their workforces, messaging becomes a key technology for staying in touch with the home office and customers.

Managing the complex messaging environment of today is a daunting task. Message stores are huge server-based multiprotocol repositories of multimedia messages accessed by different messaging clients. Aided by service-level agreements, performance measurement and application management tools, security services, and policies for retention and privacy, administrators can more easily manage highly distributed messaging systems.

Once again, the landscape of messaging is experiencing cataclysmic changes. Web-based technology is making rapid advances on client/server messaging platforms. By providing products that can be accessed from a Web browser, more users, regardless of their computing platforms, can exchange messages and files, again extending the reach of the messaging system.

Managing the messaging environment from a Web browser is also gaining prominence. Entire messaging systems are now shifting their focus from proprietary transports, protocols, and clients to Internet standards such as POP3, IMAP4, and HTTP.

What does this mean for the user and for businesses? A more open messaging environment, one that supports a variety of message types and is easily accessible from anywhere in the world — a portal to the global electronic marketplace.

## ABOUT THIS HANDBOOK

The *Handbook of Electronic Messaging* is a comprehensive guide to building, leveraging, and maintaining electronic messaging systems that have the required architecture and infrastructure to benefit organizations as they move from simple messaging to complex collaborative computing. Organized into three sections, the handbook takes the reader from an in-depth look at the current messaging environment, through using that environment to add value to organizations, and beyond to the management and protection of messaging resources. Topics range from popular messaging systems, to creating a privacy policy, to building a trusted electronic commerce infrastructure. Individual chapters are sharply attentive to what is available in the industry now in the way of messaging products and services, the impact of the Internet on corporate messaging systems, industry trends and forecasts, and real-world implementations of the technology.

Contributing authors for the handbook were selected for their high level of expertise in a particular area of messaging. The authors are actively in-

*Introduction*

volved in the messaging and electronic commerce industries as end-users, vendors, and consultants. They routinely share their insights and experience at industry events such as those sponsored by the Electronic Messaging Association (EMA).

The information contained in this handbook is beneficial for messaging managers and administrators, applications developers (especially those involved with mail-enabled applications), electronic commerce and electronic data interchange professionals, information delivery experts in various industries (e.g., law, retail, and healthcare as well as government), messaging network planners, data communications network engineers, messaging training professionals, consultants, messaging vendors, and end-users. Those who wish to distribute their products and services over the Internet as well as through traditional communications paths may also profit from the information within this handbook.

This handbook assumes that the reader has a basic familiarity with messaging applications, with client/server architectures, and with computer systems in general. After mastering the terms, concepts, and techniques described within this handbook, the reader will have the basic information necessary to construct and maintain a highly functional and serviceable messaging environment.

Nancy A. Cox

September 1997

# Section I
# The Electronic Messaging Environment

The electronic messaging environment, when viewed from 50,000 feet, seems simple. The landscape is littered with users, applications, and the networks that connect them. However, as messaging professionals well know, it's just not that simple.

Where did all those E-mail users come from? In this section, we take a closer look at the staggering growth in not only the number of messaging users but in the number and size of the messages they are exchanging. How ubiquitous and mission critical messaging has become in such a short time is examined through an analysis of strategic growth and industry trends.

This section presents an overview of popular messaging applications. Local area network (LAN) E-mail products, Internet-based messaging, and mainframe systems are discussed. Specific products such as Microsoft Exchange, Novell GroupWise, and HP OpenMail are thoroughly explored for their unique features and functionality.

We delve a little deeper within two case studies that focus on how messaging products and services are deployed and managed in the healthcare and defense industries. These chapters present real-world examples of how messaging systems have been implemented to solve specific business or operational requirements.

Users have a critical need to be able to quickly and easily locate other users, to view their contact information in order to send mail to them. Users also must have the capability to locate resources such as printers and fax machines or to schedule conference rooms. Users may want to communicate with others outside their own organization and must have a simple means to do an electronic "411" to locate names and addresses.

In this section, we deal with such topics as creating and deploying an enterprise directory service and performing directory synchronization between all the disparate directories within the organization. The X.500

## THE ELECTRONIC MESSAGING ENVIRONMENT

international standard for a distributed directory is also thoroughly examined in light of the new surge in Lightweight Directory Access Protocol (LDAP) implementations.

The communications network that connects all the organization's messaging resources may support a variety of protocols such as X.400 or SMTP and may contain several gateways to connect different E-mail systems. These topics are discussed at length, as well as the importance and contributions of value-added networks, commercial online services, and Internet service providers and resources.

# Chapter 1
# Strategic Growth of Electronic Messaging

*Rhonda Delmater*

To examine the strategic growth of electronic messaging, it is appropriate to consider the messaging architecture, standards, functional ability, product and service trends, and organizational impact. In each of these areas, examining the past and present provides perspective for understanding the strategic growth which has, and will continue, to take place in electronic messaging.

In addition to the changes in technologies and capabilities, significant changes have occurred in the types and numbers of those who avail themselves of electronic messaging services. Although the first multiorganization E-mail users were those in government and academia, today's E-mail is widely used among family and friends as an efficient, cost-effective way to stay in touch. At the same time, new commercial applications (e.g., EDI, banking, and network shopping) have emerged.

## MESSAGING ARCHITECTURE

What was the first E-mail system you used? Chances are, it provided simple character-based interpersonal messaging among users on a single host or a cluster of systems based on a single architecture (e.g., IBM SNA or DEC). Of course, this was during the days of the dumb terminal. The vendors of these systems attempted to provide a full-featured product (e.g., IBM's Professional Office System — a.k.a. PROFS — or DEC's All-In-One), which encompassed electronic mail, personal and workgroup scheduling, and a directory.

### Proprietary E-Mail Applications

Some organizations were not faced with significant challenges to providing organizational E-mail. The answer was for everyone to access one E-mail system. This approach has been, and still can be, effective for organizations with a homogeneous computing architecture. However, to realize the benefit of graphical user interfaces and client/server applications, distributed computing architectures must emerge. In addition, the single-system approach is not practical for those organizations with large contingencies of mainframe (i.e., business) users and minicomputer (i.e., engineering) users.

Many engineering firms found themselves in this situation. Some thought everyone could access one E-mail system through terminal emulation. Using an ASCII terminal for 3270 emulation is ugly at best. It is slow, cumbersome, and learning the keyboard mapping can be extremely tedious. It is not likely that the engineers were that motivated to communicate with the business staff. With the flattening of organizations, however, they were forced to function in teams with their business counterparts, thereby necessitating the exchange of messages between everyone on those teams.

**Not So Simple SMTP**

In addition to the propriety E-mail applications offered by the minicomputer vendors, many also supported Simple Mail Transport Protocol (SMTP), or what is sometimes called Unix Mail. The SMTP protocol is part of a suite of protocols related to TCP/IP. Other familiar protocols in this suite include file transfer protocol (FTP), terminal emulation (Telnet), and Simple Network Monitoring Protocol (SNMP). The word *basic* could be substituted for simple in SMTP; however, it is much less simple to explain than a product.

Because SMTP is a standard protocol, it has been implemented in a myriad of ways. Typically UNIX platform vendors bundle SMTP with their operating systems. Back in the ASCII terminal era, the interfaces were simple. Now they typically have a graphical user interface (GUI) application based on MOTIF or X-windows, which provides a very similar interface to their PC Windows counterparts. Because SMTP is closely related with TCP/IP, it is widely used on the Internet, hence it has come to be known as Internet mail. Gateways to SMTP have been widely implemented by vendors for the full spectrum of computer platforms. SMTP is very prevalent and is undoubtedly used to route the largest number of electronic messages among SMTP and non-SMTP users.

**PC E-Mail Packages**

Thinking back to life before PCs really can make one feel like an old-timer! When personal computers were initially implemented, terminal emulation was used to access the organization's host-based E-mail systems. As before the advent of PCs, this may have meant having multiple accounts to be able to communicate with multiple-user communities (e.g., business and engineering). In fact, a sender might have to type in a message multiple times to get to all of the intended recipients. And keyboard mapping for terminal emulation on each host continued to be a nightmare.

Once PCs proliferated, it was only a matter of time before users wanted E-mail systems that provided benefits similar to their other desktop office automation applications. Host-based computing cannot successfully compete with an intelligent client in terms of response time, ease of use, interoperability with other desktop applications, and the ease of use provided by the graphical user interface.

The early PC E-mail packages were character-based and ranged from providing simple interpersonal messaging to workgroup coordination. Network Courier, the precursor to Microsoft Mail, was one of the early messaging products. As Microsoft Mail, this product has grown to over 6 million users. A product known as the Coordinator from Action Technologies provided robust workflow-like features, but required a steep learning curve because of its use of a work management rather than a mail-based metaphor. The underlying foundation technology was purchased by Novell, and the transport became the foundation for Novell Message Handling System (MHS). The vendor has re-emerged in the workflow arena.

Although the interfaces have changed from character-based to windows, the underlying architecture, based on LAN file sharing, has remained unchanged until now. We are on the brink of the next generation of LAN messaging systems. The vendors are moving to client/server messaging architectures. These new architectures will allow software developers to create client applications without having to develop the message store or message transfer agent.

This is somewhat analogous to business application developers having the choice of writing their own relational database management system (RDBMS), using one of the leading RDBMS products, or better yet, writing to a standard such as structured query language (SQL). Writing to the standard allows their product to be used with multiple database engine products (or messaging engine products), so their product can run in many LAN environments. There can be significant cost savings for the client through the reduction of the number of software products that must be licensed.

Application programming interfaces (APIs) provide a defined interface for an application to utilize a service such as a messaging server. The LAN E-mail API war seems to have ended with MAPI taking a clear lead over VIM (Microsoft versus Lotus). The interface function is illustrated in the LAN messaging architecture illustrated in Exhibit 1-1.

Because the new architectures allow multiple client types to communicate with a common, scalable server, we may have come nearly full-circle to a point where an organization can have a homogeneous messaging ar-

# THE ELECTRONIC MESSAGING ENVIRONMENT

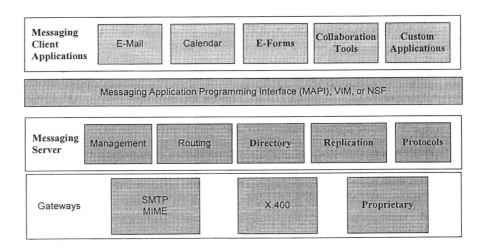

**Exhibit 1-1. LAN Messaging Architecture.**

chitecture. Now potentially many disparate client types will be able to interoperate. How things have changed!

The concept of a universal client provides a single user interface for the end-user to access all of his or her applications, including not only typical office automation and messaging, but voice, video, shared resource materials, collaboration tools, as well as proprietary business applications. In addition, the client has access to a broad range of information resources, both internally and externally through tools (e.g., Web browsers).

## MESSAGING STANDARDS

Interconnection approaches to accomplish enterprise E-mail have evolved through many phases, as discussed earlier. The gateways and messaging hubs used for interconnection are an important component of the messaging architecture. Initially, proprietary products were more frequently implemented to interconnect disparate internal E-mail systems. The applicable standards were immature, interoperability profiles (e.g., the Government Open System Interconnection Profile, known as GOSIP) had yet to be established, and there were few products available.

Early external E-mail connections were typically to a value-added network (e.g., MCI, Sprint, and ATTMail). Because the VANs were not interconnected, an organization making extensive use of external E-mail would find it necessary to connect to multiple VANs.

The international Open System Interconnection (OSI) messaging standards are known as X.400. The first set of X.400 standards was published in

1984, with updates published every 4 years. During the late 1980s, the VANs implemented X.400 networks (in addition to their proprietary ones), and established interconnections. Many Fortune 500 companies implemented X.400 connectivity through a VAN.

Some of the standards that are important to the strategic growth of electronic messaging include X.435 or Pedi. These are important for electronic commerce, multimedia, and interoperability of applications yet to be determined. At the same time the X.400 standards series has been developing, similar changes have been taking place on the Internet. The Multipurpose Internet Mail Extensions (MIME) standard has been established to support multimedia body parts. In addition, a plan is underway to expand the TCP/IP addressing space from 32 to 128 bits.

## MESSAGING FUNCTIONAL ABILITY

Basic messaging functional ability provided by today's standard applications generally includes the following features: sending mail, reading mail, file attachments, spell checking, directory of messages, searching mailbox, mailing lists, filing messages, printing messages, user lists, shared folders-bulletin boards, and online help.

In addition, most E-mail applications typically provide a set of gateways for both de jure and de facto standards (e.g., X.400, SMTP, SNADS-DISOSS, IBM PROFS, Office Vision, MHS, and VAN mail networks). Many provide electronic fax capability, although electronic fax is much more practical as a sending, rather than a receiving, technique.

Emerging functional ability at various stages of implementation includes the following:

- *Electronic forms.* E-mail provides defined formats for the exchange of information, similar to paper forms. E-forms may also include such capabilities as automatic calculations and simple validation. More advanced capabilities include DBMS interfaces and data collection.
- *Workflow.* E-mail provides process automation and tracking for each instance of a process.
- *Collaboration.* E-mail fosters workgroup communication through shared resources (e.g., forums and projects).
- *Electronic commerce.* This is used to describe a broad range of business transactions conducted over a data communications link (e.g., EDI or computer shopping).
- *Directory services.* Some features of directory services include synchronization, distributed directories, E-mail integration, distribution lists, search, administrative tools, and X.500 support.
- *Mobile and wireless messaging.* This function provides remote access to an E-mail service by dial-up or cellular access.

# THE ELECTRONIC MESSAGING ENVIRONMENT

- *Universal client.* This function provides a single graphical user interface for a full range of multimedia information, including messages and shared information resources.

## PRODUCT AND SERVICE TRENDS

The leaders of the current LAN E-mail application market include the following products (vendors in parentheses): cc:Mail (IBM-Lotus); Microsoft Mail (Microsoft), formerly Network Courier, GroupWise (Novell); Notes (Lotus); FirstClass (SoftArc); Exchange Inbox (Microsoft); and daVinci eMail (On Technology).

*Electronic Mail and Messaging Systems* reported in its January 22, 1996, issue that worldwide E-mail exceeded 90 million users at the end of 1995. Exhibit 1-2 illustrates the market share of the leading LAN messaging vendors.

These products have been available for many years, have broad installation bases, and provide commonly desired E-mail capabilities for message handling and transfer. They are based on LAN file-sharing technologies. Some of these products include one or more ancillary applications (e.g., bulletin boards, group scheduling, and electronic forms). For the functions that are not included within the E-mail applications, users have the option of implementing such products as Delrina Formflow (electronic forms), Ca-LANdar (workgroup scheduling), and Lotus Notes (collaboration).

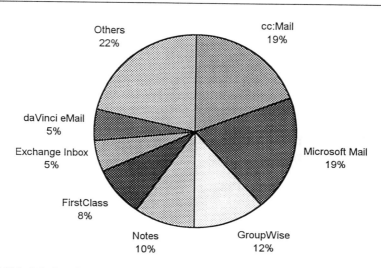

Exhibit 1-2. Leading LAN Messaging Vendors and Their Market Share.

## Client/Server Messaging Products

The next generation of LAN E-mail applications is on the horizon. These will be based on client/server technologies. They will provide tighter integration of workgroup technologies and be much more object-oriented. Client/server messaging products that have been announced in this arena include: Notes and Lotus Communications Architecture (IBM/Lotus), Exchange Clients and Back Office: Exchange Server (Microsoft), and GroupWise and Novell Communications Environment (Novell). The product formerly known as Network Courier became Microsoft Mail in the early 1990s when Microsoft purchased the company. Network Courier was one of the early LAN E-mail systems and has been widely implemented. Microsoft claims 6 million users of Microsoft Mail.

The current Microsoft Mail Server release (version 3.5) provides a multitasking message transfer agent (MMTA) for Windows NT and an electronic forms designer (which uses Visual Basic). Clients are available for Windows, MS-DOS, Macintosh, and OS/2. It also provides a gateway to SMTP, which requires a designated host on the SMTP system for routing. Microsoft Exchange will be based on a client/server rather than PC file-sharing architecture. Exchange features a single integrated inbox for multiple message types (e.g., scheduling, E-mail, fax, and voicE-mail) and cross-platform support for multiple operating systems. Product offerings from IBM/Lotus and Novell to offer similar capabilities.

Remote messaging applications will also have increased capability under the new architecture, including the capability to select which messages are downloaded from the home post office to the portable unit. CompuServe, Prodigy, America OnLine, and the Microsoft Network have joined the ranks of MCI, Sprint, and ATT in offering messaging interconnection services to Internet Mail.

## ORGANIZATIONAL IMPACT

Organizations initially implemented E-mail as a means to improve the timeliness and distribution of internal communication, and thereby increase efficiency. There is no doubt E-mail has accomplished the former. Some will argue the efficiency aspect because of the increase in sheer volume. Rules of etiquette and common sense are necessary. Electronic data interchange (EDI) has experienced steady growth, but has not yet reached its peak. Collaboration is early in its life cycle, as is workflow. Messaging is of growing commercial importance. As more and more consumers are equipped with home computers, electronic messaging will take on even greater importance.

# THE ELECTRONIC MESSAGING ENVIRONMENT

**SUMMARY**

In the past several decades, since E-mail was first used, we have seen dramatic changes in its architectures, standards, and abilities. The impact and proliferation of intelligent desktop devices has been staggering, as has been the growth in the capability of those desktop devices. The marketplace has matured somewhat through multiple acquisitions and mergers of both product companies and value-added networks. Changes in the character, functional ability, and sheer numbers on the Internet are truly amazing.

The annual conference of the Electronic Messaging Association (EMA) provides a benchmark of electronic messaging's growth. If you had attended the conference as recently as 3 years ago, you would have encountered a group of VANs and user organizations focused on migrating to OSI messaging protocols (to the near exclusion of Internet mail). You would have found primary concern on interoperability of electronic messaging systems and EDI.

If you attended again in 1995, you would have been startled by the contrast. The scope of the messaging industry has broadened to include collaboration, electronic commerce, and workflow. The tremendous growth and momentum of the Internet has caused both attempts to provide commercial offerings and interest in convergence of the Internet and X.400. Can we expect these trends to continue? Yes. Looking to the future, much work is still needed before sending multimedia objects across disparate networks to provide the capability to support such critical applications as banking and telemedicine.

# Chapter 2
# Local Area Network (LAN) Messaging
*Russell W. Chung*

Personal computer/local area network (PC/LAN)-based electronic messaging systems represent the fastest growing segment of the electronic messaging industry. This chapter discusses the reasons for the immense popularity of LAN messaging systems, provides an overview of the components and features of a LAN messaging system, and discusses some of the issues, myths, and realities regarding LAN messaging. It is not intended to be a product comparison nor a technical guide — it is a discussion of the general issues involved with planning, implementing, and managing a LAN-based messaging system.

## POPULARITY OF LAN MESSAGING SYSTEMS

The first LAN-based messaging systems made their appearance shortly after the introduction of local area networks in 1985. Some of the top-selling products in this category include Lotus cc:Mail, Lotus Notes, Microsoft Mail for PCs, Novell GroupWise, SoftArc First Class, da vinci eMail, and CE Software QuickMail. Recent trends indicate that the overall number of LAN messaging users grows at a rate of approximately 40% per year. Some of the reasons for this popularity include: ease of use, ease of setup, and low initial investment.

### Ease of Use

In the early 1980s, a typical E-mail user prepared a message on a dumb terminal connected to a mainframe- or minicomputer-based E-mail system or connected to a public value-added service (e.g., AT&T Mail, MCI Mail, CompuServe, Genie, or Delphi). If a personal computer was involved, it was used only to emulate a dumb terminal.

Users endured a featureless, character-based, command line interface in preparing, editing, addressing, and sending a message. With the command line interface, users worked on one line of the message at a time and had to invoke cryptic navigation commands to view or edit the previous lines.

# THE ELECTRONIC MESSAGING ENVIRONMENT

The single most important factor in the popularity of LAN-based messaging products is the user-friendly interface, which is made possible by the personal computer. With a personal computer, users running a LAN messaging program may prepare a message by filling in an on-screen form, edit the message by moving the cursor to the desired point on the screen, address the message by picking names from an on-screen list of users, and send the message by pressing a key or clicking on a mouse.

## Ease of Setup

The installation and setup of a LAN-based electronic messaging system on a small business or a departmental LAN is easily accomplished in a few hours by a LAN administrator. A single system can support hundreds of users. In a large company, the departmental LAN messaging systems can be linked together to form an enterprise-wide messaging system. As the number of linked systems increases, however, scalability becomes an issue.

As administrators are discovering, the management of a large, interconnected LAN-based messaging network is not a trivial task; the distributed, decentralized nature of LAN messaging systems means that a great deal of planning, training, and coordination is required to ensure reliable, timely message delivery throughout an enterprise. An electronic messaging network is like a chain; it is only as strong as its weakest link, and the distributed nature of a LAN messaging system means that it has many more links that must be managed.

## Low Initial Investment

On a per-user basis, the price of a LAN messaging system and of mainframe- or minicomputer-based messaging systems are comparable. The initial investment for software for a mainframe- or minicomputer-based messaging system may amount to hundreds of thousands of dollars, whereas the initial investment for a LAN messaging system may amount to only a few thousand dollars. The relatively low initial investment threshold makes the establishment of a LAN-based messaging system cost-effective for small businesses and for departments of large businesses.

A LAN is rarely established solely for messaging purposes. Ordinarily, a small business or a departmental workgroup makes a decision to install a LAN to facilitate access to spreadsheets, word processing documents, shared databases, or networked printers; the incremental expense of adding electronic messaging capabilities to such a LAN is relatively modest. Vendors have begun to bundle messaging client software with software office suites, further reducing the amount of the initial investment required for LAN messaging software.

## COMPONENTS OF A LAN MESSAGING SYSTEM

Developers of LAN messaging systems often use proprietary designs and architectures that may not exactly conform to the model of a messaging system defined by the International Telecommunications Union Telecommunications Standards Sector (ITU-TSS, formerly CCITT) X.400 standard or the model of a directory system defined by the ITU-TSS X.500 standard. Nevertheless, the X.400 and X.500 models provide a useful conceptual basis for describing the components of a LAN messaging system. The X.400 standard defines three components of a messaging system; user agent (UA), message store (MS), and message transfer agent (MTA).

### User Agent /User Interface (UA/UI)

The UA is the term defined in the X.400 standard for the process used to read, write, edit, file, send, and discard E-mail messages. From a user's standpoint, the look and feel and the features of the user agent are what distinguishes one vendor's product from another.

In a LAN messaging system, the user agent runs on the user's personal computer. The UA program file (or files) may be stored either on the hard drive of the user's PC, or on the hard drive of the file server and loaded in the personal computer's RAM when the user launches the program.

### Message Store (MS)

The message store is the repository for a user's messages. Messages are placed in, and retrieved from, the message store by UAs and MTAs. In a LAN environment, the message store is typically located in a file server (see Exhibit 2-1) or in a messaging server (see Exhibit 2-2). In some cases, the message store may be located in a user's PC.

Conceptually, the message store is the E-mail equivalent of an inbox, outbox, and file cabinet. As might be expected, the internal design of the message store varies from vendor to vendor. Some products use a system of hierarchical directories and separate files to store each user's messages; other products use a single shared file with a system of indexes and pointers to store all users' messages; still others use a combination of the two approaches.

### Message Transfer Agent (MTA)

The MTA is the term defined in the X.400 standard for the process that transfers messages between messaging systems. In a typical LAN messaging environment, this MTA process runs continuously on a dedicated personal computer on the LAN (Exhibit 2-1) or on a messaging server (Exhibit 2-2), regularly checking for incoming and outgoing messages and transferring them to the appropriate destinations.

## THE ELECTRONIC MESSAGING ENVIRONMENT

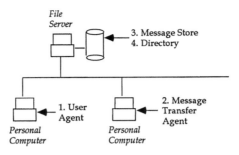

Exhibit 2-1. Layout of a Typical File Sharing-Based LAN Messaging System.

### Directories, Directory System Agent (DSA), Directory User Agent (DUA)

Messaging system directories are specialized databases that perform two functions: they provide a repository of names and addresses for the purpose of routing messages, and they provide a place to store passwords, encryption keys, and other authentication data used to control access to the messaging system.

The X.500 model contemplates decentralized, local directories. A network of distributed, cooperating directory servers running a DSA program responds to requests from users and from other directory servers for information from the directories. According to the X.500 model, the DUA uses the X.500 directory access protocol (DAP) to submit a query to a DSA when seeking information from the directory. This model requires real-time access between users and directory servers throughout the network — if real-time access is not available, this model requires periodic replication of the directory information to each server.

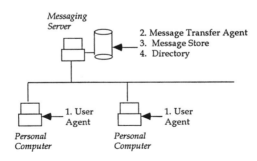

Exhibit 2-2. Layout of a Typical Client/Server Messaging System.

The architecture of a typical LAN messaging directory system does not assume real-time access to directories across the network. Instead of distributed directory servers using standards-based DAPs, current LAN messaging systems use directories that are located on each file server, messaging server, or on a user's PC. In current LAN messaging systems, the UA/UI program accesses the directory information stored on a local file server, messaging server, or on the user's PC using proprietary protocols instead of standards-based DAPs. This approach demands that each local directory contain the names and addresses of all of the users throughout the system.

To ensure that each copy of the directory contains complete, accurate information, LAN messaging vendors have developed automated processes to update, propagate, and synchronize the information contained in the directories. Typically, this information is transmitted by store-and-forward technology rather than in real-time. Despite existence of the automated processes, the maintenance and synchronization of each local directory in a large LAN messaging network continues to be one of the most challenging aspects of mail system administration.

## FEATURES OF A LAN MESSAGING SYSTEM

### File Sharing vs. Client/Server Architecture

A file sharing-based messaging system stores its messages in a LAN file server (see Exhibit 2-1). The PCs on the LAN run the UA or MTA processes to perform the work involved in managing the message store (e.g., sorting, indexing, reorganizing, or verifying internal consistency). The file server performs input, output, and storage functions for the message files, just as it would for any other data file (e.g., a spreadsheet, database, or word processing document). Otherwise, the file server does none of the processing of the messages.

Because the work of processing messages is distributed among all of the PCs, a modest file server can support many simultaneous users. The functional ability and performance of the system depends on the power of the users' personal computers. Examples of file sharing-based LAN messaging systems are Lotus cc:Mail v.6, Microsoft Mail v.3, Novell GroupWise v.4, Da Vinci eMail, and CE QuickMail.

Currently, approximately 40 million users are part of file sharing-based messaging systems, and the number of users is growing at a rate of approximately 30% per year.

A client/server-based messaging system (Exhibit 2-2) stores its messages in a messaging server. All of the work involved in managing the mes-

# THE ELECTRONIC MESSAGING ENVIRONMENT

sage store (e.g., sorting, indexing, reorganizing, or verifying internal consistency), is performed by the messaging server. The messaging server also performs the functions of the MTA. PCs on the LAN run the UA and make remote procedure calls to the server to send and retrieve messages, but the management of the message store is handled by the messaging server.

Because processing the messages is performed by the messaging server, a powerful messaging server is needed to support many simultaneous users. The functional ability and performance of the system depends on the power of the messaging server, not on the users' PCs. Examples of client/server-based messaging systems include Lotus Notes r.4, Microsoft Exchange, Novell GroupWise v.5, and SoftArc FirstClass v.3. There are approximately 14 million users of client/server-based messaging systems, growing at a rate of approximately 60% per year.

**Message Routing Topology**

A typical mainframe-based electronic messaging system needs only a single mainframe computer to support thousands of simultaneous users. As a user sends a message to another user, the message is immediately stored in the recipient's electronic mailbox on the mainframe computer. Typical LAN servers are capable of supporting a few hundred simultaneous users.

To provide electronic messaging services to a large enterprise, a network of multiple LAN file servers or messaging servers must be employed. When a user sends a message to another user on the same server, the message is immediately stored in the recipient's electronic mailbox on the server. When users send messages to users on a different server, the messages are routed between the servers within the network until they reach the electronic post office containing the user's mailbox.

LAN messaging systems typically use either a peer-to-peer topology, a hub-spoke topology, or a combination of the two to route messages between users whose mailboxes are located on different servers. In peer-to-peer routing (see Exhibit 2-3), each post office connects to every other post office to exchange messages. As the number of post offices increases, the number of possible connections grows at a geometric rate. A network of 4 post offices must maintain 12 connection paths, a network of 6 post offices must maintain 30 connection paths, and a network of 10 post offices must maintain 90 connection paths. In a system that uses peer-to-peer routing, no single point of failure would disrupt the entire messaging system. The multiplicity of routing paths is inefficient, however, for networks that have more than a few (about a half dozen) interconnected sites.

*Local Area Network (LAN) Messaging*

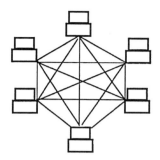

**Exhibit 2-3. Peer-to-Peer Routing.**

In a hub-spoke configuration (see Exhibit 2-4), one message store becomes the hub. Messages between the spokes are routed through the hub, and then on to their final destination. A failure at the hub would prevent messages from being routed throughout the system, so it is essential that redundant measures be taken to minimize hub downtime. A hub-spoke configuration provides an efficient means of routing messages in most situations.

Exhibit 2-5 depicts a combination configuration that uses two hubs. Messages between the spokes are routed through the hub, and then on to another spoke or to the other hub, which relays the message to the final destination. Combinations of two, three, or more interconnected hubs provide an efficient method of routing messages between large enterprise-wide messaging networks.

**Gateways**

Messaging gateways transfer messages between dissimilar messaging systems. They must convert the message contents and addressing infor-

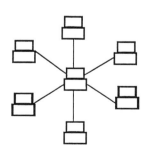

**Exhibit 2-4. Hub-Spoke Routing.**

17

# THE ELECTRONIC MESSAGING ENVIRONMENT

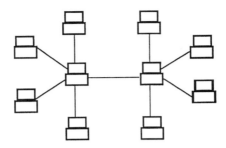

**Exhibit 2-5. Combination Topology.**

mation from one vendor's format to the other vendor's format. Because of differences between vendors, some of the features of one messaging system may not be supported by the other system, and may be lost in the conversion process.

### Application Programming Interfaces (APIs)

Messages are placed in, and retrieved from, the message store by UAs and MTAs. Until recently, the proprietary design of a message store prevented access except by UAs and MTAs from the same vendor. The development of standards-based messaging APIs permits UAs and MTAs from one vendor to access the message store of another vendor. In addition, the use of a standard API permits other applications (e.g., word processors, spreadsheets, and group calendaring programs) to become mail enabled. A mail-enabled application allows the user to send and receive messages from within the enabled application, instead of leaving the application, launching the messaging application, and attaching or detaching the information.

### Directory Updates and Synchronization

A mainframe-based messaging system needs only a single, centralized mail directory to provide name and address information to thousands of users. In a LAN-based messaging system, each server needs a complete copy of the mail directory with the correct electronic mail address for each user. When a user is added anywhere in the network, the user's entry must be propagated to each server in the network. To ensure that each copy of the directory contains complete, accurate information, LAN messaging vendors and third-party developers have produced automated processes to update, propagate, and synchronize the information contained in the directories.

### Filters, Rules, and Agents

As the volume of electronic mail increases, users begin to encounter information overload. Several LAN messaging products feature filters, rules,

## Local Area Network (LAN) Messaging

and agents to help users manage the flow of incoming mail. A filter serves as a gatekeeper; it allows messages that meet certain criteria (e.g., author, subject, or message size) to pass through to a user's mailbox, and denies passage for messages that fail. Filters are typically implemented by MTAs.

A rule examines a message that has arrived in a user's mailbox, and if the message meets certain predefined criteria (e.g., author, subject, or contents), the rule takes a predefined action on the message (e.g., filing the message, forwarding the message, or automatically replying to the message). Rules are typically implemented by a UA.

An agent acts in behalf of a user, and may operate on a local or a remote server to act on messages based on predefined criteria, even when a user is not logged in. Agents can also be used to generate messages to alert users upon the occurrence of certain predefined events.

## LAN MESSAGING SYSTEM: MYTHS AND REALITIES

### Reliability

Mainframe computers are designed with redundant hardware, and the operating systems are designed to be more failure tolerant than PCs. Therefore, mainframe based messaging systems are considered more reliable than PC/LAN-based messaging systems. Recent advances in PC technology and operating systems, however, now result in a negligible difference.

Electronic messaging system failures can be caused by several factors: hardware failures, application program failures, communications failures, and database corruption. The effect of hardware failures in file servers, messaging servers, and PCs can be reduced by the use of redundant components (e.g., redundant array of inexpensive disks, or RAID, uninterruptable power supplies, and server mirroring, or SFT level III). These measures are not inexpensive, but will result in a level of hardware reliability equivalent to mainframe hardware reliability. Even without the use of these redundant components, PC/LAN components typically operate for several years without failure.

Despite extensive testing, applications do crash. The use of multitasking operating systems (e.g., OS/2 and Windows NT) reduces the possibility that the failure of one application could cause the crash of another application on the same messaging server or PC. When a messaging application fails, the distributed nature of LAN messaging systems limits the effect of a single failure to the local workstation or messaging server, whereas a mainframe application failure could disable the entire messaging system.

The possibility that communications links can fail always exists. In a mainframe environment, a link failure may prevent users from gaining access to their mailboxes. The store and forward characteristics of a distributed PC/LAN messaging system might mean that messages would be delayed until the communications link is restored, but would not prevent local users from accessing their mailboxes. In addition, typical PC/LAN MTAs offer alternative means of forwarding messages to another site (e.g., X.25, asynchronous dial-up, or TCP/IP).

The possibility of database corruption is greater in a file sharing-based messaging system than in a client/server-based or a mainframe-based messaging system because a greater number of workstations are simultaneously writing to the message store. A UA in a client/server or mainframe messaging system makes remote procedure calls to the server, but only the server reads from and writes to the message store. Administrators of PC/LAN-based messaging systems may reduce the possibility of database corruption by performing regular maintenance on the database, and can mitigate the effect of corruption by making regular backups of the database.

**Scalability**

A single LAN serves no more than a few hundred users. To meet the needs of a large enterprise, additional LAN sites must be added to the network and messages must be routed between the sites. Although this is not a concern for small networks with only a few sites, it multiplies the amount of effort for planning, coordinating, and managing the messaging system for a large enterprise. Mail administrators are learning that the management of a large enterprise LAN-based messaging network is a nontrivial task.

**Planning**

The purpose of planning a large-scale PC/LAN-based electronic messaging network is to ensure that messages arrive at their destinations quickly, efficiently, and dependably. Planners must take into consideration the underlying LAN and WAN infrastructure, the location and number of users, and workgroup relationships in determining where to place message stores/messaging servers (i.e., post offices), and which message routing topology to use to connect them.

Generally, it makes sense to locate a post office on the same LAN segment that users are on. Ideally, to minimize network traffic, the message routing topology should follow the LAN/WAN infrastructure and be based on a hub-spoke topology. Workgroup relationships, however, must also be considered in determining message routing topology; members of workgroups that belong to the same division or department are more likely to exchange messages with each other, and for efficiency, their post offices should be connected to the same hub whenever possible.

Another consideration in the design of a large-scale, distributed messaging system is the availability and deployment of administrative and support staff. Individual sites do not require the services of a full-time mail administrator; and it usually is not cost-effective to provide extensive administrative training for duties that only require a few hours per week. Instead, it is more efficient to provide training to an administrator who will manage and support a number of sites.

**Management**

As sites are added to the network, administrators are faced with two challenges: they must monitor each and every link in the network to ensure that messages are flowing properly and without delay, and they must ensure that each local copy of the name and address book is complete and accurate.

One technique to monitor messaging networks uses SNMP to ensure that servers, routers, gateways, and workstations are operating properly. The use of SNMP management tools warns an operator that a machine or network component has failed, but does not ensure that messages are actually flowing through the network in a timely manner.

Another approach is to send small test messages from a post office to other post offices, and use an echo program to generate a response. If a response does not arrive within a predetermined time, an alert is generated that may be forwarded by E-mail or to the mail administrator's pager. Examples of products that monitor system response are Soft*Switch Mail Monitor, cc:Mail View, and Baranof MailCheck.

**SUMMARY**

Today, there is a shortage of tools for ensuring the accuracy and completeness of E-mail directories across a large organization. A few vendors have developed proprietary techniques for propagating and synchronizing directories, but they require careful configuration and monitoring for rejected entries. In a multivendor messaging environment, the problem becomes more complex because the directories must accommodate address conversion from one messaging system's format to another system's format, as well as managing additions, changes, and deletions. The X.500 standard promises to solve the problem of directory access and synchronization, but it has not yet been implemented in PC/LAN messaging systems.

# Chapter 3
# Popular E-Mail Systems
*Gary Cannon*

The Internet has grown so rapidly because of users' need to communicate and share information. Although many people were doing just that on commercial networks, the Internet offers more than just E-mail and is less expensive than commercial systems.

The number of E-mail users has grown almost 74% in the past year. More than 47 million people are using LAN E-mail systems, most of which are connected to commercial services. The larger commercial networks cannot accurately estimate how many individual users they support because most users on LANs and larger systems do not have individual accounts on the commercial networks. The majority of users access the commercial systems through corporate gateways.

This number of users continues to grow, and as commercial systems enhance their product and service offerings, there will be continued expansion on the commercial side of the market as well as the Internet.

Many business users rely on E-mail to conduct their day-to-day functions. E-mail ties together many other applications and has contributed significantly to the information explosion.

**ELECTRONIC MESSAGING: A HISTORICAL PERSPECTIVE**

For years, telex served as the only form of electronic mail and was strictly the domain of government agencies and big business. About 25 years ago, a few operating systems had rudimentary messaging capabilities. IBM Corp.'s Virtual Machine (VM) system could communicate between active terminals. Digital Equipment Corp.'s Virtual Memory System (VMS) operating system came up with the basics of what would become VMSmail.

At about the same time, General Electric's Information Services Business Division (ISBD) developed the time-sharing concept with Dartmouth University and introduced an internal system known as Cross File (XFL), which allowed employees to send messages to one another when they were working on projects together. XFL developed into a division-wide utility and each office had an address. Originally, senders wrote out their message on a piece of paper and handed it to the administrator. Sometime during the day the message would be entered into the system and the sender could expect a reply in a day or so. Functionally, the system worked fine. Practically, it took a few years before the organization accepted the application and integrated it fully into daily operation.

Over the next few years, more users would get addresses and access to the system directly via asynchronous terminals. Message traffic started to increase and ISBD offered the XFL system to other GE divisions.

Electronic mail was referred to as message switching then, which was a regulated application under the law in the U.S. and would remain so until January 1981, when it was deregulated and computer service companies entered into the E-mail market. A new commercial application was born and several companies jumped into the market, some as service providers and others as the software developers. E-mail as an industry continued to grow steadily until someone discovered the Internet — now almost everyone has an E-mail address.

**PRIMARY ELECTRONIC MESSAGING SYSTEM CATEGORIES**

Today there are four primary categories of E-mail systems and users:

- *Online services.* A relatively small number of services provide E-mail to a large number of users. Examples include CompuServe, America Online, and Prodigy. There are an estimated 12 million users of these services worldwide.
- *Commercial services.* These are traditional computer service companies with mostly corporate clients providing connectivity between companies. Examples include AT&T, GE Information Services (GEIS), MCI, and Sprint. There are an estimated 1.5 million users of these services worldwide.
- *Private E-mail.* These E-mail systems are proprietary to companies and large organizations and are maintained and operated by them. Examples include General Motors, Pfizer, and JC Penney. There are an estimated 70 million users of these services worldwide.
- The Internet. An estimated 35 million users worldwide use Internet E-mail.

**FEATURES AND FUNCTIONS**

When selecting what features an E-mail system should have, the IS department must keep the users in mind. The E-mail system must serve the users. Reliability and maintenance are also critical. There is no 800 number to call if something goes wrong with the LAN server. The LAN is a proprietary system that has to be repaired in-house. As user communities within companies expand, so does the reliability and service problem. As a company grows, so does its local networks, and soon IS and the network staff are maintaining a worldwide collection of them.

Network managers must also be concerned with connectivity. E-mail users, if they do not already, may soon need to communicate with people outside their immediate community. All the popular E-mail systems today have gateways.

## X.400 and SMTP

X.400 is the international standard for interconnecting unlike messaging systems. The X.400 recommendations were developed and continue to be upgraded by the International Telecommunications Union Telecommunications Standards Sector (ITU-TSS), an organization charted by the United Nations that represents most of the countries with modern telephone systems. Almost every E-mail vendor offers X.400 software to connect its system to the commercial world. The software is still expensive, but it is reliable and fast, handles attached files well, and offers excellent security. It does, however, have a slight problem with addressing.

Most E-mail system vendors now offer Simple Mail Transfer Protocol (SMTP) gateways with their products to connect to the Internet. SMTP is reliable, almost as fast as X.400, does an acceptable job with binary files, has its own addressing problems, and is inexpensive. There is still the directory problem.

The commercial service world also offers proprietary gateways for many private E-mail systems to their public services, which gives the corporate user a window to the E-mail world. For many private E-mail system clients who do not yet need X.400 software, the commercial services offer a gateway to the X.400 world. All of them also provide gateway services to the Internet.

## X.500 Directory Service

All of this connectivity introduces the most serious problem in E-mail today — addressing and directories. Worldwide connectivity does no good if there is no map for getting around.

X.500 is E-mail's atlas. It can interconnect distributed directories, but it is still waiting in the wings. The North American Directory Forum (NADF) has been showing a demo of interconnected X.500 for 2 years. Commercial service providers are trying to lure corporate clients into using X.500. Some large companies are even experimenting with their own in-house X.500 systems. Because there are still concerns about privacy and security with X.500, many companies are investigating alternatives. This brings ad-

# THE ELECTRONIC MESSAGING ENVIRONMENT

ditional pressures on the E-mail system vendors to define and offer competent directory services. Companies such as Hitachi are, in addition, introducing directory synchronization products such as SyncWare.

**Features and Services Checklists**

Following is a checklist of features users should look for when reviewing E-mail products:

- Editing capability
- Distribution lists
- Import/export capability
- File Transfer Body Part (FTBP) or BP-15 (X.400) ability
- A spell checker
- The ability to send forms
- Function keys
- Reply options
- A calendar/scheduler feature

Following is a checklist of services users should look for when reviewing E-mail products (several of these services are discussed in the section "E-mail Services"):

- Directories
- Ad hoc entries
- Fax output
- Fax input
- Message notification
- Delivery options
- Security
- Encryption
- Keyboard combinations
- On-screen help
- Computer-based training (CBT)
- Message storage and archive
- Storage backup
- Communications protocol
- Comm port/comm line backup
- Expansion capabilities

Following is a checklist of services users should look for to ensure sufficient connectivity in an E-mail system:

- SMTP gateway
- X.400 gateway
- Telex
- Fax
- Mail API (MAPI)

- Wireless services
- Pager services

**Features for Creating Messages**

Fortunately for users, competition in this field is intense and many of the newer E-mail systems have similar features and capabilities. Besides sending and receiving mail, creating messages is the next most important function.

A single function key should initiate this operation and prepare the user to address the message and set options. Most systems set the cursor at the TO: block, and the next keystroke should open the address book and point to the first entry, starting with the letters matching those keyed in.

Identifying the name selection with (in many cases) the return key, the user can set the address book to identify the next entry with further keystrokes. There should be no limit to the number of addresses selected. When the TO: block is filled, then a tab or another keystroke should place the user at the cc: block. This operation is processed in the same way as the TO: block. Many systems also allow for blind copies — these are addressees who receive the message but are not shown in the address block.

**Editing Capability.** After the addressing tasks are completed, the user can proceed to constructing the message text. The majority of E-mail messages are written on the fly and perhaps include some previously prepared text. Full-page editing features are a must. Cut-and-paste manipulation of the text, along with import of existing files, allows the user to create messages efficiently.

It is extremely convenient to have a spell checker in the E-mail system. If the E-mail system is part of a complete office support system such as MS Office or HP Desk, then the spell checker will be available and probably shared between the individual components. That way the user does not have to keep updating separate new-word dictionary files.

**Attaching Files.** After the user has completed the text of the message, he or she may want to include some additional files to accompany the message. These can be word processing documents, spreadsheets, drawings, charts, and graphs. Any file that can be stored on the PC, workstation, or mainframe should be able to join the message in transit to its destination. This is a standard capability of most E-mail systems, X.400, and the Internet, although they all handle attached files in different ways. Gateways are improving at allowing attached files to cross these boundaries.

Clicking on the attached file icon or function key should place the user in the directory reserved for attached files. Selecting each with a single keystroke should add the attachments to the file list for this message. Again, there may be no technical limit to the number or size of files, but

# THE ELECTRONIC MESSAGING ENVIRONMENT

the speed and reliability of transit may be affected if the files are too large in size.

**File Compression.** Many organizations struggle with whether or not to offer file compression. On a commercial service, the cost of the transfer is always important. Equally important is the time of transit. Then the question arises, "does the receiver on the other end have the same compression algorithm?"

Within the corporate environment, file compression should be easier as long as all employees have moved up to the same version of the software. The cost savings of closing the data center and putting every location on LANs is countered with the problem that most LANs operate independently and with differing versions of the software, making E-mail and file transfers more of a challenge.

**Distribution Lists.** Associated with addressing are distribution lists. Some systems cannot handle lengthy address lists. This is not necessarily a design flaw; more often, it represents memory limitations of intermediate or receiving systems. A single address entry can add a few hundred bytes to the header; therefore, distribution lists are highly recommended.

Distribution lists can be thought of as header compression techniques. Most commercial services charge for additional message copies, whether they are TO:s or cc:s. Therefore, the use of distribution lists reduces invoices as well as transit times.

In many systems, distribution lists can only be created by the administrator. Other systems allow the user to create and maintain their own distribution lists. In either case, their use is recommended for efficiency. However, many people overuse the distribution list by, for example, sending everyone a message that not necessarily everyone needs to see.

**Importing Text.** The importing feature can be used during message creation to save portions of messages for future use. IS, for example, can use this feature to explain aspects of messaging to users. An IS staffer can save many previously used answers in files on his or her PC and incorporate them in messages to current queries. This expedites the job of IS and ensures that each user gets a complete and accurate response each time. If a new query comes up, IS can use the export feature to save that response for future use.

When importing text, users may also want to search for a particular string to verify that they brought in the correct file. The ability to search the text of the message being created is a handy feature. For frequent E-mail users, it is almost a necessary tool.

**Signature Files.** Many users frequently import their signature file, which may consist of, for example, the user's X.400 address, Internet address, and telephone number. Sometimes addresses in the FROM: block of a message get scrambled or expanded by gateways in transit. A signature file contains the correct version of the sender's address. With all the gateways around the world, this is a highly recommended practice.

**Forms.** Forms are a special type of import. Some E-mail systems allow forms to be generated that are partially filled out by the sender, with the intent that the recipient fill in the rest and return it or pass it on. Forms usually have special fields that are reserved for alpha or numeric import to assist the user in entering required data. The field size can be specified, and on more sophisticated systems, only certain users can enter particular fields. Most systems with forms are restricted to sharing the forms among users of the same system. Going outside that system usually requires that the form be sent as an attached file, if it is possible at all.

## E-MAIL SERVICES

Various services such as directories, fax gateways, message notification, security, and connectivity are provided with modern electronic messaging systems.

### Directories

Directories are one of the most critical and complex problems facing the messaging industry today. The E-mail population is hovering around 100 million users. This population requires an extensive directory, to say the least.

The X.500 solution, however, has been around for about 8 years and acceptance has been very slow. In the short term, most companies are investigating interim solutions.

The bottom line is that the E-mail system should have a flexible directory service that can handle local and remote addresses. Users should be able to enter ad hoc addresses as well as access centrally administered entries from anywhere on the LAN or mainframe E-mail system. When multiple E-mail systems are involved, the gateway system or hub service should contain all entries. Control Data, Soft*Switch, DEC, and Microsoft Exchange all offer this capability. These systems are also X.500 compatible.

**System Directories vs. Local Address Books.** The difference between a system directory and a local address book is that the directory contains all the addresses for an entire system that may include other E-mail systems connected via gateways. The address book is what each user maintains on his

# THE ELECTRONIC MESSAGING ENVIRONMENT

or her individual PC or workstation. When users need the address for someone not in their address book, they search the system directory.

The directory should be available to all users and the directories of separate post offices should be able to exchange entries. Directory synchronization packages are also starting to appear on the market. Hitachi's SyncWare interfaces with a variety of E-mail system directories and includes an X.500 gateway. Individual users should also be able to load their address book from the system directory.

One feature very helpful to users is the ability to cut and paste into an address book from text. Frequently, users get messages with long cc: lists and would like to be able to copy one or more entries into their address books rather than retype the entry. Sometimes messages come through a number of gateways and the FROM: address is about three times as long as when it started out. If the sender includes his or her original address in the text, the receiver can extract it and simply transfer it to an address book.

Another feature that should be required by network-based directories is the ability to handle queries and updates by mail. This feature allows users with the proper access to send queries to the directory to search for particular entries or names, preferably with a mask character. Updating the directory by mail is also a feature needed by remote administrative users, again with the proper security permissions. This feature is not a requirement obvious to LAN users because everyone is connected and can access the directory. However, when there are a variety of systems and directories interconnected via commercial networks or the Internet, query and update by mail is a time saver.

The directory should have a local find capability that allows the user to search either on address or name for an entry. As directories and address books grow in size and scope, these features will be required by all users. Eventually, users will be able to query an X.500 directory for any entry they need.

## Fax Gateways

Even before the Internet, there was the fax. Fax gateways have existed on commercial E-mail services for more than a decade. PCs have added fax gateways within the past 7 or 8 years. Recently fax modem prices have fallen, so it is affordable for almost every PC and LAN to send and receive faxes. One major convenience of a fax modem for travelers is that they do not have to carry a printer on the road with them.

## Message Notification

PCs now have literally hundreds of applications for users, and most people only spend a short time each day on E-mail. Therefore, when a message arrives, the user may want to be interrupted and notified. Some systems provide a capability that informs the user of new mail. The form of notification, either a flashing indication in the corner of the screen or a simple beep, should be set by the user. This capability should also include a message when the PC is turned on that there is mail waiting.

## Security

As more critical business information is transported via E-mail, security options have become more important to system implementers. Many of these have been a standard part of electronic data interchange (EDI) for years and are starting to show up in the E-mail side of the industry. As the cost for sending files via E-mail decreases, the need for additional security increases.

## Gateways

This may be the most often used capability when selecting E-mail systems. Some companies feel they have to decide between either X.400 or the Internet. Most E-mail systems now come standard with a simple mail transport protocol (SMTP) gateway for Internet, and almost every E-mail system on the market has an X.400 gateway.

For an E-mail system or service to survive, it must provide access to the Internet at least via E-mail. The standard that is quoted most often for E-mail access to the Internet is RFC-822, which specifies the rules for SMTP. This is a democratic procedure for posting proposed specifications on the Internet and allowing people to debate the pros and cons of all or part of the specifications. After a proper time period, the request for comment (RFC) committee decides to make the RFC part of the standing rules of the Internet.

**X.400 Software.** The X.400 standard is administered by committee. The ITU-TSS has standing committees that create and maintain the recommendations for telecommunications. The most familiar of these are the X series, including X.25, X.400, and X.500. Each of these has committees made up of representatives from the international telephone companies, U.S. phone companies, and software companies involved in the telecommunications industry. These committees meet periodically to review the status of their efforts and between meetings usually share information via E-mail. When they complete a new version of the recommendations, they gather in a plenary session for approval from the ruling committees and new final versions are published and announced.

The meetings used to occur on a regular 4-year cycle, but have recently been changed to an as-needed basis. The recommendations can be purchased from government offices, the UN, or companies associated with the

# THE ELECTRONIC MESSAGING ENVIRONMENT

ITU-TSS. There are also supplementary documents available, such as the implementer's guide. It still helps to have a resident expert when designing and writing an X.400 gateway.

Approximately 18 companies around the world actually offer X.400 software. In the U.S., Digital Equipment Corp. and Hewlett-Packard supply X.400 software for their systems. ISOCOR and OSIware (Infonet Software Services) offer X.400 systems of a more generic nature to interconnect systems from other vendors. Europe has a number of suppliers of X.400 systems, such as Marben and Net-Tel.

**Telex, Fax, and Wireless.** Other gateways often required by E-mail systems are telex, fax, and wireless. Telex is still used extensively around the world. All service providers offer a gateway to telex. Some private E-mail systems offer a telex gateway, but this requires a separate service agreement with a telex service provider. This gateway must work both ways, unlike most fax gateways.

ISOCOR offers a fax gateway that permits fax into X.400. The fax user sends a document and, after the connect indication, keys in a code that corresponds to a directory entry in the ISOCOR software. The incoming fax is converted to text and routed to the address in the directory. This is a handy capability for medium-to-large scale service.

With the increase in the use of wireless communications for PCs and the ever-popular pager, many E-mail systems are starting to incorporate gateways for wireless services. This requires a third-party provider, but it does offer the user that last-ditch method for reaching someone away from the office.

## POPULAR MESSAGING SYSTEMS

The following sections focus on the pros and cons of the leading messaging systems on the market today that can be installed in companies.

### Lotus cc:Mail

Lotus cc:Mail has dominated LAN-based systems since its introduction a few years ago. cc:Mail provides efficient directory services and most of the features anyone would need. It handles many of the common APIs available and operates on most platforms. Many other E-mail systems imitate cc:Mail, but it has remained the market leader.

### IBM Office Vision 400

IBM Office Vision had difficult beginnings. It survived mainly on the strength of IBM in the mainframe market. Distributed Office Support System (DISOSS), which eventually became Office Vision/Multiple Virtual Storage (OV/MVS), is a very large system requiring a well-trained, knowledgeable staff. It is not the most user-friendly system and has the limited 8×8 (DGN.DEN) addressing common to SNA Distribution Services (SNADS)-based systems. PROFS, the precursor of OV/VM, is a much more user-friendly system that can use nickname files, includes a calendar feature, and has a more flexible directory system.

The most popular entry in the Office Vision stable is OV/400. The AS/400 platform is quite possibly the biggest seller after the PC. AS/400 is extremely popular in Europe and the U.S. Although OV/400 still uses the SNADS 8×8 addressing, it does have the personal address book feature popular in other systems. OV/400 also has the calendar capability found in OV/VM. Gateways are available for X.400 and the more popular APIs.

## DEC All-in-One

Digital Equipment Corp. offers two of the older and most popular E-mail systems today. VMSmail, like Unix Sendmail, is a utility feature in the VMS operating system. It is a command-line-oriented system that is strongly enhanced by the availability of DEC's Message Router system. Even with the overwhelming popularity of full-screen systems, there are still many VMS-mail systems active today.

The mainstay for DEC E-mail systems is All-in-One. This is a full-service, full-screen presentation E-mail system. Along with MAILbus, DEC All-in-One supports X.400 and SMTP as well as X.500 directory protocols.

All-in-One can operate on a stand-alone VAX as well as an entire network of VAXs interconnected via DECnet. With MAILbus, All-in-One interfaces with LANs and other E-mail systems, including OV/VM and OV/400 via SNADS. Many companies with multiple E-mail systems use All-in-One and MAILbus as their central hub system. Distributed directory services (DDS) capability, combined with the X.500 protocols, ensure that this system will be around for a while.

## HP Open Desk Manager

HP Open Desk Manager is the premier system for Unix-based E-mail on midrange systems. HP mail is based on SMTP and therefore readily interfaces with the Internet. Hewlett-Packard also offers a full X.400 system, which allows it to communicate with commercial service providers. The flexibility of HP Open Desk Manager includes interfaces to cc:Mail and Microsoft Mail, as well as Wang and Office Vision systems. The directory service is extremely flexible and allows for local and remote entries in various formats. The directory allows the user to search on a number of fields in the database, making this a very useful tool.

# THE ELECTRONIC MESSAGING ENVIRONMENT

## Lotus Notes

Compared with the other products, Lotus Notes is a fairly new arrival. However, with the strength of Lotus, now backed by IBM, Notes will most likely be around for a long time.

Notes is a true workgroup system that incorporates the spreadsheet capabilities of Lotus 1-2-3 with a database/foldering capability that has made it instantly popular. Many people wonder how the two products — Lotus cc:Mail and Lotus Notes — will develop. cc:Mail will probably take over as the E-mail engine for Notes eventually.

## Microsoft Exchange

The dominance of Microsoft in the computer industry has almost guaranteed success for the various E-mail products they offer. The original MSmail for the Macintosh is still popular and one of the most widely used. MSmail for the PC is the old Network Courier, acquired by Microsoft years ago.

Microsoft's new Microsoft Exchange is based on the Windows NT server. It interfaces the older products into the NT systems and includes interfaces to cc:Mail, Lotus Notes, and others. Microsoft Mail uses the X.400 gateway. Other X.400 software vendors also offer gateways for Microsoft Mail. Standard with Microsoft Mail Exchange is the SMTP gateway, which interfaces to the Internet for E-mail. Microsoft is also offering software for direct access to the Internet.

## Fisher TAO

Fisher International has been a very strong player in the E-mail market with EMCEE. This is a mainframe-based system that runs on the VMS platform. The newer version, EMCEE/TAO, incorporates paging facilities along with calendar and X.400 gateways. It offers the usual SNADS gateway as well as SMTP and LAN message handling system (MHS) connectivity. The company plans to add an X.500 capability, which should significantly enhance its market share.

## Memo

This mainframe E-mail system was developed by Verimation in Europe for a single client. It became so popular there that they started marketing Memo in the U.S. Two years ago they added X.400, and more recently an SMTP gateway, and the product is still selling. Memo owns a sizeable share of the messaging market, primarily in Europe.

## BeyondMail

Banyan Systems' BeyondMail does not command a large portion of the E-mail market, but it is important to mention because it is the guidepost against which E-mail systems should be measured. Since its introduction about 5 years ago, BeyondMail has had more functions and features than any of its competitors. It easily accesses documents from other applications such as MS Word, WordPerfect, and Lotus 1-2-3. It has a very flexible and powerful directory service based on Novell MHS. It includes an SMTP gateway and runs on Unix, which makes it a natural for linking with the Internet. BeyondMail also runs on Windows, DOS, and Mac platforms. There is also a link to calendar and scheduling systems and a rules-based scripting language that helps the user interface with other applications on the LAN.

### QuickMail

CE Software QuickMail is probably the best known of the Macintosh-based E-mail systems. There are more than 2.5 million users. This product is very popular in the advertising and publishing industries. It also runs on Windows and DOS platforms; however, the Mac is where it shines. QuickMail interfaces well with word processing packages on the Mac and is capable of sending attached files, including drawings. There are many gateways to commercial systems and the Internet, making this a popular and versatile system.

### GroupWise

Novell is the newest entry into the E-mail market. However, due to the strength of Novell in the industry, GroupWise ranks about third in the LAN market with more than 11% of the mailboxes. GroupWise can link to other systems via MHS and has an SMTP gateway to access E-mail on the Internet. Several X.400 software vendors offer gateways for GroupWise, making the worldwide reach of this system impressive.

### SUMMARY

This chapter has discussed all of the necessary features to look for in an E-mail package for corporate use. In summary, the leading E-mail packages are described and compared. Most of the different E-mail systems use gateways between each other, so there is little overlap of user populations, and almost everyone is able to communicate with each other. Unfortunately, it is still difficult to find someone's address. Among all the E-mail systems on the market, cc:Mail leads the LAN systems with number of mailboxes installed, and IBM Office Vision is on top of the list of midsize and mainframe systems.

# Chapter 4
# An Introduction to Microsoft Exchange Server

*Lee Benjamin*

Microsoft Exchange Server embraces Internet standards and extends rich messaging and collaboration solutions to businesses of all sizes. As such it is the first client/server messaging system to integrate E-mail, group scheduling, Internet access, discussion groups, rules, electronic forms, and groupware in a single system with centralized management capabilities.

Microsoft Exchange Server provides a complete and scalable messaging infrastructure. It provides a solid foundation for building client/server solutions that gather, organize, share, and deliver information virtually anyway users want it. Microsoft Exchange Server was designed from the ground up to provide users and administrators with unmatched open and secure access to the Internet. Native SMTP support, support for MIME for reliable delivery of Internet mail attachments, and support for Web browsers and HTML ensures seamless Internet connectivity.

Since its introduction in the spring of 1996, customers have been evaluating and deploying Microsoft Exchange Server along with Microsoft Outlook and other clients in record numbers. This chapter offers an overview of how Microsoft Exchange Server can help organizations improve their business processes, work smarter, and increase profits through improved communication. The topics to be covered include:

- Trends in messaging and collaboration
- Infrastructure for messaging and collaboration
- Redefining groupware
- Internet connectivity
- Easy and powerful administration
- Building a business strategy around Microsoft Exchange Server

THE ELECTRONIC MESSAGING ENVIRONMENT

Microsoft Exchange Server is part of the Microsoft BackOffice integrated family of server products, which are designed to make it easier for organizations to improve decision making and streamline business processes with client/server solutions. The Microsoft BackOffice family includes the Microsoft Windows NT Server network operating system, Microsoft Internet Information Server, Microsoft Exchange Server, Microsoft SQL Server, Microsoft SNA Server, and Microsoft Systems Management Server.

## TRENDS IN MESSAGING AND COLLABORATION

Information, both from within organizations and from outside sources, is becoming one of the most valuable commodities in business today. Never before has so much information been so readily available. Nor have there been such high expectations for how much individuals will be able to accomplish with this information. To take advantage of this information, businesses are rethinking every aspect of their operations and reengineering business processes to react more quickly, become more responsive, provide better service, and unify teams separated by thousands of miles and multiple time zones.

Until now, organizations looking for a messaging system had two choices: either a host-based system that provided beneficial administrative capabilities but was costly and did not integrate well with PC-based desktop applications, or a LAN-based system that integrated well with PC-based desktop applications but was not scalable and was less reliable than host systems.

### Unifying LAN- and Host-Based E-Mail

Microsoft Exchange is not a response to any one single product. Rather, it is the evolution of messaging products in general. For the past 10 years, Microsoft has been a leader in LAN messaging solutions. In 1987, Microsoft released the first version of Microsoft Mail. This product was significant in two ways.

First, as a client/server implementation of messaging, Microsoft Mail was a test platform of what Microsoft Exchange Server would become. Second, Microsoft added a programming layer (API) to the product. One might say that this was the grandfather of MAPI, the Messaging Application Programming Interface upon which Exchange is built. Microsoft Mail for PC Networks now has an installed base of well over 10 million copies. Over the years, customers have told Microsoft what they wanted in their next-generation messaging system. Microsoft Exchange is that product.

Microsoft Exchange Server delivers the benefits of both LAN-based and host-based E-mail systems and eliminates the shortcomings of each approach. It integrates E-mail, group scheduling, electronic forms, rules,

## An Introduction to Microsoft Exchange Server

groupware, and built-in support for the Internet on a single platform with centralized management capabilities. Microsoft Exchange Server can provide everyone in the organization, from professional developers to administrators to end-users, with a single point of access to critical business information. It makes messaging easier, more reliable, and more scalable for organizations of all sizes.

Technology is not only changing how businesses process and assimilate information, it is affecting how this information is transferred, viewed, and acted upon. Electronic messaging plays a pivotal role in this process. The annual growth in individual electronic mailboxes is phenomenal. It has been estimated that there was an installed base of more than 100 million mailboxes worldwide at the end of 1996. Five key trends have led to this growth:

- *Growth in PC use.* It has been estimated that an installed base of more than 200 million personal computers exists worldwide. Performance increases and price decreases have expanded the demand for PCs both at home and in the workplace. In addition, the rapid acceptance of the Internet by companies as a marketing vehicle indicates that the PC has reached the mainstream for more than just recreational or business uses.
- *Adoption of the graphical user interface.* The intuitive, icon-based graphical user client interface has made E-mail applications easier to use and has made it possible to create more sophisticated messages than with previous MS-DOS and host-based E-mail applications.
- *Integration of messaging in the operating system.* As messaging functionality has been integrated into the operating system, every application has become "mail-enabled." Users are able to easily distribute documents and data from within applications without having to switch to a dedicated E-mail inbox. Messaging functionality in the operating system also provides a platform for critical new business applications such as forms routing and electronic collaboration.
- *Client/server computing.* Client/server computing combines the flexibility of LAN-based systems — for easier management and extensibility — with the power and security of mainframe host-based systems. In addition, client/server technology has made electronic messaging systems "smarter" so they can anticipate problems that previously required human intervention, thus reducing overall support costs.
- *Growth of the Internet.* The growth of the Internet is perhaps the most important platform shift to hit the computing industry since the introduction of the IBM personal computer in 1981. The most explosive expansion is expected to be in the use of Internet technologies to improve communication within organizations.

## THE ELECTRONIC MESSAGING ENVIRONMENT

Anticipating these trends, Microsoft Exchange Server was developed to unify host-based and LAN-based environments that have historically been separate. Microsoft Exchange Server incorporates both messaging and information sharing in a unified product architecture. By taking advantage of client/server technology, organizations receive the scalability benefits of host-based environments and the flexibility of LAN-based environments.

**The Microsoft Exchange Product Family**

The Microsoft Exchange product family consists of:

- The Microsoft Outlook family of clients (Outlook, Outlook Express, and Outlook Web Access) for Microsoft Exchange Server, which includes E-mail, rules, public folders, sample applications, native Internet support, rapid and easy electronic forms, and an application design environment that makes it easy for users to create groupware applications without programming. These clients are based on protocols such as MAPI, IMAP4, POP3, and LDAP. Outlook is the premier client for Microsoft Exchange Server and is available for Windows 95 and Windows NT. A version of Outlook is also available for the Windows 3.x and Apple Macintosh System 7.x operating systems. Other clients such as POP3 and Web-based clients can also access Microsoft Exchange Server.
- Microsoft Exchange Server, which consists of core components that provide the main messaging services — message transfer and delivery, message storage, directory services, and a centralized administration program. Optional server components provide seamless connectivity and directory exchange between Microsoft Exchange Server sites linked over the Internet, via X.400 or other messaging systems. Microsoft Exchange Server supports SMTP and X.400 as standards to ensure reliable message transfer for systems backboned over the Internet or other systems. It also provides outstanding NNTP (Network News Transport Protocol) and Web access and integration, enabling customers to easily access all types of Internet information.

Microsoft Exchange Server must also be a platform for an assortment of business solutions, which organizations of all sizes can implement to meet a wide range of key challenges, including:

- Making it easy for salespeople and support technicians to find product information
- Improving an organization's access to market information
- Allowing users to receive timely and accurate information regarding sales and product activity
- Making individual and group scheduling fast and easy
- Improving customer tracking

- Allowing all technicians and engineers to share common customer technical issues
- Continuing to expand the network and add new applications while maintaining complete compatibility with existing systems

## INFRASTRUCTURE FOR MESSAGING AND COLLABORATION

Microsoft Exchange Server combines the best features of both host-based and LAN-based E-mail systems with some additional benefits all its own. The result is a messaging system that is easy to use and manage and that moves messages and files through the system quickly, securely, and reliably, regardless of how many users or servers the organization has.

### Universal Inbox

The Universal Inbox in the Microsoft Outlook Client lets users keep all messages, forms, faxes, and meeting requests in one location, where they can be easily accessed. Users can search and sort these items using a wide range of criteria — such as addressee, topic, or date of receipt — to quickly locate the information they need.

In addition, server-based rules automatically process incoming messages, including those from the Internet, even when the user is out of the office. These rules can be configured to file incoming messages in appropriate folders or to respond immediately with specified actions, such as forwarding messages to another person, flagging them for special attention, or generating a reply automatically.

### Tight Integration with Desktop Applications

Because Microsoft Outlook is tightly integrated with the Microsoft Windows operating system and the Microsoft Office family of products, it is easy for users to learn and use. Microsoft Outlook actually ships with both Exchange Server and Microsoft Office 97. With new features such as "journaling" (which allows a user to find a file based on when that file was used, rather than by file name), Outlook can keep track of what users do every day.

### Fast, Secure, and Reliable

Microsoft Exchange Server takes full advantage of the robust client/server architecture in Windows NT Server to get messages to their destinations quickly, whether across the hall or around the world. It also provides tools for easily tracking messages sent to other users of Microsoft Exchange Server and via the Internet to users on other systems, to confirm that they arrived and that they were read. Support for digital encryption allows users to automatically secure messages against unauthorized access, and digital signatures guarantee that messages get to their recipients without modification.

In addition to these security features, Microsoft Exchange Server also takes advantage of the security features built into Windows NT Server to prevent unauthorized users — inside or outside the organization — from accessing corporate data.

### Remote Client Access

Local replication is the ability to do two-way synchronization between a server folder and a copy of that folder on a local or portable machine. Local replication is initiated by creating an offline folder — a snapshot or "replica" — of the server-based folder the user wishes to use while disconnected from the server. (The use of offline folder synchronization is discussed further in a subsequent section.)

### Scalable

Built on the scalable Windows NT Server architecture that supports the full array of Intel and Digital Alpha-based servers, Microsoft Exchange Server scales to meet a range of requirements — from those of a small, growing office to those of a multinational corporation.

It is easy to add users to existing servers and new servers to an organization as it grows. Routing and directory replication occur automatically between the new and existing servers at each site. Plus, optional connectors are available to connect computers running Microsoft Exchange Server to the Internet and X.400 systems.

### REDEFINING GROUPWARE

In addition to E-mail, which allows users to send information to each other, Microsoft Exchange Server and Microsoft Outlook support groupware applications that help users share information by retrieving it wherever it might be — without the traditional complexities of navigating through a maze of network servers or jumping between multiple screens and applications.

A built-in suite of groupware applications in Microsoft Exchange Server gives users a headstart with group scheduling, bulletin boards, task management, and customer tracking. Because these applications are designed to integrate tightly with the Windows operating system and Microsoft Office, Microsoft Exchange Server provides an ideal platform for integrating business solutions with desktop applications.

While some Microsoft Exchange Server applications are ready to go right out of the box, and many more are also available on the Web, you can also easily customize them using Microsoft Outlook and extend them using popular development tools such as the Visual Basic programming system, the Visual C++ development system, Java, and ActiveX components.

The concept of discussion groups and bulletin boards are nothing new to Internet users. With the Microsoft Exchange Internet News Service, the complete set of Internet newsgroups are easily available to users through public folders. Organizations can make public folder information available to internal or external users of the Web without storing information in redundant locations or manually reformatting information into hypertext markup language (HTML) format, via Outlook, Outlook Web Access, or any NNTP newsreader client. Users can also access newsgroups hosted on Exchange Server using the native NNTP protocol.

Microsoft Exchange Server also allows users to communicate with each other and to share information from any time zone or location. This is especially important for mobile users, who need to break through traditional organizational boundaries to communicate with the enterprise. Group scheduling and public folders help users work together more effectively, whether they are across the hall, across the country, or around the globe.

**Group Scheduling**

A full-featured personal calendar, task manager, and group scheduler in its own right, Microsoft Outlook has been incorporated into Microsoft Exchange Server to provide a fully extensible system that can act as a rich foundation for business-specific, activity-management applications. It takes full advantage of the advanced client/server architecture and centralized management features in Microsoft Exchange Server.

Microsoft Outlook is a tool for scheduling group meetings, rooms, and resources. To schedule a meeting, users can overlay the busy times of all the attendees in a single calendar to automatically schedule a meeting, conference room, and any other resources required. The Microsoft Outlook contact-management features provide users with easy access to the names and phone numbers that are part of their daily work.

**Public Folders**

Public folders make it easy for users to access information on a related topic all in one place. Documents can be stored in public folders for easy access by users inside and outside an organization. These folders are easy to set up without programming; relevant documents can be dragged into the folder. Microsoft Exchange Server uses these public folders as containers for groupware and custom applications.

# THE ELECTRONIC MESSAGING ENVIRONMENT

## Bulletin Boards

Support for bulletin boards enables organizations to easily share information throughout the enterprise. Information is organized so that users can easily find what they need, leave messages, and communicate about the topic.

It is interesting to note that Internet users have been working with bulletin boards for many years using Usenet Newsgroups. Microsoft Exchange Server uses the same Internet Standard, called NNTP. Sample bulletin board folders, which are easily customizable, are included with Microsoft Exchange Server.

## Outlook Forms

Electronic forms are easy to create and modify in Outlook so users can send and receive structured information. Traditionally advanced features such as drop-down lists and validation formulas are easy to get at and use. More sophisticated capabilities are accessible from Outlook's rich programming extensibility interfaces. In addition, Outlook forms are automatically rendered to the Web so any user can get to them.

## Public Folder Replication

One of the key strengths of Microsoft Exchange Server is its ability to distribute and synchronize shared information through the Microsoft Exchange Server replication system. It is possible to have multiple synchronized copies of folders in different locations regardless of whether users are connected over a LAN or WAN, or the Internet or X.400 backbone.

Replicating information in this way means that synchronized copies of a public folder can reside on multiple servers, distributing the processing load and improving response time for users accessing information within the folder. It also means synchronized copies of a public folder can reside at several geographically separated sites, significantly reducing the amount of long-distance WAN traffic necessary to access information. If a server holding one copy of a public folder becomes unavailable, other servers holding synchronized copies of the same folder can be accessed transparently, greatly increasing the availability of information for users and resulting in a highly reliable system.

Microsoft Exchange Server offers users the unique benefit of location-independent access to shared information. With replication, the physical location of folders is irrelevant to users, and Microsoft Exchange Server hides the sophistication of public folder replication. Users need not be aware of where replicated folders are located, the number of replicated copies, or even that replication occurs at all. They simply find information more easily than ever before.

## An Introduction to Microsoft Exchange Server

With the Microsoft Exchange Internet Mail Service users can replicate public folders and groupware applications throughout a distributed organization, even if they do not have a wide area network. Managing public folder replication is very easy. Using the graphical Microsoft Exchange Server Administrator program, system managers need only select the servers that will receive replicas of the public folders.

**Offline Folder Synchronization**

Microsoft Exchange Server allows users to automatically perform two-way synchronization between a server folder and a copy of that folder on a local PC. For example, a user can create an offline folder — a snapshot or "replica" — of a customer-tracking application to take on a business trip and update it based on interactions with customers during the trip. Then, when the user reconnects to the server — either remotely by modem or by connecting to the LAN upon returning to the office — the folders can be bi-directionally synchronized with the server. Changes, including forms and views, made on the local machine are updated to the server, and changes to the server-based folders automatically show up on the user's PC. Offline folder synchronization lets users maintain up-to-date information without having to be continuously connected to the network.

Creating an offline folder is different from simply copying a server folder to the hard disk, because an offline folder remembers its relationship with the server folder and uses that relationship to perform the bidirectional update. Only changes — not the whole folder — are copied, which helps minimize network traffic.

Microsoft Exchange Server supports multiple simultaneous offline folder synchronization sessions from many different locations. Built-in conflict resolution for public folders ensures that all the changes are added. The owner of the folder is notified if there is a conflict and can choose which version to keep. With the powerful server-to-server replication technology in Microsoft Exchange Server, this information can then be automatically replicated to users of your system around the world.

**Easy-to-Create Groupware Applications**

Microsoft Exchange Server delivers a scalable set of tools that lets almost anyone — even users who have never programmed — develop custom groupware applications. It also gives professional programmers all the power they need to build advanced business software systems. Microsoft Exchange Server includes these key development features for both users and programmers:

- *Fast applications development without programming.* Users can build complete groupware applications — such as a customer-tracking system or an electronic discussion forum — without programming. As-

## THE ELECTRONIC MESSAGING ENVIRONMENT

suming they have the appropriate permissions, users can simply copy an existing application (including forms, views, permissions, and rules) and modify it as they wish with the functionality available in the Microsoft Outlook Client. They can easily modify existing forms or create new ones with a menu choice to Design Outlook Form, which requires no programming knowledge.
- *Central application management.* Once users complete an application, they will usually hand it off to the Microsoft Exchange Server administrator for further testing or distribution to others within the organization. The Microsoft Exchange Server replication engine manages the distribution of the application or any new forms that may have been revised or created for existing applications. One can also replicate these applications from one Microsoft Exchange Server site to another over the Internet by using the Microsoft Exchange Internet Mail Service.

Both of these capabilities translate into reduced cycles for creating, modifying, and distributing groupware applications. That means end-users can build applications that are valuable to them without having to wait for a response from their IS departmetns. Even if an application turns out to be less useful than the creator hoped it would be, the development cost is minor.

The IS department can also benefit because it can customize those applications that do turn out to be worthwhile, since forms created or modified with Microsoft Outlook are extensible with Visual Basic Script programming system. In addition, Outlook Forms can be further extended with other programming tools such as Visual C++, ActiveX Controls, and Java. By using the Microsoft Exchange Server replication engine, revisions and new applications can be deployed inexpensively as well.

The speedy application design and delivery process made possible by Microsoft Exchange Server enables the people who have the best understanding of the functionality needed to respond quickly to market requirements. As a result, an organization can reduce the costs of adapting and rolling out those applications. Whenever an application is rolled out within an organization, it is usually only a matter of time before the applications developer hears from users about how it could be improved. Many applications provide limited functionality — once the barrier is reached, they cannot be customized any further and require redesign from scratch with a more powerful tool.

Thus, forms designed by end-users can be customized by professional developers using the full power of Visual Basic Script. Other workgroup application design tools either require a high degree of programming skill or quickly run out of steam, as a particularly useful application requires additional functionality. Microsoft Exchange Server opens the door between end-user application design and the full power of the Windows APIs available through more powerful programming languages.

Exchange Server takes these forms even further by automatically rendering them to the Web as HTML forms. By leveraging a technology known as ActiveServer Pages, included with the Microsoft Internet Information Server, forms and the information in them can be seen by any user accessing Exchange from a browser anywhere on the World Wide Web (if they have the appropriate permissions of course).

**MAPI: Messaging Application Programming Interface**

The MAPI subsystem is the infrastructure on which Microsoft Exchange Server is built. Messaging client applications communicate with service providers running on the server through the MAPI subsystem. Through broad publication of Microsoft messaging APIs, and because of the robust messaging and workgroup functionality defined in them, MAPI has become a widely used standard throughout the industry for messaging and groupware clients and providers.

MAPI-compliant clients span a variety of messaging- and workgroup-based applications and support either Windows 32-bit MAPI applications on Windows 95 or Windows NT, and 16-bit MAPI applications running on Windows 3.x. Each of these types of applications can access the service provider functionality needed without requiring a specific interface for each provider. This is similar to the situation where applications that use the Microsoft Windows printing subsystem do not need drivers for every available printer.

Messaging applications that require messaging services can access them through any of five programming interfaces:

- Simple MAPI (sMAPI)
- Common Messaging Calls (CMC)
- ActiveMessaging (formerly known as OLE Messaging and OLE Scheduling)
- MAPI itself
- (In the near future) Internet Mail Access Protocol, or IMAP

Client requests for messaging services are processed by the MAPI subsystem — either as function interface calls (for sMAPI or CMC) or as manipulations of MAPI objects (for OLE Messaging or MAPI itself) — and are passed on to the appropriate MAPI-compliant service provider. The MAPI service providers then perform the requested actions for the client and pass back the action through the MAPI subsystem to the MAPI client.

Third-party programming interfaces that can be built upon MAPI are frequently employed. Because MAPI is an open and well-defined interface, a

THE ELECTRONIC MESSAGING ENVIRONMENT

proprietary third-party API can be implemented on top of MAPI without having to revise the MAPI subsystem itself. Thus, customers and vendors can implement their own MAPI solutions that meet their particular needs without incurring the development costs that would otherwise accrue on other messaging infrastructures.

## INTERNET CONNECTIVITY

Extensive built-in support for the Internet in the Microsoft Outlook Clients, as well as the Microsoft Exchange Internet Mail Service, Microsoft Exchange Internet News Service, and Outlook Web Access, makes it easy for organizations to use the Internet as a communications backbone, to make Internet newsgroup data available to their users through public folders, and to make messaging and public folder information available to the ever-growing numbers of Internet Web users.

The Microsoft Exchange Internet Mail Service provides high-performance multithreaded connectivity between Microsoft Exchange Server sites and the Internet. It also supports MIME and UUENCODE (and BINHEX for Macintosh) to ensure that attachments arrive at their destinations intact. Built-in message tracking helps ensure message delivery. Standards-based digital encryption and digital signatures ensure message security.

These capabilities make it possible for organizations to use the Internet as a virtual private network to connect Microsoft Exchange Server sites over the Internet and to route messages using the TCP/IP SMTP or X.400 protocols. You can easily control who sends and receives Internet mail by rejecting or accepting messages on a per-host basis.

### Integrated Internet Support

The Microsoft Outlook Clients include built-in Internet mail standards to allow users, connected locally or remotely, to reach other Microsoft Exchange Server sites and virtually anyone else using any Internet service provider. Native MIME support allows files to be transported reliably over the Internet. Support for Post Office Protocol, Version 3 (POP3), PPP, and IMAP4 ensures compatibility with all SMTP E-mail systems.

The Microsoft Exchange Inbox — a version of Microsoft Exchange Client that does not include Microsoft Exchange Server–specific functionality — is built into the Windows 95 operating system. This feature makes Internet mail easy to set up and access. Any user with an Internet mailbox via POP3 can use the Internet Mail Driver for Windows 95 in the Microsoft Exchange Inbox. Similarly, any client that supports POP3 can connect to a Microsoft Exchange Server. Outlook Express is an Internet Mail and News Client that ships with Microsoft Internet Explorer 4.0 and supports the SMTP/POP3, LDAP, NNTP, and IMAP4 protocols.

## Direct Connections over the Internet for Mobile Users

The Microsoft Exchange Inbox and Microsoft Outlook clients can also leverage the Internet in another way — as an alternative to dialup connections.

Outlook clients and Microsoft Exchange Server both have built-in support to connect to each other securely over the Internet. Mobile users can use a local Internet service provider (ISP) to connect to the Microsoft Exchange Server site located back in their organizational headquarters. Once this connection is established, users have full access to all server-based functionality, including directory services, digital signature and encryption, group scheduling, free/busy checking, and public-folder applications.

## Support for Internet Newsgroups and Discussion Groups

As previously mentioned, the Microsoft Exchange Internet News Service can bring a Usenet news feed to Microsoft Exchange Server, from which administrators can distribute the feed to users through the public folder interface in Microsoft Exchange Server. Items within a newsgroup are assembled by conversation topic — the view preferred by most discussion group users. Users can then read the articles and post replies to be sent back to the Internet newsgroup.

Using the standard Microsoft Exchange Client Post Note feature, users can post a new article or a follow-up to an article or send a reply to the author of an article. Users have all the composition features of the Microsoft Exchange Inbox for composing posts to discussion groups. As with E-mail, however, the extent to which these composition features can be viewed by other users depends on the encoding format used.

The Internet News Connector automatically uses UUENCODE or MIME to encode outgoing and decode incoming post attachments. Thus, when users see an attachment in a post, they need only double-click and watch the attachment pop up. There is no waiting for the decoder to process the file.

## Outlook Web Access

This capability provides a different, but equally important, kind of integration with the Internet. Outlook Web Service translates the information stored in Microsoft Exchange Server folders into HTML and makes it available — at the document or item level — as a uniform resource locator (URL) to any user with a Web browser. This capability teams up with the Microsoft Internet Information Server (IIS), which hosts the URL. As a result, organizations with documents or discussions they want to make available to Web users inside or outside their organization can accomplish this without storing the information in two different places, manually changing its format into HTML, or requiring that everyone use the same kind of client.

# THE ELECTRONIC MESSAGING ENVIRONMENT

## EASY AND POWERFUL CENTRALIZED ADMINISTRATION

While Microsoft Exchange Server offers the tight integration with desktop applications previously available only with LAN-based E-mail systems, it also offers the centralized administrative capabilities previously available only with host-based systems. Its easy-to-manage, reliable messaging infrastructure gives administrators a single view of the entire enterprise.

### Easy-to-Use Graphical Administration Program

Microsoft Exchange Server includes a number of tools that help administrators reduce administration time while keeping the system running at peak performance. The graphical Administrator program lets administrators manage all components of the system, either remotely or locally from a single desktop. Built-in intelligent monitoring tools automatically notify the administrator of a problem with any of the servers and can restart the service or the server if necessary. Microsoft Exchange Server integrates tightly with Windows NT Server monitoring tools as well, so administrators can even create new user accounts and new mailboxes in one simple step for those users.

### Information Moves Reliably

To keep the right information flowing to the right people, users need to be able to count on reliable message delivery. Using powerful monitoring and management tools, Microsoft Exchange Server helps ensure that the entire organization enjoys uninterrupted service. It even seeks out and corrects problems based on administrator guidelines. If a connection goes down, Microsoft Exchange Server automatically reroutes messages as well as public folder and directory changes, balancing them over the remaining connections. This greatly simplifies administration and ensures reliable and efficient communication.

### Microsoft Exchange Server Components

Let's take a closer look at Microsoft Exchange Server components: private folder, public information store, directory, directory synchronization agent (DXA), and message transfer agent (MTA) objects reside in the server container on Microsoft Exchange Server.

Each server installation of Microsoft Exchange Server automatically contains an instance of the directory, the information store, and the MTA. These Windows NT-based services control directory replication and mail connectivity within a site. Directory and public folder replication between sites, as well as mail connectivity between sites and with other mail systems, are controlled through the Administrator program.

## An Introduction to Microsoft Exchange Server

**Private Folders: Central Storage for Private User Data.** Private folders provide central storage for all the mailboxes that exist on that server. Users have the option to store messages locally, but server-based private folders are recommended for security, management, and backup purposes. Synchronizing server folders to the local machine is the best of both worlds and is the default configuration for people who travel with their computer.

**Public Information Store: Centrally Replicating Global Access Store.** On each server, the public information store houses data that can be replicated throughout the organization. Using the Administrator program to customize this replication, you can allow some data to be replicated everywhere, while other data is replicated only to key servers in each site. Data replication can be tightly controlled because rich status screens, available at all times, enable the administrator to track the replication of data throughout the enterprise.

**Exchange Diretory Replication and Synchronization.** The Microsoft Exchange Server directory provides a wealth of customizable end-user information and covers all the routing information required by a server. Automatic replication of directory information between servers in a site eliminates the need to configure servers.

Directory synchronization has been perceived as the single biggest weakness in LAN-based messaging. Microsoft Exchange Server changes this perception with a process that keeps directories automatically synchronized on a daily basis. This makes it possible to communicate quickly and easily with users on a wide range of messaging systems such as Microsoft Mail for PC Networks, Microsoft Mail for AppleTalk Networks, Lotus cc:Mail, and optionally other messaging systems.

**Message Transfer Agent.** The MTA delivers all data between two servers in a site and between two bridgehead servers in different sites. The MTA is standards-based and can use client/server remote procedure calls (RPCs), Internet Mail (SMTP), or X.400 to communicate between sites.

All transport objects that enable connectivity to other sites and other mail systems reside in the Connections Container in Microsoft Exchange Server. These objects can be accessed directly through the Administrator program. The Connections Container on a Microsoft Exchange Server site houses four objects that enable site-to-site connectivity: Microsoft Exchange Site Connector, Microsoft Exchange Internet Mail Connector, Microsoft Exchange X.400 Connector, and the Remote Access Service (RAS) Connector.

### Single Interface for Global Management

All objects are created and managed through the Administrator program using the same commands. A Microsoft Exchange Server installation

# THE ELECTRONIC MESSAGING ENVIRONMENT

can be implemented using a wide range of connectivity options that are all managed through a single interface. The exchange of all site-to-site information — from user-to-user messaging to data replication to route monitoring — is handled through mail messages. This single administration infrastructure greatly simplifies management of the rich functionality of Microsoft Exchange Server.

**Microsoft Mail Connector.** The Microsoft Mail Connector, included standard with Microsoft Exchange Server, provides seamless connectivity to Microsoft Mail Server for PC Networks, Microsoft Mail Server for AppleTalk Networks, and Microsoft Mail Server for PC Networks gateways. It uses a "connector" post office that is structured as a Microsoft Mail 3.x post office. Each Microsoft Exchange Server site appears to Microsoft Mail Server as another Microsoft Mail post office. A Microsoft Exchange Server site can connect directly to an existing Microsoft Mail post office, allowing you to replace — not just supplement — an existing Microsoft Mail MTA. No additional software is required.

**Lotus cc:Mail Connector.** The Microsoft Exchange Connector for Lotus cc:Mail also provides messaging connectivity and directory synchronization. Customers can co-exist and send information easily between these E-mail systems, and then later migrate when they are ready with the Lotus cc:Mail migration tools that are also included.

**Microsoft Exchange Internet Mail Service.** The Internet has long used several E-mail standards. RFC 821 (also known as Simple Message Transfer Protocol, or SMTP) defines how Internet mail is transferred, while RFC 822 defines the message content for plain-text messages. RFC 1521 (Multipurpose Internet Mail Extensions, or MIME) supports rich attachments such as documents, images, sound, and video. Microsoft Exchange Internet Mail Connector supports all three standards. It also enables backboning between two remote Microsoft Exchange Server sites using the Internet or other SMTP systems, making it an important component for customers who rely on SMTP connectivity to communicate with members of their own organization, as well as other organizations.

**Microsoft Exchange X.400 Connector.** The X.400 Connector supports three different connectivity options — TCP/IP, TP4, and X.25 — between Microsoft Exchange Server and other X.400-compliant mail systems, and between two different Microsoft Exchange Server sites over an X.400 backbone. The X.400 Connector supports both 1984 and 1988 X.400 communication and includes support for the latest X.400 protocol, File Transfer Body Part (FTBP). The X.400 Connector also enables backboning between two remote sites using a public X.400 services, such as MCI or Sprint.

# An Introduction to Microsoft Exchange Server

**Dynamic Dialup (RAS) Connector.** The RAS Connector object is a special-case site connector. It uses dial-up networking (also known as RAS, or Remote Access Services) instead of a permanent network connection, thereby enabling dial-up connectivity between two Microsoft Exchange Server sites. The administrator configures when the connections should be made, and Microsoft Exchange Server connects to the other site at that time. This connector is also standards-based, using the Internet Point-to-Point Protocol (PPP). The Dynamic Dialup Connector can be automatically invoked by the Internet Mail Service and the Internet News Service so companies can participate in the Internet without the added cost of a permanent connection.

## Client Support

Microsoft Exchange Server supports clients running the Windows NT and Windows 95, Windows 3.1, Windows for Workgroups, MS-DOS, and Macintosh System 7 operating systems so that users work within a familiar environment. It uses the built-in network protocol support of Windows NT Server, specifically TCP/IP and NetBEUI. In addition, its network-independent messaging protocol enables Microsoft Exchange Server to work cooperatively with existing network systems such as Novell NetWare.

You can also install and use the Microsoft Outlook Client for Windows 3.x on a Novell NetWare 3.x client running a monolithic IPX/SPX NETx, ODI/NETx, ODI/VLM, or LAN Workplace for DOS (version 4.2 or later) with no modification to the client. Microsoft Exchange clients communicate with the Microsoft Exchange Server computer by using DCE-compatible remote procedure calls, which are forwarded within an IP or SPX packet using the Windows Sockets interface.

## Manage all Components from a Single Seat

Because the connectors for Microsoft Mail, Lotus cc:Mail, the Internet, and X.400 systems all function as core parts of Microsoft Exchange Server rather than as add-on applications, they take advantage of the message routing, management, and monitoring features built into Microsoft Exchange Server. They also integrate with the administrative tools provided in Windows NT Server. By using and extending tools found in Windows NT Server, Microsoft Exchange makes use of strong authentication, provides an easy-to-use backup facility that does not require the system to be shut down to save data, and features an extensive dial-in facility that can manage up to 256 connections on a single server.

Monitoring tools include extensions to Windows NT's Performance Monitor, as well as both Server and Link Monitors that inform network administrators when there is a problem or delay in the system. Microsoft Exchange Server makes use of the Windows NT Event Log to store all types of information on the operating status of the system. This monitoring capability lets the administrator set up an automatic escalation process if a service stops. For example, if the MTA service stops, the monitoring system can be configured to automatically restart it or to notify specific individuals who can determine an appropriate action.

**Easy Migration**

Built-in migration tools make it easy to convert user accounts to Microsoft Exchange Server. These tools work with the existing system and the Administrator program to copy and import addresses, mailboxes, and scheduling information from existing systems. It is also easy to automatically upgrade client software from the server. Migration tools are included for Microsoft Mail for PC Networks, Microsoft Mail for AppleTalk Networks, Lotus cc:Mail, Digital All-in-One, IBM PROFS/OV, Verimation MEMO, Collabra Share, and Novell Groupwise.

## BUILDING A BUSINESS STRATEGY AROUND MICROSOFT EXCHANGE SERVER

Businesses of all types and sizes can implement Microsoft Exchange Server as their information infrastructure. It supports all E-mail, information exchange, and line-of-business applications that help organizations use information to greater business advantage. Microsoft has worked closely with customers throughout the development of Microsoft Exchange Server to help ensure that it meets the needs of even the largest and most complex systems. The following are some common examples of how customers are implementing Microsoft Exchange Server.

**Downsizing**

Many large organizations will migrate their E-mail systems from a host mainframe to a client/server system based on Microsoft Exchange Server. Microsoft Exchange Server provides the security and robust operations capabilities of the mainframe in a more flexible, inexpensive, scalable, and manageable implementation. It also includes migration tools that make it easy to move users from existing LAN-based and host-based E-mail systems.

Customers who are downsizing operations can develop applications for Microsoft Exchange Server using popular languages and development tools not applicable for mainframe computers. These customers require the flexibility that only a family of clients such as Outlook can offer.

**Connecting Multisystem Environments**

Customers with multiple personal computing and network platforms can use Microsoft Exchange Server to link all their users together. Organizations can benefit from the simplified administration of having just one

*An Introduction to Microsoft Exchange Server*

server and a single client interface that supports all popular computing platforms. In addition, Microsoft Exchange Server allows organizations to use the Internet as a communications backbone to connect to and share information with other geographic locations of their own organization as well as with other companies.

## Upgrading Current Microsoft Mail Systems

Many customers have built powerful messaging systems — including electronic forms and mail-enabled applications — with Microsoft Mail. All of their existing messaging investments will seamlessly migrate to Microsoft Exchange Server, allowing them to gain the new capabilities that Microsoft Exchange Server offers without losing access to their mission-critical applications already in place. Customers of other LAN shared-file system based E-mail systems will enjoy the same benefits.

The real test of Microsoft Exchange Server capabilities is in real-life business solutions. The following are just a few of the solutions that can be implemented using the Microsoft Exchange product family.

**Customer-Support Systems.** Organizations have always struggled with the costly problem of duplicating efforts because individuals do not know that others have already tackled the same issues. A customer-support system can remedy this problem by allowing support technicians to document and share their experiences and acquired knowledge with their colleagues in other support centers. This sharing helps keep organizations from "reinventing the wheel," because all employees can see and use the information and ideas generated by others. It also allows technicians to automatically route product bug reports to the engineering staff at the home office.

**Customer Account Tracking.** Providing superior customer service with distributed sales teams requires excellent communication among all team members and a shared history of customer contact. Inconsistent communication with customers is one of the main reasons companies lose customers to competitors.

An account-tracking system improves the management of customers by enabling account managers to see at a glance whenever anyone in the company has made contact with a customer account. A customer-tracking system also helps identify solid new sales opportunities and pinpoint customer problems that require immediate attention. Because many account managers travel extensively, this information must be accessible both from the office network and from remote locations such as hotel rooms, airports, or home.

**Sales Tracking.** Today, every organization that manufactures a product worries about the high cost of carrying large inventories of finished goods and supplies. A sales-tracking application can help businesses make better manufacturing planning decisions by helping sales managers and marketing executives get up-to-the-minute information, including sales volumes by region, product, and customer. This information makes it possible to identify regions or products that require special attention and to make more informed projections of demand for each product.

**Product Information Libraries.** The key to excellent customer service is providing customers with the right information, right now. A product information library application can help organizations improve customer service by providing salespeople with up-to-date, correct information.

This online library must contain a variety of interrelated information, including word-processing documents, spreadsheets, presentation graphics slide shows, E-mail messages from product managers, and, increasingly, multimedia elements such as images, sound clips, and videos. Sales reps can have read-only access to this library, while product managers at any location can change and modify those items that pertain to their particular products.

Such an electronic library of product information eliminates the need to continually distribute new printed product information to the sales force, which in turn eliminates the problem of disposing of expensive inventories of obsolete brochures and data sheets when products change.

**A Market and General Information Newswire.** Today's rapid business pace requires that managers stay in constant touch with business trends that will affect their markets and customers. A newswire application provides an easy way for employees to stay in touch with important trends, the needs of customers, and their competitors without a separate specialized application.

## SUMMARY

By integrating a powerful E-mail system, group scheduling, groupware applications, Internet connectivity, and centralized administrative tools all on a single platform, Microsoft Exchange Server makes messaging easier, more reliable, and more scalable for organizations of all sizes. Microsoft Exchange Server is also a highly extensible and programmable product that allows organizations to build more advanced information-sharing applications or extend existing applications easily, based on existing knowledge. In addition, it provides the centralized administrative tools to keep the enterprise running securely behind the scenes.

As a result, Microsoft Exchange Server can help organizations save time and improve all forms of business communications, both within and beyond the enterprise. By the time this chapter goes to press, the next version of Microsoft Exchange Server will already be available. New functionality in its clients, more integration with the Internet and Web, and greater scalability and performance are just a few of the improvements in store for customers.

Microsoft Exchange Server was designed to handle today's messaging and collaboration requirements. It is built on existing Internet standards and is designed to easily adopt new and emerging technologies to provide the best platform to its customers. Messaging is an evolutionary technology and Microsoft Exchange Server provides the foundation for any organization's messaging and collaboration growth.

# Chapter 5
# Novell Messaging Products

*Bruce Greenblatt*

Novell, Inc., is the world's leading network software provider, connecting people to other people and the information they need and enabling them to act on it anytime, any place. The company's software products provide the distributed infrastructure, network services, and advanced network access required to make networked information and pervasive computing an integral part of everyone's daily life.

Today's Novell was born from a dying hardware manufacturer called Novell Data Systems. When Novell Data Systems experienced problems with its workstation hardware business during the early 1980s, its source of venture capital, Safeguard Scientific, began looking for someone to salvage the company. Ray Noorda was invited to take a look at the company. He was not impressed by the hardware products, but he did take an interest in the ideas of SuperSet, a foursome of software development consultants that Novell Data Systems had hired in October 1981 to network its CPM Z80 microprocessors. SuperSet and Ray shared the same networking vision. Distributed computing systems could make inroads into applications traditionally implemented on minicomputers and mainframes.

SuperSet demonstrated their newly created file server operating system at COMDEX '82 using nine networked PCs. Ray Noorda stopped by their booth to take a look. By early 1983, Novell Data Systems was reorganized as Novell, Inc., and Ray Noorda joined the company as president. It did not take Ray long to change the focus of the new company to networking. Novell introduced what would become its flagship product later in 1983. The first version of NetWare was network file server software based on the Intel 8086 microprocessor. The company quickly began porting NetWare to other vendors' hardware. SuperSet, with the help of Novell engineers, soon rewrote·NetWare in 286 mode and added a variety of revolutionary system fault-tolerant features such as disk mirroring. The revamped product was demonstrated at COMDEX '85 — the same year the company went public.

## DEFINING THE FUTURE OF NETWORKING

In 1994 Ray Noorda stepped down as Novell's president and CEO, naming Bob Frankenberg as his successor. Bob articulated Novell's pervasive computing vision, defining the business of Novell as connecting people with other people and the information they need, enabling them to act on it any time, any place. This vision centers on the idea that soon the smart global network will be as indispensable and easy to access as a telephone providing a near-instant digital link to global network resources. It will be as if all businesses, from the smallest mom-and-pop shop to the largest enterprise, are branch offices of one gigantic worldwide virtual corporation.

Novell is working with other leading technology companies to build the infrastructure to fulfill this global networking vision. As the world leader in networking software, Novell has the technology, partnerships, and strategy to make the networked world a reality. Novell's strategy for forging the future of networking rests on three basic principles: build smart networks; give users network access any time, from any place; and enable heterogeneous networks to be easily connected.

To make the vision real, Novell is now creating systems and services that build the infrastructure for the smart global network; open programmer interfaces that will enable the network to integrate heterogeneous applications, systems, and platforms from all types of developers; and products that enable universal access to the global network. At the core of Novell's strategy for building a smart global network is today's NetWare 4.1, the fastest, most reliable, most scalable, and best-supported networking environment available. NetWare is the ideal platform on which to build smart network services, and the company has started with NetWare Directory Services (NDS). With more than 10 million users, NDS is quickly emerging as the de facto standard for networks worldwide.

Other smart networking services are available alongside NDS, including file and print, security, messaging, transaction processing, licensing, management, remote access, host connectivity, database, routing, and telephony. More services will follow soon, including distributed objects and electronic commerce services. Linking today's 2.5 million separate networks into one global network community is the role of NetWare Connect Services (NCS), which became available in late 1995. Through Novell's partnerships with communications providers like AT&T, NCS will enable NetWare sites to connect their NetWare LANs securely to any other NetWare LANs, virtually anywhere in the world. Integrated Internet connections will effectively eliminate the distinction between the net and the user's LAN.

To put NetWare networks directly onto the Internet, customers are able to add a robust NetWare Internet Server that is integrated with NetWare services for the most manageable Internet connection and Web publishing system available. Today, it is hard to develop software that takes full advantage of the network and almost impossible to develop software that takes advantage of a heterogeneous network. Net2000, Novell's revolutionary new set of APIs, will remove these obstacles by giving developers a simple, universal programming interface to the diverse global network. Compatible with developers' favorite toolsets, Net2000 will make building solutions for a global network as easy as building stand-alone desktop solutions today.

As powerful as network applications will be, many people today do not need or want to use a PC. Novell's Nested NetWare extends network access and services to any device with an embedded microprocessor, regardless of its architecture. For instance, one day soon the Nested NetWare computer inside a car will notify the driver by E-mail when it is time for a tune-up. A Nested NetWare vending machine will report its daily stock levels electronically. A Nested NetWare interactive television will connect viewers to the Internet.

With the next century less than 3 years away, networks will become the principal transport medium for global commerce, content, and ideas. By the turn of the century, Novell technology will enable 1 billion connections around the world. And we will all see the biggest boom in productivity since the invention of the personal computer. Clearly, the era of the network is upon us. And Novell is already well down the road toward a world of pervasive computing that connects people with other people and the information they need — enabling them to act on it any time, from any place.

## MESSAGE HANDLING SERVICES

### Global Message Handling Services

Novell's NetWare Message Handling Services (MHS) product family provides a messaging infrastructure service. Because messaging is, by nature, a fully distributed service used on a networkwide basis, the concept of a directory is very important. NetWare Global MHS is Novell's NetWare Loadable Module (NLM)-based messaging product that includes messaging-specific directory support designed to: service environments without NetWare 4.0, integrate NetWare 3.x messaging environments with NetWare 4.0, and propagate message routing information. This approach allows NetWare Global MHS to provide the complete messaging solution customers and MHS application developers need.

NetWare Global MHS is a store-and-forward messaging technology that provides messaging and directory services to any desktop that has file access to a NetWare server (e.g., DOS, Windows, Macintosh, Unix, and OS/2), and to disconnected laptops. MHS is typically used in conjunction with such messaging applications as E-mail, calendaring, and network fax applications. Commercial third-party products include da vinci's eMAIL and Co-ordinator, Coordinate.com's BeyondMail, Reach's MailMAN and WorkMAN, Powercore's WinMail, Infinite's ExpressIT!, Notework's Notework, Futurus' Team, MicroSystems Software's CaLANdar, Campbell Services' OnTime, Castelle's FaxPress, Optus' FACSys, CE Software's QuickMail, Transend's CompletE-mail, and many others. In addition, MHS comes with a starter E-mail package, FirstMail, to enable users to get started with messaging.

Submitting a message to MHS is as easy as creating a text file with appropriate headers and giving MHS access to the new file. The simplicity of this process makes it easy for third parties to develop applications and for system integrators and corporate developers to use the messaging system. For example, an E-mail message may look like this:

```
smf-71
To: Bob Smith@marketing.acme
From: Tim Johnson@engineering.acme
Subject: Q1 Results?

Are the quarterly results in yet?
```

Once a message has been submitted, MHS determines how to route it through the messaging system. Global MHS implements various messaging protocols, including standard MHS protocols, as well as other industry standards (e.g., SMTP, SNADS, and X.400). After sending a message through the messaging system, MHS delivers the message into a file that the recipient's application may access. In addition to the messaging service, Global MHS provides a directory system for use by MHS and its applications. Access to directory information is gained by opening a shared file on the NetWare server containing information about individual users on the messaging system. E-mail applications typically use this information to provide point-and-click lists to the users. The naming scheme for MHS is hierarchical, as shown in Exhibit 5-1.

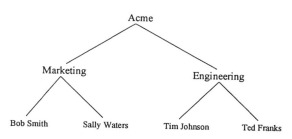

**Exhibit 5-1. MHS Naming Scheme.**

For example, Bob Smith's MHS name is formed from this tree: Bob-Smith@marketing.acme. The nodes in the naming tree that are not leaf nodes are called workgroups. Therefore, Acme is a workgroup that contains two other workgroups: marketing.acme, and engineering.acme. This hierarchical naming scheme allows a single unique global name for all MHS messaging users to ensure that no two MHS messaging users in the world have the same MHS name. In addition, this naming scheme is structurally identical to that in NetWare directory services (NDS) and X.500. Applications can also gain access to the directory information. They typically use this information to provide point-and-click lists to users. MHS provides this information in the form of an extract file that contains sorted records in a fixed-length format. Each record contains information about a user of the messaging system (e.g., mail address, phone number, title, and department). Directory information is shared between servers through a subscription mechanism. This mechanism is another characteristic of the NetWare Global MHS directory that logically relates to NDS functions.

The NetWare Global MHS product provides scalable, fully integrated MHS services to NetWare 3.x users. Implemented as a set of NLMs for NetWare 3.x and NetWare 4.x, NetWare Global MHS enables the network operating system to support a complete messaging infrastructure. This approach allows messaging services to be easily installed and also capitalizes on existing NetWare investments. NetWare Global MHS provides built-in directory support, routing, and workgroup-to-workgroup connectivity and it offers optional protocol modules for messaging interoperability with users on SMTP, SNADS, and X.400 systems.

Users and their associated mailboxes can be added to a server in a NetWare 3.x environment through the administrative utility in Global MHS. That server becomes the owner of that user object, and it commands the right to modify or delete the object. MHS propagates the fact that this object exists to other MHS systems by creating MHS directory synchronization messages. In addition to user information, distribution lists and workgroup information are also propagated the same way. To minimize synchronization traffic, only changes in the directory are propagated immediately. MHS also periodically synchronizes the entire directory to ensure that all servers have the same directory information (this is useful to guard against undelivered directory synchronization messages that may have resulted from servers being down for a long period of time or a loss of a communications link). Global MHS also uses directory synchronization messages to synchronize routing information, including server connectivity.

Global MHS uses the user information in the directory for routing purposes. Global MHS routing is a two-step process. The first step is to determine on which server the recipient's mailbox is located. The second step is to examine the connectivity of the network to determine the best path to that server. Messages are then routed accordingly.

NetWare 4.1 implements a variety of new services and functions. The feature that is particularly relevant to MHS is NDS. NDS implements a distributed directory database that takes over the role performed by the Bindery in previous NetWare releases. NDS's schema conforms to the X.500 international standard and is similar to the NetWare Global MHS directory. Names in the directory are hierarchical as shown in the diagram in Exibit 5-1. This is very similar to the MHS naming scheme, with the exception that NDS permits nodes in the directory tree to have a type associated with them (e.g., organization or common name).

Application programming interface (API) access to the NDS directory is broad, providing read and write access to a large number of objects, allowing for yellow-page search operations, comparison of attributes, modification to the schema of the directory, and partition management. In addition to information about users, the directory can contain information about many other types of objects, printers, devices, queues, file volumes, and more. Because the schema is extensible, other types of information can also be added.

To facilitate management of the directory, NDS divides the tree into logical divisions or partitions. The partitions may not overlap, and each node in the tree falls into a partition. Information about each partition is kept in a file on a server. The administrator determines where the master copy of the partition data resides and where replicas of each partition (if any) are to be kept. Making replicas of partitions increases the reliability of the directory and can also increase its performance. If a query to the directory is made and the currently attached server does not have the information locally, it will refer the requester to a server that does have the information. The effect of replicated partitions is similar to that of the Global MHS subscription mechanism.

## MHS Services for NetWare 4.1

Messaging is a network service that provides the basis for automatic data transfer across the network. Novell bundles its industry-standard back-end messaging server, NetWare messaging services, in NetWare 4 as a core network service. Because NetWare 4.1 messaging services are integrated with Novell's NDS, administration of the messaging environment is now accomplished using the same directory for both the network users and messaging users. The same GUI utility, NWAdmin, is used to administer both users in the NetWare 4 environment. This reduces administrative costs and overhead by simplifying management and administration of the messaging system.

MHS Services for NetWare 4.1 is an open platform supporting many popular E-mail client applications: Microsoft Mail, NetWare MHS client applications, and mail-enabled applications based on Common Mail Call (CMC), Vendor Independent Messaging (VIM), and Simple MAPI APIs. MHS Services for NetWare 4.1 is a service that exploits NetWare and Novell's other connectivity products, including TCP/IP, NetWare for SAA, NetWare Connect, and NetWare MultiProtocol Router (MPR).

NetWare MHS offers server-to-server connectivity over IPX and TCP/IP connections, plus asynchronous support for remote servers and dial-in laptops by making using of NetWare Connect clients and servers. Multithreaded operations within NetWare MHS support as many as eight concurrent asynchronous sessions.

The MHS Services for NetWare 4.1 solution exploits the multithreaded nature of NetWare for higher throughput. MHS Services for NetWare 4.1 has very low incremental memory requirements on the NetWare server (250K). Local groups of users are serviced without adding load to the backbone network. The NetWare 4.1 messaging services are fully backward compatible with the SMF v71 API, and ship integrated with NetWare 4.1 at no additional cost.

**GROUPWISE**

Not long ago, a messaging system was seen simply as a means for people to exchange E-mail messages. As workgroup computing evolves, however, many agree that the right messaging system can also erase the traditional boundaries of disparate applications, multiple operating system platforms, and multiple geographical locations, thereby extending the scope of workgroup computing and making the virtual workgroup a reality. For a messaging system to provide a solid yet flexible foundation for workgroup computing solutions, the messaging system architecture must be broad enough to support both an organization's current needs as well as its future needs for the next 5 to 10 years. Its architecture must therefore embody five important design principles. The architectural foundation must include the following attributes:

- It must be elegantly simple, yet powerful enough to handle very complex solutions.
- It must be flexible enough to change and expand without affecting the workgroup applications and solutions already dependent on it.
- It cannot be dependent on a specific operating system or environment, but must be open, portable, and full-functioned while operating on multiple existing and new operating systems and environments.
- It must be powerful enough to work under multiple client/server operating models concurrently to provide the best for process load-balancing and data security needs.

- It must be robust enough to support Novell's distributed document processing architecture, as well as open enough to support other industry applications and standards.

The overall effectiveness of workgroup solutions depends in large measure on how much the underlying messaging system architecture embodies these principles. As organizations look for workgroup computing solutions, they should carefully consider both the workgroup applications and the messaging system architecture that will provide the workgroup-enabling foundation for those applications.

## GroupWise Positioning

Before explaining the details of the GroupWise messaging system architecture, it is important to position GroupWise in the workgroup software industry. Workgroup software represents a broad category of applications and services that help groups of people work together more efficiently. The depth of GroupWise encompasses several groupware categories, including electronic messaging, calendaring and scheduling, and office automation. For today's workgroup communication needs, simple E-mail is not enough anymore. GroupWise provides the messaging services required for a strong workgroup computing foundation, including the following:

- Support for multiple message types (e.g., mail messages, schedule requests, task assignments, scheduled calendar notes, phone messages, and custom messages)
- Workflow routing for collaborative work
- Full message status tracking for knowing when messages are delivered, opened, deleted, accepted, declined, or delegated, and for tracking the progress of routed messages

The open messaging environment (OME) electronic messaging strategy guides the development direction of GroupWise. The OME strategy incorporates strengths from today's GroupWise products and services and those services available in MHS. The convergence of these technologies will strengthen the already robust messaging services in GroupWise.

## Integrated Messaging Services

At the heart of GroupWise's unique design is the integration of E-mail, scheduling, calendaring, and task management services into a single application. The strategy behind this messaging service integration is twofold. First, integration of the messaging services for E-mail and scheduling eliminates duplication of back-end messaging resources, including user directories, message transport, and especially the message stores. Second, combining all of these services together under a single interface provides more consistent ease of use and reduced training costs for the end-users. GroupWise is designed to work as seamlessly in heterogeneous computing environments as in those organizations that have standardized on a single OS platform.

Bringing the messaging system to desktop applications is a major objective within the Novell electronic messaging strategy. E-mail APIs (e.g., MAPI and CMC) allow desktop applications such as word processors and spreadsheet programs to make calls directly into an API-compliant E-mail system. In other words, the desktop application has become mail-enabled. For end-users, mail-enabling means the ability to mail the documents, spreadsheets, or graphics they are working on without leaving their current applications. GroupWise brings more than just E-mail functions to the desktop application. GroupWise can message-enable an application by bringing the combined ability of E-mail, calendaring, scheduling, and task management into the applications that people use most. Message-enabling means attaching an agenda to a meeting request, routing a document for review, and checking the calendar, all without leaving the applications where users do most of their work. Message-enabling means bringing the GroupWise collaborative services (e.g., workflow routing, status tracking, and task management) to desktop applications. GroupWise does not to force people into an unfamiliar work environment, but adds messaging ability to the applications they currently use.

In addition to message-enabling applications, GroupWise provides a messaging system on which message-aware applications can be built. Although a message-enabled application uses the messaging system client to perform messaging functions, a message-aware application can directly use the message transport, user directory, and message store of the underlying messaging system. GroupWise is open enough to support the message-aware applications of the future. As organizations move up to the advanced messaging capabilities of GroupWise, they must be able to provide interoperability with existing E-mail systems (e.g., Lotus cc:Mail, NetWare Global MHS, or IBM OfficeVision/VM), including both message exchange and directory synchronization.

GroupWise gateways also provide connectivity to public and standards-based messaging systems (e.g., X.400 and SMTP/MIME) for communication outside the organization. GroupWise is designed to fit seamlessly into an organization's overall messaging system and to work well with all of its components.

As today's organizations become more mobile, people want the flexibility of not being tied down to the office workstation. An important part of the Novell GroupWare strategy is to offer the widest range of mobile computing options, giving people the freedom to access their messages and information from whatever device best suits their needs. Laptop computers, telephones, and pagers can all act as interfaces to the GroupWise messaging system.

## THE ELECTRONIC MESSAGING ENVIRONMENT

Directory synchronization is completely automatic and an integral component of the administration program. GroupWise also supports directory exchange (import/export) with major network directories from within the administration program, as well as custom directory exchange capabilities through the GroupWise API Gateway. Administration and maintenance typically constitute a large portion of the overall cost of operating a messaging system. As additional improvements in administration and management services are developed, all GroupWare products will eventually be supported by the same administration and management platform.

**GroupWise 4.1 Electronic Messaging Architecture**

This section provides a technical overview of the core engine technology and administration services that bring the GroupWise 4.1 electronic messaging system to life. Before moving on to these subjects, however, it is important to understand the basic components of the GroupWise messaging system. GroupWise is a store and forward-based messaging system. The major message handling and administrative components within the GroupWise system are the client, the post office, the Message Server, the gateways, and the domain.

The client is the end-user application. The Post Office is the directory structure on a network file server that provides the message storage area for a specific group of users. The message server is the message transport agent (MTA) within the GroupWise system that provides routing services among post offices and through gateways. The gateway is the connection and translation software for communicating between GroupWise and other messaging systems. The domain is the basic administration unit consisting of the post offices and gateways directly serviced by a message server (see Exhibit 5-2).

The GroupWise core engine (shared code) technology lies at the heart of the GroupWise system's simple yet powerful architecture. All GroupWise clients and message servers share the same base code or core engine. The engine, which defines the core messaging functional ability and services, was created using object-oriented design and coding methods. The core engine's object-oriented design provides unique benefits at several levels throughout the GroupWise messaging system architecture.

All GroupWise message types, including mail messages, calendar items, meeting requests, and task assignments, are defined within the core engine. In other words, all GroupWise clients and message servers can handle multiple message types, making GroupWise able to truly combine E-mail, calendaring, scheduling, and task management into the same LAN-based messaging system. Not only are the various message types defined within the engine, but so are the GroupWise messaging services (e.g., folders, rules, workflow routing, and file attachments). What this means is that all

## Novell Messaging Products

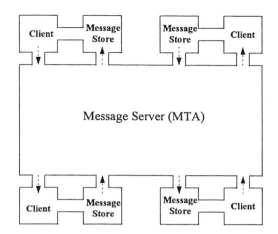

**Exhibit 5-2. GroupWise Message Server MTA.**

message types share the same functional ability. Some of the unique functions shared by all message types in GroupWise include the following:

- Rules-based message management applies to meeting requests, task assignments, and calendar notes as well as mail messages.
- Messages of all types are stored in the same folder structure.
- Workflow routing applies to both mail messages and task assignments.
- Files can be attached to any message type, including meeting requests and task assignments.

For example, the cross-message rules function allows users to auto-forward mail and auto-delegate meetings and tasks while they are out of the office.

### GroupWise 4.1 Message Server

The GroupWise 4.1 Message Server is the message transport agent used to distribute messages between post offices (i.e., directories and databases for message storage), domains (i.e., groups of one or more post offices), and gateways to external messaging systems. These messages can be E-mail, calendaring and scheduling requests, and message tracking. The message server also handles updates to domains and to databases within domains. The GroupWise 4.1 Message Server is available for Unix, OS/2, NetWare Loadable Modules (NLMs), and DOS. The message server is necessary for GroupWise customers who want to do the following:

- Connect two or more GroupWise post offices
- Connect two or more GroupWise domains

## THE ELECTRONIC MESSAGING ENVIRONMENT

- Install one or more GroupWise gateways to external messaging systems
- Support GroupWise remote users or connect to remote GroupWise sites
- Provide maximum security for message transfer

Features of the message server include compatibility with all major networks, automated maintenance for UNIX, OS/2, and NLM message servers, electronic messaging capabilities, security, flexible configuration, and directory synchronization. The UNIX, OS/2, and NLM message servers automatically clean up problems encountered in the message database without any effort on the part of the administrator. When the message server finds a problem in the database, it notifies the administrator with an E-mail message describing the problem and the actions taken to resolve the problem. Because additional server threads are used for maintenance, messages continue to flow through the system while the automatic maintenance is active. The message server also provides error checking to ensure that messages are not lost in transit.

With the message server, users can communicate with other users in their organization, whether or not they are using GroupWise. They can send messages to other users in their department, building, or a distant office within a matter of seconds. With GroupWise's connectivity solutions, they can also communicate with almost any other E-mail user in the world. The message server adds security to a messaging system. If users specify secure mode, the message server will take control of all database updates. The GroupWise 4.1 Client will still have read access to databases, but will not be able to write to or delete databases.

GroupWise lets users configure their messaging system flexibly. GroupWise provides direct links, in which message servers in two domains can directly exchange messages, and a gateway link, in which two domains are linked through another messaging system. With the message server, users can integrate the GroupWise messaging system with other messaging systems and enable communication between local and wide area networks. The message server provides simple, effective directory synchronization. When users add, modify, or delete other users, groups, or resources in one domain, the message server sends a message to all other domains informing them of the change. The other domains then handle updates to each directory. Directory synchronization is done continuously, without any effort on the part of the administrator.

The capability for both the client and message server to perform the same processing functions also offers unique load-balancing ability. The GroupWise message server includes a threshold feature that lets the administrator determine what level of processing the client should handle before turning over processing duties to the message server. For example, the client can process small transactions quickly and efficiently with no perceptible performance degradation for the end-user, while freeing the message server to process other transactions (e.g., message routing).

At the same time, a user does not want to wait for the client to process a message to all the people in a department or organization. The administrator can set a threshold that lets the client process small transactions (e.g., message delivery to one or two recipients), while passing all large transactions to the message server for processing. This load-balancing function keeps the client and message server running at maximum efficiency and improves the overall message processing speed.

An aspect of the core engine technology is the ability to support multiple processing models concurrently. For example, for single-post office systems that do not require a server for message routing, the client can perform all message-processing functions. For systems with message servers, the client and message server can share processing responsibilities, or all processing can be forced to the message server for security reasons. Each domain (i.e., message server and associated post offices) within the GroupWise system can use the client and server model that best suits that domain's needs.

Communication between the GroupWise clients and message servers is message-based (i.e., store and forward). Client/server interprocess communication (IPC) support will be implemented with OME. Although some messaging systems limit the client/server connection to one method or the other, GroupWise will offer the flexibility of using either or both. The message-based communication model provides a fundamentally simple, yet powerful design for cross-platform interconnectivity. Transaction messages are placed in a queue relative to the client's post office. The message server, which may or may not be running on the same platform as the client, can then pick up and process the transaction. All the client and message server need to work under this transaction model is network file access to the post office.

The client/server engine will also support IPC connections for those network configurations that support direct client/server protocols. In this configuration, the client will pass all transaction requests directly to the message server, which will process the requests on behalf of the client.

**GroupWise Gateways**

How servers communicate and distribute messages among similar and dissimilar messaging systems is becoming a critical issue for many organizations. The need to communicate with people outside an organization's network or messaging system is driving organizations to look for far-reaching connectivity solutions. GroupWise gateways are designed specifically to provide reliable and consistent connectivity and interoperability solutions. All GroupWise gateways provide the same level of basic services and support multiple message types. In addition, most GroupWise gateways provide pass-through messaging and directory synchronization services.

Not only must a gateway connect GroupWise with another messaging system, it must also preserve the integrity of the information flowing between the two systems. GroupWise gateways must effectively handle messages that are not natively supported in other systems. For example, if a gateway connects to an E-mail-only system, the gateway is intelligent enough to convert all meeting requests and task assignments into the other system's E-mail message format without losing any of the scheduling or task information.

If the gateway is connecting to a combined E-mail and scheduling system (e.g., IBM OfficeVision or DEC All-In-One), the gateway can convert GroupWise meeting requests into a message format that the other system understands. The gateway can also receive meeting requests from the other system. Not only is the meeting information preserved, but so is the ability to accept or decline the request. The OfficeVision Gateway also supports busy searches between the GroupWise and OfficeVision calendaring systems.

Pass-through messaging lets GroupWise connect with another physically separate GroupWise system, using a different intermediary messaging system as a transport. Pass-through messages are encapsulated in the other system's message format and then unencapsulated when they reach the remote GroupWise system. This encapsulation or tunneling preserves all messaging functions (e.g., message status tracking, attachments, priorities, and calendar-busy searches). It also maintains administrative functions (e.g., directory synchronization) between the two GroupWise systems.

Many GroupWise gateways also provide directory synchronization between GroupWise and a different messaging system. For example, the GroupWise cc:Mail Gateway automatically updates the cc:Mail directory with changes made in the GroupWise directory, and vice versa. A long list of GroupWise gateways built on the same core engine technology provides consistency of gateway functional ability, integrity of exchanged messages, and extensive connectivity to other messaging systems.

**GroupWise Administration**

Administration of GroupWise software in NetWare 4.1 environments is achieved using the native NWADMIN program provided with NetWare. NWADMIN plug-in modules allow seamless administration of the Group-

Wise users and objects. The GroupWise directory will first be synchronized with, and then eventually replaced by, NDS as the directory service in NetWare environments. The GroupWise administration architecture supports both a central and distributed administration model. All GroupWise administration can be done centrally by a single person, regardless of system size and the location of the various system components. Administration responsibilities can also be divided among multiple administrators, each with responsibility for a specific domain.

The ability to monitor and diagnose the system is currently missing in most LAN-based messaging products. Novell GroupWise offers support for SNMP management of its message server programs. GroupWise also offers diagnostic and status reporting services with the Admin program. The administrator can monitor the heartbeat, or message flow, of the system and automatically be notified of system problems via mail message, pager, or voice mail.

**GroupWise Telephony Access Server**

The GroupWise Telephony Access Server (TAS) enables users to access their personal GroupWise mailboxes remotely without using a computer. As with many messaging systems, GroupWise allows a user to gain access to his or her mailbox through a remote computer, modem, and telephone line. Although remote access to a messaging system by modem helps users be more productive and accessible, it can be limiting because not every user has access to a computer outside the workplace. Now GroupWise is leading the way into the next generation of mobile computing by giving every GroupWise user the ability to receive and send messages anywhere, anytime, using a standard touch-tone telephone.

After accessing TAS, GroupWise users can read and send messages and listen to calendar information. Users can also take advantage of extensive message search capabilities, and can even have messages or calendar information sent to a fax machine. To access TAS, a user simply calls the TAS telephone number supplied by the GroupWise administrator. TAS then prompts the user for his or her personal TAS ID and password. The GroupWise administrator assigns the TAS ID and a numerical password. If desired, the user can change the password after accessing TAS using the originally assigned password. If a user forgets the correct TAS ID, inputting his or her first and last name will prompt TAS to give the correct ID (the user must, of course, also know the password to gain access). Once the ID and password are verified, the TAS voice presents the user with a menu of options. If the user does not respond within a specific amount of time, TAS automatically repeats instructions and gives additional assistance.

If the user chooses to hear new messages from the inbox, TAS first tells the user how many new messages there are. For each message, TAS indi-

## THE ELECTRONIC MESSAGING ENVIRONMENT

cates who sent the message and reads the subject line. The user can then listen to the message or skip to the next one. When listening to a message, the user has the option of listening to attachments as well. Before reading an attachment, TAS indicates approximately how long it will take to read it. The user can reply to, forward, or delete any message. For scheduled meetings, tasks, and notes, the user can also accept, decline, or delegate the message. If the user asks to listen to calendar information, TAS prompts the user to input the day and then gives the user the option to listen to the day's appointments, notes, or tasks, or to listen to all three categories combined.

If the optional GroupWise Fax/Print Gateway is installed at the user's master system, the user can instruct TAS to send any message, along with its attachments, to a fax machine. The user can also choose to have his or her personal calendar faxed in one of three formats: planner day view, trifold day view, or week view.

A user can search for a specific message by indicating the message location (e.g., in-box or out-box), the message type (e.g., mail, calendar, or both), and a key word. TAS looks for the key word in the subject, to, and from lines of all messages meeting the search criteria and reports how many matches it finds. The user can then choose to listen to just the subject or to the entire text of the matching messages.

When sending a message, replying to a message, or forwarding a message, the user simply talks into the telephone handset. TAS records the user's voice and appends the sound file to the message. To address the message, the user can input the recipient's TAS ID or the recipient's first and last name. The subject line of a TAS message is always Audio Telephone Message. A recipient can listen to a TAS message on his or her PC if it has a sound card installed, or by dialing into TAS. Users can find out if and when messages have been delivered, opened, and deleted. They can also check to see whether meetings they have scheduled, tasks they have assigned, or calendar notes they have sent have been accepted, declined, or delegated.

Administering the telephone access server is simple and straightforward. After installing the TAS software, the administrator sets up the server from within the GroupWise Admin program by providing the necessary information for software and database location, telephone line setup, and user access. To set up the telephone lines, the administrator indicates how many lines will be used and selects the language for each line. After setup and activation of the server, the administrator can disable a specific telephone line at any time to control access and perform maintenance.

The administrator can also control who can use the TAS service. Access to TAS can be granted to individual users, to an entire post office, or to a whole domain. The administrator can also specify a pattern for automatically assigning TAS IDs as users are granted access. Systems requirements users can access TAS through any standard touch-tone telephone. The TAS software should be run as an OS/2 process on the network where the master GroupWise system is installed. Although it is possible to run TAS on the same machine as the message server, it is recommended that TAS be run on a dedicated machine.

As with all GroupWise gateways, TAS requires a GroupWise message server in the master GroupWise system. The message server is the message transport agent within GroupWise and provides message routing services among post offices, gateways, and services like TAS.

## GROUPWISE 5

GroupWise 5 is a powerful new groupware product that integrates a variety of groupware functions (e.g., messaging, calendaring and scheduling, online discussions, information storage, and retrieval and workflow) into a single environment. These functions are built on a proven set of information and communications management services that provide all users in the office, on the road, or at home access to the information they need to collaborate and make decisions.

Today's organizations face an environment that grows more complex and competitive with each passing year. Organizations must do more and more work with fewer resources than ever before. As competition increases, profits and budgets come under pressure. To remain competitive and thrive in this environment, many organizations are looking for software solutions that can increase efficiency while better leveraging one of the organization's most vital resources: information.

During the last decade, software companies have done much to increase the efficiency of individuals by providing applications that let them create and process information at the desktop. Now organizations are looking to expand individual efficiency to teams and workgroups through networked applications. People and organizations are looking for improved access to information and better ways to collaborate anytime, anywhere.

### Empowering People to Act on Information

It is a fundamental truth in business: people need information to do their jobs effectively. For example, they need to know answers to the following types of questions.

- How many of part A are currently in inventory?
- What does sales think of the new marketing plan?
- How has this problem been solved in the past?

# THE ELECTRONIC MESSAGING ENVIRONMENT

The frustrating part for many organizations is that the information people need to make sound, informed decisions usually exists somewhere in the organization if only the right people could find it, process it, understand it, and use it when they needed to. Novell GroupWare is helping organizations take advantage of their networking infrastructure to empower people to act on information anytime, anywhere. Novell GroupWare does this today by offering a family of leading-edge collaborative computing products, including GroupWise, InForms, SoftSolutions, MHS, and Collabra Share for GroupWise. Building upon this foundation of proven technology, Novell is delivering its next-generation groupware solution, called GroupWise 5. GroupWise 5 has helped individuals, teams, and workgroups in an organization better share and act on information, simultaneously shortening and improving the decision-making cycle.

A typical workday consists of a flood of notes and faxes, voice mail and E-mail, each demanding some action be taken. Users drown in the sheer volume of information that comes in multiple forms and from multiple channels. Today's typical PC user faces a daunting information feast-or-famine situation. On the one hand, an explosion in computer use has created information overload, where users routinely face mountains of raw data without structure or context for that data. As a result, people spend too much time sorting, sifting, and processing information and not enough time acting on it.

With all of this seemingly available information, however, people at all levels of the organization are starved for the vital information they need to act to make a decision. Users must ask where the important items are located. How do users access these pieces of information, store them, communicate and manage them so they can make effective business decisions in a timely fashion? How do messaging systems increase the productivity of individuals and workgroups? And how do these systems decrease the time it takes to execute a business process? GroupWise 5 addresses this dilemma.

## Solutions to the Problem

GroupWise 5 solves this dual problem of data overload and information starvation by providing the following groupware solutions:

- A work management framework. The GroupWise 5 Desktop provides an optimized and flexible framework in which users interact with other members of a team or workgroup. The desktop offers new and powerful tools to profile, organize, share, communicate, and act on information.
- A Universal In Box to manage all kinds of incoming messages and data.
- A Universal Out Box to track all outbound messages and data.

- Shared folders to provide central repositories for all kinds of information.

From this desktop, users can access each of GroupWise 5's groupware functions, including messaging, scheduling, calendaring, documents management, electronic forms, workflow, and online discussions. Information management services of GroupWise 5 provide new and flexible tools to view corporate data, as well as a method for storing information that leads to high availability and usability. GroupWise 5 leverages existing corporate data by enabling timely access to it.

Communications management services of GroupWise 5 facilitate the movement of information among members of a workgroup without corrupting or compromising the information. Common administration and management of GroupWise 5 reduces the cost of ownership and leverages the existing network services by letting organizations manage their groupware applications and their network from a common point of administration.

**Business Solutions**

GroupWise 5 reduces the cost of ownership and leverages the existing network services by letting organizations manage their groupware applications and their network from a common point of administration. GroupWise 5 empowers both the individual and the team or workgroup. Specifically, GroupWise 5 enables the following:

- Knowledge workers to accelerate their decision-making processes
- Enterprisewide and departmental workgroups to accelerate the execution of tasks and business processes
- Power users and applications developers to create custom groupware applications tailored to solve unique business or reengineering objectives
- System administrators to easily manage their groupware services, custom applications, and networks at the same time

GroupWise 5 can be used to solve a wide variety of information and communication challenges an organization might face, including the following:

- Leveraging existing information currently tied up in company databases and electronic files anywhere on the network
- Automating business processes (e.g., lead tracking, sales, expense reporting, and purchasing)
- Making individual and group scheduling faster and easier
- Accelerating information access for sales or support personnel
- Replacing less effective meetings with online discussions

The first components of GroupWise 5 were available in the second half of 1996; however, the foundation of GroupWise 5 is available today in the

# THE ELECTRONIC MESSAGING ENVIRONMENT

following products that are market-tested: GroupWise, InForms, SoftSolutions, MHS, and Collabra Share for GroupWise. In fact, current customers of these products have driven the requirements for GroupWise 5 described in the following sections.

## The GroupWise 5 Desktop

The GroupWise 5 Desktop gives users new tools to help them more efficiently collaborate with others, manage information, and make decisions. These new tools include a Universal In Box, a Universal Out Box, and shared folders. The revolutionary GroupWise 5 Desktop will be available on a variety of client platforms, including MS Windows, Windows 95, Macintosh, Power Macintosh, and UNIX.

At the heart of the GroupWise 5 Desktop is the Universal In Box. This in-box replaces the many competing in-boxes a typical user manages (e.g., voice mailbox, E-mail in-box, fax machine, pager) with a single in-box for all types of information. As a result, users spend less time processing in-box contents and only have to learn one interface to manage all types of incoming information. The GroupWise 5 Desktop provides a framework for how work gets done in an organization.

GroupWise 5 can help users effectively manage day-to-day work, as well as automate multistep business processes. The GroupWise 5 Desktop provides streamlined access to a complete set of groupware features (e.g., messaging, calendaring and scheduling, information storage and retrieval, and custom groupware applications). The GroupWise 5 Universal In Box accepts E-mail messages, schedule requests, delegated tasks, voice mail, faxes, pages, electronic forms, and many other types of data. The GroupWise 5 Desktop is truly unique because of the various forms of information that the Universal In Box can handle. For each of these data types, GroupWise 5 does the following:

- Maintains native attributes. For example, a voice message appears in the Universal In Box as a voice message and maintains its native attributes (e.g., caller identification)
- Profiles and manages information in the same way regardless of data type

Users can apply GroupWise 5's message-handling functions and server-based rules not only to E-mail but to all data types supported by the Universal In Box (e.g., a voice message can be sorted, delegated, routed, and copied). The GroupWise 5 Universal Out Box provides complete message tracking and user accountability. GroupWise 5 Shared Folders lets members of a team or workgroup easily share information related to a particular topic or project. At any time, users can see if messages have been received, opened, deleted, or delegated. They can even retract messages that have not already been opened. The message tracking offered with GroupWise 5 is especially important as groupware solutions are expanded to transport delegated tasks, appointments, and other workgroup activities.

Through the GroupWise 5 Desktop, users can also access information stored in shared folders. These folders let members of a team or workgroup easily share information related to a particular topic or project. Any type of information supported by the Universal In Box (e.g., messages, voice mail, documents, or faxes) can be stored in shared folders. Through the GroupWise 5 Desktop, users can access all their groupware functions. Messaging, calendaring and scheduling, task management, workflow, electronic forms, document management, online discussions, and even custom groupware applications all are available from one screen, with a consistent interface.

At the heart of the GroupWise 5 work management solution is the Universal In Box. It combines many in-boxes into a single point of entry for all types of information. GroupWise 5 dynamically links together objects of any data type received through the Universal In Box with actions taken on those objects. These links form context trails, a powerful way to preserve the relationship between information and actions. With context trails, users can act on information and relate it to tasks, scheduled items, messages, people, and documents in an intelligent way.

GroupWise 5 provides tools that enable users to streamline, prioritize, manage, and automate responses to incoming information. For example, a product development meeting scheduled in GroupWise 5 might be linked to one of the following:

- The original schedule request to attend the meeting
- An E-mail message from the vice-president citing a specific product problem
- Several complaint forms filled out by technical support
- An online discussion for where solutions for the problem have been discussed
- A voice mail recording from a particularly irate customer

Users rushing into the meeting could quickly navigate through these context trails to construct a full understanding of the situation. The GroupWise 5 Desktop provides a framework for how work gets done in your organization. GroupWise 5 can help users effectively manage day-to-day work, as well as automate multistep business processes, such as expense reporting and customer tracking. In a typical workday, users react to information as it arrives on their desktop, responding to a voice mail message, returning a response to an E-mail, delegating a task, or accepting an invitation to a meeting. The number of these demands is often so overwhelming that a day at work is not a set of planned activities but a series of interruptions. GroupWise 5 provides tools that enable users to streamline, prioritize, manage, and automate responses to incoming information, thereby allowing users to better manage interruptions and execute planned activities.

In addition, GroupWise 5 can automate the repetitive tasks associated with certain business processes (e.g., expense reporting). For example, a common expense report might flow from an employee to a supervisor for review and approval, then route to accounting for reimbursement. In this scenario, GroupWise 5 automates and expedites each individual's step in the process, as well as the process as a whole. From the GroupWise 5 Desktop, a user would fill out an intelligent expense report form that performs all calculations. The user would then drag the form onto a workflow icon that would launch the approval and reimbursement process.

The power of the GroupWise 5 Desktop applied to a typical new product development team should be considered. From a single window GroupWise 5 can automate a wide variety of business processes (e.g., sales tracking, customer support, and expense reporting) to obtain the following.

- A shared folder containing all of the E-mail and voice-mail messages, faxes, memos, and other inbound information on the current product development cycle
- An online discussion in which the new product's feature set is discussed with team members in sales, engineering, marketing, and manufacturing from facilities around the world
- Sales history information for previous products in the line

All the information needed to make product development decisions would be readily available without having to switch between programs or dig through a wide variety of sources. The GroupWise 5 Desktop gives individuals and workgroups a powerful work management model designed from the ground up with collaborative computing in mind. It does not dictate how users must go about their work. It is completely customizable and extensible to third-party applications so that users are free to choose the applications they need to shorten decision processes.

Supporting the GroupWise 5 Desktop are information management services that offer users access to all kinds of corporate data and information. The GroupWise 5 information management services allow users to view, query, store, and move corporate data while maintaining its integrity, version control, and security. Even though users are often faced with information overload, it often seems impossible to find the right information on demand — regardless of where it might be stored. In fact, estimates show that the average executive spends approximately 1 month per year waiting for information. Therefore, information management services that speed the process of finding crucial information can save time and money.

## Novell Messaging Products

GroupWise 5 accelerates decision-making by helping users access vital information, no matter where or how it is stored. Powerful document management features based on SoftSolutions technology can search and retrieve information from native application files (e.g., word processor documents, presentations, and spreadsheets), and integrated InForms technology allows GroupWise 5 users to view and extract information stored in company databases using a familiar forms metaphor. Users can view, collect, circulate, and route information from their GroupWise 5 Desktop.

Effective communications are strategic to collaborative computing. With this in mind, Novell is delivering a robust communications infrastructure that is modular and based on industry standards. GroupWise 5 offers a wide range of background communications services, from one-to-one communications to one-to-many, to many-to-many. These services:

- Offer complete support for remote and mobile users
- Intelligently support the many types of data workgroups need to share
- Connect to external and legacy systems
- Can be scaled to meet the needs of workgroups, departments, or entire enterprises
- Feature a true client/server architecture

Like the other services provided by GroupWise 5, the communications management services support multiple data types, including not only E-mail, documents, and forms, but also voice mail, images, and faxes. And GroupWise 5 optimizes the movement of information over a network. For example, these services give users the option of downloading files or simply moving references or pointers around the network, avoiding the consumption of network bandwidth by large files (e.g., sound or video files).

GroupWise 5 communications services enable today's road warriors to take their offices with them. GroupWise 5 provides mobile users access to their in-boxes and out-boxes by way of remote client software or a standard touch-tone telephone, so they can continue to participate in the workflow of a business process when they are away from the office. In addition, remote client software lets users access data stores and participate in conferences — it even lets them take snapshots of data or forms when on the road and synchronize it when they return to the office. A user is never out of the loop, and a business process does not have to come to a screeching halt.

**SUMMARY**

GroupWise 5 represents the culmination of the integration of Novell's MHS product line and its GroupWise product line. At the heart of communications services is the GroupWise 5 message transfer agent. It supports many desktop clients, including MHS SMF70/71 clients, GroupWise 4.1 clients, MAPI 1.0 clients, CMC 2.0 clients, and GroupWise 5 clients. Current

# THE ELECTRONIC MESSAGING ENVIRONMENT

MHS and GroupWise users can easily migrate to GroupWise 5 by replacing their MTA with the GroupWise 5 MTA, moving individual mailboxes to GroupWise 5 one at a time or upgrading entire post offices.

GroupWise 5 communications management services also support popular connectivity standards (e.g., MHS, X.400, and SMTP/MIME). GroupWise 5 provides a host of gateways to a wide variety of existing messaging systems. Users can protect their investment in legacy messaging and data systems as well as connect to the external world of private and public data networks.

The foundation of Novell's GroupWise 5 product family resides in market-tested products that are available today, including InForms, SoftSolutions, MHS, and Collabra Share for GroupWise. Workgroups using these products are already on the path to GroupWise 5. They offer immediate, real-world benefits to organizations by leveraging the value of NetWare by implementing GroupWise and MHS — robust messaging technologies that capitalize on NetWare's system administration and management.

# Chapter 6
# HP OpenMail
*Alex Morgan*

Powerful business forces are working to ensure that enterprises will embrace a wider range of technologies to achieve competitive advantage. Business has turned to the Internet and corporate intranets for business communications, sparking a tremendous growth in communications applications. Hewlett-Packard's OpenMail provides the means to integrate these disparate end-user technology choices. As the enterprise post office, OpenMail provides tremendous economies of scale for lowering the cost of commercial messaging.

OpenMail is HP's enterprise communications technology. OpenMail provides guaranteed message delivery, low-cost management, and high-end scalability. It is based on Internet standards and commercial standards (e.g., X.400 and X.500). It is the end-user post office as well the backbone, supporting personal messaging, workgroup collaboration, and application messaging applications.

OpenMail allows business units to select the end-user technologies (e.g., exchange client, cc:Mail client, Web browser) that best meet their needs. OpenMail also integrates existing investments in departmental or single-platform groupware and messaging (e.g., Lotus Notes, Exchange Server, and IBM/PROFS) for seamless enterprise communications. OpenMail was developed by Hewlett-Packard's Enterprise Messaging Operation (EMO). EMO provides messaging and process management solutions for commercial customers (see Exhibit 6-1).

## MARKET POSITIONING AND ADVANTAGES

OpenMail has the following four advantages over solutions based on proprietary technologies:

- *It works.* OpenMail is the only proven client/server messaging system for large, global enterprises. It is not new or untried: OpenMail is now in its fourth major release. More than 900 enterprises, including most Fortune 1000 corporations, are using OpenMail today — simply because it is a high-performance, low-risk, low-cost strategy.

# THE ELECTRONIC MESSAGING ENVIRONMENT

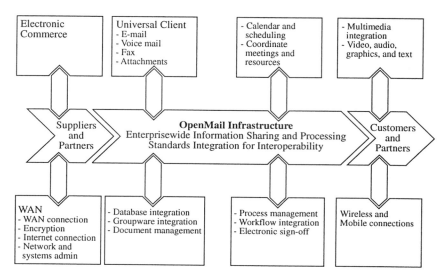

Exhibit 6-1. Support for Enterprise Communications.

- *It scales easily.* OpenMail is based on an architecture that involves a few, very powerful servers providing services to a large number of users. A single OpenMail server can support 5000 users.
- *It costs less.* The management cost of a messaging system can amount to two thirds of the cost of ownership. OpenMail's scalability, together with its close integration with OpenView, offers centralized control over a global messaging system, neatly integrated into a framework that also manages the networks, systems, applications, and databases on which the enterprise depends.
- *It thrives on diversity.* OpenMail addresses the natural diversity of the real-world messaging environment, rather than attempting to impose a single proprietary solution on it. OpenMail is based on openness, interoperability, and choice. OpenMail embraces new tools and standards, protects existing investments, and improves overall corporate communications.

## Value to Customers

By opening up choices, and by focusing on working with rather than replacing other systems, OpenMail equips users to deliver solutions that actively address the complexity of their enterprise and the world outside. Choice also protects IT investments. By integrating existing choices rather than replacing them, users can invest selectively in new technology to increase competitive advantage and reduce the risks of all-or-nothing decisions.

*HP OpenMail*

## Product Strategy

OpenMail starts with an enterprise-scale approach to directory services, management, and security. OpenMail's design principles call for these fundamental services to be embedded in the infrastructure rather than in specific end-user applications and toolsets. This approach allows applications to share such services as security. It drives down costs by leveraging resources and expertise. This modular approach has allowed OpenMail to support multiple transports (e.g., X.400 and SMTP), operating environments (e.g., NT and UNIX), and end-user client interfaces (e.g., MAPI1, POP3, and IMAP4).

Global businesses are endorsing open standards for directories (e.g., LDAP and X.500) and security (e.g., S/MIME and public key). Additional standards for multimedia integration, voice-mail/E-mail interoperability will also be supported by OpenMail as they emerge.

## ARCHITECTURE AND FUNCTION

OpenMail has been constructed to support a wide variety of services through a store and forward process. Each component of the architecture has been designed to support an open-ended set of customer requirements. This flexibility has allowed OpenMail to thrive in the corporate environment and continually add support for new messaging applications, data types, and services.

### Server Architecture

OpenMail uses a server-centric messaging model. Based around a few high-performance servers or a fully distributed architecture, each OpenMail server provides mailbox, directory, message store, and routing services (see Exhibit 6-2). This flat architecture model reduces management tasks while improving throughput between servers. Each user and application has a separate mailbox.

### Messaging

OpenMail's store- and -forward architecture supports personal messaging and application messaging. Application messaging relies on the same APIs and message tracking methods as personal messaging. Applications range from calendar scheduling and discussion groups to custom applications for supply chain integration.

**Client Access Layer.** OpenMail provides a rich client API, the universal agent layer (UAL). Drivers interface the UAL to each client type. This approach allows OpenMail to directly connect a wide variety of open (e.g.,

# THE ELECTRONIC MESSAGING ENVIRONMENT

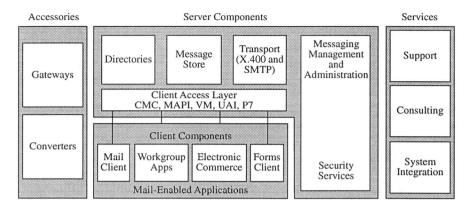

Exhibit 6-2. OpenMail Server Architecture.

POP3 and IMAP4) and proprietary (e.g., MAPI0 and MAPI1) client standards as well as custom end-user clients. The UAL also provides the connect API for message-oriented applications. All mailboxes provide full OpenMail services. The service level provided to the end-user or application depends on the richness of the client application.

**Message Stores.** OpenMail provides a full-featured message store that supports server-based and download-oriented messaging clients. Security features allow users access to public and private folders. Database design has been optimized to support such services as shared directory lists and broadcast messages. These features greatly reduce storage requirements on the server without reducing performance for individual users.

**Transports.** OpenMail rigorously supports the X.400 and SMTP standards. These transports support communications between OpenMail servers and servers from other vendors. Because these transports are native interfaces of OpenMail, each is engineered for high throughput. This contrasts with messaging servers that support these commercial standards through gateways or third-party products.

**Security.** Security is provided in a modular approach, allowing customers to choose the types of security required for each messaging application, including the following:

- *Client.* For clients, OpenMail relies on the security contained on the client. For Windows NT, this includes support for the single sign-on provided by the NT server registry. Recent advances in Web browser technologies provide for strong user authentication at the client (see

*HP OpenMail*

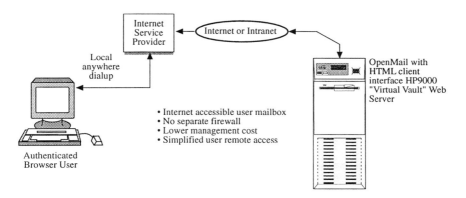

**Exhibit 6-3. Secure Dial-Up Internet Access.**

---

Exhibit 6-3). OpenMail supports vouchers from certificate authorities (e.g., Nortel Entrust).

- *Server mailbox.* Each user or application has a unique mailbox on the OpenMail server. For authentication at the server, these are initially secured according to the approach of the host operating system. For UNIX systems, user security references entries in the /etc./password security structures. For NT systems, mailbox security is managed through the registry. For trusted systems (e.g., HP's Virtual Vault Platform), mailbox security conforms to user security practices on those systems.
- *Client/server download encryption.* OpenMail supports encryption of messages during download through protocols (e.g., SSL 3.0).
- *Client information display.* OpenMail supports SHTTP — secure hypertext transfer protocol — for Web browser access to message store information. Other security protocols can be implemented for communication between the OpenMail server and clients as specified by the interface (e.g., MAPI1).
- *Message store.* The OpenMail message store places all messages, folders, and distribution lists and directory information in a private store, inaccessible from the standard file system. Access to information in the store requires user or administrator permissions.
- *Message transport.* OpenMail supports several message transport privacy approaches (see Exhibit 6-4). First, OpenMail can secure transport between OpenMail servers through an optional security product. Second, OpenMail supports secure MIME (SMIME) to encrypt message attachments within the SMTP environment. Third, OpenMail supports Nortel's Entrust technology. OpenMail clients can be made Entrust-aware and users can select to encrypt messages using public

## THE ELECTRONIC MESSAGING ENVIRONMENT

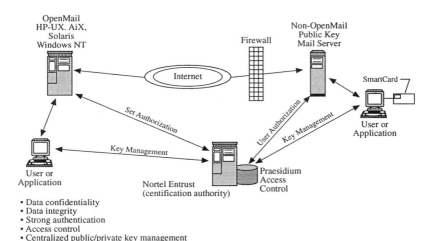

- Data confidentiality
- Data integrity
- Strong authentication
- Access control
- Centralized public/private key management

**Exhibit 6-4. Secure Document Transmittal.**

---

keys stored in the OpenMail directory. These are typically X.509v3 certificates, and therefore provide nonrepudiation and digital signature as well as encryption. Capabilities vary based on the nation or nations that are part of the network.

- *Trusted operating systems.* OpenMail is being ported to HP's Virtual Vault Platform, which many of HP's financial services customers are using to make customer account information available on the World Wide Web. In complement with strongly authenticated clients and privacy-ensured client-to-server communications, this platform solution makes enterprise E-mail available, in a secure manner, through Internet service providers (see Exhibit 6-5).

**Directories.** OpenMail provides a robust directory that can link with X.500 corporate directories as well as synchronizing with application-specific directories for Lotus Notes, Microsoft Exchange, and Netscape Directory Server. OpenMail directory services (see Exhibit 6-6) will support the lightweight directory access protocol (LDAP) to ensure synchronization with and access to other applications as well as enterprise directories. This will enable customers to leverage their investment in legacy directories (e.g., NDS) when deploying new applications. OpenMail, as the backbone messaging technology, can provide replication services to local directories supporting business units and applications.

Government and telecom suppliers have announced intentions to provide X.500 services in concert with certificate authority services. HP will

*HP OpenMail*

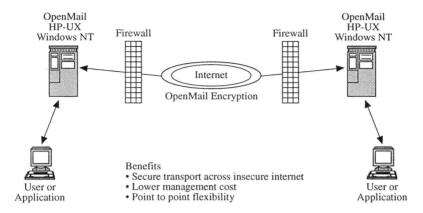

**Exhibit 6-5. Data Stream Encryption.**

work closely with these service providers to deliver complete solutions. This will support customers as they deploy secure global solutions rapidly. This means that customers will be able to leverage those services with minimal short-term investments and realize significant long-term payback.

**Calendaring.** HP OpenTime delivers real-time scheduling across enterprise servers. End-users can schedule meetings or reserve corporate resources through a single interface.

OpenTime supports free-time mapping across all servers, allowing meetings to be scheduled regardless of geography or time zone. OpenTime provides direct alerts to users whose calendars are being queried and also

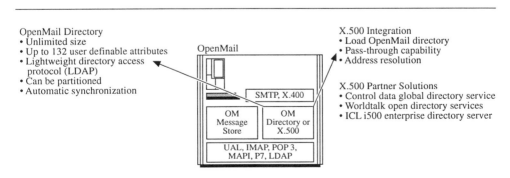

**Exhibit 6-6. OpenMail Directory Services.**

89

# THE ELECTRONIC MESSAGING ENVIRONMENT

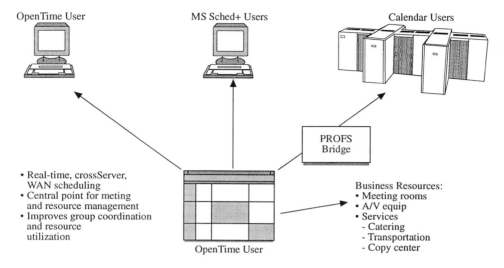

**Exhibit 6-7. OpenTime: Real-Time Scheduling.**

sends E-mail notifications to users and to others outside the OpenTime community.

OpenTime supports Windows, UNIX/Motif, and Macintosh users. A Web browser interface became available in late 1996. OpenTime's gateway to PROFS calendaring systems simplifies migration to client/server messaging. Now, PROFS users can check free-time maps of OpenTime users, and vice versa, in real time (see Exhibit 6-7). For Microsoft environments, Schedule+ integration is under development.

## Forms and Workflow

Workflow applications can be easily integrated within OpenMail. Forms packages can take advantage of OpenMail features (e.g., return receipt and time opened) to support workflow process flags. Forms-based products, and database-centric workflow technologies can leverage OpenMail's message distribution and management capabilities.

## Public Folders

The OpenMail message store provides shared folder services. Users can post information and comments in a threaded discussion group format. Feeds from external sources can be delivered through network news transport protocol (NNTP). Information in the folders can be distributed to other servers through NNTP, OpenMail message forwarding, or alternate

*HP OpenMail*

mechanisms. Through OpenMail's HTML browser interface, shared folders becomes an inexpensive, simple Web publishing mechanism. Users can post information to the discussion groups and it is automatically available to the browser interface.

### EDI/Electronic Commerce

OpenMail supports electronic data interchange (EDI) through X.400 and X.435. Many customers using X.435 also integrate OpenMail with their X.500 directories to provide enterprise routing and resource information. OpenMail's UAL provides support for application messaging. Support for X.509v3 certificates provides message privacy, integrity, and nonrepudiation.

### Management

OpenMail equips administrators with a rich and powerful set of management utilities (see Exhibit 6-8). Trouble-shooting, configuration, remote task automation, directory synchronization, error logging, and access control for all enterprise servers can be done from a single console. Besides providing a simple point-and-click, Windows-based administrator console, OpenMail provides a rich scripting facility to automate routine user and message store maintenance. Through this combination, IT can equip help-desk staff with easy-to-use tools and also offer administrators powerful tools to reduce maintenance costs.

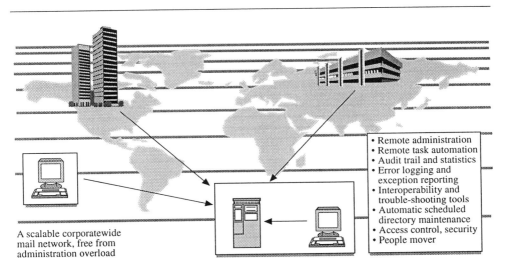

**Exhibit 6-8. Management and Administration.**

# THE ELECTRONIC MESSAGING ENVIRONMENT

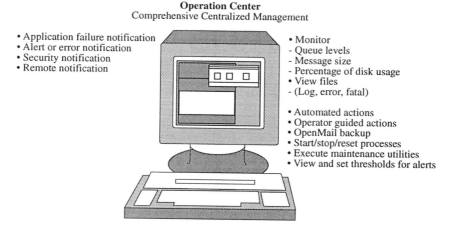

Exhibit 6-9. **OpenMail and HP Operations Center.**

OpenMail's tight integration with HP OpenView Operations Center brings oversight to the enterprise level. Intelligent agents on all messaging applications can flag error conditions and make them visible at a single console (see Exhibit 6-9). Corrective action can be taken automatically, based on Operations Center scripts, or human intervention can be initiated. This puts messaging management into the data center environment, reducing costs at the business unit level.

## Performance and Reliability

The OpenMail architecture enables consolidation of messaging mailbox users to a few, very large servers. This approach reduces the number of managed components and reduces the amount of network traffic required to deliver a message. With 5000 concurrent users on a single large server, most messages can be delivered without any network traffic. Network data transfers are generally only required to download messages or view contents of the message store.

On HP-UX systems, OpenMail integrates with MC/ServiceGuard, HP's high-availability clustering technology. With MC/ServiceGuard, the messaging server obtains maximum protection against downtime, even when hardware faults occur. Within a cluster of servers running OpenMail, MC/ServiceGuard monitors processors, disk drives, and other components. Should a failure occur, OpenMail services are handed off to another system on the cluster, allowing resumption of service in a matter of minutes.

## Platforms

OpenMail is available on UNIX and NT operating system platforms. On UNIX, OpenMail runs on platforms from Hewlett-Packard, Sun, IBM, and DEC. Other platforms can be supported based on customer demand. Besides HP-UX, OpenMail will also be available on HP's Virtual Vault Platform. This unique trusted system is effectively a second-generation firewall. When used with this platform, OpenMail can provide mailboxes directly on the open Internet.

## Complementary Technologies

Each global enterprise has a unique set of challenges. OpenMail extends its own capabilities with complementary technologies from solution partners. Solutions cover the following areas: calendaring, conferencing, consultants and integrators, document management, electronic commerce, facsimile solutions, forms and workflow, gateway solutions, management, utilities and toolkits, and X.400 and X.500.

## SOLUTIONS

Large organizations must address the unique needs of individual business units while reducing the number of technologies that are centrally managed. OpenMail is the only solution that provides support for new technologies and investment protection for prior choices. OpenMail support for the most common business collaboration technologies is described in the following sections.

## Intranet

OpenMail supports all Internet and intranet messaging standards. Any HTML Web browser can connect to OpenMail through a Web page interface. Internet mail clients can connect through POP3 and IMAP4 protocols. Publicly available and commercial single-function servers (e.g., mail, directory, and news) can communicate with OpenMail through Internet standard protocols (e.g., SMTP, LDAP, and NNTP). For customers migrating to intranet messaging, OpenMail can link their users to other environments (e.g., Exchange, Notes, and X.400).

OpenMail provides much greater message management and delivery reliability than the basic Internet standards. For instance, server-based message storage increases reliability over POP3 and IMAP4 delivery methods because messages can be stored on the server as well as on each individual's personal computer. Because POP3 and IMAP4 are download protocols, using them exclusively forces message send (including public key encryption) to originate at the desktop. Messages cannot be tracked until they reach the first managed server. This gap can generate unacceptable rates of message loss.

# THE ELECTRONIC MESSAGING ENVIRONMENT

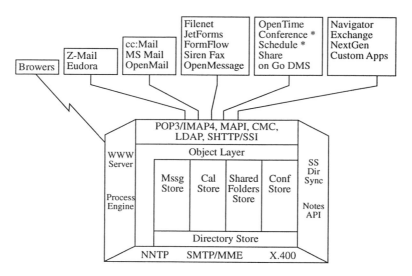

**Exhibit 6-10. Internet Collaboration Services.**

Because OpenMail supports both a browser interface and POP3/IMAP4, users can take full advantage of the full-featured browser interface and download messages when required (e.g., when traveling). OpenMail permits users to keep a copy on the server after downloading, providing an added measure of safety and control. The browser interface also provides access to OpenMail public folders.

OpenMail's Internet services architecture is illustrated in Exhibit 6-10.

**OpenMail and POP3/IMAP4.** OpenMail provides full support for these standards, which are very popular for personal messaging. Neither protocol provides the robustness required for commercial messaging, because they lack logging and guaranteed delivery processes. Users that prefer a more rigorous messaging environment within an Internet/intranet end-user toolset should use the HTML browser interface to OpenMail.

**Browser Access to Mail.** OpenMail's complete message store facility can be viewed through a Web page interface (see Exhibit 6-11). This option provides the lowest-cost deployment and highest degree of messaging management on the intranet.

All messages, distribution lists, and public folders remain on the server to ensure consistent management and backup. Users require no special software on their client. The OpenMail message log provides detailed records to resolve missing message questions. Any browser platform, in-

## HP OpenMail

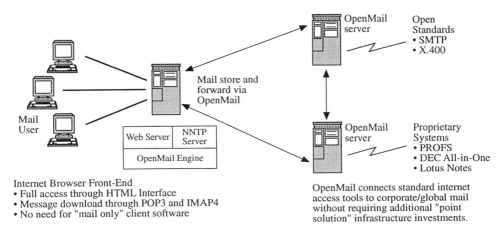

**Exhibit 6-11. Intranet Mail Client Solution.**

cluding text-only browsers, can access the message store. This model is also the ideal architecture for message-based workflow applications.

**Discussion Groups and Public Folders.** OpenMail provides public folders that are viewable through the Web browser interface. These newsgroups can be synchronized internally through OpenMail replication or can be linked to external newsgroups through OpenMail's NNTP interface (see Exhibit 6-12).

**Electronic Commerce and the Internet.** OpenMail's modular, standards-based architecture supports workflow and application messaging applications between customers and institutions as well as between business partners. As mentioned, integration with HP's Virtual Vault Platform enables secure mailboxes to be placed directly on the Internet, outside a corporate firewall. In combination with public key services (e.g., Nortel Entrust), workflow applications based on Entrust-aware forms packages can be sent to applications on any server with access to the appropriate public key certificates. OpenMail's UAL allows applications to act as their own clients and therefore to act on forms whenever they appear in the application mailbox.

These functions can be combined to provide secure, personal kiosks on the Internet. Through a standard Web browser, customers can obtain information, initiate services, and communicate with staff. This model provides high flexibility and ease of use, as well as simple integration with all exist-

# THE ELECTRONIC MESSAGING ENVIRONMENT

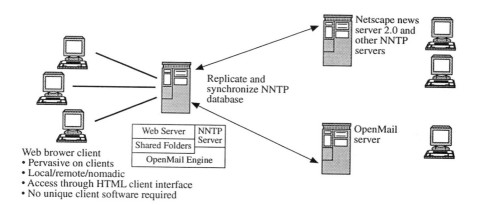

Exhibit 6-12. Intranet Shared Folders and Discussion Group Solution.

ing customer applications. It also enables institutions to market directly to individual customers and monitor their subsequent actions.

Exhibit 6-13 depicts the solution architecture for this service. As Internet Service Providers (ISPs) assist businesses to expand their presence on the Internet, OpenMail will develop additional solutions to extend electronic commerce.

## Microsoft Environments

Microsoft's family of operating systems provides a unique challenge to customers. Many customers find themselves supporting MS-DOS, Win 3.1, Windows for Workgroups, LanManager, Windows 95, Windows NT client and Windows NT server, as well as Novell NetWare. OpenMail's UAL driver approach to supporting messaging clients makes it an ideal engine for customers that must support multiple Microsoft environments.

**Microsoft Exchange.** The OpenMail server provides full support for MAPI1 clients (e.g., Exchange and cc:Mail 7.0), making the Exchange server unnecessary. For OpenMail customers that do have requirements for Exchange server, OpenMail provides complete server-to-server integration, including choice of transports (X.400 or SMTP), directory synchronization, and shared folder synchronization.

**MS-Mail and MS-Explorer.** The OpenMail server provides full support for MS-Mail 3.x clients. OpenMail is the only mail server that can support MS-Mail and Exchange from the same server, enabling customers to retain Win 3.1 or DOS clients as is. This allows customers to consolidate MS-Mail post

*HP OpenMail*

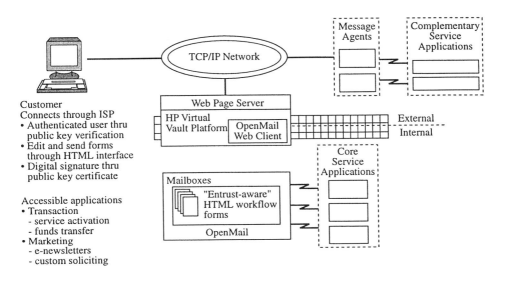

**Exhibit 6-13. Internet Customer Service Solution.**

offices to a single large OpenMail server to reduce operating costs (see Exhibit 6-14).

Once on OpenMail, customers have the choice of moving directly to browser-based mail with Microsoft Explorer, accessing the OpenMail mes-

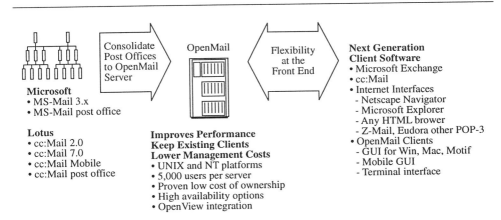

**Exhibit 6-14. LAN E-Mail Post Office Consolidation.**

97

## THE ELECTRONIC MESSAGING ENVIRONMENT

sage store directly. Customers that intend to deploy NT-based Intranet solutions can take advantage of OpenMail for NT to support Microsoft clients as well as browser clients and POP3/IMAP4 clients.

### cc:Mail Environments

OpenMail provides complete post office services for cc:Mail 2.0 clients. For example, within Hewlett Packard's internal mail system, 50 OpenMail servers support over 70,000 cc:Mail clients. Through MAPI1, OpenMail provides complete post office services for cc:Mail 7.0.

OpenMail is the only solution that provides integrated post office services for cc:Mail 2.0 and 7.0. OpenMail's scalability and low cost of ownership make it the ideal consolidation strategy for distributed cc:Mail post offices.

### Lotus Notes

OpenMail provides mail delivery services for Lotus Notes database servers. Through OpenMail, Notes mail end-users can take advantage of robust message delivery services to the other environments supported by OpenMail, including Exchange users, cc:Mail users, PROFS users, and intranet users. OpenMail's high-fidelity gateway allows OpenMail to provide message transport between isolated Notes servers.

In addition to Notes to Notes transport, the gateway converts Notes messages to MIME format, allowing them to be interpreted by non-Notes clients, including Web browser clients. This includes conversion of Notes rich text format to plain text for viewing by other clients. MIME conversion also is applied to embedded binary files, which are converted to attachments. The gateway also enables two-way directory synchronization between Notes and OpenMail. Exhibit 6-15 shows the OpenMail-Notes interface.

Investigation is underway to link OpenMail and Notes discussion databases. Including Notes mail within the OpenMail environment brings Notes mail into the same management and tracking infrastructure as the rest of the enterprise, reducing costs and lost messages.

### Host-Based E-mail Interoperability

OpenMail provides interoperability for host-based E-mail and calendaring (see Exhibit 6-16). Gateways are available to support PROFS, DISOSS, All-in-One, and similar environments. These products allow for directory synchronization as well as mail address conversion.

For instance, OpenMail can act as the enterprise directory and synchronize mainframe directories and groupware directories (e.g., Notes). Cen-

*HP OpenMail*

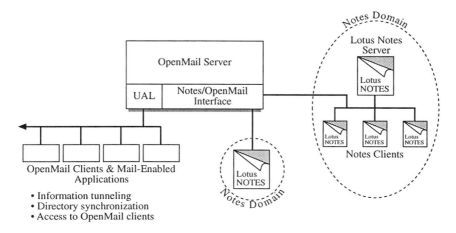

**Exhibit 6-15. Lotus Notes Integration.**

tralized directory management reduces message loss and increases overall messaging reliability. Direct integration with UNIX environments is provided by OpenMail's SMTP transport. Calendar integration with PROFS is supplied by HP OpenTime.

### Application Messaging

OpenMail supports multiple approaches to application to application messaging, including EDI X.435 and SMTP.

**EDI (X.435).** X.400 provides a robust transport standard but does not store messages on behalf of the user. OpenMail provides this capability,

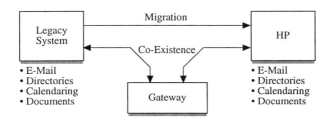

**Exhibit 6-16. Support for Host-Based Environments.**

99

# THE ELECTRONIC MESSAGING ENVIRONMENT

along with directory services and native support for industry standard E-mail clients. OpenMail's message store stores EDI messages coming from X.400, waiting for the EDI recipient application to request them. This EDI application calls the OpenMail message store API to list and retrieve messages.

For example, a customer (e.g., a large government agency) uses EDI to exchange tax information with large corporations. The application runs on an IBM CICS mainframe with an HP9000 server as an EDI front-end. The IBM connects to the HP9000 via LU6.2 and exchanges EDI messages with the OpenMail message store through remote procedure calls (RPCs). OpenMail then uses HP X.400 to send the EDI messages. The EDIFACT translator is also located on the HP9000, and submission or retrieval of EDI messages calls this translator.

**SMTP.** OpenMail supports X.400 levels of message tracking within the SMTP environment. Applications are written to the OpenMail UAL and each endpoint is assigned an endpoint. Sending applications post messages and addresses through the UAL. Receiving applications poll the mailbox periodically and then act on messages stored in the in-tray. OpenMail provides the same level of directory services, server storage, and message security to application messages as to personal messages. This makes application messaging within a public key environment an ideal method for secure document transmittal. The solution architecture used for the personal kiosk example described earlier can also be applied to application messaging. This allows "any-to-any" linking of applications across the Internet with privacy, guaranteed receipt, and nonrepudiation.

## SUMMARY

For businesses that need to integrate disparate end-user technology choices, OpenMail is a proven product that scales easily and manages the networks, systems, applications, and databases on which business enterprises depend. As the enterprise post office, OpenMail provides tremendous economies of scale for lowering the cost of commercial messaging. For more information and updates on OpenMail, users can look at the following site on the Internet: http://www.openmail.external.hp.com/OpenMail.

# Chapter 7
# The Defense Message System
*Christopher J. "Kit" Lueder*

The Defense Message System (DMS) program of the U.S. Department of Defense (DOD) upgrades the existing DOD E-mail systems, improves security, and provides consistent, interoperable messaging and directory services to all DMS users. The Defense Information Systems Agency (DISA) DMS Program Management Office has responsibility for program management and for deployment and implementation of the DMS infrastructure for DOD.

DMS provides secure, accountable, and reliable messaging services, fully integrated with a global DOD directory service, based on Joint Staff validated requirements. All components of DMS consist of commercial off-the-shelf products, provided by leaders in the electronic mail industry. DMS implements a set of international, open system standards that provide full interoperability from writer to reader. In addition, DMS will provide interfaces to, and interoperability with, other federal agencies, U.S. allies, the commercial sector, and the public at large.

DMS will support E-mail and provide store-and-forward file transfer services for such applications as nonreal-time audio, video imagery, and binary file exchange. Interactive applications (e.g., video teleconferencing or voice) will not be supported by DMS.

The DMS messaging and directory services are based on a suite of standards (recommendations) by the International Telecommunication Union-Telecommunications Standards Sector (ITU-TSS), jointly published with the International Standards Organization (ISO)/International Electrotechnical Association (IEC). The DMS electronic mail system is based on the 1988 edition of the X.400 series of standards, and the DMS electronic directory system is based on the 1993 edition of the X.500 series of standards. A new military message content type, referred to as P772, is defined in Allied Communications Publication (ACP) 123 and the U.S. Supplement to ACP 123, to enhance the services provided by the X.400 messaging system and to meet the DMS military and business-grade messaging requirements. DMS security services are then provided by placing the P772 message in an additional message envelope using the P42 Message Security Protocol (MSP), defined in Secure Data Network (SDN) SDN.701.

In addition to supporting the X.400 suite of protocols, the DMS message transfer agents (MTAs) and user agents (UAs) are integrated with the Microsoft Exchange and Lotus Notes user agent software for support of the common desktop user messaging environments for the PC and Macintosh environments. The UNIX environment is also supported. This chapter describes the Defense Message System program and the operational environment and the protocols, standards, security approach, and directory schema used in DMS.

## DMS PROGRAM OVERVIEW

### DMS Background

DMS began in 1989 as an effort by the U.S. DOD to transition the many noninteroperable messaging systems to a common X.400-based messaging system, using the Defense Information Systems Network (DISN) as the transport mechanism. The baseline DOD messaging environment consists of two main categories of systems: the Automatic Digital Network (AUTODIN) and the Defense Data Network (DDN).

**Organizational Messaging.** AUTODIN provides classified and unclassified organizational (i.e., office to office) messaging services. AUTODIN uses a message format defined in ACP 127 and Joint Army, Navy, Air Force Publication (JANAP) 128. AUTODIN is an expensive system to maintain (i.e., very labor intensive), because many aspects are not automated (e.g., release authority processing, paper-based directories). Although AUTODIN has some valuable services (e.g., reliability, security, and integrity), it has slow throughput, limited capacity, and long message delays, and it lacks support for binary data.

**Individual Messaging.** The DDN conveys individual (i.e., person to person) messaging services, based on the simple mail transfer protocol (SMTP, defined in Request for Comments 821 and 822), as well as other proprietary messaging systems by several dozen different vendors (e.g., IBM PROFS, Novell MHS, Lotus cc:Mail, Wang Mail, Microsoft Mail, MMDF-II, and UNISYS Mail).

These E-mail systems have many limitations, including a lack of interoperability among systems, inadequate or nonexistent security services (especially for messages passed between these environments), lack of enterprisewide management, complex support requirements, message accountability and reliability problems, inconsistent user addressing, no central user directory, and inconsistent messaging functional ability. The

*The Defense Message System*

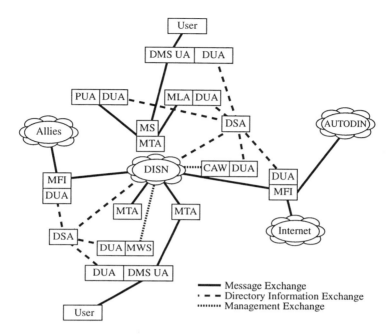

**Exhibit 7-1. DMS Functional Model.**

E-mail systems were experiencing explosive uncontrolled growth, as well as widespread and growing tactical use. The DOD found it could save money by deploying DMS and improve services in every respect.

### DMS Components

Exhibit 7-1, the Defense Message System functional model, depicts the DMS software components. The MTAs are interconnected to provide the backbone message transfer service. Some MTAs have collocated message stores (MSs) for serving end-users.

The DMS user agents (UAs) communicate with the MTAs. A directory user agent (DUA) is located with each DMS UA, multifunction interpreter (MFI), mail list agent (MLA), management workstation (MWS), certification authority workstation (CAW), and profiling user agent (PUA). The DISN network is used for the Defense Message System network infrastructure. DMS also interconnects with other messaging domains (e.g., the AUTODIN network, the Internet network, and the Allies' E-mail and directory systems). The software components used in DMS are described in the following sections.

103

## THE ELECTRONIC MESSAGING ENVIRONMENT

**DMS User Agent.** The DMS UA, a client in the messaging environment, provides the user interface and support for message preparation, submission, release, reception, and distribution determination. The DMS UA is specifically designed to support the DOD operational environment (e.g., authorized release of messages) and uses the military message content type P772. Most security functions are implemented in the UA, using the Message Security Protocol, referred to as P42. The UA uses the FORTEZZA PCMCIA card to sign, hash, and encrypt messages, and to authenticate connections (e.g., with the MTA) and directory queries.

**Message Transfer Agent.** The MTA, the server in the messaging environment, provides message submission, delivery, and relay services. No special enhancements are required for the DMS MTAs. In the future, DMS MTAs will support the X.400 standard capability for strong authentication of other entities with which the MTA communicates.

**Multifunction Interpreter (MFI).** The MFI provides gateway functions to allow DMS-compliant users to communicate with other messaging communities (e.g., vanilla X.400 systems, or commercial value-added networks, or DOD X.400 systems that are not yet DMS compliant), commercial and research networks, and AUTODIN. It converts messages between the DMS P42/P772 message content type and other messaging protocols, including P2 (1984) X.400, SMTP, and AUTODIN. It may also provide conversion of security services for communications with allies that support X.400 but use different security mechanisms from the P42/P772 messages.

**DMS Mail List Agent (MLA).** The DMS MLA is used to expand a collective name into a list of mail recipients and to forward the DMS P42/P772 message to the list recipients. The MLA is a special-purpose UA.

Mail lists are similar to the AUTODIN address indicator groups (AIGs), which may contain thousands of addresses in a single AIG. The MLA is necessary for the Defense Message System, because each recipient has a different security certificate. X.400 distribution list expansion by an MTA is not possible because the MTA cannot change the security tokens within the P42 message contents.

**Directory System Agent (DSA).** The DSA provides the directory services to the directory users. All DMS components that perform security processing (e.g., encryption of a message, validation of an electronic signature, and authentication of a connection with another system) require access to the DMS directory to obtain the necessary security certificate information and to check the certificate revocation lists. The DMS directory services are based on the 1993 edition of the X.500 series of ITU-TSS standards.

**DMS Directory.** The Defense Message System directory is a distributed hierarchical database that contains information about end-users (e.g., the E-mail address, security certificates, users' preferred E-mail delivery method, and telephone and facsimile numbers). It also contains information about computer systems, including presentation addresses and Internet protocol (IP) host addresses. The DMS directory will be expanded into a general-purpose DOD directory by integrating data requirements for other DOD applications.

**Directory User Agent (DUA).** The DUA provides the X.500 directory services to the user (a human user or a DMS messaging component).

**DMS Profiling User Agent (PUA).** The DMS PUA is a special kind of DMS UA that can automatically perform onward distribution of received military messages (forwarding the military message from an organizational recipient to action officers or other individual recipients within an organization). The PUA receives and decrypts a message, and resubmits the message to the MTA encrypted with new tokens generated by the PUA for the new recipients of the message. The originator's digital signature is retained, if present.

**Message Store (MS).** The MS is optionally used for individual messaging and is located with an MTA. The MS provides the mailbox where messages are delivered for later access by the user. It automatically sends an alert to the UA if a high-precedence message arrives; if the UA is not connected, the MS can auto-forward the message to an alternate UA that can receive and act on the high-precedence message.

**DMS Certification Authority Workstation (CAW).** The DMS CAW is used to provide security support to the DMS users, including the following:

- Generate security certificates for users by combining security keying and user privilege information with the user's X.500 directory name and adding a signature of a superior security authority.
- Program FORTEZZA PCMCIA cards with the user's private key information and authorizations (e.g., security classification, precedence level, and organizational releases).
- Update the security information held in the X.500 directory with the user's public certificate and the user's access control rights.

**DMS Management and Control Function.** This set of components is used to maintain and operate the DMS infrastructure. Included is a management workstation (MWS), the service manager application, and the DMS administrative directory UA. The management workstation consists of a suite of

## THE ELECTRONIC MESSAGING ENVIRONMENT

products, which includes a network management application, relational database management system, and trouble ticketing system.

### DMS Operational Environment

DMS components will communicate using the Defense Information Systems Network, which primarily uses the Transmission Control Protocol/Internet Protocol (TCP/IP) as the transport mechanism. The X.400/X.500 Open Systems Interconnection (OSI) application layer uses the OSI presentation and session layers. The RFC 1006 standard is used by the OSI applications to interoperate with the TCP/IP lower layers of the communications stack. OSI transport and lower layers may also be used in some situations.

Exhibit 7-2 depicts these two protocol stacks as used by DMS in the DISN environment; the left side of the exhibit shows the pure OSI stack, and the right side shows the mixed stack with OSI upper layers and TCP/IP lower layers. The standards references are also included, although some are an example of cases where more than one standard may be used. The leftmost column in the exhibit is the seven-layer OSI reference model, and the next column lists the protocols: P42, P772, P1, Transport Protocol Class 4 (TP4), Connectionless Network Protocol (CLNP), Logical Link Control (LLC), and Media Access Control (MAC), and the standards where the protocols are defined. The IS standards are defined by ISO/IEC.

The DMS backbone messaging infrastructure will be deployed by the Defense Information Systems Agency (DISA). The backbone includes MTAs, DSAs, MLAs, MWSs, and MFIs required to provide the core messaging transport services and directory service. The backbone infrastructure maintained by DISA will also include the interconnections to non-DMS user communities. Non-DMS user communities include the various SMTP-based or proprietary DOD systems that are used in a tactical environment, other federal government agencies, the commercial and research environment (e.g., the Internet), and by allies.

End systems on government sites (e.g., bases) will be deployed and operated by the services and agencies. The end systems include local MTAs, UAs, DUAs, and MWSs to maintain the local environment. The Agency is planning to operate the DMS messaging system as an administrative management domain (ADMD) and to interconnect with other commercial ADMDs as a peer messaging network.

DMS may initially subscribe as a private management domain (PRMD) to a commercial ADMD until it is able to become established as an ADMD. It is choosing an X.400 originator/recipient (O/R) addressing approach where it is named as if it were an ADMD. Users would have an O/R address

## The Defense Message System

|  | Pure OSI Stack | | | Mixed OSI/TCP Stack | | |
|---|---|---|---|---|---|---|
| Layers | Examples | | Layers | Examples | | |
| Message Format | P42, P772 | [SDN.701, ACP123] | Message Format | P42, P772 | [SDN.701, ACP123] | |
| Application | P1 | [X.411] | Application | P1 | [X.411] | |
| Presentation | | [IS 8823] | Presentation | | [IS 8823] | |
| Session | | [IS 8327] | Session | | [IS 8327] | |
| Transport | TP4 | [IS 8073] | Transport | TP0 | [IS 8073] | |
|  |  |  |  |  | [RFC 1006] | |
|  |  |  |  | TCP | [RFC 793] | |
| Network | CLNP | [IS 8473] | Internet | IP | [RFC 791] | |
|  |  |  | Network Interface | IP over | IEEE 802 [RFC 1042] | |
| Data Link | LLC MAC | [IS 8802-2] [IS 8802-3] |  | LLC MAC | [IS 8802-2] [IS 8802-3] | |
| Physical | Baseband | | Hardware | Baseband | | |

**Exhibit 7-2. DMS Protocol Stacks.**

that includes the following: Country = US and ADMD = DMS, Organization = [organization name], Organizational Unit = [location name], and Personal Name = [name]. A PRMD value would be included in the O/R address for non-DOD government agencies that wish to use the DMS environment.

**Unique DMS Characteristics**

Some characteristics listed here distinguish the DMS messaging approach from that of other X.400-based messaging environments. In general, the additional messaging requirements that the DOD has determined to be necessary for secure and reliable messaging communications have been isolated to enhancements for the DMS UA. For example, in the list of enhancements for the ACP 123 and the ACP 123 U.S. Supplement (described later), virtually all are implemented by the DMS UA. The message transfer service complies with the existing international standards for X.400, which eases the interoperability with other messaging domains.

As another example of enhancements being placed in the UA, DMS security is implemented in the UA. The security certificates are stored on a FORTEZZA PCMCIA card. Although the security protocol is standardized

## THE ELECTRONIC MESSAGING ENVIRONMENT

between the allies, the MFI gateways will still be needed between countries to convert between different cryptographic mechanisms. The DMS security approach and protocols are described in the following sections.

DMS distinguishes between individual and organizational messaging. An individual message is sent by an individual user and is addressed directly to the intended recipient; the message is not sent on behalf of an organization and no authorization is required for release of the message. An organizational message is sent as official correspondence between organizations and requires a message release authority. The DMS UA is able to differentiate between individual and organizational messages and can enforce the procedures that are required for the handling of organizational messaging.

**X.400 Elements of Service.** DMS makes use of the following four mechanisms, or elements of service, available in X.400 for alternative delivery, to ensure that the message can be routed around inoperative systems or other problems:

- Originator-requested alternate recipient is used by the sender to specify an alternate address to which the message is sent if the preferred recipient is not available. The originator includes a security token for the alternate recipient as well as for the intended recipient.
- With redirection of incoming messages, a user requests that the MTA redirect all incoming messages to an alternate address (e.g., if the user is about to power-down the workstation). The user's security certificate may have to be shared with the designated alternate recipient.
- Alternate recipient assignment provides a default or dead-letter recipient address (e.g., postmaster) to be used if a message cannot be routed as addressed; the message originator receives a nondelivery notice, so that corrective action may be taken by either the originator or the actual recipient. Encrypted information may not be readable by the alternate recipient, unless a certificate is shared between the alternate recipient and the intended recipient of the message.
- Autoforwarding may be performed by the MS or UA after the message has been delivered. For example, the DMS profiling user agent employs autoforwarding based on the received message's characteristics. Encrypted information may not be readable by the forwarded recipient unless a certificate is shared between the forwarded recipient and the original recipient of the message.

To inform the sender whether the message is received by the intended recipient in a timely manner, DMS messages use notification of delivery and nondelivery by the X.400 message transfer service. Receipt notification, with nonrepudiation and proof security services, is also required for DMS messages and is performed using the Message Security Protocol (MSP) P42.

*The Defense Message System*

DMS supports an approach called MINIMIZE to reduce message traffic during critical situations, putting less burden on communications resources and message recipients in the affected area. When a location or region is operating during a crisis, a MINIMIZE condition can be asserted. When MINIMIZE is in effect, the originator is responsible for ensuring that no unnecessary messages are sent to recipients in the affected area.

**Rationale for X.400-Based DMS**

Because the decision for DMS to be based on X.400 rather than SMTP or Multipurpose Internet Mail Extensions (MIME, defined in RFCs 1521 and 1522) has been contentious in some user communities, a rationale for DMS is provided here. The Department of Defense needed a common E-mail standard for interoperability within DOD and with other agencies of the federal government and allies. Reasons for selecting X.400 as the basis for DMS include the following:

- The international military community, including all of the U.S.'s closest allies, planned to use X.400 as the basis for future messaging systems. DMS benefited from existing NATO development of extensions to X.400 for military messaging services.
- In the federal agencies (outside of the DOD), more than 100 X.400-based messaging domains are in place that subscribe to the Federal Telecommunications System (FTS) 2000. The DMS Program Management Office expects many of these to migrate to DMS-compliant X.400 products and use the DISN network.
- X.400 is a robust messaging protocol with many services that are necessary for DMS. SMTP/MIME are known for their simplicity and do not provide sufficient features. The elements of service (e.g., the four alternative delivery and alternative recipient services) are crucial to reliable operation in DMS, but are not supported by SMTP/MIME and would be difficult to add to SMTP/MIME implementations. Other X.400 services not provided by SMTP/MIME include: authorizing user, autoforwarded indication, cross-reference, deferred delivery, directory name use, delivery or receipt notification (positive or negative), expire time, globally unique message ID, grade of delivery, incomplete copy, obsolete message, reply time, and UA capabilities registration. In addition, other elements of service defined in ACP 123 are not supported by SMTP/MIME.
- Extensibility is part of the X.400 standard. New message content types can be defined, and new EOS and protocol elements can be defined and used. A critical mechanism is provided so that the recipient system knows whether it can ignore or must reject a message that contains an extension that the recipient system does not support.
- X.400 is an international standard, with robust implementations available and with the support of many E-mail service providers. Although

# THE ELECTRONIC MESSAGING ENVIRONMENT

use of SMTP/MIME is predominant in the research and engineering environment (e.g., universities), many commercial organizations use X.400 for crucial business messaging.
- X.400 is fully integrated with the X.500 directory and supports the use of an X.500 directory name. X.500 is crucial to the DMS security infrastructure, with the Directory Name being used to identify the user in the security certificate, and the X.500 directory providing the repository for the users' public certificates.
- In X.400, message routing is explicitly defined within the messaging application layer. SMTP/MIME relies on routing by the underlying layers, based on the IP address of the recipient's host system. Although the latter is a simpler approach to configure and maintain, it does not allow the user to control message routing. Frequently on the Internet, an SMTP/MIME message will take many hours or days to transit the Internet network, and users cannot override the IP routing to bypass the problem areas in the network. In addition, message tracing on the Internet is difficult for diagnosing the cause of the delays in the environment.
- X.400 has well-defined messaging components and interfaces between the components (e.g., MTAs, MSs, and UAs). The X.400 standard defines a mechanism for strong authentication between components using certificates. SMTP has no authentication between host servers, so it cannot prevent connections from unauthorized systems — making SMTP hosts vulnerable to malicious flooding. SMTP user-authentication mechanisms, performed by the computer operating system, network file system, or the Kerberos security system, are based on passwords, not certificates. The SMTP Post Office Protocol authentication is also based on passwords.
- SMTP does not have a component comparable to the X.400 message store (MS). The MS allows a user's mail to accumulate at the server when the user's UA is turned off. It also allows a user to access the mail at the MS from more than one UA.
- The Message Security Protocol (MSP) has been defined to provide messaging security for DMS users. The available security approaches in SMTP (e.g., Privacy Enhanced Mail, or PEM, defined in RFCs 1421-1424) and MIME (MIME Object Security Service, or MOSS, defined in RFCs 1847-8) have not been accepted for use by the DOD because the security mechanisms are not strong enough and not all security services are provided (e.g., nonrepudiation of receipt).
- X.400 has better overall performance than SMTP/MIME for the conveyance of binary messages; most (if not all) DMS messages will be encrypted and therefore binary. SMTP/MIME does not directly support binary data, so all encrypted attachments must be translated (e.g., using the UUENCODE or Base64 encoding mechanism), incurring a 33%

overhead. At MITRE Corp., a preliminary measurement of the protocol overheads showed that X.400 (using RFC 1006 and TCP/IP lower layers) was as much as 13% larger than SMTP for text messages, but SMTP was roughly 26% larger than X.400 for messages with large binary attachments. MIME was not measured, but it has a higher overhead than SMTP.

## DMS Deployment Schedule

The Defense Message System Request for Proposal (RFP) was released to the public on March 16, 1994. (Section C of the RFP is the Statement of Work, which describes the requirements for the DMS components.) The DMS contract was awarded to Lockheed Martin Federal Systems (formerly Loral Federal Systems) on May 1, 1995.

DMS-complaint products were commercially released for Lotus Notes and Microsoft Exchange in April 1996. The initial nine sites were activated for the continental U.S. (CONUS), Pacific, and Europe in September 1996.

DISA initiated the infrastructure laydown and ordered infrastructure components in December 1996. The schedule was not finalized at the time of publication for the DMS Initial Operational Capability (IOC) for unclassified but sensitive messaging. At that point, the DMS infrastructure will be operational, supporting user messaging and directory traffic.

The Initial Operational Test and Evaluation (IOT&E) is scheduled for June 1997. IOT&E includes acceptance testing of the messaging, directory, security, and management components provided by Lockheed and its subcontractors.

AUTODIN is scheduled to be closed down by December 31, 1999. Funding for its maintenance and operation is only budgeted through that date. If AUTODIN continues beyond that date, enhancements would be required, such as to handle the date wrap-around in year 2000.

## Availability of DMS-Compliant Products

DMS will maintain a DMS-compliant certified product list (CPL). At present, procedures and requirements for product compliance test and evaluation are under development. The CPL will include all products provided by the DMS contract as well as from other vendors that provide DMS-compliant products. Current information should be available on the DMS home page as the products become available. Information regarding the DMS contract with Lockheed Martin Federal Systems can be obtained from

> Contracting Officer. U.S. Air Force, HQ Standard Systems Group, Contracting Officer, Ms. Dorothy Priest, Maxwell Air Force Base, Gunter Annex, AL. Telephone: 1-334-416-3207. Fax: 1-334-416-1757.

## DMS STANDARDS AND PROTOCOLS

### ACP 123, Military Message Content Types

Allied Communications Publication (ACP) 123 provides a common messaging strategy and procedures for allied military message traffic. The bulk of the ACP 123 document consists of specifications for elements of service (EOS) for military messages, a military message content-type definition (P772), and a message content profile and conformance statement for a standard military message. The ACP 123 document also contains operational procedures and specific requirements for millitary message MTAs, UAs, and MSs.

ACP 123 was developed by the U.S., Canada, the U.K., Australia, and New Zealand. ACP 123 was validated in April 1994. It was approved by the Combined Communications Electronics Board (which is comprised of representatives of the U.S. and its allies) in June 1994. ACP 123 (in Annex A) defines the military message content type, or P772, using Abstract Syntax Notation One (ASN.1). P772 adopted the same syntax as the 1988 X.420 P22 content type, but uses a content type identifier of object identifier. "Id-nato-mmhs-cont-mm88" is used rather than integer 22. The same ASN.1 encoding as P22 is used in P772, and military message extensions are defined using the same IPMS-EXTENSION macro as in P22; these extensions correspond to element of service (EOS) definitions.

The conformance statement and profile specify for each P772 protocol element whether it is optional or mandatory. It also specifies for each element of service (EOS) whether the user must be able to select the service or be able to display the associated information.

The remainder of this subsection describes the EOS specifications of ACP 123, which are cited here to convey to the functional ability provided by Military Message (MM) systems. Four categories of EOS specifications from ACP 123 exist: X.400 EOS that are used in the MM environment, X.400 EOS that are not needed in the MM environment, new EOS that are required for MM, and transitional EOS that are required to allow AUTODIN users to communicate with ACP 123-compliant users.

**ACP Elements of Service.** Most of the elements of service specified in the X.400 standard are used by ACP 123 without change, with the following exceptions:

- *Forwarded MM indication.* This EOS is similar to the X.400 forwarded interpersonal message indication, except that the message contains a forwarded MM (P772).
- *Hold for delivery.* When this EOS is supported, the MTA must support a method for ensuring delivery to a backup recipient all messages with

## The Defense Message System

a priority value of urgent (e.g., flash and override precedence) to meet delivery time requirements.
- *Military message identification.* This EOS is the same as the X.400 Interpersonal message identification, except that the O/R name of the originator is mandatory in ACP 123.
- *Receipt notification request indication.* The recipient's UA must not generate a receipt notification until the recipient has viewed the entire message indicating acceptance of responsibility.

The following elements of service are specified in X.400, but are not used in ACP 123:

- *Importance indication.* Importance is not used because it is redundant with the message precedence specifications and its use would cause confusion to the user.
- *Probe.* The invocation of the probe EOS is prohibited for security reasons, to prevent unauthorized users from sending probes to query about user addresses and characteristics without detection by the UA or user (because the service is performed by the MTA). Submission controls are used to prohibit UAs from originating probes within the Military Message Handling System (MMHS) environment; gateways and firewalls are used to prevent probes from entering the MMHS from other messaging domains.
- *Restricted delivery.* No requirement has been identified for this EOS, and no protocol to support this EOS is defined in X.400.
- *Return of content.* This EOS is not used for performance reasons. The originator should always keep a copy of messages that were sent; the copy can be correlated with the delivery or receipt notifications that are returned.
- *Sensitivity indication.* X.400 does not specify the actions or events to be taken based on the different values of sensitivity. This service might conflict with the message security label, which conveys more information.
- *Implicit conversion.* The ACP 123 document had stated that use of the implicit conversion EOS was prohibited. That position has since been reconsidered, however, and implicit conversion is now permitted.

The following elements of service are defined by ACP 123 for military messaging:

- *Primary precedence, copy precedence.* Military messages have six levels of precedence, and military procedures specify the handling of messages at the different precedence levels. The receipt of a higher-precedence message can preempt the transmission or processing of a lower-precedence message. Smaller message size limits and faster transfer times are specified for the higher precedence messages. A different level of precedence may be indicated to the primary recipients rather than to the copy recipients, because different actions are expected of the different recipients as a result of receipt of the message. The six precedence levels are mapped into the three levels of the message transfer service grade of delivery (or priority) EOS as follows: deferred and routine precedence are associated with nonurgent grade of delivery; priority and immediate, with normal; and flash and override, with urgent.

- *Message type.* This EOS enables the originating military message UA to distinguish messages that relate to a specific exercise, operation, drill, or project, and to provide a string to name that message type.
- *Exempted addresses.* This EOS is used to convey the names of members of an AUTODIN address list (AL) that the originator has specified to be excluded from receiving the message. The AL is a mechanism, like the X.400 distribution list, for the AUTODIN environment. Exclusion is provided at the point of AL expansion. The AL could be carried either as the AddressListIndicator heading field, as an X.400 Distribution List, or as an AL name in the primary-recipients or copy-recipients field.
- *Extended authorization information.* This EOS enables the originating military message UA to indicate the date and time when the message was officially released by the releasing officer, or it may indicate when the message was submitted to the MTS.
- *Distribution code.* The originating MM UA uses this EOS to convey distribution information to the recipient MM UA, to enable the recipient to perform distribution of the message to other persons or staff cells.
- *Message instructions.* This EOS is used to provide additional instructions about the handling of a message; specific values are not specified by ACP 123.
- *Clear service indicator.* This EOS indicates to the recipient that the message containing classified information has been transmitted over an unsecured channel, and therefore should be handled as confidential material.
- *Other recipient indicator.* This EOS enables the originator to indicate to the recipient the names of other recipients, as well as the category (primary or copy) that are intended to receive, or have received, the message by means other than the Military Message Handling System.
- *Originator reference.* This EOS enables the originator to assign a reference number to the message, which is provided to the MTS at the time of submission of the message.
- *Use of address list.* This EOS conveys the name of a predefined address list to which the originator is sending the message. The address list (AL) can be qualified as either a primary or copy recipient, and indicates if notifications or replies are requested from the AL members. Expansion of the AL entails placing the necessary address information

## The Defense Message System

for each AL member in the P1 envelope, except for exempted addresses that may be specified explicitly in the message header.

The following elements of service are defined by ACP 123 to meet transitional requirements, for interoperation with AUTODIN users that use the ACP 127/JANAP 128 message format. These transitional elements of service are used only in messages originating in AUTODIN, to ensure that all message header attributes can be conveyed to the DMS X.400 recipient. Use of these elements of service will stop when AUTODIN is completely phased out:

- *Handling instructions.* This EOS conveys local handling instructions for the message, which requires manual handling by AUTODIN operators.
- *Pilot forwarded.* This EOS enables an ACP 127 pilot message to be sent in the text body of a P772 message, and conveys the original ACP 127 header information.
- *Corrections.* This EOS indicates to the recipient that corrections (carried in an externally defined body part type corrections-body-part) to the text body are required.
- *ACP 127 message identifier.* This EOS conveys the original ACP 127 message identifier of the message.
- *Originator Plain Language Address Designator (PLAD).* This EOS indicates the PLAD of the originator of the message in AUTODIN, if the originator's PLAD cannot be translated into an X.400 O/R Name.
- *Codress message indicator.* This EOS indicates that the message is an ACP 127 codress encrypted message.
- *ACP 127 Notification Request, ACP 127 Notification Response.* These elements of service enable the originator to request and to receive a positive, negative, or transfer notification from the ACP 127 gateway.

## ACP 123 U.S. Supplement 1

The U.S. Supplement to ACP 123 provides U.S.-specific enhancements to be used for military messaging within the U.S. Department of Defense. It applies to all users of the Defense Message System, including all U.S. DOD services and agencies and other participating federal agencies. Use of these enhancements, however, may not be supported in communication with allied and NATO messaging systems unless they are included in a future version of ACP 123 or in the X.400 standard.

ACP 123 U.S. Supplement 1 is currently still in draft form. Another supplement will be developed by the U.S. intelligence community to meet intelligence security requirements. ACP 123 U.S. Supplement 1 contains the following:

- Refinement of the ACP 123 EOS specifications for use in the Defense Message System

## THE ELECTRONIC MESSAGING ENVIRONMENT

- Security services to be used by U.S. systems and procedures for security. The security services are based on use of the message security protocol (MSP)
- Procedures for message distribution, system management, and naming and addressing
- Functional requirements for the Defense Message System MTA, UA, MS, PUA, MLA, and MFI
- Additions to the ACP 123 P772 ASN.1 definitions for use in DMS
- A conformance statement and profile for the EOS, the P772 protocol, and the use of P1, P3, and P7 protocols

The EOS specifications are included here to inform the reader of the functionality provided by DMS messaging systems. The following EOS changes are made in ACP 123 U.S. Supplement 1 for use in DMS.

**Message Type.** Additional options are added to the message type EOS: other-organization, individual (working or organizational record), draft-organizational (working or organizational record), onward-distribution (working or organizational record), and forwarded-record-body.

**Primary Precedence, Copy Precedence.** The deferred and override precedence values are not used, but critical (CRITIC) and emergency command precedence (ECP) are added for use in DMS. The six precedence levels are mapped into the three levels of the Message Transfers Service (MTS) grade of delivery (or priority) EOS as shown in Exhibit 7-3. The exhibit also shows the MTS delivery time requirements, originator to recipient time of delivery, and largest message lengths for which the delivery times are guaranteed.

**Message Instructions.** Specific values for the instructions are given, as follows:

- "Minimize considered," to override filtering of the message in a minimize situation
- "Limited distribution," to only distribute to recipients who need to know
- "No distribution," to prevent further distribution of the message
- "Send to," to request that the message be provided to another user (who may not be accessible by DMS)
- "Alarm required," to request that the recipient be alerted on receipt of the message

**Multipart Body.** Externally defined X.400 message body part types are defined for use in DMS: U.S. Message Text Format (USMTF) messages, forwarded MSP message, forwarded report (e.g., delivery notifications), forwarded notification (e.g., receipt notifications), server query (e.g., to send structured query language, or SQL, commands to a server), and file transfer (e.g., for AUTODIN data pattern messages or EDI transaction sets).

## The Defense Message System

Exhibit 7-3. Message Precedence, Grade of Delivery, and Speed of Service Requirements.

| Military precedence | MTS grade of delivery | MTS time of delivery | Originator to recipient time of delivery | Assumed message length |
|---|---|---|---|---|
| CRITIC | | As fast as possible, <3 minutes | 3 Minutes | 5,400 |
| ECP | Urgent | | 3 minutes | 5,400 |
| Flash | | | 10 minutes | 7,000 |
| Immediate | Normal | <20 minutes | 20 minutes | 1,000,000 |
| Priority | | | 45 minutes | 2,000,000 |
| Routine | Non-urgent | <8 hours or start of next business day | <8 hours or start of next business day | 2,000,000 |

### International Standardization of ACP 123 Message Extensions

Nine of the ACP 123 elements of service have been submitted to ITU-TSS to consider including them in the X.400 series of standards. In May 1995, ITU-TSS Study Group (SG) I, Question (Q) 12 accepted the inclusion of the EOS in the next edition of Recommendation F.400/X.400. The contribution was submitted by the National Institute of Standards and Technology (NIST). Funding for NIST's participation in this activity ran out; final approval of inclusion of the EOSs by SG I Q12 is not certain because an advocate for the contribution may not be present at the meetings. The EOS submitted to ITU-TSS are as follows:

- Copy precedence
- Distribution codes
- Exempted addresses
- Extended authorization information
- Message instructions
- Message type
- Originator reference
- Other recipients indicator
- Primary precedence

The protocol elements associated with these EOS were reviewed by ITU-TSS SG VII Q14 in June 1995, but further development of the protocol specifications and review is required at subsequent meetings of SG VII Q14. The protocol enhancements would be defined as interpersonal message header extensions in the P22 content type, in the X.420 Recommendation. One EOS, exempted addresses, would require a protocol extension to the P1 and P3 envelopes as well, to inform the MTA of exempted addresses when the MTA performs distribution list expansion.

# THE ELECTRONIC MESSAGING ENVIRONMENT

## DMS Security

Security mechanisms are needed to protect against possible threats to the DMS messaging and directory environment (e.g., unauthorized disclosure, modification, deletion, insertion, replay, or redirection of DMS information, or denial of service for DMS users). The security mechanisms implemented include:

- Message confidentiality, by encrypting the message and including a message token for each recipient, using the connectionless confidentiality service of Message Security Protocol (MSP).
- Message integrity and data origin authentication, by including a digital signature in the MSP message using the MSP integrity and authentication services.
- Message nonrepudiation with proof of origin, by including the originator's digital signature, using the nonrepudiation service of MSP.
- Message nonrepudiation with proof of delivery, by including the recipient's digital signature on a return receipt, using the nonrepudiation service of MSP.
- Receipt authentication, to provide nonrepudiation of receipt of a message, using the MSP signed receipt function.
- Message security labeling, to convey the security classification (e.g., unclassified or secret) and, optionally, a security policy identifier (e.g., DMS, Secure Data Network System, or NATO), security category marks (e.g., a list of permissible nations), and a privacy mark (e.g., clear service indication).
- Peer entity authentication, by using the strong form of Association Control Service Element (ACSE) peer entity authentication during the bind operation (association establishment).
- Access control, based on privileges set in the user's FORTEZZA PCMCIA card, to control what information or services are available to a user.
- Firewall systems, to prevent unauthorized access from unsecured networks.

**Multilevel Security.** DMS security is based on the DOD Multilevel Information Systems Security Initiative (MISSI) program, which was designed by the National Security Agency to support programs such as DMS. The FORTEZZA crypto card, a credit-card-sized Type-II PCMCIA card, is an implementation of the MISSI security. FORTEZZA is from the Italian word meaning *fortress*; it is not an acronym. FORTEZZA supports three basic security algorithms, all of which were developed or invented by the U.S. National Security Agency (NSA). Specifically:

- *Digital Signature Algorithm is used for authentication.* It is a public-key algorithm, but can only be used for digital signatures. That algorithm is the basis for the Digital Signature Standard (DSS).

- *Secure Hash Algorithm (SHA) is used for integrity.* It performs a one-way hash computation, producing a 160-bit hash value from an arbitrary-length message. That algorithm is the basis for the Secure Hash Standard (SHS).
- *Skipjack Encryption Algorithm (Type II) is used for confidentiality.* Skipjack is a DOD-classified encryption/decryption algorithm. The algorithm includes a key escrow mechanism, where the government retains a copy of the user's key; in some cases, a government agency may be able to issue a warrant to obtain the escrowed key and be able to decrypt the user's encrypted messages.

The FORTEZZA card is used to provide security services for unclassified but sensitive messages. All messages in the initial phases of DMS are considered unclassified but sensitive. Subsequent releases of NSA's MISSI program will provide security services for higher classifications and compartments of messages.

For a DMS component to perform security processing (e.g., generating an electronic signature for a message being sent, decrypting a security token on a received message, and performing user authentication when connecting with another system), that component must have a FORTEZZA card inserted in a card reader locally on that computer system. The user must provide a personal identification number before the FORTEZZA card can be used. The same FORTEZZA crypto cards will be used by several other DOD systems, including the Defense Information Systems Network (for the identification and authentication of dial-in subscribers) and the Global Command and Control System (GCCS).

**Message Security Protocol, P42**

The Message Security Protocol (MSP) is an X.400 message content type of P42. It supports MISSI security, and the DMS implementations make use of the FORTEZZA cards to provide encryption, digital signatures, or both for DMS messages.

MSP is defined in Secure Data Network (SDN) SDN.701, developed by the U.S. National Security Agency. In DMS, an MSP P42 message carries an embedded P772 message; the MSP message acts as a security envelope for the P772 message it contains.

Exhibit 7-4 shows the format of a DMS message encoded in the P42 and P772 messaging protocols. The user's text message is contained as a message body part in a P772 military message. The P772 message is in turn encapsulated in the P42 message, for conveyance across the DMS messaging service; the P772 message is carried as an encrypted string if the content confidentiality service is used. Finally, the P42 message is encapsulated in the P1 X.400 message transfer envelope, for transfer between MTAs, or it is

# THE ELECTRONIC MESSAGING ENVIRONMENT

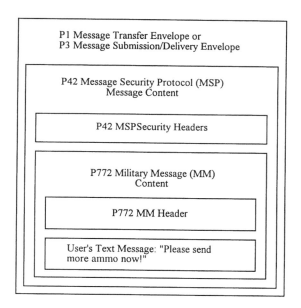

**Exhibit 7-4. DMS Message Format.**

encapsulated in the P3 X.400 message submission and delivery envelope, for submission from the DMS UA to the MTA, or for delivery from the MTA to the DMS UA.

## DMS Directory Schema

The Defense Message System directory is based on the 1993 edition of the X.500 series of recommendations. The directory can be used for authentication of users, name resolution (e.g., mapping from an X.500 directory name to an X.400 originator/recipient address), mail list expansion, and user capabilities assessment. A document specifying the DMS directory schema is still in draft form; the descriptions given here are based on the current draft specifications, though significant changes are not expected.

The directory will contain security information about users and DMS system components. The syntax for the security information stored in the directory is based on the definitions in ITU-TSS Recommendation X.509, as well as the directory object classes and attributes defined in the NSA document "Secure Data Network System (SDNS) Directory Specifications for Utilization with SDNS Message Security Protocol," SDN.702.

**Object Classes.** Exhibits 7-5, 7-6, and 7-7 list the object classes (OCs) that will be used in the DMS directory. OCs define the types of directory entries

## The Defense Message System

**Exhibit 7-5. Standard Object Classes Used Unchanged in DMS.**

| Object class | Standard reference |
|---|---|
| Country | X.521 |
| Organization | X.521 |
| Organizational unit | X.521 |
| Locality | X.521 |
| Group of names | X.521 |
| Group of unique names | X.521 |
| Device | X.521 |
| Application entity | X.521 |
| MHS user agent | X.402 |
| dsa-fortezza | SDN.702 |
| dsa-sdns | SDN.702 |

that can be stored in the directory information base and define the types of attributes or data that can be placed in those entries.

Exhibit 7-5 lists those OCs that are already defined in existing standards to be used, as is in the DMS directory. Exhibit 7-6 lists OCs, already defined in existing standards, that form the basis for new DMS OCs; one reason for defining new OCs is to add security attributes (e.g., to store certificates in the directory) to the OC definitions. Exhibit 7-7 lists the new DMS OCs that have been defined, along with a brief description.

**Exhibit 7-6. Standard Object Classes Used When Creating New DMS OCs.**

| Object class | Standard reference |
|---|---|
| Top | X.501 |
| Alias | X.501 |
| Organizational unit | X.521 |
| Organizational person | X.521 |
| Organizational role | X.521 |
| Application entity | X.521 |
| Device | X.521 |
| MHS message store | X.402 |
| MHS message transfer agent | X.402 |
| msp-user-fortezza | SDN.702 |
| mail-list | SDN.702 |
| ca-fortezza (Certification Authority) | SDN.702 |
| msp-user-sdns | SDN.702 |
| ca-sdns | SDN.702 |
| strong-authenticate-user-fortezza | SDN.702 |
| strong-authenticate-user-sdns | SDN.702 |
| ckls-sdns (Certificate Revocation Lists) | SDN.702 |

# THE ELECTRONIC MESSAGING ENVIRONMENT

**Exhibit 7-7. New DMS Directory Object Classes.**

| New DMS object class | Description |
| --- | --- |
| Certification authority | Attributes for a CA |
| Computer | Attributes for a computer system |
| ML | Mail list |
| MLA | Mail list agent |
| Messaging organizational unit | To describe an organizational user |
| MFI | Multifunction interpreter |
| Release authority | The authority who releases organizational messages |

Twelve additional DMS object classes (not shown) have been defined for transitional AUTODIN purposes, to be able to store different types of AUTODIN plain language addresses or routing indicators in the DMS directory. These AUTODIN-specific OCs will be dropped from the DMS directory after AUTODIN is phased out.

An example, the Certification Authority Object Class is shown in Exhibit 7-8 to demonstrate the kind of information that will be stored in the DMS directory. A condensed version of the planned directory information tree for the DMS directory is depicted in Exhibit 7-9. In the exhibit, C is country, O is organization, OU is organizational unit, L is locality, and CN is common name. An example of a directory name is as follows: C = us, O = u.s. government, OU = dod, OU = Army, OU = locations, L = Pentagon, OU = [office name], CN = John Q. Private.

**Exhibit 7-8. Certification Authority Object Class.**

```
           certificationAuthority  OBJECT-CLASS := {
           SUBCLASS OF      { organizationalRole | msp-user-fortezza |
                            msp-user-sdns | ca-fortezza | ca-sdns}
           MAY CONTAIN      { alternateRecipient |
                            minimize |
                            nationality |
                            preferredDelivery |
                            aliasPointer |
                            listPointer |
                            mhs-or-addresses |
                            mhs-deliverable-content-length
                            mhs-deliverable-content-types |
                            mhs-deliverable-eits |
                            mhs-undeliverable-eits |
                            mhs-or-addresses-with-capabilities |
                            mhs-message-store-dn |
                            rfc822Mailbox |
                            effectiveDate |
                            expiration Date }
           ID               Id-oc-certificationAuthority }
```

*The Defense Message System*

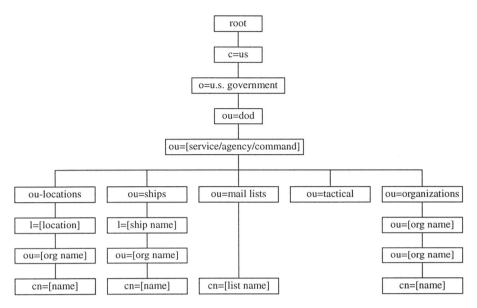

**Exhibit 7-9. DMS Directory Information Tree.**

---

A document ACP 133 is being developed to specify a consistent requirements and profile specification for X.500 directory systems for the U.S. Government and its allies. The Combined Communications Electronics Board established the Allied Message Handling (AMH) International Subject Matter Experts (ISME) forum, which formed a task force (with participation by the U.S., U.K., Canada, Australia, and New Zealand) to develop the ACP 133 document. The North Atlantic Treaty Organization is an observer for the ACP 133 Task Force.

**SUMMARY**

Information about DMS is available on the Internet. Using the World Wide Web (WWW), the DMS home page may be found at URL: http://www.itsi.disa.mil/dmshome.html. "Anonymous" FTP is also available for access to the DMS home page server. Using the Internet FTP protocol, users can connect to host ftp.itsi.disa.mil (or IP address 198.4.59.6), give a username of "anonymous," and give their E-mail address as the password value to connect to directory /pub/dms_exe.

An SMTP E-mail address is also available for users to send their comments or questions:

dmswww@ncr.disa.mil

# THE ELECTRONIC MESSAGING ENVIRONMENT

The Navy DMS Home Page URL is:

> http://www.chips.navy.mil/dms/index.html

The Army DMS Home Page URL is:

> http://www.cec.army.mil/~isma/dmsarmy.html

The Army Information Systems Management Activity (ISMA) office also has a Bulletin Board server, host IP address 134.80.3.23. For "Anonymous" FTP access, users can connect to directory /pub/dms.

The Air Force DMS home page URL is:

> http://w3.af.mil/DMS/dms.htm

These home pages also have links to other service or agency home pages, which may have additional information on DMS. Also, if one of the preceding addresses for a server changes, the new server can be found by following links from one of the other home pages. Using the WWW, two SDNS documents, SDN.701 and SDN.702, may be retrieved using the following two URLs (for SDN.701, unzip the compressed files first, which are then in PostScript format):

- ftp://acqbbs.ssc.af.mil/dms/refd/sdn701.exe
- ftp://acqbbs.ssc.af.mil/dms/refd/sdn702.ps

Internet RFCs may be retrieved using "Anonymous" FTP from a variety of FTP hosts, including ftp.psi.com and nic.ddn.mil. The U.S. Department of Commerce NIST Federal Information Processing Standard (FIPS) publications may be obtained from the Superintendent of Documents, U.S. Government Printing Office, Washington, D.C. 20402. Telephone: 1-202-512-1800. Fax: 1-202-512-2250. A separate E-mail effort by the federal government-wide E-mail Program Management Office, by the General Services Administration Office of Emerging Technology, is documented at the following WWW URL:

> http://www.fed.gov

# Chapter 8
# Electronic Messaging in the Healthcare Industry
*Rhonda Delmater*

The common objective of healthcare providers, payors, and patients is timely access to quality care in a cost-effective manner. In a survey conducted by the Community Medical Network Society, hospital management cited "communication among physicians for clinical and referral purposes" as one of the top two issues, along with "using outcome measurements to evaluate quality and performance."

As is true of most American industries (e.g., electronic messaging), the healthcare industry has been undergoing dramatic change. The traditional focus of the messaging industry has evolved from interpersonal E-mail to a much broader scope encompassing electronic messaging applications (e.g., electronic commerce and workflow).

Changes in the healthcare industry may be considered even more dramatic with emphasis shifting reactive treatment of patients to proactive wellness programs, and payment for services changing from fee for service to managed care or even capitation. Under fee for service, healthcare providers submit claim forms for services rendered, and are reimbursed to the amount considered reasonable and customary for the location where the service is rendered. Under managed care, rate schedules are prenegotiated. The providers accept a smaller fee in exchange for access to the patient population covered by an employer or healthcare administrator. Under capitation, a provider organization agrees to provide healthcare services for a monthly fee per member. In other words, if the members are healthy, the providers get to keep the fees while providing little service, but if the members are less healthy than the providers expect, they can be in a losing proposition.

# THE ELECTRONIC MESSAGING ENVIRONMENT

## EMERGENCE OF TELEMEDICINE

Information systems in healthcare have historically been focused on the financial business of obtaining reimbursement for services rendered. Under the new paradigm, increased emphasis will focus on measuring the effectiveness of treatments, or what is commonly referred to as *outcomes measurement*. This will require increased collection and dissemination of clinical information. The number of medical specialists will decrease as a result of cost restructuring, so it will be important to provide appropriate clinical information to caregivers, and to make the specialists as accessible as possible through the application of information technologies (e.g., telemedicine).

Telemedicine simply means practicing medicine across a distance. In practical application, it is often used to supplement the capabilities of the local care provider (e.g., a rural physician), or to provide specialty or even subspecialty consultation regarding a case. In the future it may even be used to practice remote surgery using robotics. Current telemedicine is almost exclusively based on point-to-point or dedicated network connections (e.g., video teleconferencing, file transfer, or direct file access), and can use nearly any network topology. Many telemedicine applications use video teleconferencing technology, while others use file transfer technologies (e.g., teleradiology) where large binary files of diagnostic images are transmitted.

In the future, as networking capabilities continue to advance and electronic messaging networks evolve from a textual medium to an object-oriented one supporting multimedia, telemedicine may take advantage of electronic messaging to provide worldwide access to medical specialists for consultations. The referring physician could send a multimedia message, including transcribed documents, video clips, portions of the patient record, dictation (audio), and so on, to whatever expert is needed for a particular case. As telemedicine becomes more common, E-mail is likely to become, not only viable, but the preferred transport vehicle for telemedicine consultations.

The current point-to-point connections support consultations by a limited, predetermined set of specialists. E-mail-based telemedicine could make a much broader field of specialists available for consultation for unusual or dramatic cases. Multimedia electronic messaging will dramatically affect healthcare delivery by transforming telemedicine from real-time point-to-point video teleconferencing, to store-and-forward video referrals composed of a full range of clinical information (e.g., diagnostic images, EKGs, and voice and video for telemedical consultations). At the same time, electronic messaging will support such administrative requirements as electronic commerce. Workflow applications are also receiving increased emphasis in healthcare, for both administrative and clinical applications.

*Electronic Messaging in the Healthcare Industry*

Today, most healthcare institutions provide some form of interpersonal messaging, or E-mail for internal staff use. Many hospitals have central host-based systems based on legacy E-mail applications (e.g., IBM PRofessional OFfice System, or PROFS). PROFS is one of the early mainframe E-mail systems. Most university hospitals and large provider organizations have more robust messaging capabilities with interconnection among two or more E-mail systems along with Internet access (e.g., SMTP).

## Expanding EDI into Healthcare

The current use of electronic data interchange (EDI) in healthcare is primarily to verify eligibility for benefits and to submit medical claims. The WorkGroup for Electronic Data Interchange (WEDI) promotes the expansion of EDI in healthcare. The WEDI Steering Committee has approximately 25 members, including several large insurance companies (e.g., Aetna and Travelers) and several prominent industry associations (e.g., the American Hospital Association, the Medical Group Management Association, the National Association for Home Care, and the American Association of Retired Persons, AARP). Medicaid and ANSI X12 are also represented.

WEDI's charter pertains exclusively to healthcare. They assert that from $8 to $20 billion could be saved in the U.S. by implementing only four core transactions, and as much as $26 billion could be saved by implementing the first 11 business transactions. There are three types of healthcare transaction sets: administrative, materials management, and patient care. WEDI claims that a net savings of more than $42 billion can be realized over the next six years if EDI is expanded to 11 types of healthcare transactions (see Exhibits 8-1 to 8-3).

---

**Exhibit 8-1. Administrative Transaction Set.**

| | |
|---|---|
| 257 | Health Care Eligibility/Benefit Inquiry |
| 258 | Health Care Eligibility/Benefit Information Immediate Response |
| 270 | Health Care Eligibility/Benefit Inquiry |
| 271 | Health Care Eligibility/Benefit Information |
| 276 | Health Care Claim Status Request |
| 277 | Health Care Claim Status Notification |
| 278 | Health Care Services Review Request |
| 279 | Health Care Services Review Information |
| 834 | Benefits Enrollment and Maintenance |
| 835 | Healthcare Claim Payment/Advice |
| 837 | Healthcare Claim |

# THE ELECTRONIC MESSAGING ENVIRONMENT

**Exhibit 8-2. Materials Management Transaction Set.**

| | |
|---|---|
| 810 | Invoice |
| 812 | Credit/Debit |
| 820 | Payment Order/Remittance Advice |
| 832 | Price Sales Catalog |
| 836 | Contract Award |
| 840 | Request for Quotation |
| 843 | Response to Request for Quotation |
| 844 | Request for Rebate |
| 845 | Price Authorization (Contract Charges/Award Confirmation) |
| 846 | Inventory Inquiry/Advice |
| 849 | Response to Request for Rebate |
| 850 | Purchase Order |
| 855 | Purchase Order Acknowledgment |
| 856 | Ship Notice |
| 867 | Product Transfer/Sales Report |
| 867 | Product Transfer/Sales Report |

## CASE STUDY

St. John Medical Center (SJMC) of Tulsa, Oklahoma, is a 750-bed acute-care facility. It is comprised of multiple entities, including the Regional Medical Lab, Cardiovascular Institute and Physicians Support Services. It is a partner in an HMO with St. Francis Hospital, St. Anthony's and Mercy Hospital in Oklahoma City, and a member of the Community Care Alliance. SJMC is part of the Sisters of the Sorrowful Mothers (SSM) Ministry that has hospitals in several locations, within and outside of the U.S. SJMC has earned a reputation for high-quality healthcare.

SJMC recognizes IT as a strategic investment. Their business requires the processing of a vast amount of data, and the availability of meaningful information to support patient care processes. They have recently undertaken a large project to upgrade their network infrastructure in preparation for implementing new distributed applications. As other healthcare organizations are looking forward to implementing a computer-based patient record (CPR), SJMC has recently implemented one to provide a clinical information resource regarding its patients.

A CPR is an information system that collects and disseminates patient data across multiple episodes of patient care, often comprised of a clinical

**Exhibit 8-3. Patient Care Transaction Set.**

| | |
|---|---|
| 274 | Request for Patient Information |
| 275 | Patient Information |

workstation (user interface), a clinical data repository (data warehouse), a master patient index (table of patient identifiers), and an interface engine (gateway to ancillary information systems — admissions, laboratory, radiology, pharmacy, dictation/transcription).

SJMC has approximately 3400 employees, with nearly 70% having E-mail accounts. Until recently, all of the accounts were on a mainframe CICS-based system called Wizard Mail. Command Information Control System (CICS) is an IBM teleprocessing monitor which is widely utilized for mainframe transactions. The system also supports E-mail accounts for affiliated organizations — Jane Phillips Hospital, Omni Medical Group, and Regional Medical Lab — from a single host.

**Anatomy of an E-Mail Network**

SJMC has made very effective use of E-mail as a communication medium among multiple sites within the organization. The center's IT management, however, recognized the gap between the capability of the mainframe-based E-mail system and what will be needed as the healthcare industry continues to experience increasing competitive pressures and the emphasis shifts from administrative to clinical information systems. It is clear that a robust electronic messaging capability, with unlimited connectivity options, will be essential for St. John Medical Center.

SJMC purchased Wizard Mail roughly 11 years ago for a very small sum. The vendor is H&W Computer Systems, Inc.(Boise, Idaho). Integration products offered by the vendor are Wizard Mail Gateways for: Message Handling Service (MHS) from Novell; IBM Mail Exchange, PROFS, and Office Vision from IBM; and SYSM, another E-mail system offered by H&W.

Wizard Mail is a host-based system with a character user interface that supports the following features and functions: sending mail, reading mail, directory of messages, mailing lists, filing messages, user lists, bulletin boards, calendars-scheduling, online help, ticklers, and common messages (i.e., structured forms).

Although enterprisewide scheduling is available through Wizard Mail, use occurs predominantly at the workgroup level, with IT making the most use of the system. SJMC has developed its own interface to accept incoming messages from both batch and online application systems for distribution to Wizard Mail users. These are primarily for notification and report distribution. Currently, more than 200 types of messages are routed in this manner. Examples include dirty room notification, error messages to computer programmer, and patient transfer. These types of notifications can continue to be routed to other E-mail systems from Wizard Mail interface, as long as Wizard Mail and the related interfaces continue to be used.

## THE ELECTRONIC MESSAGING ENVIRONMENT

SJMC has also been the premier site (installation number 001) for the Cerner PathNet system, which is one of the leading laboratory information systems. As such, SJMC desires the capability to send lab results and reports automatically through the E-mail network. The computing environment at St. John's includes an IBM mainframe, AS-400's, RS-6000's, Suns, and DEC VAX and MicroVAX machines as host computers, along with various LAN servers. Users access these hosts (or servers) through various types of terminals and both DOS and MS/Windows PCs.

Even though SJMC has a diverse computing environment, until they began their electronic messaging migration project in 1995, they had a homogeneous E-mail network. There is significant interest from the user community to migrate to Microsoft Exchange, and in connection to Internet mail. Any of the leading LAN E-mail applications would offer similar benefits over the mainframe E-mail system, including a graphical user interface featuring windows, icons, menus, and pointers, and the ability to attach binary objects (e.g., documents, spreadsheets, pictures, and graphics). Evolution of LAN E-mail to client/server architectures has provided even more advanced capabilities (e.g., increased support of collaborative features including shared folders).

Even though there was no connectivity to other E-mail systems implemented within SJMC or to external E-mail networks, SJMC made a significant investment in enhancing its networking infrastructure to support these and other information systems enhancements. An electronic messaging plan was developed to identify both tactical and strategic implementation activities. The planning methodology included a site survey of the medical center, a technical survey of current and announced messaging products, a survey of messaging applications in the healthcare industry, and development of a vision of future messaging applications for healthcare delivery systems.

SJMC selected Microsoft Exchange as its LAN E-mail system, based on earlier standardization on Microsoft Office. SJMC will continue to have a large number of mainframe Wizard Mail users during the near term, therefore interconnection between Wizard Mail and MS Exchange have been necessary to support the migration. SJMC plans to phase out Wizard Mail by the year 2000.

In addition, external E-mail connectivity is desired, with near-term emphasis on SMTP. Preference for SMTP over X.400 is based primarily on the number of users, or access to a greater E-mail population. Most X.400 users also have access to Internet mail; however, the converse is not true. X.400 may be reconsidered in the future as SJMC and other healthcare providers become more reliant on external messaging.

*Electronic Messaging in the Healthcare Industry*

## Electronic Forms and Other Goals

The SJMC Printing Services organization currently has a limited implementation of an electronic forms package, Formflow from Delrina. SJMC has had the software for 2 years and has licensing for 20 filler users. The proposed LAN messaging implementation must support transport of the Formflow electronic forms. It is desirable for the messaging system as a whole to support E-mail routing of forms created using Formflow.

Printing Services has identified four phases of electronic forms evolution: print on demand, typewriter replacement, intelligent forms (includes calculations), and data collection (includes DBMS interface). Of the 2800 forms used at SJMC, about 125 have been converted to electronic format using the Formflow developer package, with some forms at all phases of the evolution path. There is also interest in routing forms to physician offices and to Jane Phillips Hospital, and in support of electronic signatures.

The vision for the SJMC Messaging System is to take full advantage of the emerging technologies that will support multimedia clients. This is likely to become important in support of telemedicine, as the capability emerges to support store-and-forward of video transmissions for expert consultation. Exhibit 8-4 illustrates the Visionary SJMC Messaging System.

SJMC has identified the following system-level goals for E-mail:

- E-mail should be available on the users' native platform.
- Each user should be able to originate (and receive) mail from his/her local work platform to be routed to (from) the standard SJMC E-mail systems and to external users (by SMTP).

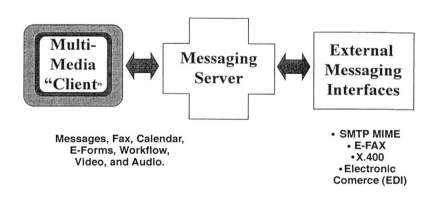

Exhibit 8-4. SJMC Electronic Messaging Vision.

# THE ELECTRONIC MESSAGING ENVIRONMENT

| ID | Task Name | Duration |
|----|-----------|----------|
| 1 | Project Planning | 60d |
| 2 | Microsoft Exchange Pilot | 108d |
| 3 | Microsoft-Wizard Connection | 23d |
| 4 | External Gateway Implementation | 86d |
| 5 | User Migration | 1088d |
| 6 | Mail Enabled Applications | 195d |
| 7 | Messaging Server Evaluation | 129d |
| 8 | Messaging Server Implementation | 133d |
| 9 | Multimedia Client Evaluation | 133d |
| 10 | Multimedia Client Implementation | 130d |
| 11 | Gateway Upgrades | 262d |
| 12 | Decommission Wizard Mail & MHS | 23d |

**Exhibit 8-5. SJMC Electronic Messaging Implementation Timeline.**

---

- Message addressing should be easy.
- Users should be able to access an accurate directory from their platform.

The electronic messaging timeline for SJMC is illustrated in Exhibit 8-5.

As the healthcare industry shifts to greater emphasis on clinical applications, the electronic messaging network will be required to provide the electronic transport for many types of objects, including audio, video, graphical representations (e.g., EKGs), and diagnostic images. It is unknown whether a single multimedia client will emerge, but it is probable, because the APIs will support a best-of-breed approach on a single messaging server, that there will be many special purpose clients as well as some general-purpose client applications.

## SUMMARY

The messaging technologies and products that are rapidly becoming available can have an exciting effect on patient care. Multimedia messaging servers and client software can support electronic delivery of a full range of clinical information (e.g., clinical graphics, diagnostic images, EKGs, and voice and video for telemedical consultations), as well as supporting administrative requirements, including electronic commerce. Electronic messaging also supports workflow that can improve administrative and clinical processes when combined with process engineering.

The vision for healthcare is to take full advantage of the emerging technologies that will support multimedia clients. This could change the face of telemedicine from real-time point-to-point video teleconferencing consultations, to store-and-forward video referrals addressed to virtually any specialist for a specific consultation. The specialist may receive a compound message that includes any or all of the following types of body parts: text audio, video, graphical representations (e.g., EKGs), and diagnostic images. The healthcare organizations who have such a capability available early may well have a competitive advantage.

# Chapter 9
# Enterprise Directory Services

*Martin Schleiff*

Many consulting organizations, trade associations, and vendors are touting directory services as the center of the communications universe. They believe that enterprises will benefit as they put increasing efforts and resources into their directory services. Application directory services (e.g., the address books in today's E-mail packages) are devoted to enhancing the functional ability of a particular product, service, or application to its user community. Even though E-mail address books are capable of displaying and possibly administering some information about E-mail users, they are not geared to manage corporate-critical information and deliver it to other applications, users, and services. The predicted benefits are not realized until a company embraces the concept of an enterprise directory service.

## APPLICATION VS. ENTERPRISE DIRECTORY SERVICES

Perhaps the simplest way to contrast application and enterprise directory services is to consider the community, or user base, being served. Application directory services typically have a well-defined user base. This may consist of the users of a particular calendaring or scheduling system, or an electronic messaging system. Where multiple, disparate E-mail systems are deployed, the community may consist of the users of all interconnected messaging systems, and the application directory service may include processes that synchronize directory information between each system.

Enterprise directory services focus on providing a core set of fundamental services that can be used by many environments and customer communities. The prime objective is to leverage the efforts of few to the benefit of many (see Exhibit 9-1).

The enterprise directory service provides an enabling infrastructure on which other technologies, applications, products, and services can build. It is wise to establish this infrastructure even before receiving hard require-

# THE ELECTRONIC MESSAGING ENVIRONMENT

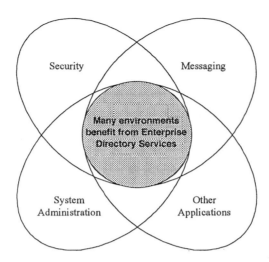

Exhibit 9-1. Enterprise Directory Services Environment.

ments from specific customer environments. Two symptoms of companies that lack enterprise directory services are hindered deployment of new technologies and applications, and redundant processes that have been built by each environment to meet its own needs.

Another way to contrast application and enterprise directory services is to consider how they evaluate the merit of a potential new service. Providers of enterprise directory services will consider the potential for their efforts to eventually be leveraged across multiple user communities. If only one community will ever use the new service, little advantage is associated with hosting that service at the enterprise level. It might be more appropriate to encourage that community to build its own service and offer to support its efforts with managed directory information. For example, if an organization wishes to track the activities of its employees on various projects, an enterprise directory service would likely decline the request to track activities, but offer to support the project by providing access to the appropriate employee information.

## WHAT GOES ON IN AN ENTERPRISE DIRECTORY SERVICE?

In a nutshell, any directory service consists of activities and processes to collect and publish information. Many directory services also add value by integrating various types and sources of information into commonly useable formats.

Services that are likely candidates to be provided as part of an enterprise directory service include the following: information solicitation, reg-

*Enterprise Directory Services*

istration, naming, authentication, directory synchronization, and coordination of publication infrastructure. Each of these services is briefly described in the following paragraphs. Consideration is given to some of the issues and challenges facing the providers of directory services.

## Information Solicitation Services

Providers of enterprise directory services can be characterized as data hungry; they realize that empty directories are worthless. They continually seek new and better sources of information to include in their directories. Frequently, these information sources are less than enthralled about expending any efforts and resources that do not directly benefit their own causes.

For example, the human resources department of a subsidiary may not see the benefit of including subsidiary employee information in the directories of the parent company. Likewise, a payroll organization may not fully appreciate that the information they maintain about who is authorized to sign an employee's time card could also be used to approximate the company's management hierarchy. Presenting this information in a directory provides visibility of the company's dynamic organizational structure.

Enterprise directory service providers tend to accept information in any way they can get it. Data submitted by E-mail, file transfer protocol (FTP), floppies, facsimile, phone calls, and even yellow sticky notes are readily accepted in any format. A service provider would rather bend over backward than impose the service's conventions on a hesitant supplier and risk losing their willingness to provide information. In fact, the most critical task in providing enterprise directory services is to build and maintain relationships with data providers.

Service providers also realize that directories with misleading information are worse than worthless. They therefore fret continually about the condition of the data they receive and publish.

At the system level, service-level agreements normally provide some level of confidence about the availability of systems and the quality of hardware and software support. The challenge for enterprise directory service providers is to expand this concept to include information.

Agreements with information sources help identify the quality and availability of the information that will appear in the directory. Information does not need to be perfect to be meaningful and useful, as long as the quality characteristics of the information are made known to users so they can make value judgments about how to appropriately use the information.

## Registration Services

Whenever possible, enterprise directory service providers attempt to acquire directory content from other sources — they prefer to be informa-

THE ELECTRONIC MESSAGING ENVIRONMENT

tion publishers rather than information owners. Sometimes, however, no managed source for a particular type of information can be identified. In such cases, building a registration service to collect the desired information may be considered. The following list offers examples of registration activities that may be included in an enterprise directory service:

- *Collection of workstation E-mail addresses.* In most messaging environments, E-mail address information can be collected from a well-defined group of administrators. Unfortunately, where UNIX and other workstations are abundant, administration tends to be much less defined and coordinated. In such areas, each workstation may have its own messaging system, and each user may be responsible for administration of their own system. There is probably not a single source from which to obtain address information for a large company's workstation community.
- *Primary address designation.* Many people have more than one E-mail mailbox. For example, engineers may prefer to use their UNIX systems for messaging, yet their managers may require them to maintain an account on the same mainframe used by all the managers. Individuals may require the flexibility to work with binary attachments offered by a LAN messaging system and still maintain a mainframe account to use a calendaring system that scales to enterprise levels. With frequent migrations from one messaging system to another, large groups of people may have multiple mailboxes for several weeks while they gain familiarity with the new system and transfer information from the old system. All these cases breed confusion among senders when it is not apparent which system holds the intended recipient's preferred mailbox. Registration services may be established to let individuals designate their preferred mailbox (or preferred document type or preferred spreadsheet type).
- *Information about non-employees.* The human resources department is an obvious source of employee information to be included in a directory. The HR department, however, probably does not track information about non-employees who may require access to a company's computing resources and who may have E-mail accounts on the company's systems. A service to register, maintain, and publish information about contractors, service personnel, and trading partners may be considered.
- *Information about nonpeople.* Other entities that may be appropriate to display in a directory include distribution lists, bulletin boards, helpdesks, list servers, applications, conference rooms, and other non-people entities.

Caution is advised when considering to offer registration services, because such services frequently require extensive manual effort and will sig-

*Enterprise Directory Services*

nificantly increase the labor required to run the enterprise directory service. Also, the enterprise directory service then becomes a data owner and can no longer defer data inconsistencies to some other responsible party. Even though resources are consumed in providing such services, the benefits can far outweigh the costs of uncoordinated, inconsistent, absent, or redundant registration activities.

**Naming/Identification Services**

A naming service simply provides alternate names, or identifiers, for such entities as people, network devices, and resources. A prime example is domain name service (DNS), which provides a user-friendly identifier (i.e., host.domain) for a networked resource and maps it to a very unfriendly network address (e.g., 130.42.14.165).

Another popular service is to provide identifiers for people in a format that can be used as login IDs or E-mail addresses, and which can be mapped back to the owners. LAN-based messaging systems frequently identify and track users by full name instead of by user ID, assuming that people's names are unique across that particular messaging system. This is one of the major hurdles in scaling LAN-based messaging systems beyond the workgroup level. A naming service could manage and provide distinct full names and alleviate a major scaling problem.

Naming services should adhere to the following principles when defining identifiers.

1. *Stable and meaningless.* Identifiers, or names, are like gossip; once they become known, they are difficult to change, recall, and correct. Inherent meanings (e.g., organizational affiliation or physical location) should not be embedded in an identifier because these values change frequently, rendering the identifier inaccurate or out of date.
2. *Uniqueness.* Identifiers must be unique within a naming context.
3. *Traceable back to an individual.* To effectively manage a system, knowledge about the system's users is required.
4. *Extensible to many environments.* Many companies are striving to minimize the number of user IDs an individual must remember to gain access to myriad accounts. Others hope to eventually implement a single logon by which their users can access any computing resource.
5. *User friendly.* Identifiers should be easy to convey, easy to type in, and easy to remember.
6. *Easy to maintain.* Algorithmically derived identifiers (e.g., surname followed by first initial) are easy to generate, but such algorithms may cause duplicate identifiers to be generated if they are not checked against some registry of previously assigned identifiers.

Some of the principles are in contention with each other, so enterprise directory service providers must identify a practical balance of desirable characteristics. A hybrid approach might incorporate some of the following rules:

- Algorithmically assign identifiers (ease of generation)
- Include a maximum of eight characters (extensible to many environments)
- Base identifiers on people's real names (semi-stable and user friendly)
- Use numbers as needed in the rightmost bytes to distinguish between similar names (uniqueness)
- Register the identifiers in a database (traceable back to owner and guaranteed uniqueness)
- Allow individuals to override the generated identifier with self-chosen vanity plate values (user friendly)

Naming an entity should occur as soon as that entity is known (e.g., on or before an employee's hire date), and the identifier should immediately be published by the enterprise directory service. Then, when system administrators establish user accounts for an entity, they can find the entity's identifier in the directory and use that identifier as the user ID. Thereafter, the identifier can be used to query the directory for a user's updated contact and status information. This approach enables system administrators to focus on managing user accounts instead of employee information.

**Authentication/Confidentiality Services**

The swelling interest in electronic commerce has brought much attention to the need for encryption and electronic signature capabilities. The most promising technologies are referred to as public-key cryptographic systems (PKCS), which provide two keys for an individual — one a private key and the other a public key (see Exhibit 9-2). A private key must remain known only to its owner. An individual's public key must be widely published and made easily available to the community with which that person does business.

Information encrypted with an individual's public key can only be decrypted with the associated private key. Therefore, information can be sent confidentially to its recipient. Information encrypted with an individual's private key can only be decrypted with the associated public key. Therefore, a recipient can authenticate the origin of the information.

Companies are now grappling with such PKCS deployment issues as which organization will manage keys, which algorithms will be used to generate keys, will private keys be held in escrow, and will employees be allowed to use company generated keysets for personal use. The one issue that seems clear is that public keys should be published in a directory

*Enterprise Directory Services*

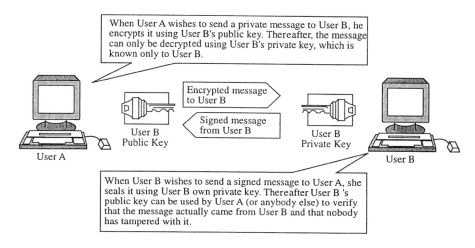

**Exhibit 9-2. Public Key Encryption: How It Works.**

---

service. Soon, enterprise directory service providers will be grappling with their own issues (e.g., when should the directory actually begin to carry public keys?, how to encourage vendors to build products that can query a directory for somebody's public key, how many keys should be carried for an individual?, and how to make these keys accessible to trading partners and other areas beyond corporate boundaries).

**Directory Synchronization Services**

The press routinely represents cross-platform directory synchronization as the most difficult task in a directory service. Depending on the approach taken, this situation may be true. Several messaging system vendors provide products that can synchronize non-native address books into their own. If a company attempts to use such products to synchronize address books from more than two vendors, a true mess will result.

A much more effective manner is to synchronize each address book with a central (probably vendor neutral) bi-directional synchronization service. The central synchronization service can pool the addresses from all participating messaging communities, translate between native and common address formats, and finally make the pooled addresses available to each community in its respective native format. The advantages of a central service over two-way address book synchronizers can be likened to the advantages of providing a messaging backbone service instead of two-way E-mail gateways. Large companies taking the centralized approach have been able to provide cross-platform synchronization services with minimal resources (e.g., one allocated headcount).

Directory standards (e.g., X.500) have the potential to further simplify cross-platform directory synchronization. If vendors ever half-embrace the standard to the point that their products actually query an X.500 service instead of (or in addition to) their local address books, directory synchronization can become a uni-directional process. Each messaging community would still submit the addresses of its users to the enterprise directory service. However, there would be no need to bring other addresses back into their local address book. If vendors ever fully embrace the standard to the point that their products actually manage their own address information right in an X.500 directory, processes for cross-platform directory synchronization are altogether unnecessary. Enterprise directory service providers ought to encourage their vendors to include these capabilities in their products.

With growing interest in electronic commerce, enterprise directory service providers are being asked to carry trading partner information in local directories. In response to such requests, and in the absence of better solutions, service providers commonly exchange files containing address information with trading partner companies and then incorporate the address information into their own processes.

A better approach would be to link the directories of the trading partner companies so that each company is responsible for its own information; there would be no need for a company to obtain a file of trading partner addresses and manage the information locally. This is yet another area where directory standards provide hope for a better future. The X.500 standards specify the technologies and protocols required for such an interenterprise directory. Before participating in an interenterprise X.500 directory, service providers will need to evaluate their company policies on external publication of directories, access control to sensitive information, and the readiness of trading partners to participate.

**What Is X.500?**

X.500 is a set of international standards jointly developed by the International Telecommunications Union-Telecommunications Standards Sector (ITU-TSS) and the International Standards Organization (ISO) that specify standards-based directory components and protocols. Some of the major components and protocols include the following:

- *Directory System Agent (DSA)*. This is where information resides. Multiple DSAs can replicate and distribute information among themselves. They can communicate among themselves to resolve directory queries.
- *Directory User Agent (DUA)*. This enables a user (or application) to access information stored in DSAs. DUAs may be stand-alone programs or incorporated into other applications (e.g., an E-mail user agent).

## Enterprise Directory Services

- *Directory System Protocol (DSP)*. This protocol is used by DSAs to communicate among themselves to resolve queries.
- *Directory Access Protocol (DAP)*. This protocol is used to communicate between DUAs and DSAs. A Lightweight Directory Access Protocol (LDAP) has been defined that is less bulky than DAP and that is appropriate for desktop devices.

### X.500 Infrastructure Coordination Services

Various approaches can be taken for deployment of X.500 technology. Frequently, X.500 is included as a component of some other product the company wishes to deploy. For example, a company may purchase a new message transfer agent (MTA) that includes an X.500 DSA to manage message routing information; a company can also acquire a public-key crypto system product that includes X.500 to publish public keys, or a company may use X.500 as a distributed computing environment (DCE) global directory.

Another approach is for a company to first deploy a stand-alone X.500-based directory service, and then tie in messaging, security, and other products as they are acquired. In a large company, some combination of these approaches may occur simultaneously. Setting up X.500 for a particular application is much easier than setting up a general X.500 infrastructure, but it is not the preferred approach.

It is likely that various computing communities will acquire their own X.500 directory system agents to meet their own special needs. Unless the efforts of these groups are coordinated, conflicting definitions, incompatibilities, and redundant efforts and data will emerge — foiling the potential for the efforts of the few to be leveraged to the benefit of the many.

Even though the enterprise directory service may not own all the directory components, it is the best place to organize the activities of these components. Coordination activities include acting as the company's registration authority, managing the root DSA within the company, providing a map of the company's directory information tree, registration of new DSAs, schema management, registration of object classes and attribute types, controlling where new information is incorporated into the directory, and managing replication agreements.

### A SERVICE PROVIDER'S PERSPECTIVE ON X.500

These days, technologists promote technologies as directory services, vendors market products as directory services, and standards bodies represent specifications as directory services. Enterprise directory service providers must remember that all these are just tools by which to provide a directory service, and it is their responsibility to select the appropriate tools to best deliver their services. They must discern between what works in theory, what works in principle, and what works in practice.

X.500 technology is sound, but until vendors actually deliver products that use the technology, it will remain difficult to justify a serious investment in establishing X.500 infrastructure. Home-grown applications are driving X.500 deployment in some companies, but even in these companies a stronger vendor commitment would ease the burden of justification.

### Information Management Issues

X.500 is frequently cast as a panacea for directory services. This is not true; the biggest challenges for enterprise directory service providers are information management issues.

X.500's strength is in information publication, not information management. In some companies, X.500 receives more attention from the security and systems administration communities than from messaging organizations. These communities have much more stringent requirements for timely and accurate data; service providers will need to revamp processes and clean up content to meet these requirements. Remembering that directories with bad information are worse than worthless, service providers will need to give as much attention to information processes and relationships with data providers as they do to X.500 technologies.

### Database Issues

X.500 is often described as a type of distributed DBMS. This can be misleading, and some applications that began building on X.500 have had to back off and use conventional database products. X.500 is well-suited for information that is structured, frequently accessed by a heterogeneous community, primarily read only, and latency tolerant. X.500, however, lacks database capabilities (e.g., referential integrity or the ability to update groups of information in a single operation). A common debate concerns whether data should be managed in X.500, or if it should be managed in a database and then published in X.500.

### Directory Information Tree Structure

Most literature suggests that there are only two basic models to follow when designing the X.500 directory information tree (DIT): the DIT should reflect either a company's organizational structure or its geographical structure.

In fact, both the organizational and geographical approaches violate the first principle of naming (stable and meaningless identifiers). Frequent reorganizations and relocations will cause frequent changes to distinguished names (DNs) if either the organizational or the geographical model is fol-

## Enterprise Directory Services

lowed. When designing a DIT, sources of information must be considered. For example, in companies where all employee data are managed by a central human resources organization, it may not make sense to artificially divide the information so that it fits a distributed model.

These observations are not intended to discourage use of X.500-based directory services; rather, they are intended to encourage cautious deployment. Reckless and inappropriate activities with X.500 will likely damage its chances for eventual success in a company, and this technology has far too much potential to carelessly squander away. The previously lamented lack of vendor products that use X.500 can also be taken as a window of opportunity for companies to gain X.500 experience in a controlled manner before some critical application demands knee-jerk deployment. An ideal way to start (assuming that the information is already available) is to build an inexpensive and low-risk White Pages directory service using the following components:

- *A single DSA and LDAP server.* Inexpensive products are available today; a desktop system can provide good response for thousands of White Pages users. Worries about replication agreements and getting different vendors' directory system agents to interoperate can be postponed until the service grows beyond the capabilities of a single DSA.
- *A simple Windows DUA.* Vendor-provided directory user agents can be purchased. Unfortunately, many vendor products tend to be so full-featured that they risk becoming overly complex. An in-house-developed DUA can provide basic functions, can be easily enhanced, can be optimized for the company's directory information tree, and can be freely distributed. Freeware Macintosh DUAs are also available.
- *A Web-to-X.500 gateway.* Freeware software is available.
- *A White Pages Home Page that resolves queries through the Web-to-X.500 gateway.* This brings the service to a large heterogeneous user community.

A precautionary note about deploying such a service is that users will consider it a production service even if it is intended only as a pilot or prototype. Support and backup capabilities may be demanded earlier than expected.

## ROLES AND RESPONSIBILITIES

As companies attempt to categorize enterprise directory services, especially those that include X.500, they begin to realize the difficulty in identifying a service provider organization. An enterprise directory service is as much a networking technology as it is an information management service, or a communications enabler, or a foundation for security services. One cannot definitely predict that messaging will be the biggest customer of the service or that HR information and E-mail addresses will be the most important information handled by the service. Possibly the best approach is to create a new organization to provide the enterprise directory service.

To build and provide effective directory services, service providers must be intimately familiar with the data so they can witness the changing characteristics of the information, recognize anomalies, and appropriately relate various types of data. Rather than assign an individual to a single role, it is preferable to assign each individual to multiple roles so that nobody is too far removed from the data and so that no role is left without a backup.

Roles and responsibilities fall into the following categories: operations management, product management, project management, and service management.

**Operations Management.** Directory operations management focuses on quality and availability of directory data as specified in service-level agreements. Responsibilities include the following:

1. Operate day-to-day directory processes and provide on-time availability of deliverables.
2. Ensure that the directory service is accessible by supported user agents.
3. Register information that is not available from other sources.
4. Interface with existing data suppliers to resolve inaccurate information.
5. Validate data, report exceptions, and track trends.
6. Staff a user support helpdesk.

**Product Management.** Directory product management focuses on the type of products and services offered to directory customers, as well as tools and processes to best provide the services. Responsibilities include the following:

1. Acquire, build, and maintain tools that enable or improve day-to-day operation of directory processes. Appropriate skills include database development, programming in various environments, and familiarity with the company's heterogeneous computing environment.
2. Establish and administer service-level agreements with information providers and customers. Negotiate formats, schedules, and methods of information exchange between the directory service and its suppliers and customers.
3. Provide and maintain approved directory user agents to the user community.

4. Promote current capabilities of directory services to potential customers.
5. Build directory products to meet customer needs.

**Project Management.** Directory project management focuses on the near-term future of directory services. Responsibilities include the following:

1. Conduct proof-of-concept projects to explore new uses of directory services.
2. Coordinate activities to put new uses of directory services into production.
3. Conduct proof-of-concept projects to explore the use of new technologies in providing directory services.
4. Coordinate activities to deploy new directory technologies into production.
5. Consult and assist customer environments in incorporating directory services into their applications and services (e.g., assist with LDAP coding, assemble toolkits to ease coding against directories, train application developers in use of toolkits, and provide orientation to the enterprise directory service).
6. Ascertain which projects are most needed. Possible projects include inter-enterprise directories, participation in global directories, authentication and access control, public keys in directories, electronic Yellow Pages, combining disparate registration services, achieving DNS functional ability with X.500, explore the use of whois++ and finger as directory user agents, partner with other groups to achieve synergy with WWW and other technologies, and assist corporate communications organizations to optimize electronic distribution of information bulletins.
7. Participate with appropriate forums to define and establish international standards.

**Service Management.** Directory service management focuses on the soundness of current and future directory services. Responsibilities include the following:

1. Coordinate project management, product management, and operations management efforts.
2. Author and maintain service descriptions.
3. Run change boards and advisory boards.
4. Coordinate X.500 infrastructure (registration point for DSAs, object classes, and attributes).
5. Develop and gain consensus on position statements and direction statements.
6. Formulate business plans and justify required resources.

# THE ELECTRONIC MESSAGING ENVIRONMENT

7. Act as focal point for major customer communities; collect requirements, assess customer satisfaction, and gain consensus on direction statements.
8. Represent the company's interests at external forums and to vendors.

## SUMMARY

It is important to have realistic expectations of a directory service. Some who expect that a directory will provide huge payback as a single data repository and source of all corporate information may be disappointed. Others who consider directories to be of limited use beyond messaging will reap only a portion of directory benefits. It is difficult to predict which visionary ideas will become practical applications in the near future.

Probably the most vital point to consider when planning for an enterprise directory service is flexibility. In light of dynamic customer demands, new sources and types of information, increasingly stringent requirements for quality data, and emerging technologies, many service providers have elected to use a database to manage directory information (exploiting its data management capabilities) and a directory to publish managed data (capitalizing on its publication strengths).

As buzzwords proliferate, it is easy to get caught up in the glitter and pomp of new technologies. Wise service providers will realize that the important part of directory services is service, and that the role of new technologies is simply to better enable the providing of a service.

# Chapter 10
# Directory Synchronization

*Sathvik Krishnamurthy*

What is a directory? Simply put, a directory is a repository of information designed to be used by applications or people. Directories are everywhere. People use them every day. Often referred to as databases, directories are typically implemented with an underlying database that manages low-level storage and retrieval of raw data.

Directories, however, are much more than a low-level database. They provide mechanisms (i.e., directory services) for access, management, distribution, and control over the data. The data stored in directories have evolved from simple textual information into complex objects containing multimedia information (e.g., photographs).

One of the significant challenges facing today's IT managers is in learning how to provide consistent directory services across the plethora of incompatible directory implementations that exist across the enterprise. Creating such a "meta-directory" can be achieved with the right approach and tools for directory synchronization, which are explored in this chapter.

## ELEMENTS OF A DIRECTORY

A general-purpose directory consists of four basic elements: service agents, user agents, storage, and various protocols. This nomenclature is derived from the ITU X.500 directory recommendations, thereby paralleling the role that the ITU X.400 recommendations took in defining basic elements of electronic messaging systems (e.g., message transfer agent, message user agent, and message store) over the last decade.

### Directory Service Agent

As shown in Exhibit 10-1, the directory service agent (DSA) can be considered the central component. It interfaces with the storage elements to store and retrieve information, services requests from directory user agents (DUAs), and in the general case, communicates with neighboring

# THE ELECTRONIC MESSAGING ENVIRONMENT

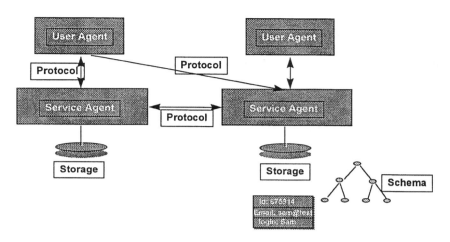

**Exhibit 10-1. Elements of a Directory.**

---

DSAs for the purpose of sharing directory information. The protocols used between DSAs are not necessarily the ones used between DUA and DSA, and most often are not.

## Directory User Agent

The DUA is the component that users or applications use to access the directory service. It may be as straightforward as a stand-alone application that gives users access to directory information as it appears in storage.

This approach, while academically feasible, does not lend itself to simplicity — users or applications must sift through potentially large amounts of irrelevant information before finding what they really want.

A common alternative is to create directory-enabled applications that only have access to well-defined sets of relevant information. For example, an enterprisewide Yellow Pages service might only need to access user names, departments, E-mail addresses, and phone extensions. Such a directory-enabled application would only be granted permission (or access control) to retrieve that specific information.

## Storage

Any way you look at it, the storage within a directory is some sort of database. How the information is organized at the underlying database level (i.e., physical schema) is implementation specific, and is not generally exposed.

At the directory level, however, the logical information schema is used to organize the information for purposes of providing access, management,

*Directory Synchronization*

distribution, and control. For instance, in X.500, systems managers describe the DSA schema with a directory information tree (DIT). This tree contains objects (e.g., users), which in turn contain attributes (e.g., E-mail address, phone extension, title, and photo ID). The DSA keeps track of not only the directory information tree, but also enforces who is allowed to access which subtrees.

**Protocol**

The protocols used between two DSAs, and between DUA and DSA to intercommunicate, will vary, depending on system. In X.500, the directory service protocol (DSP) is used to transmit and fulfill real-time requests for information from one DSA to another.

The 1993 X.500 recommendations added another protocol, called the directory information shadowing protocol (DISP), that allows for batch-level replication of information from one DSA to another. This addition was necessary to reduce search times across the network for nonlocal information. DUA-to-DSA communication is achieved with the directory access protocol (DAP). To take advantage of the ubiquity of TCP/IP for underlying transports however, a new standard called lightweight directory access protocol (LDAP) that runs over TCP/IP instead of an Open Systems Interconnection (OSI) transport protocol, was introduced.

**SO, WHAT IS THE PROBLEM?**

If all applications used X.500 for their directory needs, there would not be a problem. All information would be accessible by the appropriate applications or users across the enterprise (indeed, across the globe).

The problem is, the number of product implementations that contain information that needs to be accessed, managed, distributed, and controlled (i.e., introduced into a directory service) is tremendous. Roughly 38 million users exist in the network name services directories throughout the enterprise, including the NetWare Bindery, DEC DDS, IBM CallUp, and the NT Name Services. In addition, most electronic mail systems use limited forms of directories (more aptly called name and address books) that contain names, E-mail addresses, and limited templates of user information, including phone numbers, departments, and the like.

Roughly 56 million users reside in these systems, including Lotus cc:Mail, Lotus Notes, Microsoft Exchange, IBM PROFS, and DEC All-In-One. Add to this the 3 million users in the network directory service systems (e.g., Novell NDS, Banyan StreetTalk, and native X.500 directories) and one begins to see the widespread diversity of directories in today's enterprise.

Not only do these directories have proprietary, incompatible implementations, but the information they contain (e.g., names and E-mail address-

## THE ELECTRONIC MESSAGING ENVIRONMENT

es) is incompatible and continues to diverge. Consider the examples of the various different formats in Exhibit 10-2 used for E-mail addresses in some typical E-mail name and address books:

**Exhibit 10-2. Divergence of E-mail Addresses.**

| E-mail system | Example name | Example E-mail address |
|---|---|---|
| Lotus cc:Mail | Clark Kent | Clark Kent at Planet |
| Microsoft Mail | Kent M, Clark | us/planet/KClark |
| X.400/X.500 | /us/planet/sales/Clark | /c=us/a=telemail/o=planet/ /ou=sales/g=Kent/s=Clark |
| QuickMail | Kclark | Kent_Clark@planet |
| Microsoft Exchange | Clark Kent | sales/int'l/server/KClark |
| SMTP/MIME | Kent.C | Kent.C@planet.com |

This lack of consolidation of directory information across the enterprise is an expensive problem; indeed the cost of user information management can exceed $4000 per user, per year when considering down-time and other factors that exist when users cannot get to the information they need to access.

### THE META-DIRECTORY

The approach described here for solving the problem of providing universal access, management, distribution, and control for divergent directory implementations, is to create a meta-directory. X.500 is well-suited for the task of providing the backbone service for management and distribution of data (see Exhibit 10-3). Its DSP and DISP protocols can easily be used to securely propagate information in a distributed environment as previously described. The issues of information access and control, however, can only be solved with specific tools that are designed to synchronize the various network name services, name and address books, and network directory services, with the X.500 Meta-Directory.

### Elements of Directory Synchronization

A directory synchronization solution is composed of many different elements. Among them, the following are explored: protocol, transport mechanism, name and object mappings, scheduling, filtering, and management.

Unlike the areas of directory services and electronic messaging, there are no ITU recommendations for directory synchronization. Each proprietary directory tends to have its own mechanisms for distributing information within its own system; however, it cannot interface with other systems such as X.500. Multivendor directory synchronization solutions are rare but are in high demand to solve the problem of creating the enterprisewide meta-directory.

## Directory Synchronization

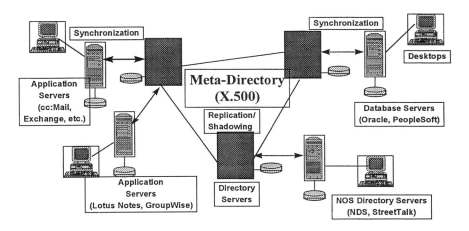

Exhibit 10-3. The Meta-Directory.

### Protocol

As previously mentioned, there are no well-adopted public domain protocols for directory synchronization. Although there have been some attempts to introduce them (e.g., the Retix Directory Exchange initiative), they have not received wide adoption because of their lack of flexibility. As a result, vendors that provide directory synchronization solutions have created their own proprietary protocols for this purpose.

The typical solution defines a directory synchronization cycle, where multiple alternate storage (see Exhibit 10-4) entities have the opportunity to download and upload directory updates from centralized storage (e.g., an X.500 directory). The protocol used usually requires a bulk load initially, and reverts to only propagating directory updates as necessary. Of course, the protocol itself must be integrated into an overall solution that performs name and object mapping, scheduling, filtering, and management.

### Transport Mechanism

Underlying the directory synchronization protocol, there must exist a transport that can reliably transfer information between the alternate storage entities and centralized storage. One popular alternative that vendors have taken is to utilize E-mail as this transport.

This approach, although convenient for synchronizing E-mail name and address books, limits the solution to only being able to synchronize information that resides on a machine that has access to an E-mail system. Although E-mail is increasingly ubiquitous, this solution does not lend itself

# THE ELECTRONIC MESSAGING ENVIRONMENT

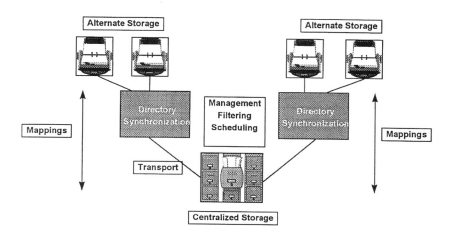

Exhibit 10-4. Elements of Directory Synchronization.

well to synchronization with legacy systems (e.g., human resource databases) or dedicated hardware directories (e.g., PBX databases) where E-mail is not available. In those cases, vendors must provide an entire E-mail implementation, which can be quite an overhead.

Another common approach is to use a file-transfer type of transport, by way of such standards as FTP or NFS, which is perhaps even more ubiquitous than E-mail. In general, this approach is stronger for companies that need to introduce non-E-mail information stores into the synchronization cycle.

## Name and Object Mapping

At the core of directory synchronization are the mechanisms by which information can be introduced into the meta-directory. The process of taking an object out of a proprietary directory and mapping its various attributes into a common object format is the first step.

As shown in Exhibit 10-5, this critical procedure must occur without any loss of the context in which the information originally occurred. Then, when the same object is synchronized with another alternate storage entity, a similar attribute mapping must occur so that the context of the attribute is preserved. The process of altering the format of information as it travels through this cycle can be quite complex and is typically handled with regular expression-based engines that can cope with various string patterns and perform character-by-character alterations as necessary.

Two of the most crucial aspects to directory synchronization are: ensuring that the names and other objects are correctly transformed or mapped from one format to another, and maintaining the correct context of the at-

## Directory Synchronization

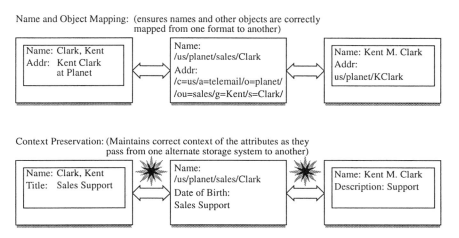

Exhibit 10-5. Name and Object Mapping with Context Preservation.

tributes as they pass from one alternate storage system to another. In Exhibit 10-5, there would be terrible consequences if the Title attribute was mapped to the Date of Birth in the X.500 directory, and subsequently mapped to the Description attribute in the destination directory. Without context preservation, this would happen unless the schemas for all directories were the same.

### Filtering

The ability to control what information gets propagated between alternate storage entities is an important component of a directory synchronization solution. Without such controls, all information would presumably be available to everyone.

For example, when dealing with E-mail name and address books, it may be desirable for certain secretive organizations within a company to view other organizations' E-mail addresses, but not to propagate their own addresses to others. Or perhaps a sales department may only be given the E-mail addresses of other sales and marketing departments, but not the R&D groups. Such examples are commonplace in today's enterprise, and a directory synchronization solution must provide the ability to filter on various attribute criteria, including departments, names, E-mail addresses, domains, and titles.

### Error Alerts and Recovery

One thing is certain: a directory synchronization system cannot assume that its operations are always successful. All too often, an alternate storage

## THE ELECTRONIC MESSAGING ENVIRONMENT

entity is not available, either through network failure, power outages, or system crashes. Even worse, if a synchronization cycle fails in the midst of doing an update, the alternate storage can be left in an indeterminate state that is difficult to fix. In such cases, it is extremely important that the errors are logged, and that there is an automated mechanism that backs out any changes made in the erroneous cycle. Such measures are extremely important and contribute directly toward reducing information management costs.

**CASE STUDY: WORLDWIDE BANKING INSTITUTION IMPLEMENTING A DIRECTORY BACKBONE**

Achieving success with a directory backbone is definitely possible. For example, a major U.S.-based financial institution with over 60,000 employees is using NetJunction products from Worldtalk Corporation to provide an E-mail and directory services backbone. The institution has E-mail and groupware systems from Hewlett-Packard and Lotus.

Before implementing a Worldtalk solution, directories of these E-mail and groupware systems were not synchronized. Changes made to the directory entries in one system were not reflected in others. These changes were primarily employee name and address modifications resulting from mergers and acquisitions, and employees moving from one division to another. Failure to maintain current directory information resulted in attempted message routing to users who no longer worked in, or had moved within, the institution.

Today, however, with the Worldtalk solution acting as an E-mail and directory backbone, the institution is able to synchronize its E-mail address books. Changes to the directory entries in one system are automatically reflected in all other systems.

The institution has deployed an X.500-based NetJunction network application router in the U.S. serving the Americas and Europe. This network application router is connected to the institution's E-mail systems and provides E-mail and directory services connectivity. The network application router also acts as the master of all directory entries. It collects individual changes from the institution's E-mail systems, applies them locally, and creates a master change list that is distributed back to E-mail systems. E-mail traffic between different E-mail systems is also routed through the NetJunction network application router.

**SUMMARY**

So how do you know if you need directory synchronization solutions? Ask yourself the following questions:

## Directory Synchronization

- *How many different E-mail systems do you have currently?* If the answer is more than two, you probably need to synchronize their directories regularly.
- *Do you have any kind of automation to do directory synchronization?* Directory synchronization will almost always save you money in the long run by reducing the amount of time you need to manage your directories.
- *What are the soft-costs that you pay as a result of the lack of up-to-date directories?* You can further justify the costs of directory synchronization solutions from the downtime, lost productivity, and lost revenues that your company experiences. These costs will likely far outweigh the cost of your own time.

Even with the basics in mind, other issues arise when attempting to deploy directory synchronization solutions. For instance, the scalability of a system is an issue to consider. Directory synchronization solutions that are suitable for enterprisewide deployment typically involve significant installation time and management overhead on high-end UNIX machines. Smaller organizations may require lower-end systems that are easy to install, configure, and maintain, on such platforms as Windows NT.

In enterprises with smaller satellite offices, a combination of high-end and low-end solutions may be appropriate. The ability to manage these remote systems, however, then becomes important. Robust management tools that allow system managers to view log files and alter configurations are commonplace and should not be compromised.

Finally, give yourself time to set up a pilot network. Directory synchronization solutions rarely work out of the box and require some handholding; but once you are up and running, you can realize substantial savings for your company with your very own meta-directory.

# Chapter 11
# X.500 Directory Services: A Business Process Enabler

*Roger Mizumori*

---

Virtual organizations, electronic commerce, business process reengineering. These are the prominent buzzwords of the day. The way organizations do business is changing dramatically. Managers are as likely to outsource a business or process function as they are to run it within the company. New-world markets are becoming more accessible. The World Wide Web has shown that global marketing is a reality.

Yet, when it comes down to actually conducting business, there are still fundamental needs — like knowing the mailing addresses and telephone numbers of your trading partners and customers. There is a requirement for understanding where business associates are and how to communicate with them. There is the basic need for an address book.

Once an address book is established, it needs to be maintained and sometimes shared with other co-workers. As more businesses move to the world of electronic data interchange (EDI) and online commerce, where computing applications interact directly with the business applications of other companies, it becomes more critical for these applications to programmatically "know" where to send transactions, under what conditions, and with what security mechanisms.

Administratively, there is a need to know where things are in a distributed environment. This includes network components, telephones and voice messaging systems, and fax machines. The ideal is to have all that information in a common directory and have the information stay current.

Clearly, there are a number of business processes that need to be put into place to support these opportunities. This chapter describes a fundamental tool to enable implementation of business processes. That tool is

THE ELECTRONIC MESSAGING ENVIRONMENT

X.500. With its facilitation of communication, X.500 is allowing the information age to mature into the information-sharing age.

## A JOINT EFFORT

Since the mid-1980s, people have been developing a vision for a global directory service, capable of working on any hardware platform and interoperating with any vendor's software implementation. This work emerged as a joint effort between the International Telecommunications Union-Telecommunications Standards Sector (ITU-TSS, formerly the CCITT) and the International Standards Organization (ISO). Each standard body has designated the series of work with a document identifier; the ITU-TSS called it X.500, whereas the ISO called it ISO 9594.

This joint effort resulted in a single set of documents representing both standards bodies. The current edition of the standard is called the 1993 version of X.500. The first edition, called the 1988 X.500, was approved in 1990 and, due to time constraints, did not include several key features. However, the standard defined sufficient functionality that resulted in many implementations. The current edition was approved in 1994 and added key functionality in the areas of replication, access control, distributed operations, and schema.

In this document, the word *Directory* is capitalized when referring to the standard or standard-based implementations.

## DRIVING FORCES

Over the past 30 years, there have been significant advances in computing technology that have resulted in increased performance and reduced prices. This has made it possible for communications protocols to be developed to connect multiple processors together over a wide area network.

The popularity of the PC alone has been monumental. Most schools have incorporated their use into the classroom. Many families own one or more at home, and many people have access to personal computers at their place of employment. At the same time, this new class of microprocessor technology is beginning to make its way into many other aspects of our daily lives. Most new cars are now microprocessor controlled; most cash registers contain microprocessors; and it is virtually impossible to make a telephone call without many computers being involved.

At the same time, the software industry has made software very affordable. Just 10 years ago, it was highly unusual for a business to purchase a piece of software that cost less than $10,000. Today, most corporate software purchases are under $500, and very often under $100.

Why have prices dropped so dramatically? Most analysts agree that competition has been a significant factor in this equation. Also, the price reductions have opened up many new markets that did not exist 20 years ago, especially the home market. This volume has made it possible for software and hardware vendors to cut their profit margins to near zero, but because of the volume, continue to make significant financial gains.

Today, you can configure a Pentium-based processor with adequate memory and software to build an application server for under $5000. You can build a client system for under $2000, with the same processor, less memory and disk, and the related client software.

In summary, the systems are readily available and affordable to move to client/server computing.

**TECHNOLOGICAL READINESS**

Computers are ready and available for many client/server applications. To build a large-scale client/server computing environment, it is impractical to maintain software and network configuration information on every system for every application. A distributed directory technology is necessary.

Most corporations have found that they can no longer purchase the necessary software for all line-of-business applications from the same vendor. Because of the diversity of requirements, corporations are bringing in many hardware and software solutions from different vendors, and facing significant interoperability problems.

For example, one organization may find that an E-mail system from one vendor solves its needs, while another organization finds that an E-mail system from a different vendor solves its needs. Each of these environments requires a different proprietary directory system. If employees are loaned to the other organization, it becomes very awkward for them. They must suddenly use both E-mail systems, since the desktop system at the temporary assignment location is loaded with the other user agent. This has led to the need for establishing gateways between the two messaging systems and synchronizing the two proprietary directories.

Information systems managers have realized the need for standards-based solutions that would eliminate the need for gateways and directory synchronization. A common directory infrastructure could be deployed that everyone could use. The only standards-based directory technology at this time is X.500.

**Performance in Prime Time**

The two important questions that remain to be answered are whether or not the technology is capable of providing the functionality and performance that would meet current directory requirements, and whether implementations are available in the marketplace to deploy.

Over the past few years, numerous vendors have participated in conformance and interoperability testing. New versions have continued to improve on performance and functionality. It is important to note that the X.500 Directory is typically implemented using agreements that have been harmonized internationally. These agreements specify mandatory functionality, which is the basis of the initial releases. These same agreements also specify optional functionality. This is where we are seeing product growth today.

Many companies have deployed X.500 in both pilot and production environments, often in support of messaging. The Internet community has deployed an X.500 Directory, called the Paradise Project, with approximately 2 million entries that are distributed around the globe with close to 1000 organizations participating.

There are very many instances of private X.500 Directories that are deployed within a single enterprise. For example, WalMart has deployed an X.500 Directory Service with a server in each of its close to 3000 stores. Another vendor, Control Data Systems, currently has more than 3 million X.500 user licenses sold. The technology has proven that it is stable and deployable. Many other vendors have large amounts of users using Directory in intranet installations.

## ORGANIZATIONAL READINESS

Another real question is whether or not the enterprise is ready to consider a Directory infrastructure that would help enable large-scale client/server applications. Many companies have not considered these implications yet. Many are currently just beginning to consider their first large-scale client/server deployment — a messaging system.

Corporations should consider their strategic direction here. If a Directory service is deployed that cannot be easily grown into a full-service infrastructure, it will be much more difficult to reengineer the Directory service to meet future requirements.

## THE X.500 STANDARD

As more and more countries, public electronic mail carriers, organizations, computer processes, and users established electronic messaging capabilities, the need emerged for a means to locate information about subscribers of the service. A name-to-address translation service would be required to facilitate communications between open systems. The service would provide simple naming, name-to-address mapping, and be acces-

## X.500 Directory Services: A Business Process Enabler

**Exhibit 11-1. X.500 Recommendation.**

| Recommendations | | Description |
|---|---|---|
| X.500 | ISO 9594-1 | Overview of concepts, models, and services |
| X.501 | ISO 9594-2 | Models |
| X.509 | ISO 9594-8 | Authentication framework |
| X.511 | ISO 9594-3 | Abstract service definition |
| X.518 | ISO 9594-4 | Procedures for distributed operations |
| X.519 | ISO 9594-5 | Protocol specifications |
| X.520 | ISO 9594-6 | Selected attribute types |
| X.521 | ISO 9594-7 | Selected object classes |

sible to subscribers of the service without regard to where the request originated.

Documents for both the ITU-TSS's X.500-X.521 standard and for the ISO's 9594 standard describe directory services, concepts, and models necessary to facilitate the use of electronic store-and-forward messaging systems within international data communications networks. The X.500 recommendation (the 1998 standard) is composed of several integral parts shown in Exhibit 11-1.

The X.500 standard describes the structure of the directory information model, specifies the protocols required for interoperability between open systems with directory information, and specifies procedures to operate distributed information directories across multiple open systems. Collectively, this set of X.500 recommendations enables the development of a distributed global directory service (see Exhibit 11-2) that can be accessed by users and computers to obtain the information and addresses required to exchange electronic messages within the international X.400 community. As stated in the standard (from Recommendation X.500, page 6), "The *Directory* is a collection of open systems which cooperate to hold a logical database of information about a set of objects in the real world. The *users* of the Directory, including people and computer programs, can read or modify the information, or parts of it, subject to having permission to do so."

The standard is updated every 4 years to accommodate changes in the global computing environment and to add items that were not included in the previous edition. The resulting recommendations are called *extensions* to the base standard (i.e., extensions to the original 1988 X.500 standard).

The next round of imminent changes is being published in 1997 and was not available at the time this chapter went to press. However, the 1997 extension is expected to be the definitive set of recommendations to complete a directory service description. Subsequent changes are expected to be maintenance enhancements.

# THE ELECTRONIC MESSAGING ENVIRONMENT

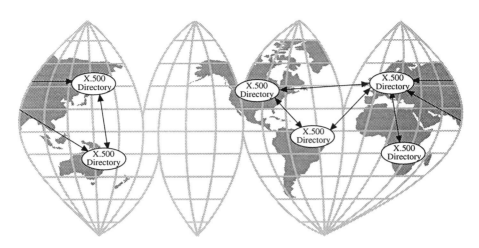

**Exhibit 11-2. X.500 Directory Worldview.**

## PURPOSE OF THE X.500 DIRECTORY

The intent of the X.500 Directory specification is to provide for the uniform deployment of physically distributed but logically centralized information databases maintained separately by their prospective open system owners. These information bases, by their open accessibility, are essential in the deployment and interoperability of multivendor information processing systems. The distributed, standards-based directories will then form one logical, "universal" directory enabling users to reach anyone, anywhere, anytime.

The X.500 Directory specification, with its user-friendly naming, common address format, and distributed operation, was developed to permit information directories to reside as close as possible to the electronic mail system they serve and to let different databases communicate.

### Business Benefits

The principal business benefit of using an X.500-based directory system is to enable the owner to leverage existing investments made in disparate information processing systems. These systems may be from different vendors, of different ages, and administered by different groups.

X.500 reduces the cost of procuring new standards-compliant systems by providing a standardized interface. It permits global access to people, applications, computer processes, and information. In addition, X.500 provides the following benefits:

- Promotes ease of use because of its single name space

- Separates the object's name from the network address
- Shields the user from changes made to the network
- Simplifies additions, deletions, and modifications
- Provides the capability to locate all network objects such as people, devices, computer applications, files, document types
- Facilitates electronic communication with business partners, vendors, and customers in a global marketplace
- Provides a "clearinghouse" in multivendor environments, enabling a single master for information
- Reduces "telephone tag" and accommodates different time zones

## X.500 DIRECTORY ARCHITECTURE

The X.500 Directory architecture is based on a client/server relationship in which the users represent the clients and the Directory represents the server. Typically, directory systems include a user interface (input screen), an access method, a database containing information about users and devices, and a response method to return the located information to the user.

The standard presents several X.500 Directory models, including the Directory Information Model, the Directory Functional Model, and the Security Model. The X.500 Directory model is composed of various components — for example, the Directory User Agent (DUA), the Directory Service Agent (DSA), the Directory Access Protocol (DAP), the Directory System Protocol (DSP), the Directory Information Base (DIB), the Directory Information Tree (DIT), directory services, directory management domains, and directory security mechanisms. Each of these Directory components, as depicted in Exhibit 11-3 will be discussed in detail in the following sections.

The X.500 Directory functions by means of the interactions between the user and the DUA, between the DUA and the DSA(s), and between the DSA(s) and the Directory Information Base. This basic functionality permits a user to request and obtain information regardless of where the request was submitted.

### Directory User Agent (DUA)

Users gain access to the Directory and its services through a Directory User Agent (DUA). X.500 does not specify any formats for the DUA.

In a typical DUA, the user would see an input screen with choices of directory services to request, such as "search" or "compare." The user enters the information required by the service (e.g., a common name, company name, and country). The DUA is an abstract service and represents a single user that can either be a person or a computer process that is permitted to access the Directory, request Directory services, and receive a response from the Directory.

# THE ELECTRONIC MESSAGING ENVIRONMENT

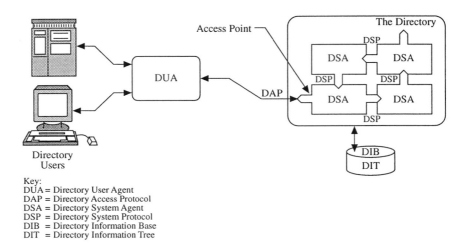

Key:
DUA = Directory User Agent
DAP = Directory Access Protocol
DSA = Directory System Agent
DSP = Directory System Protocol
DIB = Directory Information Base
DIT = Directory Information Tree

**Exhibit 11-3. Directory System Model.**

All components of the Directory can be encoded in a form called Abstract Syntax Notation (ASN.1), defined in Recommendation X.407. For example, the DUA encoded in ASN.1 is displayed as follows:

> dua
>
> OBJECT
>
> PORTS { readPort [C],
>
> searchPort [C],
>
> modifyPort [D]}
>
> ::=  id-ot-dua

The DUA acts on behalf of the user in interacting with the Directory. The X.500 standard does not specify any recommendations as to the design of facilities within the DUA to enable a user to easily communicate requests to the Directory or interpret responses from the Directory. For example, the standard does not specify how a query should be composed or how the response from the Directory will appear to the user.

The DUA, using the Directory Access Protocol (DAP), reaches the Directory by means of an access point. The Directory may provide one or more access points. The DUA concept is illustrated in Exhibit 11-3.

The DUA initiates a bind operation with the Directory System Agent (DSA) in order to gain access to the Directory and make requests for Directory services. When the Directory either returns the results of the specified service operation or an error, the DUA initiates an unbind operation to release itself from Directory access.

Typical implementations distinguish user agent functionality into a user's DUA (focused on search and compare functions) and an Administrative DUA (ADUA) that is used for maintaining the Directory itself and the information contained in it.

**Directory System Agent (DSA)**

The Directory is composed of one or more Directory System Agents (DSA), each containing one or more access points. The DUA is permitted to access one or more DSAs.

A Directory containing one DSA is referred to as a *centralized directory* while a Directory containing multiple DSAs is called a *distributed directory*. These multiple DSAs may be found on numerous OSI systems and cooperate to form the logical, worldwide directory. The DSA permits access to the Directory Information Base in order to fulfill requests made by the Directory User Agent on behalf of the user. To satisfy a request, one DSA may communicate with one other DSA or may require interaction with several other DSAs.

**How X.500 Works with X.400**

The X.500 Directory standard and the X.400 Message Handling System specification recommend that each user (object) of the service have an unambiguous name. A name identifies an object or entity that resides on or uses the communications network. The name must be unique within the global X.400 and OSI community.

A user can be an originator of a message or a recipient of a message and can be a person, a distribution list, a document type, a device, a directory component, or a computer application. According to Recommendation X.400, an originator should be able to provide a descriptive name for each recipient of a message using information commonly known about that user. The Originator/Recipient (O/R) name will contain the user's country, administration domain (e.g., public electronic mail carrier), and other information about the user, such as the company or organization name and personal or common name.

The country name and administration domain name are required for the successful exchange of messages. At least one of the remaining attributes must also be cited to form the O/R name and ensure the global uniqueness of the name.

The O/R name must then be mapped to a network address in order to permit the exchange of electronic messages. The address identifies an object/entity location in a computer system. The route of the message can be determined by the attributes in the O/R name. The route identifies the path or part of a path to the location specified in the address.

There is an Internet Request for Comment (RFC) that defines the use of an X.500 Directory by the X.400 Messaging Service. This is called the MHS-DS standard.

One objective of both the X.500 and X.400 standards is to shield the user from the complexities and constant modifications of the data communications network. The resolution of an O/R name to an address and the resolution of an address to a route is most commonly accomplished by using a translation table. These tables are generally proprietary in nature and closed to inquiries outside the particular system.

The Directory separates the name from the route. The user can then search the Directory to obtain another user's unique name and successfully exchange X.400 messages without having to know or include any complex routing instructions. When an X.500 Directory is integrated with an X.400 electronic mail system, the user simply specifies the unique name without having to cite the E-mail address. The originator is thereby isolated from changes made in the location of the recipient on the network. The properly addressed electronic message is then sent to the recipient using the X.400 message transfer system. The relationship between X.400 and X.500 is illustrated in Exhibit 11-4.

## X.500 DIRECTORY PROTOCOLS

Both the Directory User Agent and the Directory Service Agent(s) communicate by means of protocols defined in the X.500 Directory standard.

### Directory Access Protocol (DAP)

A Directory User Agent communicates with a DSA in another open system by means of the Directory Access Protocol (DAP). DAP is an OSI Layer 7 application protocol that enables communication between two application layer processes.

DAP is employed when the DUA requests a Directory service such as "read" or "search." DAP is also used by the applications to obtain from the Directory the user's credentials for the purpose of authentication.

### Directory System Protocol (DSP)

Cooperating DSAs communicate with each other by means of another OSI Layer 7 application protocol known as the Directory System Protocol (DSP). This protocol is required to provide distributed X.500 Directory services.

Interactions between a DUA and a DSA are always originated by the DUA, and only the DUA can release the interaction. Therefore, if the communica-

# X.500 Directory Services: A Business Process Enabler

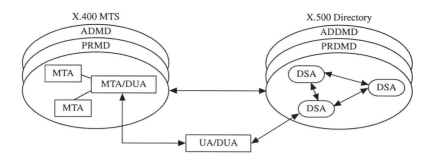

Key:
MTS = Message Transfer System
ADMD = Administrative Management Domain
PRMD = Private Management Domain
MTA = Message Transfer Agent
ADDMD = Administrative Directory Management Domain
PRDMD = Private Directory Management Domain
UA = User Agent
DUA = Directory User Agent
DSA = Directory System Agent

**Exhibit 11-4. X.400/X.500 Relationship.**

---

tions fail due to a problem in the DUA, only the local user will be affected. When communications with a DSA fails, however, other DSAs that have knowledge references to the failing DSA and any DUAs associated with the failing DSA will be affected.

## Replication Protocols

Replication of Directory information is specified in a Recommendation called X.525. Replication is the process of distributing reproductions of the Directory Information Base held by a particular DSA to other cooperating DSAs. The DSA holding the original information is called the master DSA.

Replication is accomplished either by holding cache copies of information that resulted from a Directory operation such as "search" or by holding shadow DIT copies obtained from another DSA. The Directory user will be informed that the information requested was obtained from a copy.

The designers of the 1993 extensions have defined replication as a Directory operational service and have specified replication services to meet the following objectives:

- Enable users to locate information more quickly by moving the information closer to the users
- Reduce the time and costs associated with Directory service requests
- Protect the Directory Information Tree (DIT) in the event of a failure and provide for redundancy
- Force cooperating DSAs to establish scheduled updates to guarantee consistency
- Enable the dissemination of knowledge and access control information

# THE ELECTRONIC MESSAGING ENVIRONMENT

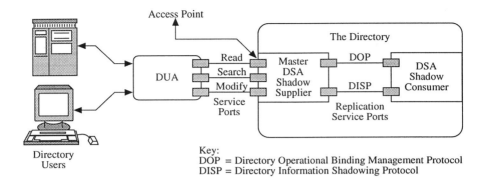

**Exhibit 11-5. Replication Service (1993 Extension).**

---

DSAs participating in full replication should support both the Directory Information Shadowing Protocol (DISP) and the Directory Operational Binding Management Protocol (DOP), as shown in Exhibit 11-5. DISP is required to perform shadowing. A DSA shadow supplier sends a copy of its DIT to a DSA shadow consumer who receives the duplicate information. DOP is used to establish and terminate an interaction (binding) between two cooperating DSAs.

## X.500 DIRECTORY INFORMATION MODEL

Organized hierarchically, the X.500 Directory information model permits each nation to define the configuration of its own data nationwide. The standard also permits each company, organization, or entity within the particular country to internally define its own data format.

This proved to be extremely straightforward in its initial implementations. Since then, the world has changed. Organizations and business processes are now extremely multinational and mandate collaboration across national boundaries to ensure some level of consistency for how information is kept.

### Directory Information Tree (DIT)

The directory information is organized in a hierarchical tree structure called the Directory Information Tree (DIT). Immediately below the root of the directory tree are located the country names formed from the vertices or branches of the information tree.

Each country has a name composed of a two-character abbreviation. For example, the United States is abbreviated US and the United Kingdom, GB. All X.500 directory entries in the United States will begin with the country displayed as US (C = US) below the root, as shown in Exhibit 11-6.

## X.500 Directory Services: A Business Process Enabler

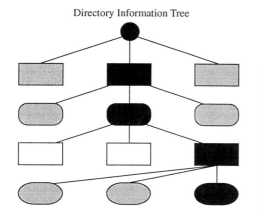

| Root | Null |
|---|---|
| Countries | C = US |
| Organizations | O = Acme Parts |
| Organization Units | OU = Sales |
| People, Devices, Applications, etc. | CN = Smith |

**Exhibit 11-6. Directory Information Structure.**

Companies and other organizations reside immediately below the country level. The X.500 standard takes advantage of the hierarchical structure of most organizations. These organizations are further divided into organization units such as sales, manufacturing, and purchasing. Organization units are then subdivided into attributes for people, devices, applications, document types, and distribution lists. These attributes and their values then form an X.500 directory entry. Each complete entry contains all the information about one (and only one) object and can be referred to as a *directory entry*.

Currently there is work in progress to establish how multinational organizations such as NATO can be represented in the DIT without putting it under a given country.

### Directory Information Base (DIB)

The collection of entries describing single objects as specified in the Directory Information Tree form the database known as the Directory Information Base (DIB). The DIB can contain all the object entries held under the root such as a country. The DIB may also contain only portions or fragments of the logical, "universal" directory, as in the case of a firm operating within one country or a division operating within a multidivision company.

The Directory Information Base contains "the complete set of information to which the Directory provides access.... All of the pieces of information which can be read or manipulated by the operations of the Directory are considered to be included in the DIB" (according to Recommendation X.501, page 25). The standard does not specify the type of database to be used.

# THE ELECTRONIC MESSAGING ENVIRONMENT

## Information Objects

Within the context of the X.500 recommendations, an object is a user, client, or consumer of the services that the Directory provides. An information object can be named uniquely within the global computing environment. An information object may be a person, a computer application, a document type, a device such as a printer, or a distribution list composed of the names of other information objects. The structured description of the information object is the object entry that resides within the X.500 Directory Information Base.

## Object Classes

Objects are organized into classes. An object class is a set of objects having at least one attribute in common. An attribute is a characteristic or quality of a particular object.

For example, all countries would compose an object class, as would all residents of Orlando or all printers on a particular network. In this example, countries, location, and devices would be the common attribute for each object class. Exhibit 11-7 offers a list of selected object classes that are available in the X.521 standard.

## Attributes

An attribute is composed of an attribute type and attribute value pair. The collection of attribute pairs pertaining to a single information object forms the X.500 structured directory entry.

An attribute type such as "country" would, for example, have a corresponding attribute value of "US" or "GB." An attribute type may have more than one attribute value, such as a type of "telephone" and values of home, business, fax, or cellular phone.

**Exhibit 11-7. Selected X.521 Object Classes.**

| Class number | Object class | Class number | Object class |
|---|---|---|---|
| 0 | Top | 9 | Group of names |
| 1 | Alias | 10 | Residential person |
| 2 | Country | 11 | Application process |
| 3 | Locality | 12 | Application entry |
| 4 | Organization | 13 | DSA |
| 5 | Organizational unit | 14 | Device |
| 6 | Person | 15 | Strong authentication user |
| 7 | Organizational person | 16 | Certification authority |
| 8 | Organizational role | | |

### Exhibit 11-8. Selected Attribute Types.

| Type number | Attribute type | Type number | Attribute type |
|---|---|---|---|
| 0 | Object class | 21 | Telex number |
| 1 | Aliased object name | 22 | Telex terminal identifier |
| 2 | Knowledge information | 23 | Fascimile telephone number |
| 3 | Common name | 24 | X.121 address |
| 4 | Surname | 25 | International ISDN number |
| 5 | Serial number | 26 | Registered address |
| 6 | Country name | 27 | Destination indicator |
| 7 | Locality name | 28 | Preferred delivery method |
| 8 | State/province name | 29 | Presentation address |
| 9 | Street address | 30 | Supported application context |
| 10 | Organization name | 31 | Member |
| 11 | Organizational unit name | 32 | Owner |
| 12 | Title | 33 | Role occupant |
| 13 | Description | 34 | See Also |
| 14 | Search guide | 35 | User password |
| 15 | Business category | 36 | User certificate |
| 16 | Postal address | 37 | CA certificate |
| 17 | Postal code | 38 | Authority revocation list |
| 18 | Post office box | 39 | Certificate revocation list |
| 19 | Physical delivery office name | 40 | Cross certificate pair |
| 20 | Telephone number | | |

In addition, various object entries with the same attribute type can have different attribute values. For example, the attribute type of "address" can be paired with one attribute value for the physical address for a corporation and another attribute value for a different physical address for the corporate divisions. If no attribute value is present, the attribute type is considered to be nonexistent and will not be relayed by the DSA. Selected attribute types within Recommendation X.520 are featured in Exhibit 11-8.

The attribute type and attribute value pair (e.g., Country = US) is interpreted as an Attribute Value Assertion (AVA). An AVA proposes that the attribute pair is either "true, false, or undefined." For example, the attribute pair O = Acme Parts is true, while the pair O = Akme Parts may not be true.

## X.500 Directory Names

The structure and content of Directory names must balance the competing aims of simple names and manageable data (which implies a deeper hierarchical tree and longer names). Users who interact frequently with objects should be able to intuitively determine or predict the structure of those object names. Directory names should be simple enough for a person to understand and recall. Names should not contain complex machine-readable character strings more useful to a computer than to people.

One of the goals of the X.500 standard is to specify globally simple and consistent object names and to separate the object's name from the object's route or location on the network. Therefore, the only criterion for designing Directory names cited within the X.500 standard is that of user-friendliness.

**User-Friendliness.** Annex E of the X.501 Recommendation suggests that a name can be deemed user-friendly if it meets the following criteria. First, a human being (can) correctly guess an object's ... name on the basis of information about the object that he naturally possesses. This information would be obtained during the normal course of business interaction such as acquiring a business card or reading a name tag.

Second, if the object is an ambiguous name, the Directory should recognize the fact rather than conclude that the name identifies one particular object. Consider the following: The user specified "Chris Smythe," an ambiguous name. The Directory presented two Smythe entries, requiring the user to specify another attribute such as location or image to conclusively identify the object entry.

In addition, user-friendly name design should drive toward the simplest multipart name possible, exclude computer addresses, employ aliases to limit the number of entries searched, and provide for abbreviations. The name should also be the same across all instances of use on the open data communications system.

## Information Model Updates

The 1988 X.500 Directory model was composed of an entry and an attribute. The 1993 model extends this scheme to include Administrative Points (AP). These APs will be immediate subordinates of the root of the Directory Information Tree but can also be located at the organization level or the organization unit level.

At each AP resides a Directory entry called an Administrative Entry (AE). These AEs have operational attributes, which contain information regarding the way the Directory will function within this particular domain, such as access control information, subschema rules, replication, and access point information.

## X.500 DIRECTORY ORGANIZATIONAL MODEL

The Directory is organized into Directory Management Domains (DMDs). These DMDs are composed of one or more Directory System Agents under the control of a single entity such as an organization, company, or public electronic mail carrier. Domains operated by public carriers or other telecommunications service providers are called Administrative Directory Management Domains (ADDMDs); those operated by an organization or company are called Private Directory Management Domains (PRDMDs), as depicted in Exhibit 11-4. These names and the management of the domains closely parallel ADMD and PRMD entities as described within the X.400 Message Handling System standard.

The way that a PRDMD operates internally, the services that are provided by the PRDMD, and the configuration of the PRDMD are not within the scope of the X.500 standard.

**Administrative Authority**

A Directory Management Domain is responsible for the services offered by its DSA(s) and for ongoing maintenance of the DSA(s). The DMD empowers an Administrative Authority to manage the infrastructure of each DSA. This authority administers the object and alias entries listed within the Directory System Agent and controls the origination and ongoing changes made to the entries.

A Naming Authority within the management domain assigns directory names within the framework of a directory name format. The Administrative Authority then inserts this naming structure within the DSA and is responsible for the ongoing maintenance of the structure in accordance with the rules of the Directory Information Base.

The Administrative Authority manages and maintains the entries within the portion of the DIB held by the DSA. That portion or fragment is composed of one or multiple naming contexts. A naming context is a partial subtree of the Directory Information Tree defined as starting at a vertex and extending downward to leaf or nonleaf vertices. Each naming context must also include the path from the vertex at the beginning of the context to the root of the DIT.

Illustrated in Exhibit 11-9 are five naming contexts managed by three DSAs. It is preferable to have an Administrative Authority manage as few naming contexts as possible to reduce the time associated with locating information and maintenance and administrative duties.

**Naming Authority**

The assignment of names within an object class such as "country" or "organization" is conducted by a Naming Authority. This entity is tasked with the creation and allocation of names immediately subordinate to its own object class.

The Naming Authority must take care not to issue duplicate names. The Naming Authority should select stable, user-friendly names and seek to

# THE ELECTRONIC MESSAGING ENVIRONMENT

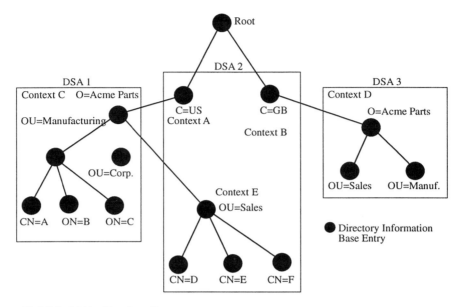

**Exhibit 11-9. Naming Contexts.**

minimize the number of valid Attribute Value Assertions required to form the Distinguished Name of the object.

The Naming Authority allocates valid names with a consistent format and submits them to the Directory Management Domain's Administrative Authority to be placed into the DIB. Then the organization delegates the allocation of subordinate names such as Organization Unit or Common Name to the subordinate Naming Authority within the Organization.

For example, the sales division (Organization Unit) of Acme Parts would allocate common names within its domain, and the Organization of Acme Parts would allocate Organization Unit names such as Sales or Manufacturing within its domain.

## USES AND APPLICATIONS OF AN X.500 DIRECTORY

The X.500 Directory provides the necessary mapping between an object and its location within the data communications environment. This mapping isolates the user from the complexities and constant changes in a typical network and allows the user to reference objects by a user-friendly name. The X.500 Directory may be used within open systems for a variety of uses and applications.

## Initial Planning

Typically, an organization will begin to consider the improved functionality that a directory infrastructure can provide. A small project will be started to better understand the feasibility and impact of deploying the directory infrastructure. Very early in this process it will become evident that the only technically reasonable solution is an X.500 Directory.

**Building a Team.** It is important to build a small core team from the different disciplines that will make the effort a success. This includes people with the following responsibilities:

- E-mail administration
- Data management (suppliers)
- Network management
- System management
- Application programming
- Customer support

**Defining Activities.** The team needs to begin discussions on the requirements of the Directory service. These discussions will be very preliminary, but will serve as a motivator and starting point for building a level set.

**Technology Introduction.** Users should organize a workshop that provides base-level training on X.500 technology. This introductory workshop should provide a reasonable overview of all the necessary components needed to develop an initial approach to Directory architecture. This workshop should be used as an opportunity to leverage the level-setting exercise to also build a preliminary design of the Directory solution.

Typically, if 1 week is put aside for this training and design effort, significant progress and agreement can be reached to have a published baseline that the project team can use to build a production pilot. This effort will also lead to a better set of expectations that would help keep user expectations in check.

**Understanding Potential Technology Consumers.** The preliminary design effort will initiate independent thinking on the possibilities of a Directory infrastructure. Some of the more common ideas are the replacement of the corporate paper phone book and department-level phone lists.

For example, a simple spreadsheet macro can be developed that will query the Directory and return information into the cells of a spreadsheet in the format of the department-level phone list. This demonstration tool could serve as a significant way to build support for the project and identify new champions.

THE ELECTRONIC MESSAGING ENVIRONMENT

As more of these demonstration applications are built, dependence on these tools will grow. In fact, if carefully managed, infiltration of a large cross-section of the enterprise can be made. The benefits are significant. The Directory team will better understand user requirements, users will better understand the potential of the technology, and the momentum will be difficult to stop.

**Generic Directory Uses**

The X.500 Directory may be used as a generic framework to support X.400 Message Handling Systems, other OSI applications, management processes, and telecommunications services. The following high-level uses can be supported by the Directory.

**Simple Lookup.** Simple lookup is the most frequently used type of query. Lookup uses the "read" interrogation operation. The user submits a Distinguished Name and an attribute type to the Directory. The Directory returns the attribute values associated with the DN. The user can also employ aliases when performing lookups and can request lookups on multiple attribute types.

**Browsing.** The browsing feature lets a Directory user request information about a number of entries by using the list and search operations in conjunction with the read operation.

**Yellow Pages.** The Gartner Group predicts that an X.500 "business-to-business" Yellow Page service will be one of the very first applications of the Directory standard. Companies will use filtering and special DIT structures to form listings grouped by attribute values within certain categories such as the "Business Category" attribute. The business requirement is to provide trading partners with addressing information for all current and potential business contacts internally (i.e., persons, devices, and applications), while protecting proprietary information that may currently reside in the same directory. Businesses would use aliases to create a subtree or subset of information that would be made available to their trading partners.

**Groups.** Directory users may query the Directory to ascertain if a particular entity is a member of a group and to list the group's membership. Groups may be formed by using the "member" attribute type. Compare and read functions may be employed to determine the group and the membership therein.

**Authentication.** Various applications mandate that the participants prove their identities prior to carrying out the functions associated with the application. The X.500 Directory supports these authentication requirements by holding a value in the attribute type of User Password suitable for the simple authentication mechanism. The Directory may also hold the user's public keys in the strong authentication scheme.

## DIRECTORY APPLICATIONS

The X.500 Directory may be used in a variety of applications ranging from a global electronic directory to a repository for printer addresses. Even in its most basic form, that of support for X.400 message handling, the user-friendly naming of X.500 Directory facilitates the exchange of electronic messages. The list of applications discussed here is by no means complete as new applications for the X.500 standard emerge daily.

### Messaging

An important initial consumer of Directory services, at least from most typical deployments, is the messaging system infrastructure. There are two major aspects that must be considered here.

**The Backbone.** The first is the deployment of a messaging backbone. The backbone messaging infrastructure must be capable of resolving mail addresses and delivering them to the proper post office or gateway.

A Directory service significantly simplifies this task as well as reduces overall system costs. As additional message relay systems are added to the backbone, they can take advantage of the existing directory infrastructure and not be required to replicate the routing tables and subscriber information that is already in place.

**Desktop Access.** The second area where messaging takes advantage of the Directory technology is making the information available to desktop messaging users. Traditionally, every proprietary messaging system has provided a proprietary address book function. In most organizations, this required a central organization to collate information from many different sources, such as human resources and teleprocessing organizations, and generate custom directory feeds to each of these environments. As update frequencies varied, the data was often incorrect. For example, at The Boeing Company, a typical week would result in 2000 to 6000 directory changes.

An even more difficult management issue is that not all address books are created equally. Some address books have severe limitations to the number of attributes that they may store about an individual entry and the level of coordination between post offices from the same vendor. Even more severe is the fact that numerous implementations cannot scale to large enterprises with large numbers of entries.

A current trend by many proprietary messaging systems is to provide X.500 address book access, either directly embedded in the products or soon to be embedded in an upcoming future release. With the emergence of MAPI 1.0 interfaces becoming more common in the market, a number of vendors have begun to develop Address Book Service Providers based on MAPI 1.0. This means that a vendor should be able to use the same service provider with different products, such as Microsoft Exchange and Lotus Notes. One such vendor to demonstrate this functionality is Enterprise Solutions Limited.

The address book requirement is very important to most large installations. For example, the Defense Messaging System (DMS) is expecting to deploy 2 million desktop messaging users. It would be extremely impractical and expensive to attempt to synchronize proprietary directories with the DMS X.500 Directory. In fact, it would be difficult to synchronize 2 million directory users in any proprietary directory product currently on the market.

**Phone Books**

Most organizations currently produce corporate phone books using a proven technology that has been available for many years — paper. Unfortunately, paper is quite expensive and typically produces significant amounts of errors between the time of going to print and the time the phone books are distributed to the users.

As an example, in 1993, The Boeing Company was producing company telephone books four times per year. Each production cycle generated 97,000 phone books at a printing cost of $155,000. Interestingly enough, it was accepted that approximately 10% of the printed books would be removed from Boeing facilities. The staff necessary to produce the phone book consisted of four and a half people (180 hours per week). It required 11 trucks to pick up the printed material and deliver it to the Boeing mailrooms. At that point, the phone books were bundled into odd-size lots and addressed to a secretary or secretaries and distributed to every mail-stop in the company.

The amount of work to deliver these books was very significant, yet the costs were considered overhead to the mailroom and never accounted for. Another cost was the disposing and recycling of the used books. In other words, the cost to Boeing for printing the corporate phone book was over $1 million per year. And people still didn't have current information.

Most companies are in the same predicament, but even worse. Most departments and projects generate their own custom phone list to overcome numerous factors about corporate phone books. Those factors include:

- Timeliness of information (people are constantly moving and changing)
- Organization of the data (cannot search by organization, function, or location)

- Quantity of information (too many entries and not enough information)

Companies have determined that moving information to a single source that is widely available is extremely desirable. For example, as part of a research project on X.500 at Boeing, a spreadsheet was developed using Microsoft Excel and its macro language called Visual Basic for Applications (VBA). Using VBA, it was possible to identify search criteria, request the information from the X.500 Directory, and place it into cells of a spreadsheet.

Imagine providing all department secretaries with the capability of printing their local phone list with the touch of a button — no more time wasted in maintaining employee data. When you multiply the numbers, it becomes very significant. For example, if a company of 100,000 employees has a secretary for every 25 people, then there are 4000 custom phone lists being generated every week or two. Every time it takes approximately 30 minutes to generate the phone list and another 30 minutes for fine-tuning, ongoing maintenance, and distribution. This comes out to anywhere from 48,000 to 192,000 hours per year just to get around the ineffectiveness of a corporate phone book.

Telephone directories are an area where current technologies are more than ready to provide solutions and cost savings. The single source of data can be published using X.500. The delivery mechanisms for this information can be either integrated into messaging user agents or delivered via a stand-alone user agent.

One such stand-alone user agent can be a Web browser, such as Netscape, Mosaic, or Microsoft Internet Explorer. The interface technology is readily available using a Web-to-X.500 gateway. This has been proven and demonstrated around the Internet and is currently in production at many companies.

**Business Tools**

The area of business tools is probably the largest untapped resource. The availability of a single directory infrastructure that can be shared by an entire enterprise has such significant impact to business processes that it cannot be understated.

Early entries in the area of business tools are workflow applications. As you design workflow applications, it becomes evident that current business processes are typically very ineffective and costly. Many studies have shown that the amount of process reengineering necessary to implement effective workflow can result in tremendous cost savings and major reductions in cycle time.

## THE ELECTRONIC MESSAGING ENVIRONMENT

For example, one large retail chain in the U.S. converted its human resources processes from a paper-based system to a message-enabled application. This new process reduced the amount of time to completely process new hire information from over 2 weeks to just a few minutes. The human resources organization that processed this information was reduced from 130 people down to 4 people. At the same time, management had much more timely information on current staffing, which resulted in better planning.

As dramatic as this example has been to this point, there is more to the story. The same retail chain also developed message-enabled applications that reported sales and inventory information from each store and distribution center. In the past, a subset of this information was manually sent back to a central area for data entry, processing, and later analysis. Suddenly, it was possible to generate overnight or even on demand business-trend analysis.

With this level of analytical information, coupled with improved staffing information, the retailer was able to quickly react to market trends and manage costs. Inventory management becomes more effective. The retailer has embarked on communicating transactions with its suppliers over an X.400 messaging system using EDI technology. The directory infrastructure has made it possible for these applications to drastically improve the competitive advantages that this retailer enjoys.

There are many other business tools that are possible when an infrastructure is in place. It is expected that over time, all client/server applications will utilize directory infrastructures.

**Emerging Network-Related Applications**

One untapped application relates to network-related infrastructures, such as routing. Several years ago, a project called the Internet Soft Pages took advantage of knowing the network infrastructure in locating duplicate objects on the network. Namely, the Soft Pages application was able to query the X.500 Directory for the closest copy of a document that was stored in multiple locations around the world. Users were able to retrieve the nearest, most available copy of documents that were requested. The benefit here stems from reducing network resource requirements. The application itself was transparent to the user.

Other areas of network applications that can have significant impact on business processes are printing and name services such as the TCP/IP Domain Name Service. Imagine being able to print a document at the nearest location to the desired destination. For example, you may want to print a document on a two-sided postscript printer that is nearest to the location of a business colleague that is currently located in a different facility. You would have the ability to query the directory for the information.

**Global Electronic Directory.** The X.500 Directory standard will be used by organizations worldwide to hold White and Yellow Page information about residents and organizations within countries. The Directory represents a standardized means to improve the efficiency of the global telephone network by storing user telephone numbers.

**National Electronic Directory (NADF).** The NADF is currently working to establish a national electronic directory based on X.500. The directory will be accessible by public mail carriers, value-added networks (VANs), and end-users. This directory will be an electronic version of the nationwide "411" Directory Information service provided by the telephone companies.

**Corporate Directory.** Corporations will initially use X.500 to provide messaging support for their EDI service. Thereafter, corporations typically will use X.500 to replace proprietary directories to facilitate the interconnection of diverse E-mail platforms. The Directory will contain electronic mail users, devices such as printers, facsimile and telex machine addresses, hard copy (paper) mail addresses, and application and file names. The Directory may then be expanded to include such information as employee skills and education levels, vendor profiles, corporate assets (buildings, real estate, etc.), passwords, employee pictures, public keys, and inventories, among other information.

**Network Management.** As X.500 provides a repository for electronic mail addresses to facilitate X.400 message handling, the Directory system can similarly provide a repository of network information, notably presentation addresses. In this manner, the Directory can facilitate intersystem communications. Users will have a common base for connecting different vendors' networks and mail systems without having to rely on inefficient techniques such as gateways.

**Library Services.** Library systems within local governments and within university systems could use the X.500 Directory standard to create a distributed repository for author, title, and subject classifications. The user could query the Directory to determine whether the entire system has a particular book and pinpoint its location within a distant library.

**Government Services.** Local, state, and federal governments could use the X.500 Directory standard to establish listings of various government agencies, services, and organizations. Users would then query the Directory, for example, to determine what services are available within a particular classification within their geographical area.

# THE ELECTRONIC MESSAGING ENVIRONMENT

**Third-Party Directory Services.** Some corporations and organizations may not wish to have external users access their X.500 Directory directly for security or competitive reasons. Those entities that form an industry group such as the Aerospace Industries Association or the American Petroleum Institute may want to employ a third-party X.500 Directory service whose access would be limited to other firms within the same industry group.

**Other Uses.** The usage and application of the X.500 Directory is limited only by the imagination of the users, developers, and strategic planners within organizations. Such esoteric uses as a national auto parts directory, a stolen property registry, a directory for fax and cellular phone numbers, a national employment directory, or a national organ donor directory are possible.

## SUMMARY

In businesses' quest for an electronic, global, cross-platform directory system, the X.500 Directory standard is key. X.500 is the only standards-based directory technology at this time that is proven to be stable and deployable. Deployment is often in support of messaging.

This chapter has described the principal business benefits and possible applications of an X.500-based directory system, as well as the X.500 Directory's client/server-style architecture. Chapter 12, "X.500 Directory Services: Under the Covers," goes more in-depth describing how to use an X.500 Directory to find and house information, as well as secure it.

# Chapter 12
# X.500 Directory Services: Under the Covers

*Roger Mizumori*

The X.500 Directory is useful for finding and housing information. This chapter describes the directory entries and explains how they are stored in a schema. Next, it discusses the naming structure used by X.500 to index information. The Directory services and principal mechanisms for conducting a search are also explained, along with access controls and security services. Lightweight Directory Access Protocol (LDAP), which is emerging as a standard way to access a directory service, is also covered in this chapter. Finally, there is a description of the distributed model and how the many Directory Systems Agents coordinate their services. Throughout the chapter, the word *Directory* is capitalized when referring to the X.500 standard or standard-based implementations.

## UNDERSTANDING DELIVERY MECHANISMS

It is important to understand how people and applications access information. To do this, you must be able to profile your current environment and have a reasonable understanding of future directions. Here are some questions to answer.

- What platforms do line-of-business applications run on?
- Is the enterprise moving to client/server computing?
- What role do desktops play in the computing environment?
- Is there an existing network infrastructure?
- How do users and applications get to directory data today?
- How often and to how many groups do data owners provide their information?
- How often is data synchronized with current corporate information?
- How quickly can you react to disable a user from further access to company resources?

- How do you control access to company information?
- Do you have requirements to share certain information with external trading partners?

### Building a Vision

It is important to build a vision of the potential impact of a directory infrastructure on an organization. This vision will help you set a direction for future activities, as well as make it easier to explain your work to current and potential customers.

### CHOOSING LOW-HANGING FRUIT

Early in the effort, it is very important not to get overly enthusiastic about the breadth of available information and applications that are possible with this technology. Instead, carefully select your initial targets for both customers and potential data. A wise decision early in the cycle will lead to very quick acceptance of the technology and the potential for rapid demands for additional services. Concentrate initially on small, meaningful, quick successes.

### Understanding Data Management Issues

As with any large database of information, data must be carefully managed for consistency, currency, and accuracy. Typically, when people begin to investigate initiating a pilot with X.500, they quickly realize that many of the desired data sources are not available or that they need significant work to make them useful. Early involvement by experienced data management personnel is very essential. A Directory is useless without accurate information.

### DIRECTORY ENTRIES

The complete information about a single object in the Directory Information Base (DIB) is represented by a directory entry. The entry is composed of attribute and value pairs. The hierarchical nature of the directory entry illustrates object classes residing above it and subclasses residing below it.

The structure of the directory entry is illustrated in Exhibit 12-1. A directory entry for the organization of Acme Parts might show, for example, "country" as a superclass or superior, and "sales" as a subclass or subordinate. Entries depict superior and subordinate class references as well.

### X.500 Directory Schema

The Directory Information Tree (DIT) is composed of directory entries, attributes, and values (see Exhibit 12-2). Before the creation of the DIT, a

# X.500 Directory Services: Under the Covers

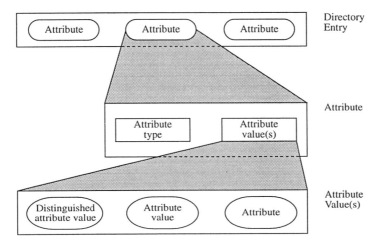

**Exhibit 12-1. Structure of a Directory Entry.**

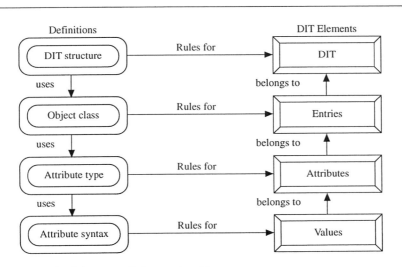

**Exhibit 12-2. Directory Information Tree.**

# THE ELECTRONIC MESSAGING ENVIRONMENT

foundation for the structure and content of the DIT elements must be established.

The schema of the X.500 Directory specifies "a set of definitions and constraints (i.e., rules) concerning the structure of the DIT and the possible ways entries are named, the information that can be held in an entry, and the attributes used to represent that information" (according to Recommendation X.501). This set of rules contains the structure of the DIT, object classes, and attribute types and their syntax. These definitions and rules specify the relationship between Distinguished Names, the mandatory and optional attributes in an object class, and whether the attribute type may have multiple values.

The Directory schema enables the creation of object classes composed of valid entries, the correct assignment of attribute types to an entry, and the use of the correct syntax for the content of the attribute values. According to the 1993 extensions to the X.500 Directory standard, the X.500 Directory schema has been expanded to include structure rules that define the form of the Distinguished Name and the relationships that names have to each other within the DIT. The 1993 extensions also specify subschemas that are created by the Administrative Authority and that contain the rules, definitions, and constraints for DIB entries located within a single administrative area. The collection of subschemas is then known as the Directory Schema.

## NAMING STRUCTURE

### Relative Distinguished Name (RDN)

The separate pairs of true attributes of the unique directory entry are also known as the Relative Distinguished Name (RDN). In Exhibit 12-3, the RDN is represented by the attribute type and attribute value of one object class.

The RDN is valid and unique within the superior object class. As in Exhibit 12-3, an organization of "Acme Parts" is unique within the country of

---

Exhibit 12-3. Example of a Relative Distinguished Name.

| Attribute | RDN | Distinguished Name |
|---|---|---|
| Root | Null | { } |
| Country | C = US | {C = US} |
| Organization | O = Acme Parts | {C = US, O = Acme Parts} |
| Organization Units | OU = Sales | {C = US, O = Acme Parts, OU = Sales} |
| Common Name | CN = Smith | {C = US, O = Acme Parts, OU = Sales, CN = Smith} |

## X.500 Directory Services: Under the Covers

"US." If the attribute value of "Common Name = Smith" is not unique within the organization, then additional information must be supplied, such as a first name and middle initial or an Organizational Unit attribute type and value.

### Distinguished Name (DN)

When the set of separate valid Attribute Value Assertions in the RDN are arranged in a hierarchical sequence, the name becomes a Distinguished Name. In Exhibit 12-3, the entire DN for the Directory entry is C = US, O = Acme Parts, OU = Sales, CN = Smith. The DN sequence depicts the object and all of its superiors. Aliases cannot be DNs because the sequence of RDNs is in descending order from the root.

### Alias Names

Aliases are object classes that provide a different name and a different path for the same object. An alias is a pointer to a different hierarchical name sequence for the same object. An object may have none or multiple names, as well as none or multiple alias entries in the Directory Information Base. According to Recommendation X.501, aliases permit object entries to achieve the effect of having multiple immediate superiors. Furthermore, an alias entry can have no subordinates.

Aliases are useful for reducing the time required to search the Directory, containing the size of the Directory response, and limiting the scope of the search. An alias entry is depicted in Exhibit 12-4.

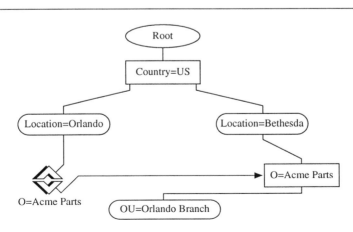

**Exhibit 12-4. Structure of an Alias Directory Entry.**

## Name Registration

To ensure unique and unambiguous names within the global X.500 community and within OSI, names must be recorded, collected in a repository, and published. Provisions must also be made to handle conflicts and challenges based on duplicate names.

A naming authority must exist to perform these administrative duties on behalf of all X.500 consumers and providers. For the U.S., the American National Standards Institute (ANSI) has assumed the role of naming authority for OSI organization names under the joint ITU-TSS/ISO arc and is registering, collecting, and publishing those names. Similarly, in Canada, the Canadian OSI Registration Authority (COSIRA) has been established for this function. Registration Authorities are being established in a number of other countries as well. However, to achieve a global directory service, these registration authorities will ultimately need to coordinate the registrations so that multinational entities, such as NATO or the United Nations, can be effectively registered without having to individually register in every country.

The process in the U.S. is that the requester of the organization name pays ANSI a fee to have the name registered, published, and stored in a database. ANSI has a procedure to handle disputes on naming rights, should they arise in assigning names. The requester receives a numeric name or object identifier (OID) and, for an additional fee, may also request and receive the alphanumeric registered name, both of which are guaranteed to be unique within ANSI.

## X.500 DIRECTORY SERVICES

The X.500 Recommendations offer a basic set of services for obtaining and manipulating the information within the Directory. Also known as the *abstract service,* users access the Directory services through an abstract service interface. This interface relates requested services to ports within the Directory User Agent (DUA).

Requests flow through the DUA ports to the DSA and beyond to the Directory Information Base. Users may interrogate and search the Directory as well as update directory entries. The specification also recommends procedures for distributed directory operations services and service controls. The Directory will either return a confirmation of the success of the particular operation requested or an error message will indicate that the operation has not taken place. In the latter case, no changes will be made to the Directory Information Tree.

### Directory Interrogation Operations

Within the X.500 standard, users are afforded several ways to interrogate or query the Directory. The user requests information based on selected attributes and the Directory returns the results of the operation. The user may initiate a read operation such as "read" or "compare." The Directory also permits two search operations: "list" and "search." The user may elect to limit the extent of the search by specifying a "filter" that removes entries from the search that are not of interest to the user. The abandon service operation is used to inform the Directory that the user is no longer interested in continuing the outstanding interrogation request.

### Directory Update Operations

The X.500 Directory contains four directory update operations that may be performed by the end-user, if authorized to do so, or by the Administrative Authority of the Directory Service Agent. The modify services are add entry, remove entry, modify entry, and modify Relative Distinguished Name.

**Add Entry.** The user may add an object entry or an alias entry to the Directory Information Tree within the same DSA as the superior entry by selecting the "add entry" operation. A user would create a valid object entry with the appropriate Attribute Value Assertions and add the entry to the DIT. The Directory ensures that the entry to be added conforms to the Directory rules and constraints as defined in the Directory schema.

**Remove Entry.** Entries in the DIT within the same DSA as the superior entry may be deleted using the "remove entry" operation. The entry may be either an object or an alias entry.

**Modify Entry.** The object entries within the DIT of a specific DSA may be changed. The user selects one or more changes to a particular entry that will not affect the Relative Distinguished Name. The user can add or remove an attribute; add, remove, or replace attribute values; and modify an alias entry. The Directory processes the request for modification in the sequence specified. As with the addition of a new entry, the modification of an existing entry must not violate the rules and constraints of the Directory schema.

**Modify Relative Distinguished Name.** The "modify RDN" update operation allows a user to change the Relative Distinguished Name of the object or alias entry within the DIT. For example, a company may have consolidated its sales operation, necessitating the deletion of the Organization Unit names of "Eastern" or "Western" from the structure of the company's RDN. Any attributes not in the existing RDN will be automatically added in the new RDN. If the new attributes cannot be added, for example, due to a rule violation of the schema, the Directory will return an error to the user.

### DISTRIBUTED X.500 DIRECTORY OPERATIONS SERVICES

The X.500 standard provides powerful search capabilities for distributed Open System Interconnection (OSI) directories. The Directory is poten-

# THE ELECTRONIC MESSAGING ENVIRONMENT

tially able to satisfy a user's request regardless of the particular access point used. The original DSA invokes service operations within the adjacent DSAs. For a DSA to satisfy the user's request, the DSA must have "knowledge" of the location (i.e., address) of other DSAs as well as "knowledge" of the information (portion of the Directory Information Tree) contained within the DIB of the other DSAs. The distributed Directory is a logical construction of multiple Directory Service Agents (see Exhibit 11-2 in the related Chapter 11, "X.500 Directory Services: A Business Process Enabler").

**Referrals**

If a Directory User Agent, on behalf of a user, requests a particular operation of its own DSA and the DSA cannot fulfill the request, the DSA will return a referral to the user. If the referral received is the whole response, it is dubbed a referral error.

The referral will contain a knowledge reference indicating which DSA within the distributed Directory may have the information sought. The DUA can then use the referral to contact other DSAs to continue the search.

The Directory User Agent of one DSA may choose to interact directly with the other DSAs on a one-to-one basis. This interaction is required when the involved DSAs do not have chained service ports, as shown in Exhibit 12-5.

Alternatively, the DUA may choose to have its own DSA gather all the information first, acting as a central collection point, before reporting the results of the search back to the user by means of the DUA. This action is possible when all the affected DSAs have chained service ports as depicted in Exhibit 12-6.

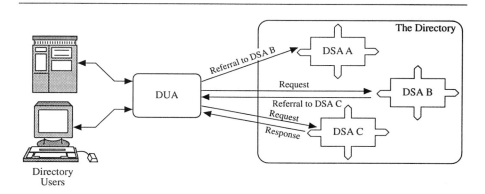

Exhibit 12-5. Referral Mode: DSA with No Chained Ports.

*X.500 Directory Services: Under the Covers*

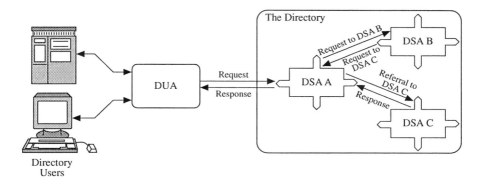

**Exhibit 12-6. Referral Mode: DSA with Chained Port.**

## Chaining

The chaining mode of operation is used by a particular DSA when it has some knowledge of what is contained within another DSA's Directory Information Tree. The originating DSA, when unable to satisfy a service request, will use the knowledge obtained in the Directory's response to contact a single DSA specified in the knowledge reference. Exhibit 12-7 depicts the chaining mode of operation.

## Multicasting

If the DSA cannot fulfill the user's request and the DSA has no specific knowledge of the naming scheme of any other DSAs, the DSA may route the

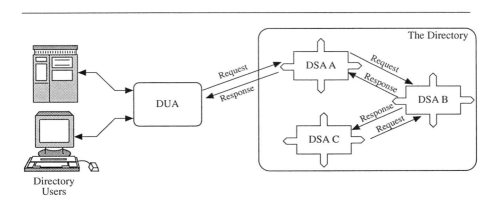

**Exhibit 12-7. Chaining Mode.**

193

# THE ELECTRONIC MESSAGING ENVIRONMENT

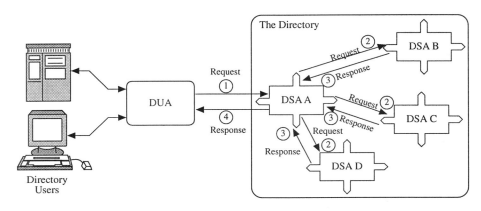

**Exhibit 12-8. Multicasting Mode.**

same request in a broadcast manner to the other DSAs as illustrated in Exhibit 12-8. This operation, known as multicasting, is a form of chaining.

Since the contents of the Directory Information Tree of the other DSAs may not be known completely to the original DSA, this feature represents the "fuzzy" search capabilities of the X.500 Directory. If a remote DSA is unable to continue processing the request or resolve the name, the DSA will return an "unable to proceed" service error to the original DSA.

## Search

The original 1988 X.500 search capability has been enhanced to include an extended filtering arrangement whereby the user may specify an alternative filter to the original filter entered to reduce the scope of the search. The search result may also indicate which attributes have been matched by the search. The search operation is subject to access controls and as such the Directory will not return any information within the search results that will compromise the security of any entry.

## ACCESS CONTROL

The current 1993 X.500 Directory extensions provide for a Basic Access Control scheme within the Security Model along with authentication, as depicted in Exhibit 12-9. This scheme contains a more detailed level of access control than that suggested within the 1988 standard.

When a user requests a Directory service (e.g., read, add, or compare), the Directory must make a decision to grant or deny permission to the requester. This decision, rendered by the Access Control Decision Function (ACDF), is based on the security policy associated with the component of

# X.500 Directory Services: Under the Covers

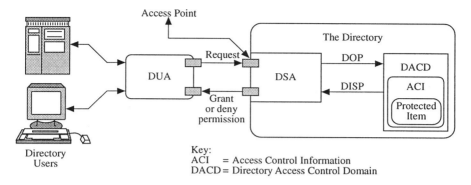

**Exhibit 12-9. Basic Access Control Scheme.**

the Directory being accessed. The component is also known as the protected item.

Protected items may be entries, attributes and values, and Directory names. These items may be collectively controlled within the scope of a particular Administrative Point. The requester will attempt to gain access to either an entry or to an attribute whose values contain user and operational information. In the absence of any access control information associated with the protected item, the ACDF will deny access.

Access Control Information (ACI) is stored within the Directory as an operational attribute of entries and subentries. An ACI item may be contained within a collection of entries that form a Directory Access Control Domain (DACD). The higher-level DACD may set overall access permissions for the entire administrative area beneath an Administrative Point. Each subordinate DACD may then create access controls on an increasingly more detailed level. The subordinate ACI may override the defaults of the DACD.

## X.500 DIRECTORY SERVICE CONTROLS

Service controls apply certain constraints, rules, and preferences to optionally limit the Directory services provided to a user. The provision of these service controls is a matter for the local administration of the Directory service.

The default set of service control options includes no preference for chaining (although chaining is not prohibited), no limit on the scope of the operation, the use of copy permitted (i.e., cached information gleaned from previous operations), and the requirement that aliases be dereferenced. (Dereferencing means to substitute the object's Distinguished Name for the alias.)

# THE ELECTRONIC MESSAGING ENVIRONMENT

The user may elect to use this default set of options or may individually select the options to be used. For example, the user may prefer to use the cached information while interrogating the Directory. The user may also prefer that the scope of the search not extend beyond the local Directory Management Domain.

Other controls available are:

- Priority service controls
- Time limitation service controls
- Size limitation service control
- Scope of referral service control
- Filters
- Entry information selection service control
- Security parameters

## X.500 DIRECTORY SECURITY

One of the goals of an open system is to permit quick and user-friendly access to information requested by the user. This openness also carries with it two inherent security risks.

Readily accessible objects must be protected from the threat of unauthorized use and abuse (authorization) by comprehensive security measures. The open system also must be able to ascertain the true identity (authentication) of the user requesting an operation or service before the system acts on the request.

Security measures can either be integrated with the communication or occur as a separate operation. The X.500 standard addresses the accessibility of the Directory in the form of guidelines for authorization, access rights, and access control. The standard leaves specific recommendations on authorization to the 1993 extensions. The X.500 Directory standard specifies several recommendations for authentication in detail in X.509.

A related recommendation is the X.800 Security Architecture for Open Systems Interconnection for ITU-TSS Applications. This standard is an architectural document or master plan describing security services and specifying security mechanisms and at which layers of the OSI Reference Model these services may be provided.

### Directory Security Services and Mechanisms

The X.500 Directory may encounter any of several widely known threats to the security of the service. Potential threats applicable to the Directory service include identity interception, masquerade, replay, data interception, manipulation, repudiation, denial of service, misrouting, and traffic analysis.

The X.500 standard suggests several security services as effective in preventing these threats and protecting areas within the Directory vulnerable to encroachment. Security services such as entity authentication, data confidentiality, data integrity, and nonrepudiation are discussed later in this chapter.

Generally, Directory security involves two primary areas. A user must first be able to prove conclusively that his/her access to the Directory is permitted. This is the *authorization security service*. A user must also be able to prove conclusively his/her true identity. This is known as the *authentication security service*.

Often, a user must present proof of identity before authorization to access the system is granted. Three additional services are also recommended: nonrepudiation, data confidentiality, and data integrity.

**Authorization.** The particulars of authorization, the granting of permission to use the Directory, are not specified in the 1988 X.500 standard. This is a major omission. The lack of an authorization specification forces developers of X.500 Directory products and services to create their own proprietary authorization mechanisms that may be in conflict with mechanisms used by other open systems. When users of these disparate open systems attempt to access them, the authorization mechanisms will be in conflict. This conflict may result in the inability to access the Directory system.

The X.500 standards committee left the bulk of the authorization specification (in particular, the access control recommendations) to the 1993 extensions. Many vendors have already implemented their own version of authorization and require modifications to be compliant with X.500. The authorization Directory security mechanism is composed of access rights and access control.

**Access Rights.** Access rights are those privileges and permissions granted to a user of the Directory that let the user issue requests for Directory services such as "read," "compare," or "modify entry." The information required to determine a user's right to perform a given operation must be available to the DSA(s) involved in performing the operation in order to avoid further remote operations solely to determine these rights.

A collection of access rights may form a list that associates each user with a certain access category as depicted in Exhibit 12-10. The Directory should be able to ascertain the access category of the user, either by having the category supplied with the initial request (within the BIND operation) for Directory services or within the Directory operation requested.

Security policies should clearly specify all levels of access rights and the maintenance of these lists of access categories. The policy should also en-

## THE ELECTRONIC MESSAGING ENVIRONMENT

**Exhibit 12-10. X.501 Access Categories.**

| Category | Items[a] | Description |
|---|---|---|
| Detect | A | Allows the protected item to be detected |
| Compare | A | Allows a presented value to be compared to the protected item |
| Read | A | Allows the protected item to be read |
| Modify | A | Allows the protected item to be updated |
| Add/delete | AE | Allows the creation and deletion of new components (attributes or values) within the protected item |
| Naming | E | Allows the modification of the RDN of, and the creation and deletion of, entries which are immediately subordinate to the protected entity |

[a] A = attribute, E = entry, AE = both.

force the user's access rights by specifying access control mechanisms as detailed below.

**Access Control.** Access control may be used to protect directory objects from unauthorized use and to specify the information that will be made available to requesters. Access control mechanisms serve to guard the vulnerable areas of the Directory Information Base from unauthorized inspection, manipulation, and modification. Items that are to be protected include an entire subtree of the Directory Information Tree, an individual entry, an entire attribute within an entry, and selected instances of attribute values.

Access controls enforce the rights and privileges granted to a particular user or group of users. Access control mechanisms use the authenticated identity of an entity, its capabilities, or its credentials to determine and enforce the access rights of that entity. An entity must first prove its identity and be authenticated before access privileges to the Directory are granted.

## AUTHENTICATION

Authentication involves ascertaining the true identity of an entity requesting services provided by the Directory. The Directory must have the capability to confirm the identity of Directory System Agents, users, and access points within the Directory structure. The Directory and its users must also have the capability to determine the identity of the origin of any information received. The X.500 Directory provides two levels of authentication: simple and strong.

# X.500 Directory Services: Under the Covers

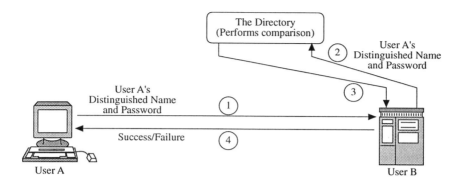

**Exhibit 12-11. Unprotected Simple Authentication.**

---

Because most secure data communications systems rely on authentication, the X.500 Directory is an ideal location to store authentication and authorization information.

**Simple Authentication**

Simple authentication, in its most basic form, requires that entities use an uncomplicated password arrangement. This bilateral arrangement must be mutually understood and uniformly handled by the parties involved.

The entity's Distinguished Name would accompany the password. Additional information such as a time stamp or a random number may also accompany the DN and the password. This set of information is termed the user's credentials.

Passwords and DNs should be protected to provide more communications. If the simple authentication scheme is not protected, the Directory must be involved, as shown in Exhibit 12-11. The DUA sends its DN and password to the DSA. The DSA sends the DN and the password to the Directory. The Directory then uses the compare operation to ascertain a match between the password attribute in the user's Directory entry and the DN and password relayed by the DSA. If the two match, the Directory notifies the DSA that the DN and password are valid. If they do not match, the Directory sends a denial notification to the DSA. The DSA may then notify the DUA that the authentication is confirmed or denied.

In a protected simple authentication scheme, the recipient confirms or denies the credentials of the originator. User A in Exhibit 12-12 sends its DN and password, together with a time stamp and a random number used to hide the password, to User B. The DSA uses the DUA's credentials to gener-

## THE ELECTRONIC MESSAGING ENVIRONMENT

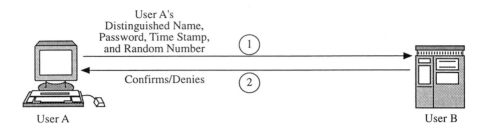

**Exhibit 12-12. Protected Simple Authentication.**

ate a local version of the DUA's information. The DSA then compares its own copy with that sent by the DUA. The DSA confirms or denies the validity of the DUA's supplied credentials, thereby eliminating the step of submitting the DN and the password to the Directory for comparison and verification. In effect, the originator and recipient peer entities confirm or deny each other's credentials.

Simple authentication is recommended for local use between such entities as a DUA and a DSA, and between cooperating DSAs. Simple authentication inhibits the masquerade threat.

**Strong Authentication**

Strong authentication mechanisms employ complex cryptographic and encipherment techniques to ensure the integrity of the user and Directory data and to secure authentication. Cryptography provides mathematical means and methods to transform data in order to hide its information content, prevent alteration, disguise it, and prevent its unauthorized use. Again, it is advantageous that the strong authentication information can be stored by the Directory as attributes.

The X.500 Directory authentication framework does not specify the particular cryptographic system, hash functions, or algorithms to be used. However, cooperating entities within the same community or users of the same application may want to have bilateral agreements as to the system or functions used in an effort to maximize the number of entities using secure communications.

**Strong Authentication Procedures.** The X.500 Directory standard specifies three strong authentication procedures. The simplest is one-way authentication in which an originator authenticates itself to a recipient. Two-way authentication requires the same procedure plus an additional one — that the recipient generate a reply to the originator, which further authen-

## X.500 Directory Services: Under the Covers

ticates the pair of users. Three-way authentication uses the same procedure as two-way with the need for checking the time stamps present in the other two procedures.

*Public Key Cryptosystem.* The strong authentication framework uses a cryptographic system called a Public Key Cryptosystem (PKCS). This system converts human-readable text into encrypted or enciphered text by the use of mathematical algorithms.

PKCS, in contrast to customary single key symmetric systems, is an asymmetric system requiring both parties to the communication to possess two keys (see Exhibit 12-13). Each user has a public key that can be freely disseminated and a private key known only to the particular user. Both the public and the private keys (a key pair) can be used to encipher and decipher text. If the public key is used to encrypt the text, the private key is used to decrypt and vice versa.

Once an originator is in possession of the recipient's DN and public key, the originator can detect the presence of the recipient's private key.

Users may generate their own key pairs, a third party may generate the pairs and relinquish them, in a secure manner, to the user, or the key pair can be generated by a Certification Authority.

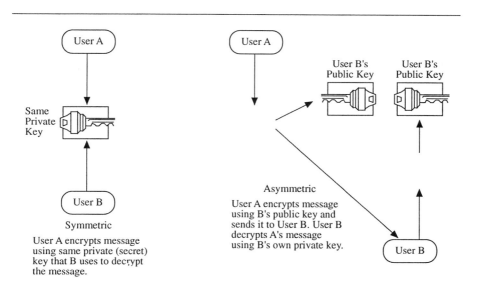

**Exhibit 12-13. Symmetric and Asymmetric Cryptosystems.**

Within an application, the mathematical algorithm used for the encryption should be identical to enable a larger community of users to communicate and authenticate.

*Certification Authority.* Authentication takes place when both users have Distinguished Names and public keys. This collection of information or credentials forms the user's certificate, which is created, signed, and issued by a Certification Authority (CA).

The certificate is nonmodifiable, cannot be forged, and can be published in the Directory as part of a user's entry. The CA is also an entity in the Directory Information Tree. Any user possessing the public key of the CA can obtain the public keys of any other entities in the DIT. CAs may exist at various points within the DIT hierarchy and, together with a superior CA reference, form a "certification path" from one user to another. The Directory provides both the certificates and the reverse certification path for users to authenticate each other.

The CA must validate a user's identity before generating the user's certificate. The CA must not issue certificates for duplicate names and must produce certificates offline before issuing a copy to the user. The CA generates certificates with serial numbers and time stamps for a specified duration and may revoke a certificate prior to its expiration. The Directory contains lists of revoked certificates as attributes.

*Digital Signatures.* Digital signatures use a two-part message (see Exhibit 12-14). The first part is the plain text information to be transmitted to the recipient. In the second part, the originator uses a one-way hash function to reduce the size of the information, thus forming a summary. (Hashing is a mathematical function that transforms a potentially large value into a smaller one.) Then the originator encrypts the summary using his/her private key. The encrypted and hashed summary of information is known as a certification token, digital signature, or seal.

The originator then sends the plain text information and the seal to the recipient. The recipient applies the same hash function to the summary and then decrypts the information using the originator's public key. The recipient validates the user's digital signature by comparing the results of the decryption to that of the plain text information received. If the two results are identical, then the information that was received is the same as that sent. The process also authenticates the originator.

Digital signatures are effective in preventing the modification of the data and in proving the identity of the originator.

*X.500 Directory Services: Under the Covers*

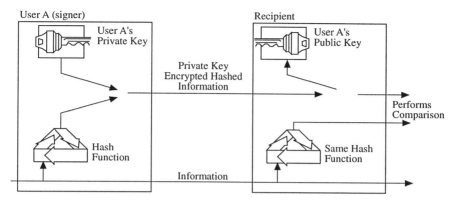

**Exhibit 12-14. Digital Signature.**

*RSA Encryption.* Developed by Rivest, Shamir, and Adleman at MIT in the late 1970s, the RSA cryptosystem is used to generate keys by employing algorithms involving very large prime numbers. This system is gaining acceptance in the field of cryptography. The system has been supported as a standard by Digital Equipment Corp. and the Internet Activities Board. IBM Corp. has licensed it for use in product development. RSA may be used to generate the public and private keys required by the X.500 strong authentication framework.

*Electronic Funds Transfer (EFT).* Strong authentication mechanisms such as digital signatures, public, and private keys are applicable to the transfer of secret information such as electronic funds. As X.400 matures and is used as the transport medium for electronic data interchange (EDI) transactions such as invoices, interest in using X.400 for the transfer of funds will increase. The X.500 Directory can assist in these transfers by providing strong authentication of the parties and applications involved.

### Nonrepudiation

Nonrepudiation means that neither a sender nor a receiver can deny that they sent or received a particular message or its contents. This prevents the sender and the receiver from making false claims about or from refusing to recognize their participation in the exchange.

### Data Confidentiality

Data confidentiality can be used to prevent an unauthorized entity from intercepting or observing user data during a Directory operation. This security

## Data Integrity

This security service protects the data in the Directory and user data from manipulation or modification. The service maintains the integrity of the user data by guaranteeing that the data has not been modified or ruined.

## LIGHTWEIGHT DIRECTORY ACCESS PROTOCOL (LDAP)

LDAP is currently known as RFC 1777 and is on the standards track of the Internet Engineering Task Force (IETF) in a version 2 draft state. Version 3 edition is currently under development by the IETF.

As stated in RFC 1777, the protocol is designed to provide access to the X.500 Directory while not incurring the resource requirements of the Directory Access Protocol (DAP). This protocol is specifically targeted at simple management applications and browser applications that provide simple read/write interactive access to the X.500 Directory and is intended to be a complement to DAP.

Key aspects of LDAP are:

- Protocol elements are carried directly over TCP or other transport, thereby bypassing much of the session/presentation overhead.
- Many protocol data elements are encoded as ordinary strings (e.g., Distinguished Names).
- A lightweight BER encoding is used to encode all protocol elements.

### Significance of LDAP

LDAP is currently being adopted by most vendors in the computing industry, including Microsoft, Lotus Development Corp., IBM, Novell, Netscape, Worldtalk, Banyan, Hewlett-Packard, and many others. It provides for a single standard interface to directory information. Any application written is able to interwork with systems from any vendor that conforms to the standard. This is important because it has started the momentum for a directory infrastructure that allows applications to be written without knowledge of specific database interface requirements.

### LDAP Uses and Limitations

LDAP is not currently suited for a number of applications. The version 3 effort is trying to address these issues.

Perhaps the most difficult issue is that schema definitions do not necessarily match between LDAP servers. This means that an attribute may have different names but the same value, depending on which server one connects to. It is not possible to refer to attributes via a unique object identifier (OID), as can be done with DAP. In addition, new matching rules are not possible, which makes it difficult to use X.509 security certificates. Version 2 of LDAP cannot proceed without work in the areas of authentication and internationalization.

There has been significant and somewhat-successful effort in using LDAP as a simple interface to local databases, known as SLDAP. Companies such as Netscape have built products that store information locally within a database and can be accessed using LDAP. While this mechanism is attractive at the department or small company level, it is clear that this solution is lacking many features that have been well developed by the X.500 community. These are:

- Replication
- Distributed operations
- Access control
- Scalability
- Multivendor interoperability

Each of these features is critical to building a common directory service. To compensate for this shortfall, numerous efforts have begun to define metadirectory technologies where directories become well integrated and access to them is seamless.

The SLDAP approach is increasing the number of LDAP-based applications being developed worldwide and encouraging vendors to provide LDAP interfaces to their proprietary directories. However, as users deploy these technologies, it is apparent that this solution does not get them to a simple, single-directory service across multiple enterprises and nations.

## X.500 DISTRIBUTED DIRECTORY SYSTEM MODEL

The realization of the vision of a global directory service can only be feasible if the Directory is widely distributed. The Directory is distributed if it is composed of more than one Directory Service Agent (DSA). This distribution becomes possible when one DSA knows the existence of other DSAs, knows the information contained in the DSAs, knows how to reach them, and cooperates with the other DSAs to satisfy requests from a Directory User Agent in a uniform and user-friendly manner.

Users interact with the distributed directory by initiating a request through the DUA as in Exhibit 12-15. The DUA funnels the user's request through service ports to an access point dedicated to a particular directory function such as "search." The Directory has two types of ports: service ports and chained service ports.

# THE ELECTRONIC MESSAGING ENVIRONMENT

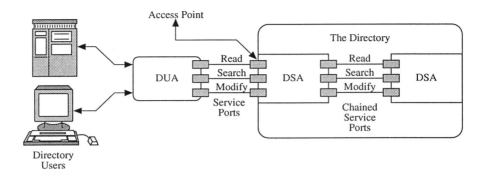

**Exhibit 12-15. Distributed Directory Model.**

The service ports provide access to the read, search, and modify functions of the Directory Service Agent. Chained service ports are used when the DSA cannot provide the requested information and must communicate with another cooperating DSA to fulfill the DUA's request. Chained service ports are the key to a global, distributed Directory service.

Ideally, this model should permit an authorized user to gain access to the distributed DIB, regardless of where the request was originally made. The DSAs cooperate with each other by either referring the user to another DSA or by interacting directly with another DSA using the multicasting or chaining modes (as discussed previously in the section "Distributed X.500 Directory Operations Services").

## DISTRIBUTED DIRECTORY KNOWLEDGE REFERENCES

One of the goals of the X.500 Directory is the ability to transparently satisfy a user's request for information regardless of where the request originated. To realize the potential of the X.500 Directory, each Directory Service Agent must contain both directory and knowledge information.

The directory information is that portion of the Directory Information Tree managed by the DSA's own Administrative Authority. The knowledge information illustrates how each particular DSA of the distributed Directory fits into the global, distributed DIT hierarchy.

Knowledge is the mapping of a name to its location within a fragment of the DIT (according to the Recommendation X.518 definition). The DSA, when given a request to provide information about a particular entry, uses its knowledge information to determine the location of the remote DSA(s) that contain the particular entry specified by the user's request. Each DSA also must contain knowledge of superior and subordinates within the DIT and their DSA's location (presentation address). This information provides

# X.500 Directory Services: Under the Covers

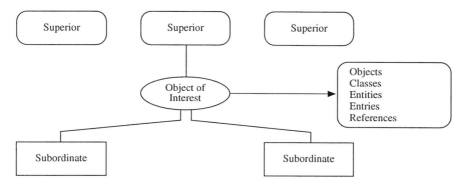

**Exhibit 12-16. Superior and Subordinate Hierarchical Concept.**

a physical view of the logical structure of the DIT. The concept of superior and subordinate references is depicted in Exhibit 12-16.

## Root Context

The global hierarchy of the Directory Information Tree refers to the "root" of the X.500 Directory. Due to the distributed nature of the Directory and the well-defined borders of the nations or international organizations whose DSAs comprise it, no one DSA will hold the root. Therefore, each DSA with entries directly below the root, such as Country = US, must function as if they are holding the root. These are called first-level DSAs.

Each first-level DSA must have knowledge of all the other first-level DSAs. This knowledge implies a reference path to the other first-level DSAs. All non-first-level DSAs must have a single superior reference that implies a reference path to any first-level DSA.

## Internal Reference

Within the Directory Information Base held by the specific DSA are located Relative Distinguished Names representing DIB entries. This information represents the knowledge that a DSA has of its own entries and how they fit into the overall hierarchy of the global Directory Information Tree. The DSA also uses a pointer to indicate the location of each entry within its own DIB. The Administrative Authority of the DSA manages the set of internal references.

## Superior Reference

A superior reference contains the name of the DSA holding the superior entry. Each DSA has only one superior reference in its knowledge information. This superior reference forms part of a reference path to the root. For example, the DSA for Acme Parts would have a superior reference of Country = US. References are useful in navigating through the hierarchy of the global Directory Information Tree.

### Subordinate Reference

A subordinate reference consists of the Relative Distinguished Name (RDN) for the immediate subordinate entry and the name of the DSA (access point) holding the entry. A company with multiple divisions and multiple DSAs managed by those divisions would have an organization name of Acme Parts, for example, and subordinate references to the divisions of Sales or Manufacturing in the DSA.

### Nonspecific Subordinate Reference

If a DSA is known to the originating DSA, but the extent of the knowledge is limited to some subordinate entries, the originating DSA would contain nonspecific subordinate references to the remote DSA. These references are likened to a "fuzzy" search operation.

The reference contains the name of the DSA holding the immediate subordinate entry. The DSA contacted based on a nonspecific subordinate reference must either be able to satisfy the request or return an "unable to proceed" service error. The referenced DSA must resolve the request directly and cannot use a referral or chain the request to another DSA.

### Cross-References

Cross-references contain the Distinguished Name for an entry and the name of the DSA holding the entry. Cross-references are used to optimize the facility of name resolution within the Directory. There is no constraint as to the number of cross-references in any one DSA. For example, one DSA may cross-reference its entries to another DSA to enable faster interrogation of the Directory.

## THE WEMA 97 DIRECTORY CHALLENGE

In July 1996, the 97 Directory Challenge was initiated with the express purpose of setting up the infrastructure for a global public directory service. This global initiative is supported by the World Electronic Messaging Association (WEMA) and has showcased four directory-enabled applications: open EDI, secure messaging, colored pages, and voice messaging. An April 1997 show in Philadelphia demonstrated a global infrastructure with international colored pages. A June show focused on secure messaging. An October 1997 show in Australia focused on electronic commerce.

## Directory-Enabled Applications

**Open EDI.** The open EDI Application is essentially a sophisticated Yellow Pages application. Businesses list themselves, contact information, trading partner profiles, and even catalog information. The intent is to facilitate the initial search for electronic trading partners and for vendors of particular products.

**Secure Messaging.** In most approaches to secure electronic messaging, encryption is a significant part of the solution. A direct implication is that a mechanism is required for managing use and access of the encryption keys. It is the objective of this application to demonstrate how X.500 can be used to accomplish this function.

**Colored Pages.** This application provides search access to government agencies just like the Blue Pages in the telephone book. In the 97 Directory Challenge, the U.S., Canadian, and Australian national governments are posting Blue Pages listings. The state of Texas is also expected to participate.

**Voice Messaging.** To facilitate interoperability of disparate voice messaging systems, a common repository is needed to house routing information and, potentially, account information. This demonstration also shows how the VPIM standard is used to share messages across disparate systems.

## Operating Parameters

Operating parameters are the rules developed by the Directory infrastructure team, in consensus, in order to effectively share knowledge references. They may or may not be formally adopted subsequently. Rules have been identified for:

- Registration
- Root DSA
- Schema
- NameFlow coordination
- Master DSAs
- Participating DSAs
- Access
- Service providers
- Corporate participants
- Access control
- Network
- Time service
- Naming convention

## THE ELECTRONIC MESSAGING ENVIRONMENT

- Certification Authority and PKI approach
- Formalized DIT maintenance
- Directory management
- LDAP
- Integrating DNS

Several of these operating parameters are described in more detail in the following paragraphs.

**Registration.** All U.S. members must register with ANSI and all Canadian participants must register with COSIRA or, to operate under the O = EMA branch on a temporary basis, with the Electronic Messaging Association (EMA) acting as registrar. Multivalued Distinguished Names are discouraged. Registration for all participating countries is a local issue.

**Root DSA.** For the purposes of the EMA demonstration, GSA is the Authoritative Directory System Agent for C = US and will replicate to all participating service providers in the Directory Challenge. The government of Canada is the Authoritative DSA for C = CA and replicates to all participating service providers. Shadowing of the root is supported using either DISP (preferred) or a flat text file of knowledge to be posted on the EMA server for interested parties. The root DSA for all interoperating countries must at minimum support the 1993 X.500 Directory System Protocol.

**Schema.** All efforts are oriented toward a single common schema for all WEMA demonstrations. The initial baseline schema is the GSA-developed U.S. Government-wide schema. Enhancements will be adopted as needed to support directory-enabled applications. Because variations will no doubt surface by countries, leading to national DITs that are unique, it is the responsibility of each country to establish its specific DIT. It is hoped that, where possible, commonality will be leveraged.

**NameFlow Coordination.** NameFlow is a global initiative in process on the Internet to demonstrate the scalability of X.500. The Directory Challenge Infrastructure team collaborates with the effort; however, within the time frame available, this connection has been declared technically infeasible for the Challenge.

**Master DSA(s).** Initially, the well-known entry point is the GSA DSA. As it becomes feasible to connect the NameFlow project, PSINet will be supported as an alternative entry point. Each participating entry must maintain at least one entry to contain name, address, and contact point. This is necessary to avoid responses indicating lookup failures, as might happen when the entity actually exists and participates but chooses not to expose information. If a query is made against a nonparticipant, it is appropriate to respond with an error message. If an object identifier is sought in both the NameFlow and WEMA context, the formally registered entity takes precedence.

For example, in the U.S., ANSI registration would establish ownership of the identifier. This approach leverages the conflict resolution mechanisms established by the formal registration authorities. For the purposes of the Directory Challenge, O = US government registers aliases. Alternatively, an organization may register a proxy under the EMA branch. The form and format will be user-defined. The worldwide public directory service begins with the first two DSAs being 1993 X.500-compliant with X.525 replication via DISP. All approaches seek maximum flexibility.

**Participating DSAs.** Participating X.500 DSAs must support at least the 1993 X.500 DSP. The Directory infrastructure will use the baseline schema and application schema designed by the workgroup, and all participating public X.500 DSAs must support the entire schema. X.500 DSAs supporting an application only must support the application-specific schema.

**Network.** All network connections will use TCP/IP with RFC 1006 support with connections through the Internet.

**Naming Convention.** Among the requirements for a naming convention are the need for uniqueness, the use of multiple character sets, and a user-friendly look and feel. It is understood, however, that legacy environments cannot be expected to change.

It is recommended that, for the Directory Challenge, the principal instance of the common name be surname, given name. Other variations will be additional instances for the multivalued attribute.

**Certificate Authority.** A Certificate Authority (CA) using Nortel Entrust 2.1v3 is supposed to support cross-certification across other Entrust sites, but without interoperating with others. The secure messaging application is to establish the S/MIME demonstration. WEMA has taken a Privacy Enhanced Mail (PEM) approach, which opposes the Entrust approach. There is, consequently, concern that interoperability may not be achievable soon.

**Directory Management.** Optionally, DSAs will present RFC 1567 support for management consoles to show directory monitoring. Validating this capability is part of the testing checklist.

## SUMMARY

The X.500 Directory standard has no counterpart and no real competition within the proprietary directory arena. The pilots now under way or planned indicate that X.500 is gaining wide acceptance by disparate users of the Internet, X.400, and proprietary E-mail systems.

## THE ELECTRONIC MESSAGING ENVIRONMENT

The vision of an electronic, global directory system is becoming a reality. The 1993 extensions include finely tuned recommendations for the replication of the Directory and for access control — items missing in the 1988 version. The Directory and its models for information, organization, functionality, distribution, and security allow the user to reach anyone or anything listed in the Directory, from anywhere, at any time.

X.500, with its facilitation of communication, is allowing the information age to mature into the information-sharing age.

# Chapter 13
# Messaging Gateways
*Peter M. Beck*

Gateways are a necessary evil in today's world of diverse electronic messaging systems and networks. They serve the important function of linking together users of different environments which, without a gateway, would not be able to exchange messages effectively.

This chapter explores the many facets of gateway technology. The topic is timely and vast, but also somewhat confusing as a result of a lack of standardization, unique jargon, and a moving-target quality.

## ASSESSING THE MESSAGING ENVIRONMENT'S COMPLEXITY AND COSTS

At first glance, the concept of an electronic messaging gateway seems straightforward, particularly to the nontechnical observer. On closer examination, however, gateway issues can be quite complex, controversial, and even costly if mishandled.

The fact is that a very wide range of different messaging environments exist today inside and around the fringes of almost any large organization. These varieties of messaging can range from legacy mainframe systems to local area network (LAN)-based packages to extremely rich value-added networks (VANs) and the Internet. On the horizon in many organizations is new or increased use of groupware and a shift to client/server products for messaging. The protocols, addressing methods, data formats, and capabilities in each of these environments are very different.

In a perfect world, gateways would be unnecessary. Users would simply standardize on one messaging type and discontinue the use of everything else. Unfortunately, it is not quite that simple. An example helps illustrate the point (see Exhibit 13-1).

A large, multinational corporation uses several types of electronic messaging systems internally. The firm's U.S. headquarters is using IBM's Professional Office System (PROFS) product on a centralized mainframe. Alternatively, the corporation's engineering and production divisions recently deployed a large new network and began using Microsoft Mail. Because of an acquisition,

# THE ELECTRONIC MESSAGING ENVIRONMENT

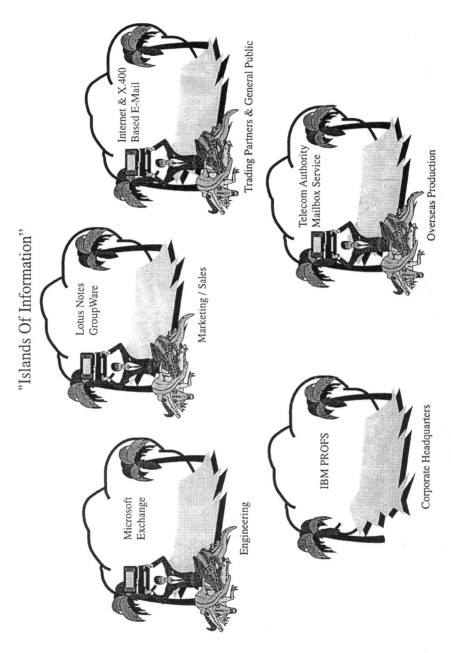

Exhibit 13-1. Multinational Corporation with Many Different Messaging Environments.

many factories have been added overseas in a country where the local communications authority provides an E-mail service.

Each of the groups is fairly satisfied as long as they stay on their island of information. Users cannot easily exchange messages with users in other divisions, however, because the systems essentially speak different languages. To make matters more complicated, employees must exchange messages with external trading partners and the general public by way of several different types of public networks, including the Internet. Finally, senior management would like to see fax, groupware, and possibly electronic commerce considered in any long-term solution.

Although this scenario sounds like enough to perplex even the most savvy of data communications experts, it is quite typical in today's fast-moving world. Several high-level questions come to mind immediately, including:

- How can gateway technology be used to link all of these messaging environments together?
- In what particular areas are careful analysis and planning especially important before starting a project?
- How much will this sort of connectivity cost to put in service, and how can these costs be recovered or justified?
- What level of expertise is required to put gateway technology to work successfully?
- What sorts of products, facilities or features, and services are available?
- What are some of the issues, pitfalls, and long-term hidden costs?

**BASIC GATEWAY FUNCTIONS**

Before embarking into the more detailed aspects of gateways, it is important to discuss functions in somewhat simple terms. In other words, what does a gateway system really do?

At the most basic level, a gateway must be able to act as a middle man, facilitating the exchange of messages between the two (or more) different environments it links together. As shown in Exhibit 13-2, each message received by the gateway from environment A must be evaluated in terms of its sender, format, content, and intended destinations.

Next, the gateway must perform a translation. This essentially renders the message suitable for the destination, shown as environment B. The degree to which the message remains intact as it traverses the gateway depends on many factors, most of which will be touched on later in this chapter.

Finally, the message must be sent by the gateway into environment B for subsequent delivery to the intended recipients. During message handling, several additional functions are often performed by a gateway. These

# THE ELECTRONIC MESSAGING ENVIRONMENT

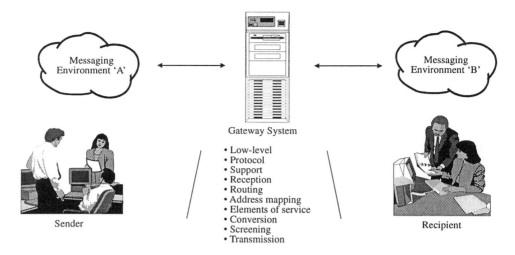

Exhibit 13-2. Basic Gateway Functions Linking Messaging Type A to Type B.

might include creating log entries for tracing purposes, archiving the message itself, and updating statistical-billing counts.

The process just described is, of course, oversimplified for the sake of illustration. Each electronic messaging environment has its own special characteristics that make it unique. Despite this, it is useful to break things down further into a set of generic functions that practically all gateway systems must provide to some degree. These functions include the following:

- Message reception processing
- Routing-address mapping
- Element of service handling
- Screening
- Message transmission processing

## Message Reception Processing

As with any technical issue, it helps to divide things into steps. The first step to consider for a gateway is the initial transfer of a message into the gateway system for processing. This is referred to as *reception*.

Externally, message reception usually works as follows: First, the sender submits a message addressed to a recipient in another environment. One of the components of the recipient's address, typically the post office or domain name, results in the message being forwarded into the gateway system for handling.

*Messaging Gateways*

The internal mechanisms actually required for reception might vary greatly depending on the situation at hand. It could range from relatively simple transfers of formatted ASCII text over a modem-telephone line connection to very complex communications protocol handshaking over a dedicated high-speed link.

When interfacing with a LAN-based messaging system, a gateway usually achieves *message exchange* by a combination of file server access and an applications programming interface (API). APIs are important for several reasons, which will be discussed later. In any case, as part of this step a gateway typically checks the message for errors, creates a log entry indicating the reception, and stores the message in preparation for the next step.

**Routing and Address Mapping**

Once a message has been successfully received, the next step is to determine where the message should be sent. This is referred to as *message routing*.

Routing can be accomplished in a variety of ways. Some low-cost gateway products do not really perform routing at all. They simply force all messages received from one environment into another environment serviced by the gateway. Other more flexible products use a combination of predefined routing rules or a directory to decide where to send a particular message.

The difference between routing and address mapping is subtle, and the two are often confused. Actually, they are two separate functions. Routing is the process of deciding where to send a message. Address mapping refers to the rendering of the recipient (and sender) addresses on transfer out of the gateway. Both are certainly among the most important functions a gateway provides.

In a simple example of address mapping in a gateway, to reach an external trading partner named John Smith (who works in the accounting department at a fictional company called ACME), a user of Lotus cc:Mail might send a message to the following address: JOHN SMITH at GATEWAY. When processing the message, the gateway system might need to render, or map, this address into the recipient's true Internet electronic mail address, which could be: SMITHJ@ACCOUNTING.ACME.COM.

Like message routing, address mapping functions can be accomplished in many different ways. These include the use of predefined rules, the use of a directory, or a combination of the two. When implemented correctly in a gateway product, the end results should always adhere to the following two basic rules: the sender should be able to easily address a message to a recipient in another environment using a familiar addressing method,

THE ELECTRONIC MESSAGING ENVIRONMENT

and the recipient of a message should be able to easily identify and reply to the sender.

**Element of Service Handling**

The term *element of service* is electronic messaging jargon for a feature offered to users. For example, the sender's ability to designate a message for high-priority delivery is an element of service.

As might be expected, it is very important for a gateway to handle each element of service in a consistent, meaningful fashion when transferring a message from one environment into another. Sometimes this involves a rather straightforward conversion or mapping; such is usually the case when handling message priority.

Often, however, the element of service is much more tricky. For example, some modern messaging environments allow users to request return of content in case an error condition prevents delivery to the intended recipient. This amounts to something like requesting a package be returned by the postal service when the mailman cannot deliver it. Electronically performing a conversion of this element of service in a gateway can be fairly involved. To illustrate, the following are some of the steps a gateway might perform:

- Recognize the reception of a non-delivery type notification message, containing returned content.
- Extract the returned content section from the notification.
- Perform a correlation to determine which recipient could not receive the message.
- Translate the often cryptic non-delivery reason code into a plain English text that the sender will hopefully understand (e.g., recipient's address incorrect).
- Render the returned content, recipient address, and the reason text into a notification message appropriate for the sender's environment.
- Transmit this notification back to the sender.

Element of service handling is where the bulk of all difficult message processing is performed in a gateway. It is therefore the one area where especially careful analysis and planning pay off, particularly when older legacy systems are involved.

Every important element of service offered in each environment linked by the gateway should be thought of in terms of how it will be handled. This is because there are often gaps or, even worse, subtle differences in the way elements of service work in one product or environment versus another. The return of content feature is a perfect example. Some products or environments support it wonderfully, but many do not.

## Screening

What happens if a user in one environment unknowingly attempts to send a very long message by way of a gateway to a user in another environment that only supports very short messages? Ideally, the message should be gracefully discarded by the destination environment and the sender should be notified. At worst, the message might cause a serious problem in the destination network. In any case, gateways are usually required to screen messages for conditions like this and take corrective action. The action taken depends on the type of error condition. For example, when detecting an overlong message, a gateway might prevent transfer or perhaps perform a splitting operation.

Other conditions gateways usually check for to prevent problems include: invalid content or data, illegal file attachment, too many recipients, or a missing or invalid sender's address. How well a gateway performs this type of screening often dictates how successful an implementation will be over the course of time. A poorly designed gateway can cause problems if it allows "bad" messages to enter into an environment not expecting to have to deal with these sorts of error conditions.

## Transmission Processing

The final step in handling any message is its transfer into the recipient's system or network, referred to as *message transmission processing*. Much like reception processing, the actual functions required can vary greatly. Most important is that the gateway adheres to the rules of the destination environment and provides for safe, reliable transfer of the message. On completion of this step, the gateway has fulfilled its main responsibility for the message.

## GATEWAY COMPONENTS

Now that basic functions have been examined, it is appropriate to take a look at the main components that comprise a gateway system. Without focusing on a single product or scenario, this discussion must be left at a high level — but it is interesting nevertheless. The typical gateway system can be broken down into both hardware and software components. Hardware is more concrete, so it is an appropriate place to start.

## Hardware Components

Generally, the minimum hardware needed for a gateway is a computer, generous amounts of main memory, and some sort of mass storage device (e.g., a hard disk). This might range from a PC with a small hard drive to a multiprocessor server with huge amounts of storage. The exact system required will depend on such factors as the operating system selected, throughput desired, fault tolerance needs, communications protocols used, and message archiving and logging requirements.

It is worth mentioning that many gateway products are what is termed *hardware model independent.* In other words, the vendor really supplies only software. If this is the case, the vendor usually indicates minimum hardware requirements and leaves it up to the person deploying the gateway to decide on details (e.g., what exact model of computer to buy). Beyond the basic computer and its storage device, hardware components are very specific to the situation. Perhaps the most frequently used is a network interface card (NIC). The two basic types of cards are Ethernet and token ring.

When a gateway must be connected to a public network (e.g., the Internet or a VAN), it is common to have a communications hardware device in the configuration. This might take the form of an X.25 adaptor card, a multiport asynchronous controller board, or modems. High-end gateway systems often use additional equipment (e.g., a CD unit or magnetic tape drive). A CD unit is essential for loading and installing software components. A tape drive is often useful for backing up the system hard drive, archiving messages, and saving billing and statistical information.

### Software Components

Gateway software components fall into one of the following three categories: operating system software, gateway applications software, and third-party support software.

**Operating System Software.** The operating system is the fundamental, general-purpose software running on the computer. It allows reading and writing to a disk, allocating memory resources, and a user interface mechanism. Examples of operating systems software supported by gateway product suppliers include UNIX, Novell NetWare, Windows NT, and MS/DOS.

**Applications Software.** Applications software is the set of programs that perform the actual message handling in the gateway. This might be as simple as one single program/file or as complex as a whole directory structure filled with dozens of software modules, each performing a special function.

**Third-Party Software.** Typically a certain number of third-party software components are necessary in a gateway system. Examples of this are device drivers, communications protocol stack components, and tape archive and backup packages.

## GATEWAY CONFIGURATION MODELS

Individual hardware and software components must be put together in a certain way to make things work. This is referred to as a *configuration.*

*Messaging Gateways*

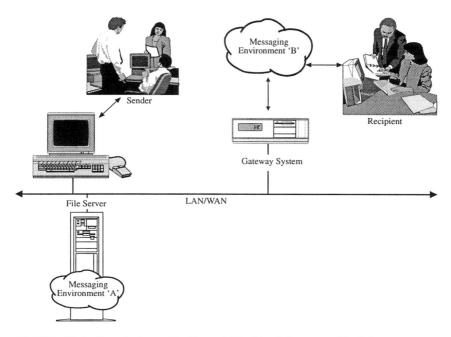

**Exhibit 13-3. Stand-alone (Dedicated Machine) Gateway Model.**

Gateway configuration models fall into three categories: stand-alone or dedicated, colocated, and distributed processing.

**Stand-Alone or Dedicated Model**

The stand-alone configuration model (see Exhibit 13-3) is typical in today's environments. For this, a separate computer is dedicated to performing the gateway functions. All hardware and software components reside in this type of single box system. In some sense, this is the easiest gateway configuration to understand and explain because messages enter the box in one form and come out in another.

This model has several other advantages. First, all or most of the computer's resources can be allocated to the job at hand. Second, very little — if any — change is required in the environments being linked together by the gateway. When considering this model, one important question to ask is, how will the gateway system be managed? This is particularly true if a number of gateway systems will be deployed across a wide geographical area.

**Colocated Model**

Colocating the gateway components in the same main computer system where one or more of the electronic messaging environments to be linked

## THE ELECTRONIC MESSAGING ENVIRONMENT

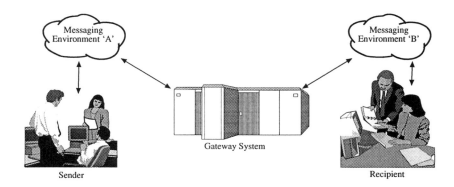

**Exhibit 13-4. Co-located Gateway Model.**

resides is also possible (see Exhibit 13-4). The disadvantage with this model is that gateway functions, which are often resource-intensive, will steal power from the main computer system. In addition, it may be necessary to add hardware and software components into a system that has been in operation and running unchanged for quite some time — something generally frowned upon.

Sometimes management advantages are associated with using this model. Often an already familiar set of commands, reports, and alarms can be used to manage the gateway.

### Distributed Processing Model

As shown in Exhibit 13-5, one of the newer and more interesting configurations now available from some product suppliers actually allows gateway functions to be distributed across multiple computers. In this model, clever software design allows for a high-speed exchange of information between dispersed gateway components. Typically, a file server acts as a central, common repository for information. This effectively forms one logical gateway system out of pieces running on several computers.

With the availability of hardware as a relatively cheap commodity (due to reduced hardware prices) today, this configuration model is worth examining. For large-scale gateways, it offers advantages in the area of scalability, load sharing, and fault tolerance.

## GATEWAY SCENARIOS

This section offers a brief summary of three fictional but realistic scenarios in which gateway technology is used. Most actual situations will match, or at least have much in common with, one of these scenarios.

*Messaging Gateways*

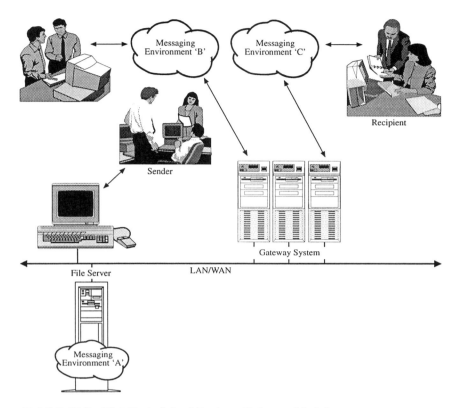

Exhibit 13-5. Distributed Architecture Gateway Model.

## Scenario One: Linking LAN Messaging to the Outside World through a Network Service Provider

This scenario is very typical today for a large organization that has deployed any one of the very popular and powerful LAN-based E-mail packages available (e.g., Lotus cc:Mail, Microsoft Exchange, or a Novell MHS-compliant product). Inevitably, users start to ask how they can exchange messages with external trading partners. Some discover on their own that if they use a somewhat crazed mix of packages, network access subscriptions and a modem, they can communicate individually. But this gets messy, expensive, hard to control, and is a bad long-term solution for any organization of size.

The best answer for business-quality messaging is to deploy what is known as an X.400 gateway (see Exhibit 13-6). This essentially links the LAN-based E-mail environment to the outside world via a network service

# THE ELECTRONIC MESSAGING ENVIRONMENT

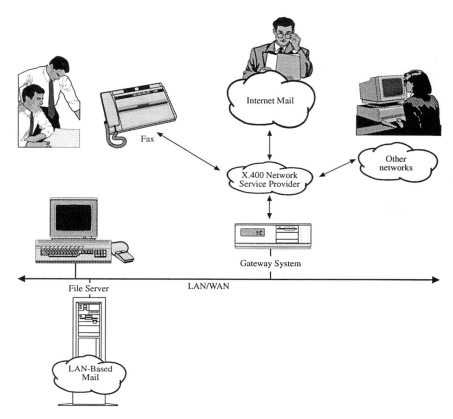

Exhibit 13-6. Scenario One: X.400 Gateway Used to Communicate with the Outside World.

---

provider. In the U.S., the more well-known service providers include MCI, AT&T, Sprint, GE Information Services, and IBM Advantis.

**X.400 Implementations.** Gateways can exchange messages with the network service provider by way of X.400. This is a very flexible, rich, and open set of standards or protocols for messaging. Lower-level communication protocols (e.g., X.25, TCP/IP and OSI transport/session) are used to ensure the reliable exchange of data.

Also important is that in this configuration the gateway forms what in X.400 terminology is referred to as a *private management domain* (PRMD) for the organization. To send a message to anyone in an organization, external trading partners must specify this PRMD by name in the address. Many network service providers offer enormous capabilities, including international access points, links to other carriers, Internet mail access, EDI capabilities, fax delivery, and detailed billing. In other words, they provide immediate connectivity and huge potential. Of course, this is not free. There is normally a charge for the communications link and for each message sent through or stored on the service.

**Benefits to Users and Administrators.** So what does having a setup like this mean to users in the organization? Now they can exchange messages with external trading partners by way of their normal desktop E-mail application. They only have one mailbox to deal with and do not have to be trained to use several packages or understand different addressing schemes. The payback in large organizations can be enormous.

What are some of the administrative advantages when using a gateway in this scenario? First, all messages exchanged with external trading partners can be logged and archived in one central point — the gateway. This is much neater, easier to manage, and more economical to the organization than allowing users to send, receive, and store correspondences in a chaotic fashion.

Second, the network service provider and gateway to a certain degree act as a security firewall between the outside world and the organization's private network.

Finally, it can be said that these companies are professionally staffed, 24-hour operations that will lend a hand with some of the issues pertaining to providing business-quality communications.

[*Author's Note:* An increasingly popular and in most cases more economical solution is to deploy an Internet mail gateway. Again, this can link your local environment to the outside world via an Internet service provider (ISP) such as UUNET, AOL, or Sprint. Direct Internet gateways present additional issues such as security, reliability of the Internet as a transport network, and, as of this writing, the lack of certain value-add features such as delivery confirmation. For example, it is usually necessary to deploy a separate firewall system to protect your local network from unauthorized access. However, because of the rapid growth of Internet E-mail as the de facto standard for messaging, a direct Internet mail gateway should be part of any large organization's strategy.]

## Scenario Two: Tying Together Legacy and Modern Messaging Environments

This scenario is also quite common as an attempt is made to tie a legacy messaging environment together with one of the newer messaging environments by gateway technology. Often a very large investment in money, personnel, and political chips has been made in the legacy network. Examples include messaging running on older mainframe hosts, specialized networks for civil aviation and banking, as well as government or military systems.

# THE ELECTRONIC MESSAGING ENVIRONMENT

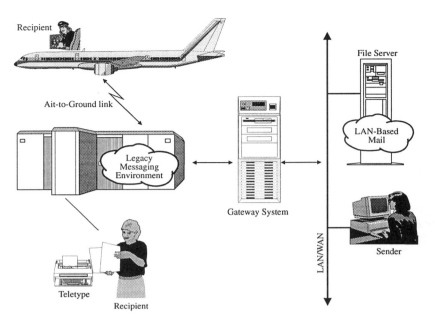

**Exhibit 13-7. Scenario Two: Tying Legacy and Modern Messaging Environments Together.**

---

Sometimes gateways are required for a very long transition period as the legacy environment is phased out. Other times, the investment in the legacy network is so large that there is no end in sight. The gateway becomes a permanent fixture, allowing for a modern user interface to an older or very specialized environment (see Exhibit 13-7). Gateway issues are sometimes difficult in this type of scenario and often require some degree of customization. This is because of the large difference in capabilities offered in the legacy environment versus the modern environment.

The legacy environment may have specialized features that do not convert unambiguously into an element of service used in the new environment. On occasion, it is even a challenge to determine enough about how a long-ago-deployed environment functions to effectively set up a gateway. Address space restrictions can also be an issue in this scenario.

For example, many older legacy networks use a simple seven- or eight-character addressing format. Given a fixed number of possible character combinations, it simply is not possible to allocate a separate address in the legacy network for each potential recipient when linking to a huge new environment (e.g., Internet mail). In cases like this, it is necessary to work with a gateway product provider to make special provisions.

*Messaging Gateways*

### Scenario Three: The Electronic Mail Hub

This scenario, in which several different noncompatible packages are being used inside a single large organization, exists frequently. It is usually the result of a merger or acquisition, a high degree of decentralization, business-unit autonomy, or just poor planning. In addition, during the past several years some organizations have stumbled into this scenario as a result of deploying groupware at the departmental or division level and beginning a slow shift toward using its built-in E-mail capabilities for forms, workflow, and electronic commerce.

Whatever the cause, multiple gateways are sometimes configured together into a master, hub-like system (see Exhibit 13-8) used to link all of the environments together. If this approach is used, gateway issues take on a new dimension of importance and complexity. The master system becomes mission-critical. For example, should this single central system break down entirely, all of the messaging environments are disconnected from each other. In many situations, a failure like this can be very costly to an organization.

A master, hub-like system can also be far more complex to run than a simple two-way gateway. It is not uncommon for administrators to request

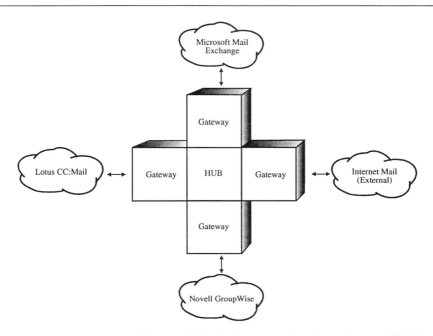

**Exhibit 13-8. Scenario Three: Multiple Gateways Configured as an E-Mail Hub.**

227

THE ELECTRONIC MESSAGING ENVIRONMENT

the use of an outside professional service organization to perform the initial design, installation, and testing of such a system.

## GATEWAY FACILITIES AND FEATURES

This is where gateway products differentiate themselves. Low-end products typically offer a minimum set of facilities and features — just enough to get the job done. On the other hand, more expensive high-end products deployed by large organizations generally supply a comprehensive set of facilities and features. These usually allow for a much better level of service. The facilities and features can be broken down into the following two categories: message handling-related facilities and features, and management or administration facilities and features.

### Message Handling Facilities and Features

Message handling-related facilities and features dictate how well a particular gateway performs its main job of reliably moving messages from one environment to another. Although usually not directly visible to users, they are the "meat and potatoes" of any gateway. What follows here is a short definition of some of the more important messaging handling facilities and features. Administrators should look for these in any products being considered and discuss them with gateway vendors/integrators.

**Support for Lower-Level Communications Protocol Layers.** Much like a river with barges floating downstream, a lower-level communications protocol provides a means of reliably transporting information, such as messages, from point A to point B. Examples include Ethernet, TCP/IP, X.25, IPX/SPX, Appletalk, and OSI Transport/Session.

Even today, the world of lower-level protocols can sometimes be complex. With regard to gateways, message reception and transmission functions essentially ride on top of these types of transport protocols. In other words, these functions cannot work without lower-level connectivity. This means a gateway product must include top-quality support for the particular flavors being used by the environments to be linked together.

**Prioritized Message Handling.** This is an important feature, especially in environments where messages contain extremely vital, time-sensitive information. Ideally, a gateway should handle messages according to a first-in, first-out (FIFO) order within priority grade. This means, for example, that a message marked as high priority will be forwarded by the gateway before a message marked as low priority.

There is a big difference between a gateway product that simply converts the priority field from one format to another versus one that actually handles messages internally based on priority. In the case of the former, an administrator might have to explain to the CEO why the junk E-mail got through the network before his or her urgent message to all vice-presidents.

**Delivery Notifications.** During the past decade, all sorts of mechanisms have sprung up in an effort to make electronic messaging more reliable and generally useful for commerce. One such mechanism is delivery notifications.

Essentially, this is an element of service that allows the sender to request a confirmation be returned by the destination system or network indicating successful delivery of a message. The delivery notification is the equivalent of requesting that the postal service return a receipt when a package is delivered by a mailman.

It is extremely important that a gateway be able to handle the delivery notification element of service. This usually refers to several capabilities, including: forwarding a sender's request for delivery notification into the destination network (typically as part of the message envelope or header), processing or translating any delivery notifications returned by the destination system or network and relaying information back to the sender, and issuing delivery notifications when appropriate.

*A word of caution about notifications*: Not all gateways handle this element of service correctly or consistently. Some gateways issue a delivery notification immediately on successful reception of a message. This is premature, and can be very misleading to the sender. Also, some gateways issue a delivery notification on transferring a message into a legacy or proprietary network; again, potentially misleading to the sender. Administrators should be aware of these issues and, at a minimum, alert users to the fact that there may be inconsistencies.

**Directory Facility.** It is fair to say that no two system or network managers ever agreed on the best way to handle electronic messaging directories. This holds especially true in the case of gateways. In any event, many gateway systems use some type of local directory facility. This directory is typically used for routing messages, mapping addresses, and sometimes even to define lower-level communications protocol-related information.

To save work for administrators and make things consistent, gateway product providers have come up with some ingenious ways to automatically populate and maintain a gateway's local directory. These include features known as directory synchronization and auto-registration.

Arguably, the best place to not have a local directory facility is within a gateway. This is because of management concerns. A large, enterprisewide messaging network with dozens of different gateways — each with its own local directory to populate, maintain, synchronize, and backup — is easy to imagine. Even with some degree of automation, this can be a nightmare.

The use of a local directory facility can be one of the first difficult issues encountered during a gateway project. Methods and technologies are available to minimize the use of local directories in gateways; however, they require the right planning, product, and people factors.

**Automatic Registration.** Automatic registration is a useful feature in some situations. To avoid the administrative work of what otherwise might amount to the defining of thousands of entries in a gateway local directory, a clever technique is used. Users essentially define themselves in the directory by way of messaging.

When a user for the first time sends a message through a gateway, his or her address is automatically registered in the directory. On responding, the recipient's address is registered. Slowly but surely, the gateway's local directory is populated as needed, allowing users to communicate easily and take advantage of advanced messaging features (e.g., replying and forwarding).

**Directory Synchronization.** When tying different messaging environments together with gateway technology, the consistency of directories is almost always a challenge. Each time a user is added or deleted from one environment, the user should also be added or deleted in a gateway's local directory and in the directory of the other environments as well. This process is called *directory synchronization.*

Most gateway product providers either include automatic directory synchronization as a feature or sell it as a separate add-on facility. Many organizations (including the U.S. Department of Defense) are looking to X.500 technology to help solve directory- and security-related issues. X.500 is actually a set of open standards that provide for defining, distributing, and accessing directory information. Many experts feel that using an X.500-based directory, in combination with synchronization tools, is an effective long-term strategy when linking diverse messaging environments.

**Wild Addressing.** Wild addressing is sometimes referred to as *in-line* or *native addressing.* It is a feature that allows users to easily send a message to an external trading partner using a public electronic mail address specified in its original format (e.g., taken right from a business card).

For wild addressing to work, the gateway must be capable of accepting and using a public electronic mail address encapsulated within a normal environment-specific recipient address. The intended recipient does not have to be registered in the gateway's directory, hence the name wild addressing.

Wild addressing is most valuable in situations where much ad hoc communications with external trading partners is required and it is impractical or cumbersome to register them in the gateway's directory. Because this is an increasingly important feature to look for in a gateway, an example is ap-

*Messaging Gateways*

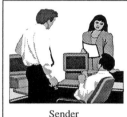

**Exhibit 13-9. Sender Uses Public E-Mail Address Taken from Business Card (via Wild Addressing).**

propriate. Exhibit 13-9 represents a typical business card for Bob Jones, a manager of a toy store chain. Bob Jones's public E-mail address is shown as: BOB_JONES@TOYSRGOOD.COM.

The president of another company, Sally Roberts, wishes to send Bob a thank-you message for the order he placed for 30,000 model airplanes. For this example, it is assumed that Sally's firm uses Lotus cc:Mail on a LAN and has an Internet mail type gateway. At the cc:Mail TO: prompt, Sally might look at Bob's business card and set the address for him as follows: BOB_JONES@TOYSRGOOD.COM at IGATEWAY.

The gateway would receive the message from the cc:Mail network as a result of Sally specifying "at IGATEWAY." Next, the gateway would perform the necessary message translation into Internet format. It would include the recipient address exactly as entered by Sally, but with the "at IGATEWAY" stripped off. The gateway in this case would not require a directory entry be defined for Bob. Nor would it perform any validation (other than perhaps basic syntax checks) on the address.

## THE ELECTRONIC MESSAGING ENVIRONMENT

**Error Handling and Message Repair.** What happens when a message is received by a gateway system with an error? Perhaps the address of the recipient is misspelled or the message is not in the expected format.

For cases like this, it is quite important that the gateway recognize the error and handle it gracefully. Sometimes it is appropriate for the gateway to automatically issue a negative notification response back to the sender. In many circumstances, however, this is not possible or not desired. The message might be from a far-away international location and contain vital, time-sensitive information. For this, a repair-correction facility in the gateway is very useful. It allows the administrator to view the message, correct it, and resubmit it for delivery.

**Transmission Retry.** Every network administrator understands Murphy's law and plans accordingly. The administrator tries to shield users from equipment failure, communications line outages, and software bugs.

Transmission retry is a gateway feature network administrators will therefore appreciate. It provides for the automatic retry by the gateway of a failed transmission attempt into the destination network. An additional capability allows the administrator to adjust the maximum number of retries, and the wait period between retries should be included as part of this feature.

**File Attachment Support.** Most newer messaging environments allow users to attach a file (e.g., a word processing document) to a message being sent. Some of the more sophisticated packages even allow for the launching of a particular application (e.g., a word processor) to view or edit a file attachment. It is therefore quite reasonable for users to expect a gateway to allow for the transfer of a message containing a file attachment from one environment into another. In fact, most gateways do allow for this.

Unfortunately, however, a few interesting pitfalls must be considered when it comes to file attachments. For example, one of the environments being linked together by a gateway may not even support file attachments. In this case, the gateway should detect any message containing an attachment, prevent its transfer into the destination network, and inform the sender.

Another fact of life is that there is often a loss of functional ability when a message traverses a gateway. For example, the file name, creation date, and access privileges are usually dropped as the message passes through a gateway into another environment. This situation has attracted much attention in recent years and may be improved in next-generation standards and products. For now, the bottom line is to use caution when planning to mix file attachments and gateways.

**Applications Programming Interface (API) Support.** An API is actually a standardized mechanism gateway that software developers use to interface with a compliant messaging environment. Examples of well-known APIs include vendor-independent messaging (VIM), the messaging applications programming interface (MAPI), common messaging calls (CMCs), and the X.400 applications programming interface (XAPI).

What do all these names and acronyms really mean in the context of gateways? First, if a gateway software product uses one of these standard mechanisms to access a messaging environment it services (versus a homemade interface), it helps ensure future compatibility. If the environment changes (e.g., a version upgrade), the gateway itself will be minimally affected. This can be very important in today's world of constant new product releases.

For example, gateway software written to use VIM for interfacing with Lotus Notes Mail generally will not be affected when a new version of Notes is released. The second reason support for an API is desirable is that larger organizations with special needs sometimes must customize or enhance a gateway product themselves. This can be much easier if a standard API has been used versus another access method.

**Management and Administration Facilities and Features**

These facilities and features allow the administrator of a gateway system to ensure a high quality of service to users. Over the long haul they are invaluable, especially in high-traffic volume situations and when message accountability is of the utmost importance.

While on the topic of management, it should be noted that this is also an area of much activity in the messaging industry. As of this writing, product vendors, network service providers, and users are working together to formulate standards for better messaging management. These efforts promise to bring about increased compatibility across different hardware and software platforms, and are certainly worth watching in the context of gateways.

**Alarms and Reports.** Many gateway products have become sophisticated enough to automatically detect problems and alert an administrator before users are affected. This might take the form of a visual/audible alarm or perhaps a report printout. In any case, the ability to detect events such as an excessive backlog of messages awaiting transfer, a failed software component, or a communications link outage is very important.

**Billing and Statistics.** A billing or statistical facility is essential in most large-scale settings. Regardless of how well a gateway is functioning and how happy users are, the issue of cost justification and recovery usually arises. In some organizations, a so-called chargeback policy that allows the cost of a gateway service to be paid for by the specific departments most

## THE ELECTRONIC MESSAGING ENVIRONMENT

frequently using the service is put in place. For this, very detailed usage, information (often in the form of counts) must be generated by the facility on a weekly or monthly basis.

Also, if the gateway is used as a link to a network provider, billing and statistical information can be used to prove that overcharging is not taking place. Finally, should there be a need to expand or enhance the gateway service, the information provided by a billing or statistical facility can help justify the cost. It is always easier to advocate something with the numbers to back it up in hand.

**Message Tracing.** Message tracing is a feature that offers the administrator the ability to determine exactly when a message was received and transmitted by a gateway. This feature is usually accomplished via the automatic generation of some sort of log entries (usually written to disk). Often, the log entries contain detailed information (e.g., the sender, the recipients, the time of reception, the time of transmission, and the message priority level).

This is a feature found in all better gateway products. It is certainly a requirement for any project involving business-quality communications.

**Message Archiving.** As its name indicates, archiving refers to the short- or long-term storage of messages handled by the gateway. Short-term storage is usually on disk; long-term storage is typically performed using tape.

Archiving can be considered a high-end facility, but not necessary in all gateway situations. This is because archiving often takes place outside the gateway in the environments being linked together.

Despite this, there are instances when this feature proves useful. For example, in a gateway used to link a private network to a public electronic messaging environment, archiving might be used to record all messages sent outside the organization.

**Message Retrieval or Resend.** This feature provides the retrieval and viewing of selected messages by the administrator from an archive. An administrator can typically pick a certain message or range of messages to be retrieved from the archive. This is done by specifying a date-time range, a particular sender, or a particular recipient.

Associated with retrieval is the ability of the administrator to actually resend a message. This is useful in environments where messages are sometimes garbled or lost on their way to an endpoint.

**Alternate Routing.** Once found only in very large-scale backbone message switching systems, alternate routing is a feature now finding its way into some gateway products. It allows the administrator to redirect messages from one destination to another during vacation periods, equipment outages, office closures, or temporary personnel shifts. Some very high-end gateway products even offer an automatic alternate routing feature, which is activated without any manual intervention when there is some sort of component failure.

**Hold and Release.** This feature is simply a way in which the administrator can instruct the gateway to temporarily stop sending messages to a particular destination environment. Messages queue up in the gateway until the administrator issues a release command.

**Remote Management.** This facility allows the administrator to manage a single gateway system or perhaps a whole network of gateway systems from a single location. In its simplest form, this feature is sometimes offered by a so-called remote control software package. This essentially allows an administrator to take control of a system from a remote location by a telephone line or a network connection. Video display, mouse, and keyboard information is exchanged across the connection, giving the effect that the administrator is actually on-site. Almost all of the features normally provided locally can then be used remotely.

More sophisticated (and expensive) management solutions are now becoming available for gateways. These go beyond simple remote control in that they provide a single console facility for monitoring, controlling, and troubleshooting multiple gateways dispersed across a network. Graphical displays, showing overall status as well as actual or potential trouble spots, are often included in such packages.

**Security.** One item sometimes overlooked by those tasked with putting an electronic messaging gateway in service is security. Perhaps this is because a messaging-only gateway, unlike a general-purpose access unit allowing file transfers and real-time login sessions, is viewed as somewhat less dangerous.

Many issues must be considered when deploying a gateway, particularly if one (or more) of the environments being linked together is considered secure. Should a particular sender be allowed to access the destination network by way of the gateway? Are messages containing file attachments screened for viruses as they transit the gateway? Is there password protection when connecting to a public network? These are all important questions that must be asked ahead of time.

## QUALITIES TO LOOK FOR IN A GATEWAY PRODUCT

Everyone, from managers to administrators to users, desires quality, especially in something that is eventually used by the entire organization.

# THE ELECTRONIC MESSAGING ENVIRONMENT

What should administrators look for when deciding on a gateway product? The answer, of course, varies depending on the type of organization, messaging environments to be linked together, and the budget available.

Certain points should be ensured before even beginning. The hardware and operating system platforms should be reliable, up-to-date, well-supported by vendors and third-party suppliers and, if possible, fit the organization's overall information technology strategy. Beyond these basics, many additional gateway-specific qualities must be checked.

### Ease of Installation and Configuration

Nothing is more frustrating than starting a project with a product that is difficult to install and configure. Gateways by their nature are not simple desktop applications — they are data communications systems, and often complex ones at that. Some vendors have done a better job than others at designing and documenting installation and configuration procedures. This is usually a positive sign when considering a product.

### Documentation

Administration manuals, utilities reference guides, cookbook-type procedures, and even release notes fall into the category of documentation. Administrators should ask to see these ahead of time to avoid any unpleasant surprises during the first month of a project.

### Reliability

Even if the installation goes well and the documentation is perfect, over the long term perhaps nothing matters as much as the reliability of the system. Design problems, not-ready-for-prime-time early product releases and nasty software bugs can be the downfall of even a carefully planned endeavor.

Administrators should check with other users of the gateway product. If not many can be found or they will not discuss their experiences, it is a bad sign. Some product suppliers can be enticed to actually help set up an evaluation system at little or no cost to allow an organization to test drive a gateway.

### Flexibility

Administrators should look for a gateway product that offers maximum flexibility in terms of its message handling and management facilities. After the gateway is in service, it should be possible to add an interface to another messaging environment, perform version upgrades, and perform maintenance easily. Is it possible to do these things while the system is running or is a complete shutdown required?

## Throughput

Some gateway systems perform incredibly poorly in this category. Message throughput rates of one message per minute are not unheard of when complex translations, protocols, and routing decisions are necessary.

In low-end situations, this may not be an issue worth spending much time and money on, but in many cases it is. Unfortunately, vendors are traditionally averse to handing out performance statistics because it involves them in a numbers game that is difficult to win. Even when they do, the numbers should be taken with a grain of salt and should be proven firsthand if possible.

## Scalabilty and Ease of Expansion

It is rare for a single, standard issue gateway system to be deployed across an entire large organization. In one location, only a small-capacity gateway might be needed. In another location, a high-capacity gateway might be necessary. If the same basic gateway product can be used in both instances, perhaps adding more powerful software or hardware components where they are needed, the gateway can be considered scaleable and easy to expand.

## Fault Tolerance

In some messaging environments, any downtime is totally unacceptable. Such is the case when messaging is used to facilitate point-of-sale transactions, air transportation, or military operations. For networks like this, gateways must be just as bulletproof as all of the other components in the network. This is referred to as fault tolerance.

Solutions that provide hardware fault tolerance include specially designed casings, redundant storage devices, and even the use of a complete backup (hot-standby) system. On the software side, some gateways include a mechanism designed to restart components automatically in the event of a failure. Features like this are usually rather expensive, but can be vital.

## Value

Getting the most for the organization's money is always important. In general, a low- to medium-end gateway software product used to link two commonly found environments (e.g., Microsoft Mail and Internet mail) can be obtained for $5000 to $10,000. This price can be deceiving, however. In larger organizations, the true cost of a project is usually in manpower (e.g., analysis, installation, configuration, testing, trial service, and support).

If the situation calling for a gateway requires customization, fault tolerance, high throughput, or special message accountability-archiving, de-

## THE ELECTRONIC MESSAGING ENVIRONMENT

ploying a gateway may amount to a much more significant investment. High-end products start at approximately $40,000 for the basic software license, with hardware and software options often increasing the cost well beyond this figure. The payback point in terms of increased productivity can be difficult to measure and prove. Most experts believe that in a large organization, where E-mail has become business critical, it is reached quite quickly.

### STEPS TO DEPLOYING A GATEWAY

The following sections detail the actual steps involved in selecting, deploying, and running an electronic messaging gateway.

#### Up-Front Analysis and Planning

During this step, which might last anywhere from several days to several months, the functional requirements for the gateway are considered, confirmed, and documented.

In a small-scale, low-budget scenario, the resulting document might amount to just a few pages of bulleted requirements. For a very large organization with complex issues, this document might be quite long. In this case, it usually serves as part of a request for proposal (RFP) sent to gateway product vendors.

The rule of thumb is that the more precise the functional specification document, the better a vendor will understand the organization's true needs and, hence, the less chance of any misunderstandings.

#### Picking a Vendor or Product

In today's business climate, giving advice on how to pick a vendor is difficult. Some household names in the computing industry seem to be doing poorly these days, while new, upstart firms blossom almost overnight (and sometimes disappear just as quickly).

It is important to know that most of the larger, more well-known messaging product vendors have in fact formed relationships with smaller firms to supply gateway technology. This is a win-win situation for the partnering firms, with the larger company rounding out its offerings and the smaller company receiving royalties. At least in theory, this makes it less risky for the buyer because the large company stands behind the gateway product.

One idea when trying to select a gateway vendor is to check the Electronic Messaging Association's (EMA) Products Service Guide. Published annually, it lists dozens of gateway providers. A few of the larger, more well-known names include Control Data Corp., Digital Equipment Corp., Infonet, Lotus, Microsoft, Novell, and Worldtalk Corp.

*Messaging Gateways*

Many smaller, less-known companies have good products. When considering these vendors, regular attendance and participation in organizations (e.g., the EMA) usually indicate a healthy company with a commitment to the messaging industry as well as a desire to listen to the needs of the user community.

**Do-It-Yourself vs. Outsourcing**

Installing, configuring, and putting in service a business-quality gateway system is often a task an organization can handle on its own. There are times to call in an expert, however, especially if the project involves customization, a legacy system no one on staff understands, links to a public network, unusual time constraints, or complicated lower-level protocol issues.

Often, even large organizations with big IS departments will rely on the product vendor, an expert consulting firm-integrator, or a network provider to supply at least a few months of planning, training, and support. Some large organizations have even gone the route of completely outsourcing the selection, deployment, administration, and maintenance of their entire messaging infrastructure, including gateways. For those wanting to focus on their core business, such companies as CSC, Control Data, Infonet, EDS, Lockheed Martin Federal Systems, and a host of smaller firms are more than willing to give birth to, rear, and even baby-sit an organization's messaging environment — for a hefty price.

**Initial Testing**

Once the gateway is installed and configured, a period of initial testing begins. During this period, the administrator checks to make sure that all the basics work well. Messages and notifications (if applicable) are exchanged, facilities and features are tried, and overall reliability is confirmed. This is an appropriate time to torture-test the gateway. Examples of "tests" to try include large messages, messages with multiple recipients, messages with file attachments, and high-priority messages. Not to be forgotten is a duration or high-load test.

Larger organizations will go so far as to write an acceptance test plan (ATP). Often a payment schedule milestone is associated with successful completion of this ATP. Then, if everything seems acceptable, it is time to begin trial operation.

**Trial Service and User Feedback**

No matter how much testing has been done, trial service is always interesting. This is a chance to see how well the gateway performs in a live environment with a small, select group of users exchanging messages by way of the gateway.

## THE ELECTRONIC MESSAGING ENVIRONMENT

Before the trial service begins, administrators must make sure of a few things. First, users should be given some advance notice, explaining why the gateway is being put in place and how it will affect them. Second, all of the parties involved (technical and nontechnical) must be sure of the role they are to play. Gateways, because of their nature of tying together different messaging turfs, are often a "political hot potato" in an organization. One single person or group must clearly be in charge of administering the gateway during the trial operation.

Another thing to carefully consider is the selection of users. It is helpful to find a small group of users who meet the following criteria:

- Users who have an interest in seeing the gateway project succeed (perhaps the department needing to exchange messages with an important external trading partner)
- Users who are somewhat computer literate
- Users who seem willing (and have the time) to offer constructive feedback

A department under unusual workload pressure or a group of users who will in any way feel threatened by the new gateway capabilities should be avoided.

### Live Service and Maintenance

Assuming that the trial service went well, the gateway is ready to be cut-over into full-scale operation. During this step, message traffic levels will typically escalate steadily until all users are aware of the gateway and are actively using it to communicate. It is during this step that administrators must reckon with the cost of providing help to users and quickly troubleshoot any problems.

For example, one or two users might discover that the gateway allows them to transport huge amounts of information to a remote database system by way of messaging — something they always wanted to do but found difficult. This may be fine, and in fact help prove the gateway useful, but perhaps it will have to be restricted to off-peak hours.

Maintenance issues become very important at this point. These include regularly scheduled hardware housekeeping and careful back-up of software, configuration information, and perhaps directory facility files. In some cases, messages must be archived to tape on a periodic basis.

### SUMMARY

Gateways are complicated beasts. In the likely continued absence of widespread standardization, however, there is often no choice but to use them to link different messaging environments together. The important steps to go through during a gateway project include an analysis phase, product selection, installation and configuration, initial testing, trial operation, and live service.

During the analysis phase, many important issues must be considered. Depending on the particular situation, these might include hardware-software choices, configurations, message routing, address mapping, directory use, element of service handling, lower-level communications protocols, and systems management.

When examining different products, certain qualities are desirable. Reliability, high throughput rates, flexibility, fault tolerance, scalability, and ease of administration fall into this category.

Gateways come in various grades, from low-end products that just get the job done to high-end systems with advanced facilities and features. Keeping in mind that the initial cost of a product can range from a few thousand to a few hundred thousand dollars, administrators should strive for the best range of features and facilities possible within their budget.

The key to success with gateways is a common-sense recipe. Ingredients include effective planning, selecting the right products, successfully dealing with integrators and network service providers, as well as a healthy dash of sheer hard work. When these are mixed together correctly, electronic messaging gateway technology can be useful and rewarding to an organization.

# Chapter 14
# X.400 vs. SMTP
*Gordon L. Preston*

X.400 or SMTP/MIME? In recent years, this question has been the subject of much canonical debate in the messaging community; comparisons of their functionality have, unfortunately, at times overshadowed more important issues of security, systems management, message management, and performance.

Significant improvements in network management tools, utilities, and techniques are essential before a large, integrated network can begin cost-effective operations. The lack of adequate management tools at the application layer is a major impediment to implementing economical enterprisewide messaging systems. Operating such networks requires highly skilled software engineers who understand large-scale global enterprise networks. No matter what your choice in E-mail protocol, the pros and cons of each technology should be evaluated in the context of the real underlying issue — keeping the network running.

This chapter covers the issues of operating a reliable messaging environment using either X.400 or SMTP/MIME as the protocol to transfer E-mail. A brief look at the features, philosophy, and development process for each protocol is included to help readers understand the differences and similarities between X.400 and SMTP.

## BACKGROUND OF X.400 AND SMPT

X.400 is based on a formal messaging model created in standardization groups in the International Telecommunications Union-Telecommunications Standards Sector (ITU-TSS). It is the international standard for message handling. It is a full-featured, store-and-forward message-handling system designed to process multimedia and complex business documents.

In particular, X.400's specification of robust message delivery and non-delivery notification schemes makes it well suited to support electronic commerce transactions. X.400 is a commercially viable and secure message-handling technology supported by a worldwide infrastructure and officially sanctioned by various governments, telecommunication vendors, and public service providers. In addition, X.400 is designed to address not only messaging, but also directory, security, and network management.

## SMTP

Transmission Control Protocol/Internet Protocol (TCP/IP) is the de facto standard network protocol offering a connectionless-mode network service in the Internet suite of protocols. Simple mail transfer protocol (SMTP) is the application-level protocol offering message-handling service. However, because SMTP has its roots in the primarily academic and research and development background of the Internet Engineering Task Force (IETF), its use has been in a relatively benign and open environment without the need for rigidly enforced network performance, security, and message-delivery criteria.

**SMTP/MIME.** Messaging by SMTP has been greatly enhanced with the development of Multipurpose Internet Mail Extensions (MIME). This is the official proposed standard format for multimedia Internet mail encapsulated inside standard Internet Request for Comment (RFC) 822 messages.

In simple terms, MIME provides a way to exchange multimedia E-mail among many different computer systems. It is a collection of specifications that describe how mail user agents (MUAs) can identify arbitrary document types and message body types so the interface can decide how best to display the incoming data to the user. All the information about the attachment is embedded in the message itself. The MUA redefines the structure and contents of RFC 822 message bodies. Users can send word processor documents, spreadsheets, audio files, images, and textual data to someone else regardless of the platform, mail transport agent (MTA), or network operational system that is used by the sender or receiver.

## STATE OF ELECTRONIC MAIL

E-mail is by far the most popular application carried over the Internet. Internet mail is based on various Requests for Comments (RFCs), including RFC 822 for SMTP.

E-mail with SMTP has become very popular in research, development, and engineering environments because of their use of UNIX. UNIX and engineering environments usually have TCP/IP and SMTP interconnection protocols bundled in with their operating systems. The Internet has proved to be eminently successful in providing information services to a widely diverse worldwide community with more than 9.4 million host computers on the Internet. One hundred and twenty nine countries now have direct connectivity to the Internet and 39 million users are reachable by E-mail. (A full report is available at http://www.nw.com/zone/WWW/report.html.)

## How X.400 and SMTP/MIME Were Developed

X.400 was designed as a total international messaging environment from the beginning, whereas SMTP developed as an outgrowth of earlier experimental work on the Defense Advanced Researched Projects Agency (DARPA) Network. X.400 provides a complete set of internationally agreed-to standards; approved SMTP/MIME RFCs do not have the same level of official international agreement and approval. The unified design of X.400 as a total messaging environment is also reflected in its clean design interfaces with other international standards required to provide a messaging service — namely, directory (X.500), management (X.700), and security (X.900) services.

SMTP is an outgrowth of the Internet and the Internet Engineering Task Force (IETF). As a quasi-official body with no set membership, the IETF is not necessarily representative of all potential customers' needs. It has primary responsibility for the development and review of potential Internet standards from all sources. The IETF's working groups pursue specific technical issues, frequently resulting in the development of one or more specifications that are proposed for adoption as Internet standards. Most IETF members agree that the greatest benefit for all Internet community members results from the cooperative development of technically superior protocols and services.

SMTP/MIME, although capable of interfacing with these other international standards, does not work with the same level of designed interoperability as does X.400. This lack of designed interoperability will directly (and negatively) impact system operational maintenance and management costs.

## THE SMTP VS. X.400 DEBATE

There is no right or wrong answer when it comes to making a choice between X.400 and SMTP. The choice of technology depends on each organization's particular needs and which strength or weakness of different technologies is most important to them. An organization that wants to share similar technology with as many people as possible can effectively use SMTP/MIME. If security, increased functionality, and operational features such as guaranteed message traceability are most important, then X.400 is the answer.

## SMTP/MIME Characteristics

SMTP/MIME supports the transmission of sophisticated information, including images and video, yet it is simple in its design and extensible in nature because of its unique content-type/subtype body part identification mechanism. In short, SMTP/MIME provides a low-cost solution for messaging backbones.

## THE ELECTRONIC MESSAGING ENVIRONMENT

Each part of a multimedia message identifies what type of information is carried in the message part. An entire MIME message, as opposed to an individual part of a multimedia message, can also have a type. For example, a message might have the type "text/plain" and consist of entirely plain text. A MIME message containing parts of different types has the umbrella type "multipart/mixed." Many types and subtypes have been defined to include audio, image, external data source reference, and partial messages.

The simplicity and flexibility of SMTP/MIME are its main strengths because it can easily be implemented on all systems. Its weaknesses include no support for non-ASCII character sets, limited header structure, and an unstructured message body.

The following table summarizes the strengths and weaknesses of SMTP/MIME:

| Strengths | Weaknesses |
|---|---|
| Very popular in the marketplace with millions of users worldwide | Lacks functionality |
| Low cost | Sendmail is free |
| Available on numerous platforms | Implementations differ |
| Text body parts keep everything simple | Too simple for some uses |
| Runs over IP, which comes with UNIX | Limited security |
| Simple addressing | Questionable for financial transactions |
| Simple message routing with DNS | Uses DNS |
| Simplicity and flexibility of the format of RFC #822 messages | Simplicity and flexibility of the format of RFC #822 messages |
| Numerous gateways available | Lack of structure |

### X.400 Characteristics

The ITU-TSS has developed an ambitious set of standards for electronic messaging called the X.400 message handling system (MHS) and the X.500 directory services standards. X.400 has a very complete set of functional characteristics and can accommodate any type of messaging from simple interpersonal text to attached graphics, voice, and video clips. X.435 is defined for electronic commerce, using basic MHS components. X.400 is based on a functional model consisting of a few main components:

- *User agents (UAs):* used on the desktop for message creation/reading.
- *Message store (MS):* stores messages until recipient chooses to read them.
- *Message transfer agents (MTAs):* stores and forwards messages within and between networks.
- *Access units (AUs):* interfaces to other messaging entities (i.e., voice, facsimile, telex, physical delivery).

## X.400 vs. SMTP

An X.500 directory is a collection of entries that contain information about things such as countries, organizations, people, computers, security, and application programs. The directory is a collection of one or more directory system agent (DSA) computers, each of which holds information for some portion of the directory.

Users access the X.500 directory via a computer process referred to as a directory user agent (DUA). Specific protocols have been developed to control directory access and the exchange of information with distributed directories.

X.500 is absolutely essential for implementing the address translation, document conversion, and sophisticated message routing needed for large-scale E-mail integration efforts. Directory synchronization is the basis for implementing transparent user addresses between systems. X.400/X.500 systems are used by most of the world's telecommunication service providers.

X.400 is the preferred technology for backbone messaging services of several large commercial companies in the U.S. and is widely used in Europe. The U.S. government is implementing a global X.400 messaging system for the military called the Defense Message System (see Chapter 7). NATO is also working toward implementing an X.400-based messaging system. This system defines a Military Message Handling System (MMHS) using X.400, similar to DMS. The militaries of Australia, Canada, the U.K., and New Zealand are also implementing X.400-based MMHS.

The strengths and weaknesses of X.400 are summarized in the following table:

| Strengths | Weaknesses |
| --- | --- |
| Rigorous standards process through ITU | Rigorous standards process through ITU |
| International standard for message handling | Expensive |
| Functionality | Complex to understand and configure |
| Robust message delivery and non-delivery schemes | Not widely accepted by commercial marketplace |
| Well suited for electronic commerce | Lack of robust user agents from popular vendors |
| Strong security standards defined | Security implementations lagging |
| Works well with X.500 directory services | X.500 complex to implement |
| Predictable performance | Overhead is significant; complex system administration |

## INDUSTRIAL-STRENGTH MESSAGING REQUIREMENTS

X.400 has superior functionality defined in the standards, although many of these enhanced functions, such as multimedia and security, have

## THE ELECTRONIC MESSAGING ENVIRONMENT

yet to be deployed commercially. Electronic commerce using X.435 is still waiting for large industry segments to take advantage of this defined standard. Additionally, features such as delivery notifications, delivery to alternate recipients, and receipt notifications are critical to "industrial-strength" messaging systems needed by large commercial organizations or a system such as DMS.

SMTP, as defined by RFC 822, lacks the functionality required for backbone messaging systems or a highly complex network such as DMS. However, several improvements have taken place over the past few years. Functionality enhancements defined under MIME to extend SMTP and provide for messages with enclosed software objects such as images, video, audio, and binary file data have greatly enhanced Internet mail use for large organizations. Privacy Enhanced Mail (PEM) RFCs address many shortcomings with regard to security. These RFCs define data confidentiality, authenticity, integrity and nonrepudiation, message encipherment, and digital signatures.

One of the questions being asked by messaging system architects is: Can SMTP/MIME meet the messaging requirements of a large global enterprise? A "qualified" yes is the answer, assuming further extensions to MIME would be required, primarily in the areas of message management. Members of the IETF have shown great resiliency to further enhancing messaging functionality over the Internet when the need arises. IETF members could develop the missing pieces to make SMTP/MIME functionally similar to X.400.

**Critical Comparison Factors**

The most important comparisons between X.400 and SMTP/MIME concern functionality, security, systems management, message management, management manpower requirements, and performance.

**Functionality.** X.400 is more advanced in this respect, but developers are working to improve SMTP/MIME to match X.400's functionality.

**Security.** Message security capabilities provided by the X.400 standards are far superior to SMTP/MIME. However, there are very few large-scale implementations that take advantage of the numerous security-related features specified within the standards.

Internet security is a major concern for many users. Although security options (such as software for trusted and privacy enhanced mail) exist, they are not widely nor uniformly deployed.

Besides lacking security, SMTP/MIME lacks reliable audit trails. Spoofing, a process by which someone masquerades as another correspondent, is easily done via Internet mail. A user is also allowed to send a message through a re-mailer service so that the original address is not attached to the message when it arrives at its final destination. It is therefore almost impossible to audit messages.

**Systems Management.** X.400 has greater potential in this respect. The entire area of management — including message management, component management, and complete MHS management — needs more attention. The experience of E-mail managers in large organizations demonstrates the need for many additional management tools for a large, complex messaging network.

Managing complex, enterprisewide messaging systems is difficult for several reasons:

- A lack of standards for message management
- Interoperability of technologies; network managers have to deal with X.400, SMTP, proprietary LAN protocols, and legacy systems
- The large number of components; an enterprisewide messaging system is composed of many different kinds of components, each with its own specific behavior characteristics (i.e., MTAs, UAs, directories, and gateways)

Much work has been done on developing standards to govern the individual components associated with X.400, although very little agreement has been reached on how to manage these various components. The network administrator must have tools and utilities available to manage day-to-day network operations. A fully deployed messaging system such as DMS with 2 million users will carry several million messages per day.

**Industry Standards for E-Mail Management.** Significant work has started on developing industry standards specific to E-mail management. A joint International Federation Information Processing (IFIP) group examined the overall problem of messaging management. A similar IETF task force led to the development of RFC 1566 (also known as the Mail and Directory Management MIB, or MADMAN MIB), which defines a class of managed objects that can be deployed within any vendor's messaging architecture. The MADMAN MIB, however, is oriented to the Internet and SMTP, and therefore lacks the ability to model some of the more complex features present in X.400-based systems.

Simple network management protocol (SNMP) and SNMP version 2, both of which are associated with the SMTP/MIME environment, are the leading protocols for managing network transport functions. However, SNMP does not work across non-TCP/IP transports. The management information base (MIB) is a definition of the managed object (i.e., what can be managed remotely). The MADMAN MIB is complete with approved standard definitions, but very limited in functionality (i.e., monitoring only). SNMP and the MIB definitions are only 5% of the puzzle, however.

## THE ELECTRONIC MESSAGING ENVIRONMENT

The ITU-TSS and the International Standards Organization/International Electrotechnical Commission (ISO/IEC) are currently working on the following MHS management documents:

- General
  - MHS management model and architecture — X.460
  - MHS management information — X.461
- Management Functions for MHS
  - Logging — X.462
  - Security — X.463
  - Configuration — X.464
  - Fault management — X.465
  - Performance management — X.466
- MHS Managed Entities
  - MTA entity — X.467
  - UA entity — X.468
  - MS entity — X.469
  - AU entity — X.470

**EMA Requirements for Messaging Management.** The Electronic Messaging Association (EMA) is working on a framework that will allow management of multivendor messaging systems. The EMA's work leverages the IFIP's work and is aligned with the MADMAN MIB definitions. The effort is broad in scope because it also addresses the area of message tracing and standardizing a set of tasks for message management across a multivendor environment.

The EMA's Messaging Management Committee has characterized requirements for messaging management in the following four major categories:

- *Operational management:* deals with finding outages and fixing them as well as doing routine maintenance. Statistical analysis of traffic and components is accomplished. *Comparison:* there is little difference between the two technologies — X.400 and SMTP/MIME — in this area.
- *Configuration management:* deals with managing the addition and deletion of components in the messaging system. It includes tasks such as dynamic updating of message-routing tables, starting and stopping messaging system components, and discovering and depicting messaging system components across the network. *Comparison:* both X.400 and SMTP/MIME are lacking in this respect.
- *Administration management:* provides a means of managing subscribers, distribution lists, and accounting information. It includes facilities for security administration. Control throughout some portions of the Internet is loose. No person or group has authority over some functional subnetworks, such as Usenet, as a whole. Every administrator

controls their own subnetwork. This is different from the X.400 assignment and demarcation of responsibilities, which are vested in management domains, with accountability for performance and control being highly defined. *Comparison:* X.400 is considered superior to SMTP in this area.
- *Network management:* the process of keeping the underlying networking layer healthy. *Comparison:* X.400 and SMTP/MIME are equal in this category.

**Message Management.** X.400 is more sophisticated than SMTP/MIME, but still needs significant improvements. The ability to track a message through messaging systems is central to the establishment of a trusted delivery infrastructure for any complex commercial usage. Maintaining unique identification of a message as it crosses intersystem boundaries represents a significant challenge that no previous standardization activity has addressed.

**Human Resources Requirements and User Support Costs.** Much work is needed in this area for both X.400 and SMTP/MIME technologies. The largest messaging networks can carry several million messages per day. Network administrators do not have time to stop and analyze trouble spots — there is too much traffic coming. They need utilities to shunt aside a problem message and let the traffic flow continue. It also takes very knowledgeable software engineers to accomplish this work, and they are expensive.

Managing these distributed messaging systems from a single, centralized, administrative control system is difficult and costly. One major Fortune 500 corporation estimates that it spends approximately $40 per user, per year to acquire messaging hardware and software versus $200–$300 per user, per year in operating costs to manage and administer the messaging network.

A study by Creative Networks, Inc. indicated messaging support costs, including end-user support, to be:

- $4,189 annual cost per desktop user
- $5,426 annual cost per mobile user

A key cost factor is the amount of end-user support required. Companies lose approximately $684 per user annually to downtime, $764 to lost productivity, and $1,198 to lost revenues due to messaging system problems. Problems with E-mail cut productivity in environments where jobs depend on computer-based information. A typical downtime incident takes 6.3 hours of staff time to resolve.

The need for a resident administrator at each major site can significantly increase the cost of managing large-scale messaging systems. In addition to being on call to deal with system failures or changes in configuration, administrators find themselves subject to normal corporate cost-containment efforts. They are called to manage high levels of ongoing expenses in training as well as in development of complex internal procedures for managing the messaging network across different departments and dissimilar platforms. A major business imperative is to improve the reliability of electronic messaging while reducing the costs of maintaining the messaging infrastructure.

**Performance.** There is no preferred technology from a performance standpoint. Engineering benchmarks are needed to demonstrate the performance of all components and the overall network. Performance bottlenecks must be identified and corrected by the system administrator, but additional tools are desperately needed. Further analysis and modeling of both X.400 and SMTP-based networks is needed.

SMTP-based networks carry large volumes of information, but with very limited functionality. X.400-based networks also carry significant traffic loads and provide very reliable service. X.400 can be engineered to deliver reliable and predictable performance. Both technologies suffer in performance when encryption is added. However, there is overhead involved that requires additional bandwidth.

**Commercial Use.** X.400 is preferred over SMTP/MIME by large organizations needing guaranteed network services. X.400 has gained international acceptance and is used by most European Postal, Telegraph, and Telephone (PTT) services and telecommunication providers throughout the world. The International Civil Aviation Organization standardized on X.400 because of the greater flexibility and enhanced features available.

The most effective means of tying messaging systems together still is the old tried-and-true X.400 backbone. Numerous large commercial and government organizations need a robust and reliable messaging network. Vendors of X.400 components have not experienced significant revenues selling X.400 components because there are still too many unresolved issues; namely, lack of management tools and utilities, plus fully developed directory services. Although SMTP/MIME vendors have made significant sales, SMTP/MIME also has the same problems with lack of management tools and directory services.

In forums such as the Electronic Messaging Association, customers repeatedly state they want the benefits and capabilities offered by X.400 messaging. It is this demand that has led to changes in the current SMTP systems in an attempt to offer the same functionality provided by X.400.

## SUMMARY

X.400 is a better protocol than SMTP/MIME for building a sophisticated network. The standards definitions are very complete for functionality and most networking requirements. X.400 has numerous security features and guaranteed message delivery and notification. These are extremely important for large, predictable, commercial messaging networks.

SMTP/MIME functionality is missing some important messaging requirements, such as delivery notifications, delivery to alternate recipients, and receipt notifications. In fairness, these elements probably could be added, but further work is needed. SMTP grew up in the UNIX environment to provide simple text messaging. Numerous features have been added over time, but the entire process has been an ad hoc development — not the planned architectural development process that the international standards bodies followed with X.400.

In the long run, X.400 and SMTP/MIME are expected to converge on a single set of standards, or at least sufficient development of bridging technology to enable the seamless coexistence of both technologies. The National Institute of Standards and Technology (NIST) has a special interest group working on coexistence and convergence profiles that will promote coexistence as a step toward convergence. This further effort strengthens the belief that the few functionality differences between X.400 and SMTP/MIME are not critical in choosing between the two.

From a standards perspective, the most critical missing ingredient in providing a robust, reliable network is the lack of management tools and utilities. This is where development attention needs to be focused, rather than on functional differences between the protocols. It does not matter which technology is chosen, nor how robust the individual components are, if the network cannot be managed. Users must be able to easily manage the overall network to provide the type of messaging environment that everyone is striving for.

# Chapter 15
# Value-Added Networks: Marketplace Trends
*David A. Zimmer*

Businesses today have a wide selection of choices for transporting information between locations and trading partners. Initially, businesses used point-to-point connections to exchange computer data. As companies expanded and the number of locations grew, the number of connections to maintain grew geometrically. Mergers added additional complexities because of differences in communication capabilities and infrastructures.

From this morass of criss-crossing, point-to-point connections, the need for centralized networks emerged. Today, we have value-added networks (VANs) comprised of business communication services, Internet service providers (ISPs), and online services. Each has its unique niche providing services supporting different needs. At the basic level, they provide interconnectivity. Businesses can use these networks for connecting locations, as well as trading partners and newly merged companies. The network's ability to support multiple connection types and transport protocols frees the internal network administrator from the geometric communication lines.

The Internet and its acceptance by businesses as a viable transport medium has caused a blurring of otherwise sharp lines of demarcation. Business communications services appealed to business needs while the online services supported individual needs. To remain competitive, the business communications services and the online services have needed to integrate tightly with the Internet. As a result, business needs and individual needs are being supported by all three types of services.

THE ELECTRONIC MESSAGING ENVIRONMENT

## UNDERSTANDING THE NEED FOR SERVICES

### Point-to-Point Connectivity Complexity

Many businesses maintain their own networks running frame relay, asynchronous transfer mode (ATM), and other high-speed protocols. These point-to-point networks provide many benefits, including secure transport of information without much security overhead, guaranteed delivery with error-correcting protocols, and logging of exact delivery time for tracking information. Unfortunately, a point-to-point connection has several drawbacks:

- It requires that both the originator and recipient use the same protocol and compatible equipment.
- It demands an expensive, expert staff to support the number of lines and complexity of technology.
- It incurs steep learning curves and additional expenses as each new technology emerges.

### Value-Added Networks Connectivity Simplicity

VANs offer a variety of connection types, services, and conveniences. Unlike point-to-point connections that provide transports only to a specific location, some VANs offer translation from one format to another, information broadcast to multiple locations, and fewer connections.

For example, a company supporting point-to-point connections to five locations must install and maintain 10 connections (see Exhibit 15-1). Broadcasting information to each location would require four separate information transfers. For each location, the data may need to be translated into the proper format before transfer. Using a VAN (Exhibit 15-2), the company simply needs one connection per location to the VAN, transfers the information to the VAN once, and the VAN can translate and broadcast the information to the recipients.

By using the VAN instead of the point-to-point connections, the company relinquishes some control over the data. For example, VANs are store-and-forward facilities, which means the information may be stored in a temporary facility before reaching its destination. As a result, the company may not know the time it reached the recipient. The time information may be important in electronic commerce applications where the time stamp may be critical in accepting a bid, as in a telex transaction. Some VANs have instituted a return receipt system so that originators know that the information has traversed the network and when the recipient received the data.

VANs support mobile workers better than point-to-point connections. VANs typically offer a user interface that runs on either a notebook PC or connects directly to a VAN's network. By using the VANs' dial-up connec-

*Value-Added Networks: Marketplace Trends*

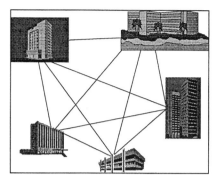

Exhibit 15-1. Five Locations, Ten Point-to-Point Connections.

tions, a company eliminates the need to support large modem banks and servers. Mobile workers can use the VANs to send and receive E-mail, electronic faxes, E-forms, and tap into informational databases.

## BUSINESS COMMUNICATION SERVICES

### Market Overview

Business communication services (BCSs) provide large, ubiquitous networks for transport of messages. Typically, these providers are used by large corporations for exchanging messages with other companies and by small businesses who do not want the expense of running an internal network for the amount of messaging performed internally.

The network providers support several types of messaging protocols connecting disparate E-mail systems permitting a user to exchange mail

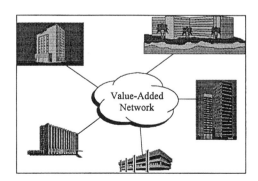

Exhibit 15-2. Five Locations, Five Connections, and One Value-Added Network.

257

with another user regardless of the recipient's messaging platform. In addition to message transport and translation, the network providers support such additional features as E-mail conversion to fax, telex, paper (regular letters), and EDI. Besides the computer-based interface, some provide access to messages by telephone interface.

The current growth of and business enthusiasm for the Internet poses a threat and an opportunity for business communications services. Many companies are beginning to use the Internet for information transport because of the low transmission cost versus the metered systems of the BCSs. By the same token, BCSs offer customer support, secured transport, and reliable delivery. By leveraging their connections to the Internet, their customer service, and reliable deliveries, BCSs will gain by the growth of the Internet.

Employee mobility has become important. Many companies are implementing telecommuting practices and encouraging customer support by face-to-face interactions. When not in the office, the mobile worker must communicate, typically through voice, fax, and E-mail. BCSs, with their multiple points-of-presence and direct support for the popular E-mail packages, make the implementation of mobile messaging more simple. Wireless connectivity eliminates the mobile worker's need to find a phone jack.

BCSs have been adding useful services and will continue to do so. The ability to receive news updates, weather, travel information, and other information directly through E-mail has made life simpler for many busy people. Additional services (e.g., information search and retrieval, online shopping, and intelligent messaging) will further add to the appeal and necessity for BCSs.

**Current Trends and Future Directions**

Four major trends are occurring in the public network market: electronic commerce, mobile messaging, Internet connectivity, and intelligent messaging.

**Electronic Commerce.** Many BCSs have been supporting electronic data interchange (EDI) for some time. Typically, EDI is implemented by larger companies for transacting business with other large companies and smaller companies using simple EDI interfaces. BCSs have provided transport and conversion utilities between companies. Smaller companies were able to use the BCSs to connect to these larger companies. Because of their solid infrastructures, BCSs have the opportunity to coalesce the fluid electronic marketplace into an offering useable by consumers. In addition to financial transactions and large intercompany purchases, BCSs can open the avenue of electronic commerce to any size player.

**Mobile Messaging.** BCSs have focused on the mobile messaging market. As a result, products and services that support the mobile professional exist. More products and services are on the way.

For example, wireless connectivity permits the mobile professional to communicate easily with co-workers and contacts. Using the same messaging infrastructure as their deskbound colleagues, mobile professionals remain productive without the need for training on new products and services. Networks supporting portable devices (e.g., personal data assistants or personal communicators) allow the mobile professional to travel with more easily, without having to pack and lug excessive baggage when on the road. BCSs will continue to add such mobile-enabling services as intelligent messaging, information services (e.g., travel-related services and maps), and ubiquitous access.

**Internet Connectivity.** The Internet certainly has gained momentum since the advent of browsers as a graphical interface to the World Wide Web (WWW). The vast resource of information and people on the Internet makes it appealing to most professionals.

At a minimum, BCSs have provided E-mail access to the Internet. Exchanging messages with someone on the Internet is as simple as adding an electronic mail address to the message. The next level of access the BCSs will add is full connectivity with such features as file transfer protocol (FTP), Telnet (the ability to connect to a remote computer as an online participant), WWW browsing, and other Internet functionalities. The major concern BCSs must overcome is security from hackers and other malicious behaviors.

BCSs offer a first-level defense against security threats to the company's internal network. Rather than connecting indiscriminately with any computer, the single connection to the BCS provider can be monitored for any malicious behavior. Some companies use the BCSs as a first-level firewall to the Internet.

Business communications services have gone to great lengths to protect themselves from unwanted activity. As a result, companies using BCS services are protected as well. (This level of protection does not preclude the need for additional protection by the individual company.)

**Intelligent Messaging.** Intelligent messaging is an exciting new concept attached to electronic messaging. User agents have had filtering of messages for a couple of years. Progress has been made by using the filters to automatically organize incoming messages.

Public networks have also begun offering filtering, intelligent agents, and infobots. For example, some BCSs have supported mobile professionals by permitting them to set the maximum message size they will accept while traveling (a filter). Messages larger than the stated size will be retained in the mailbox for later processing.

Intelligent agents, as introduced by the AT&T PersonaLink Services, provide capabilities to notify a subscriber of new messages by pager based on subject or originator. Additional agents (e.g., information searches, specific actions based on input — watching a stock ticker and selling when it hits a threshold and notifying the trader, or arranging for dinner, theater tickets, and flowers and reporting success when complete) are being developed. Infobots are used to respond to requests for information (e.g., return company or product literature).

**Why Use a BCS?**

BCSs offer a broad range of services, network access, and E-mail. They generally appeal to businesses more than consumers because of several critical functions, including connectivity, ubiquitous access, reliable delivery, message conversion, message translation, and customer support.

**Connectivity.** BCSs provide a vital link between companies. Rather than establishing a point-to-point connection with each trading partner, companies can connect themselves to a VAN. In fact, Company A may connect to VAN X, and Company B may connect to VAN Y, and yet still communicate with each other simply because the BCSs interconnect. The need to establish point-to-point connections are eliminated, saving tremendous effort, time, and money in maintaining those connections.

**Ubiquitous Access.** BCSs provide access to their networks from many places. Some provide 800-number and local area number dialup connections. For heavy traffic support, BCSs offer X.25 and other direct connects. Because BCSs cover large geographical areas, they can support companies with several, dispersed locations. Some provide international access and support.

**Reliable Delivery.** Within their domains, BCSs provide very reliable delivery. When delivery cannot be completed, a negative receipt is returned to the originator. For inter-BCS traffic, the same reliability typically exists. BCSs have negotiated non-delivery protocols between themselves so that the highest quality service can be provided, even for inter-BCS communications.

**Message Conversion.** BCSs provide additional messaging capabilities. Electronic mail messages may be converted to fax if the recipient is a fax machine. Messages may be printed on paper for delivery through the postal system. Additional conversions may be provided.

**Message Translation.** BCSs provide message translation from one messaging system to another. For example, PROFS users can easily exchange mail with HP OpenMail and Lotus cc:Mail users. The PROFS message format is translated into the appropriate format of the recipient system prior to downloading. Different formats are supported by different BCSs, so the customer must be sure the VAN supports the necessary formats.

**Customer Support.** Most major BCSs provide 24-hour customer support, 7 days a week. This added benefit permits troubles to be diagnosed and fixed at any time. This important service permits companies that rely on their messaging system to minimize downtime.

## BCS Selection Criteria

When choosing a BCS, a company must consider its needs. The following questions are representative of what a company should ask before selecting a BCS:

1. Is the company using the BCS for interconnectivity with trading partners?
2. Is the company supporting a mobile workforce with and without wireless connectivity needs?
3. Does the BCS service the geographical area needs of the company?
4. What is the pricing structure: monthly fee plus per message charge or flat rate with unlimited messaging? What types of discounts are offered?
5. Are additional services offered so that other needs of the company are met (e.g., faxing, EDI, or paper mail)?
6. What is the company's relationship to the BCS — a good working relationship or strained?
7. What are the future directions of the BCS in question? Is it along the same lines as the company's goals?
8. Are other trading partners using a particular BCS?
9. Are the messaging packages used within the corporation supported?
10. If international support is needed, are international access points available?

BCSs offer a broad array of services, network access, and E-mail connectivity. They generally appeal to businesses more than consumers. Exhibit 15-3 presents a comparison chart of the most popular BCSs — what they are, who makes them, what they offer, and how they stack up against the rest. For a list of URLs and for more information on these products, see Exhibit 15-4.

# THE ELECTRONIC MESSAGING ENVIRONMENT

| | AT&T Easy Link | GEIS Bus. Network | Harbinger EDI | IBM Global Network | network MCI | SprintMail | Sterling | WorldLinx |
|---|---|---|---|---|---|---|---|---|
| **Service Access** | | | | | | | | |
| Maximum dial-up speed | 28.8 Kbps | 28.8 Kbps | 28.8 Kbps | 28.8 Kbps | 28.8 Kbps | 28.8 Kbps | 28.8 Kbps | 28.8 Kbps |
| 800-number access | ● | ● | ● | ● | ● | ● | ● | ● |
| 56 Kbps lines | ● | ● | | | | ● | | |
| T1 | ● | ● | | ● | | ● | | |
| Frame Relay | ● | ● | | | | | | |
| ISDN | | | | | ● | | | |
| **Protocol Support** | | | | | | | | |
| Asynch | ● | ● | ● | ● | ● | ● | ● | ● |
| Bisynch | ● | ● | ● | ● | ● | ● | ● | ● |
| EDIFACT | ● | ● | ● | ● | ● | ● | ● | ● |
| ODETTE | ● | | | ● | ● | | ● | |
| SMTP | ● | ● | | ● | ● | ● | ● | ● |
| SNA/SDLC | ● | ● | ● | ● | ● | ● | ● | ● |
| TCP/IP | ● | ● | ● | ● | ● | ● | ● | ● |
| TDCC | ● | ● | ● | ● | ● | | ● | ● |
| VICS | ● | ● | ● | ● | | | ● | ● |
| WINS | ● | ● | ● | ● | | ● | ● | ● |
| Wireless | ● | | | ● | ● | | | |
| X12 | ● | ● | ● | ● | ● | ● | ● | ● |
| X.25 | ● | ● | ● | ● | ● | ● | ● | ● |
| X.400/X.435 | ● | ● | ● | ● | ● | ● | ● | ● |
| X.500 Directories | | ● | | | | | | ● |
| **E-Mail Connectivity** | | | | | | | | |
| Lotus cc:Mail | ● | ● | | ● | ● | ● | | ● |
| Microsoft Exchange | ● | ● | | ● | ● | | | ● |
| Lotus Notes | ● | ● | | ● | | | | ● |
| MHS based | ● | ● | | ● | ● | | | ● |
| AT&T EasyLink | ● | ● | ● | ● | ● | ● | ● | ● |
| CompuServe | ● | ● | ● | ● | ● | ● | ● | ● |
| Internet | ● | ● | ● | ● | ● | ● | ● | ● |
| IBM Global Network | ● | ● | ● | ● | ● | ● | ● | ● |
| networkMCI | ● | ● | ● | ● | ● | ● | ● | ● |
| PROFS | ● | ● | | ● | ● | ● | ● | ● |
| GEIS Bus. Net. | ● | ● | ● | ● | ● | ● | ● | ● |
| SprintMail | ● | ● | ● | ● | ● | ● | ● | ● |
| U.S. Postal Service | ● | | | ● | ● | | | ● |
| Fax | ● | ● | ● | ● | ● | ● | ● | ● |
| Fax Broadcast | ● | ● | ● | ● | ● | ● | ● | ● |
| Telex | ● | ● | | | ● | ● | | ● |
| X.400 | ● | ● | | ● | ● | ● | ● | ● |
| Paging | ● | | | | ● | | | |
| **News and Information** | | | | | | | | |
| News Services | ● | ● | | ● | ● | ● | | ● |
| News Clipping | ● | ● | | ● | ● | ● | | ● |
| Historical News Search | ● | ● | | ● | ● | ● | | ● |
| Other Info Services | ● | ● | | ● | ● | ● | | ● |
| **Other Services** | | | | | | | | |
| BBS | ● | ● | ● | ● | ● | | ● | ● |
| Document Conversions | | ● | ● | ● | | | ● | |
| E-Forms | ● | | | | | | ● | ● |
| Online Directories | ● | ● | ● | ● | ● | | ● | ● |
| Real-time Document Sharing | | | | | ● | | | |
| Videoconferencing | | | | | ● | | | |
| Customer Support | 7x24 | 7x24 | 7x24 | 7x24 | 7x24 | 7x24 | 7x24 | 7x24 |

Exhibit 15-3. Business Communication Service Comparison Chart.

*Value-Added Networks: Marketplace Trends*

AT&T EasyLink............http://www.att.com/easycommerce/easylink
GEIS Bus. Network......http://www.geis.com
Harbinger......................http://www.harbinger.com
IBM Global Network ..http://www.ibm.com/globalnetwork
networkMCI.................http://www.mci.com/networkmci
SprintMail...................http://www.sprintbiz.com/jproducts/jfaxamssg.html
Sterling........................http://www.sterling.com
WorldLinx ...................http://www.worldlinx.com

**Exhibit 15-4. Business Communication Services URLs.**

## INTERNET SERVICE PROVIDERS

### Market Overview

The Internet and the WWW have been likened to the revolution started by the printing press. Others have stated that it will drastically change the way we live, socialize, and work. To say that it has had an impact on society would be an understatement. Why have the Internet and the WWW caused such a stir?

Before the invention of Mosaic — the first popular Web browser — the Internet and WWW were the domain of technical people who had spent many hours learning and developing ways to ferret information stored on various computers. The interface was typically character-based with cryptic commands, and it was difficult to learn and boring to view. Until a graphical interface that supported multiple types of media was developed, the information would remain unavailable to the uninitiated. Companies were reluctant to train employees on such archaic systems, and individuals were not willing to spend the time to learn on their own.

Mosaic, a graphical interface to the WWW, supported multimedia information, made the WWW simple to use, and opened a new world of information for users. Because of its simple interface and navigational methods, the WWW has become the information resource that has been needed for a very long time. Rather than pushing tons of information on employees by way of paper manuals, directories, and other communiqués, companies simply place information pages with related topic links on internal Internets (intranets). Publishing work was always difficult because of publication and distribution costs. Now, simply generating a Web page makes the information available to anyone in the world.

Netscape fully commercialized the Web browser market with its introduction of the Navigator browser. Others followed but quickly faded once Microsoft introduced its Internet Explorer browser. Netscape's and Microsoft's browsers are the most widely used, with Netscape leading the

THE ELECTRONIC MESSAGING ENVIRONMENT

market and Microsoft gaining rapidly. Although they have slightly different functions, both access Web pages with equally agility.

Browsers run on every known hardware platform and operating system. Because the Web uses the standard hypertext transport protocol (HTTP) for transporting information and each Web page follows the hypertext markup language (HTML) for representing the information, browsers can display information universally. The Web browser has therefore been dubbed the "universal client."

**Current Trends and Future Directions**

The Internet marketplace has broken into three sections: the Internet, the World Wide Web, and intranets. The Internet is the underlying support infrastructure, providing transport and information repositories. It supports many types of programs (e.g., FTP, Whois++, and WAIS). The Internet continues to grow in number of sites at a very fast rate.

**Dynamic Web Pages.** Much excitement continues around the World Wide Web. It is the focal point for many new services, tool development, and customer interest. New programming languages (e.g., Java and ActiveX) have been developed to support even greater use of the Web. Initially, the Web was a collection of static documents with links to other related documents. It has been transformed by additional tools that let page viewers perform additional functions. Dynamic Web pages that list the latest values of a person's stock portfolio and allow users to make merchandise orders and even restaurant reservations are simply the beginning. Many companies envision full electronic commerce capabilities, virtual reality, and other exciting services in the very near future.

**Intranets.** Intranets are being deployed by companies internal to their firewalls. Firewalls are barriers from the internal network and the external Internet and WWW. Intranets are information distribution systems within corporate environments. What was once published in manuals and updated by paper distribution is now placed on intranets. Employee manuals, benefits information, and company bulletins are now published electronically by human resources departments. Companies are saving millions of dollars annually because of the electronic distribution.

Groups within companies are using intranets to distribute memos and other documents electronically. Executive reports, business cases, design documents, and others are placed on intranets instead of being mailed office-to-office in paper form. The speed of information delivery has increased so that each member of a team can obtain the information at the same time. It is expected that by the year 2000, companies will spend billions of dollars developing intranets. Although the investment seems rather high, the cost savings and additional beneficial gains far outweigh the cost of implementation.

The Internet, WWW, and intranets will continue to evolve and move further into the consumer sector. Already, several services support the individual subscriber. All socioeconomic sectors can enjoy these services. An entry-level PC (e.g., a 386 running MS-Windows) with a 14.4K or 28.8K bps modem is sufficient hardware to enjoy the WWW. Used 386 machines can be purchased for $200 or $300.

Network computers, slated to be relatively inexpensive, will eliminate the need to configure a general-purpose machine for Internet use. Ubiquitous service offerings for less than $20 per month from large telephone companies makes access simple. The electronic age is on us and the information society is here. What does the future hold? There are four areas of consideration: consolidation, enrichment of features and services, globalization, and realignment of socialization.

**Consolidation.** The Internet is supported by a backbone of Internet service providers (ISPs). These companies were initially not telephone companies, because they were not permitted to compete in this area of enhanced services. ISPs would buy communication services (i.e., telephone lines) from telephone companies for the express purpose of transporting data, unlike telephone companies that focus on voice traffic. Communications reform legislation in the U.S. has granted license for the telephone companies to compete in this area.

The current Internet market consists of local (small), regional (medium), national (large), and international providers. With the entrance of the telephone companies, some smaller and regional companies will be bought by the larger companies in an effort to gain market share. Other small- and medium-sized players will be forced from the market as customers switch to more competitive offers. In some cases, even the larger companies will succumb to the competitive pressures of the telephone companies.

AT&T WorldNet is an example of telephone company entrance into the market. Within the first several months of existence, it surpassed its goal of subscribers and continues to add more customers. Bell Atlantic has tossed its hat into the ring and has begun offering Internet access. MFS/Worldcom entered the marketplace by buying one of the largest ISPs, UUNET.

**Enrichment of Features and Services.** Content is king. Industry experts have agreed that the most profitable aspect of the WWW is the content provided. At first it was the access offer, but because of telephone company entrance, consolidation, and price-slashing competition, content will become the profit center for most players.

# THE ELECTRONIC MESSAGING ENVIRONMENT

The current number 1 use of the WWW is information searching. The number 1 use of the Internet is E-mail. Although not officially reported, the number 1 use of intranets is information distribution.

**Content Rules.** The value of the content becomes very important. People are willing to pay for valuable information and information value will continue to evolve. What was once valuable will become ordinary.

For example, people once paid to get weather and stock information. Today, both types of information are free. Other information will follow a similar trend. At one point, attending a seminar to learn about the Internet (and the presenter's consulting ability) would cost about $500 per day. Today, the same information is worth $50 to $100 per day. The consulting service can be done by several people. New programs and applets will appear as more developers write applications using Java and ActiveX. Because the Internet is global, brilliant and creative minds will continue to evolve the network for more useful endeavors.

**Globalization.** The Internet and WWW span the world. Unlike earlier, when only large companies could create business internationally, small businesses can create presence in foreign lands. A Web page, developed and stored locally, can be viewed by anyone in the world. Behind a Web page, a small company can look like a megacorporation and a large corporation can look like a small, personal service company. The Web has been called the business equalizer. This globalization trend will continue as more and more companies discover how to use the Web to its advantage. Goods sold on the Web can be both physical as well as informational.

**Realignment of Socialization.** The Internet and Web are used as social outlets. Many people would prefer to assume a persona behind a computer screen rather than meet people face-to-face. Behind the veil of the computer screen, the timid can become bold, the bored can become interesting, and the conservative can become liberal.

Most users of the Web are professionals, medium to upper class, and computer literate. Efforts are being made to ensure that the Web and its informational stores are available to all levels of the socio-economic scale. The result: equalization of information access. Those so inclined to search on a particular topic can do so without the need for expensive research tools. If it is available on the Web, information is open to all, regardless of race, creed, or intellect. (Of course, there will be those "clubs" that limit access to information by membership or exclusion, but they will be in the minority.)

Evidence of this trend has already been seen. The Internet and fax machines have been credited in part for the downfall of the Soviet Union. Soviet citizens were able to access information that existed outside of Soviet borders. They began to understand that there was more than they were being led to believe. News of internal government rumblings were relayed to the outside world, making it possible for outside diplomatic pressure to occur. Additional liberation through knowledge will continue to occur as more Internet sites and Web hosts come online. Today, a high-tech tide is rising in China.

Children today are being trained to interact with others outside of their local schools. Children from around the country and world are interacting on a regular basis. Cultural differences are being explored and accepted because of the interaction. Prejudices based on lack of familiarity are disappearing. This effect on socialization will drastically change how we view the world today.

Those who decide not to become computer literate and Web surfers will suffer. Their skills will not be sharpened. Every industry — from construction and hairstyling to high-tech industries — will be affected. New products and services will be offered electronically that will make "doing business the old way" obsolete. Banking and other financial transactions will become more electronic each year. Payments from customers will take a more electronic direction, eliminating those businesses that do not provide for electronic transfer of funds.

Companies providing access to the Internet (ISPs) will become as important as the telephone companies have become in everyday life. Their ability to provide the services needed will be critical. Unlike the telephone industry that had territories designated by legislation, ISPs have no defined domain. As a result, fierce competition for market share will happen. Those ISPs providing the socialization needs of the customer will prevail.

What does this have to do with selecting an ISP? Understanding these socioeconomic issues, network administrators and IS directors can do a better job of selecting an ISP that can support them for years to come. The process of selecting the right ISP can be time-consuming for large companies. Having to repeat the process a few years later because the current ISP is merging with another vendor could be costly.

**Why Use an ISP?**

ISPs are used by companies and individuals for similar reasons: ubiquitous access, low-cost transport of information, and information repositories.

**Ubiquitous Access.** ISPs have provided access points around the globe. By simply accessing the Internet from anywhere, mobile employees can access data back in the company database. Individuals can access the Internet from anywhere they might travel. College students can continue communicating with others whether they are at school, at home, or visiting friends. As the Internet continues to grow, access points will be added.

# THE ELECTRONIC MESSAGING ENVIRONMENT

**Low-Cost Transport.** Unlike BCSs that charge by the amount of data sent, ISPs simply charge a flat monthly fee. The fee is based on the access method (e.g., a dial-up modem or a dedicated line). This flat fee makes it more conducive to send more data by way of this connection. The low-cost transport option is not without problems. ISPs do not guarantee delivery, do not return nondelivery notifications, or provide end-to-end customer support services. Therefore, for mission-critical information transfers (e.g., electronic commerce), most businesses opt for the more expensive BCSs. Notwithstanding, the ISPs are working to build more secure and reliable networks.

**Information Repositories.** The Internet contains a wealth of information on every topic imaginable. The key problem to the Internet is finding the information. The introduction of the World Wide Web made finding the information much easier. Publishing information was made simpler as well. Although the quality of information ranges from useless to invaluable, Web surfers can easily research any topic. Unlike before, where researchers would have to search professional journals, research findings, and other hard-to-find sources, anyone can become an expert in a topic by accessing the information repositories. Companies have used this same concept internally on their intranets. Rather than use paper distribution for employee handbooks or other company materials, electronic copies are placed on intranet servers so that employees can easily access and retrieve the latest information.

## ISP Selection Criteria

When choosing an ISP, a company must consider the following points:

1. Does the ISP provide convenient access points for all company locations?
2. Does the company need an international access point?
3. Does the ISP support the connection requirements for anticipated company traffic load?
4. What has been the ISP record for up-time, and does it meet the company's service requirements?
5. What is the pricing structure for various connection types?
6. Does the ISP provide intranet capabilities or virtual private networks?
7. What is the future direction of the ISP, and does it match the company's objectives?
8. Are trading partners using the same ISP providing more secure links for applications (e.g., electronic commerce)?
9. What is the relationship between the ISP and the company — positive or strained?
10. How is the ISP positioned in the market, and what is its potential fate as industry consolidation continues?

|  | AT&T WorldNet | BBN | internet MCI | NETCOM | PSINet | UUNET |
|---|---|---|---|---|---|---|
| Connectivity Support |  |  |  |  |  |  |
| T1 | ■ | ■ | ■ | ■ | ■ | ■ |
| 10 Mbps | ■ | ■ | ■ |  | ■ | ■ |
| T3 |  | ■ | ■ | ■ | ■ | ■ |
| 56 Kbps | ■ | ■ | ■ | ■ | ■ | ■ |
| Frame Relay | ■ | ■ | ■ | ■ | ■ | ■ |
| Dial-up | ■ | ■ | ■ | ■ | ■ | ■ |
| ISDN | ■ | ■ | ■ | ■ | ■ | ■ |
| Intranet Offering | ■ |  |  | ■ | ■ |  |
| Customer Support 7x24 | ■ | ■ | ■ | ■ | ■ | ■ |
| NetNews | ■ | ■ |  | ■ | ■ | ■ |
| Personal Accounts | ■ | ■ | ■ | ■ | ■ | ■ |

Exhibit 15-5. ISP Connectivity Support.

## ISP Comparisons Chart

ISPs provide direct access to the Internet. Subscribers have access to all the Internet tools, including FTP, Telnet, Finger, WAIS, Gopher, and the World Wide Web. Subscribers have access to newsgroups, Usenet news, and electronic mail. Delivery options of E-mail is limited to E-mail and fax. A comparison chart of various ISPs and their associated connectivity support is presented in Exhibit 15-5. URLs for additional information on these ISPs can be found in Exhibit 15-6.

## COMMERICAL ONLINE SERVICES

### Market Overview

Online services have typically focused on individual subscribers. The services provide forums, discussion groups, or bulletin boards on a variety

AT&T WorldNet ................................................. http://www.att.com/worldnet
BBN ...................................................................... http://www.bbn.com
internetMCI ....................................................... http://www.internetmci.com
NETCOM ........................................................... http://www.netcom.com
PSINet ................................................................. http://www.psi.com
UUNET ............................................................... http://www.uunet.com

Exhibit 15-6. Internet Service Provider URLs.

of topics. Prodigy was initially focused on the home market. America Online was the first with an MS Windows interface and quickly gained acceptance as a consumer-based offering. CompuServe has traditionally had a business flavor to its contents.

With the phenomenal growth of the Internet, online services have added access to it. They continue to expand their Internet offering, at first focusing on E-mail and now delivering full Web browsers. Topics covered by the services include news, weather, travel, personal finance, entertainment, customer support by many companies, and chat groups. The services provide E-mail as a form of communication between subscribers and to those subscribers of other VANs.

**Current Trends and Future Directions**

Online services have traditionally focused on individual subscribers. CompuServe has complimented its individual subscriber service with connections support of greater bandwidth than typical modem speeds. These larger connections let corporations tap into the rich information stores by replicating some of the forums or using CompuServe's infrastructure for message transport. In addition, Compuserve has leveraged its infrastructure to support point-of-sale transactions, data transfer, and LAN-to-LAN connections. America Online (AOL) and Prodigy continue to focus strictly on individual users by way of their proprietary interfaces.

The Internet and WWW have affected online services, which provide tight integration with the Internet and WWW by providing Web browser functions in interfaces. Gateways to the Internet and WWW permit their subscribers to surf. Time spent on the Internet and WWW is billed at a metered rate similar to the time spent on the native system.

The systems will continue to evolve by supporting the Internet and WWW. They will continue to support the special-interest forums for which they are known. The services will continue to leverage their communities of interest. The communities may be geographically located (e.g., softball teams, church groups, and school districts) or virtually located (e.g., areas based on such specific interests as photography, fishing, and construction). Content must evolve from text-based to the graphical nature of the Web. Chat rooms with videoconferencing will be a heavily used capability.

As the Internet and WWW continue to evolve, online services will be severely threatened. The same functions that exist in online services exist on the Internet and WWW, but at lower cost from ISPs. This cost factor could shift some otherwise loyal constituents to an ISP.

Online services may be able to survive by leveraging their content. As stated earlier, content is king in the light of cheap access to the Internet and WWW. Online services currently have a great deal of content and agreement with content providers. By assuring uniqueness or freshness of information, online services will be able to charge for access to the forums. If they cannot maintain that edge, they may suffer tremendous losses.

**Why Use a Commercial Online Service?**

Commercial online services are mainly used by individuals. The most popular use of these services are chat rooms.

Chat rooms permit people to converse on a particular subject or no subject at all. It is a virtual community square where people gather to meet others of the same kind. The irony of the virtual world is that the persona of the online user may not be anything like the real person behind the screen.

Others use the online services for information research. Because most of the same information is found on the Internet or will be available soon, there will be a major shift away from the metered search time to cheaper, Internet-provided information.

**Online Service Selection Criteria**

When choosing an online service provider, individuals must consider their needs. They should start by asking the following questions:

1. Does the service have a local access point?
2. Does it host my interest areas?
3. Does it have information resources of interest to me?
4. Is the fee within my budget, and do I feel comfortable with their fee structure?
5. Am I limited to accessing on the online service or can I freely surf the Internet?
6. Does the online service restrict my access to any Internet areas? If so, are those restrictions acceptable?
7. Is the proprietary interface acceptable to me?

A comparison chart of online services is presented in Exhibit 15-7. Service access to online services (e.g., America Online and Prodigy) provide dial-up connections at speeds of 14.4K bps or 28.8K bps. CompuServe uses its network for more than just online services. It supports ISDN, frame relay, virtual private lines, 3270/5250, and X.25 packets. CompuServe uses these access methods to support point-of-sale transactions, data transfer, and LAN-to-LAN connections. Prices are negotiated for each case. For more information on online service URLs are listed in Exhibit 15-8.

**SUMMARY**

Value-added networks come in three flavors: business communication services, Internet service providers, and online services. Historically, they

## THE ELECTRONIC MESSAGING ENVIRONMENT

|  | American Online | CompuServe | Prodigy |
|---|---|---|---|
| Connectivity |  |  |  |
| Dial-up | ■ | ■ | ■ |
| ISDN |  | ■ |  |
| Frame Relay |  | ■ |  |
| Virtual Private Lines |  | ■ |  |
| 3270/5250 |  | ■ |  |
| X.25 |  | ■ |  |
| Internet/WWW | ■ | ■ | ■ |

| | American Online | CompuServe | Prodigy |
|---|---|---|---|
| E-Mail | ■ | ■ | ■ |
| Chat | ■ | ■ | ■ |
| Forums | ■ | ■ | ■ |

**Exhibit 15-7. Commercial Online Services Comparison Chart.**

provide the needs for three different markets. Each provides services and resources needed for its particular population of users.

BCSs cater to businesses with large amounts of data transmissions. They provide additional services that supported business needs (e.g., paging, E-mail-to-fax translations, E-mail-to-postal service support, EDI, and telex). In addition, they support protocol conversion so that sending electronic mail to disparate systems was easy. They eliminated the need for point-to-point connections for each branch office or trading partner. Their secure network provided the safety needed to communicate with outside organizations.

The Internet and WWW permit inexpensive access to valuable information resources. In addition, they provide lower-cost transport of data between end points. Their open nature and readily available server products let anyone become an extension of the network. Information does not flow through a homogeneous, continuous network (e.g., BCSs). As a result, the user has no guarantee of delivery, must suffer the risk of security breaches, and faces problems without any end-to-end customer support.

Online services are geared mainly toward individuals and services that support their needs (e.g., chat rooms and communities of interests). Because they focus on the individual subscriber, they are not well suited for

America Online ................................................ http://www.aol.com
CompuServe .................................................... http://www.compuserve.com
Prodigy .......................................................... http://www.prodigy.com

**Exhibit 15-8. Online Services URLs.**

large data transmissions, except for Compuserve, which has leveraged its infrastructure to support special applications (e.g., point-of-sale and WAN connections).

The Internet has become the great equalizer in many respects. Small companies can compete on a global scale as large companies because of the World Wide Web. It duplicates the massive data transport of the BCS networks and the individual services of online services. As a result, the blurring of lines between the various markets is occurring rapidly. In a very short time, those lines may be erased completely. At that point, a subscriber's loyalty will be difficult to maintain. Content will become the deciding factor for which service to use.

Deciding which service to use may continue to be based on bandwidth and connection to a particular service, but the real deciding factor will be the content forums. Content will continue to increase in value for maintaining subscriber loyalty. Telephone companies will control the access and bandwidth of the networks. Consolidation will occur in the ISP marketplace. Larger ISPs will buy smaller ISPs in an effort to gain market share. Telephone companies, in turn, will buy the larger ISPs to gain control over the transmission lines and win market share. The effect on society will be great. The way we do business now will not resemble how we will be doing business in the virtual, electronic world of the future.

# Chapter 16
# Commercial Online Services in Review
*Stewart S. Miller*

The online service industry has found its place in our emerging electronic community. Online services provide a quick and simple answer for busy corporate end-users who neither have the time nor desire to search endless sites in search of the one jewel on the information superhighway. Online services provide an effective road map that paves the way for many users. Electronic communication is reliably facilitated in the safety of an online community that provides carefully chosen links to the outside Internet via the World Wide Web.

**COMMUNICATIONS PERSONIFIED**

Online services redefined communications that had, until recently, been provided by newspapers, magazines, radio, television, and the telephone. Unlike radio and TV, online is an interactive medium in a person's house. Online services have enhanced human communication and are fast becoming a core information resource. Commercial online services provide a variety of information media, including general news and commentary, sports, business, and weather. Services also provide extended library content (e.g., encyclopedias, dictionaries, directories, maps, and guides).

Interactivity is the cornerstone to the online industry's success. The leading commercial online services — America Online, CompuServe, The Microsoft Network, Prodigy, and AT&T — allow people to get together to send messages through E-mail or bulletin boards. These discussions can take any mood; however, no other medium permits people to exchange ideas, information, and opinions in such a conducive manner.

**CREATING ONE WHOLE ONLINE PACKAGE**

The key to survival for online service providers in this information age is to seamlessly integrate Web browsers into their proprietary software. The major online services keep a comprehensive list of hot sites to visit and integrate these links into the main interactive service content. New

# THE ELECTRONIC MESSAGING ENVIRONMENT

customers are attracted by the unique content and programming online services offer.

Commercial online services provide an on-ramp to the Internet as well as individual content. This move may well position them as the pitstop of the information superhighway. Users rely on commercial services to point out the best of the Internet. Most users desire instant gratification and do not care for the hassle of spending hours on the Net to find only one valuable site or only a few interesting tidbits of data.

## AT&T Interchange

AT&T Interchange, unlike online services that organize and present their own preformatted content, operates as an online network, which waits for third-party content providers to rent cyberspace there. Providers run their own service with its own content and pricing. Hourly rates are fixed regardless of which Interchange services are used.

**A Plethora of Communication and Content.** ZD Net offers computer news, forums, software, and information. ZD Net's home page on the Internet is URL: http://www.zdnet.com/. These sites include content from *PC Week, PC Magazine,* and *Windows Sources,* plus an archive of several of Ziff's Computer Database Plus publications. Users can also retrieve technical support and software, and the service allows users to participate in discussion forums.

The AT&T Business Network offers a centralized information and business online service with business-like content. This collection of specialized business information and databases includes TRW Business Information Services, The Kiplinger Washington Editors, and Dun & Bradstreet Information.

## CompuServe

The pioneer of the online industry offers access to several research databases that offer exceptional content at a cost. Professional resources include news, background, and opinion on most topics. CompuServe allows its users to link to several databases, including Dialog, TRW, Information Access Company, and Dun & Bradstreet. These databases also offer informed topical discussion forums in which online citizens may participate. Prodigy and America Online searches often focus on the archives of one business magazine; in contrast, a CompuServe search covers numerous consumer magazines, professional journals, newsletters, books, and more.

**CompuServe's Split Regarding the Internet.** CompuServe does not wish to integrate the Web completely into its own service content. CompuServe has several key technological strengths; however, it has suffered by not providing a clear consumer identity. The service now offers aliases to overcome this deficiency by allowing users to be recognized also by a name (similar to AOL) and their ID number.

## Prodigy

Prodigy is a service that concentrates on sports, kids, and family issues. These topics generate significant interest among its online population without instigating any controversy. The online guest forums have featured such celebrities as Leonard Nimoy, Bob Saget, and Tom Selleck, who represent the pure theme Prodigy presents to its users.

Chat provides an avenue of communication that fosters discussion of family issues ranging from infertility to car buying to pets. Newsweek Interactive integrates sound, graphics, and linked news stories. Prodigy is making all of its content compatible with hypertext markup language (HTML) — the language of the Web. This will allow Prodigy to be able to send its messages across the Web.

Prodigy's effort is striving to make the Web even easier to use. In keeping with its wholesome image, Prodigy uses a software-filtering tool designed by IBM Corp. that screens electronic mail and bulletin boards for inappropriate words. Prodigy Services Co. was the first online service provider to add a proprietary Web browser to its basic subscription package. America Online Inc., CompuServe Inc., and Microsoft Network compete by offering their own Web access tools.

## America Online

America Online (AOL) offers general-interest content as well as consumer-oriented areas. It also has a wide range of content regarding the latest news, movies, Hollywood, and links to the World Wide Web.

**AOL's Impressive Content.** AOL has a large newsstand with new headlines and graphics. The broadcast media has found its niche in creating online versions of its traditional news sources. *The New York Times, ABC News On Demand, The Atlantic Monthly, Business Week, The Chicago Tribune, Congressional Quarterly, Consumer Reports, MTV, Time, Woman's Day, Omni,* and *Smithsonian* are some magazine and newspaper publishers who reproduce the full text of each issue online. In addition, these magazines encourage discussion forums pertaining to their particular topics.

AOL maintains a virtual community that includes member chats with writers, actors, and other celebrities. Their easy interface simplifies E-mail, news headlines, the World Wide Web, and USENET newsgroups. Internet content is seamlessly integrated into the proprietary GUI, to which users have become accustomed. Although business content on AOL does not compete with CompuServe's premium business databases, many areas are well stocked with daily business information that provides business people with a good perspective on the current trends.

**AOL Growth.** AOL has grown to be the market leader. It provides expanded content and has achieved universal name recognition. The amount of Web content designed around Netscape's HTML extensions will likely inspire future versions of AOL's browser to be compatible with Netscape. AOL recently launched a separate, Internet only service: Global Network Navigator.

### Microsoft Makes Its Grand Entrance

The growth of the Internet has persuaded The Microsoft Network (MSN) to open its service to the Internet instead of making it a private online service. This measure allows MSN to compete with America Online, Prodigy, and CompuServe. Software resellers have found the Internet and online services a means to transmit software applications that now ship in boxes. In the near future, when bandwidth and security issues improve, vendors can find ambitions of using cyberspace as a cost-effective means of delivery.

When users download software from any of the major commercial online services, it is slow, expensive, and often does not even work. Leaps in speed and more efficient compression technologies, however, will eliminate this drawback and provide a unique means for business vendors to relay their message throughout the online world. Commercial online service providers must now provide Internet access merely to survive. In addition, how they add value is just as important in a sector overrun with new players of every size.

The World Wide Web is changing the commercial online service industry. It continues to increase growth among commercial online service providers. In contrast, the Web cuts into the exclusive content and a large number of online service subscribers. To compete, online companies are promoting a Web-centric environment that gives the service a unique feel in cyberspace.

**Microsoft Creates Opportunities.** The Microsoft Network offers opportunities for consulting and development work in addition to a useful business tool and conduit for distributing products. MSN had originally created its service to attract individuals similar to America Online and Prodigy. Microsoft developers plan to offer several business services, however, which will also allow corporations to directly communicate and interact with both customers and staff.

CompuServe, in comparison, aggressively develops relationships with resellers as corporate customers. They have recruited about 300 value-added resellers (VARs), systems integrators, and Internet access companies to market its Internet software products. Microsoft will attract resellers with the incentive of gaining expertise in software development, which helps merchants and content providers add products and services to MSN.

Microsoft is selling itself as a network where resellers can expand sales by placing their catalog on the network and allowing customers to browse products and place orders. Customers who purchase software can download it directly off the network or have it shipped through traditional means.

## E-MAIL USES AND ABUSES

When traveling, the advantages of dial-up connections become increasingly attractive. E-mail messaging is as close as a local phone call that keeps users in touch with business contacts, friends, and family. Cybercitizens realize the versatility of dial-up online service connections. An increasing level of corporate interest surrounds dial-up connections for business travelers and telecommuters.

Staying online while traveling is easily accomplished with simple preparation. After contacting an Internet service provider for some simple directions, this one call can save users long-distance telephone charges while allowing them to access their online account through a local telephone number. Most users become readily comfortable with logging on to their E-mail account from other systems, forwarding E-mail to a new location, or automatically notifying people who send them messages that they are unavailable until a later date.

All that is needed when traveling is the service provider's local connection numbers for destination. The Internet service provider will tell users where its local dial-up or points-of-presence (POPs) are located around the country or world. National providers are advantageous as they usually have many more POP access numbers. In comparison, Netcom and UUNET have local POPs in more than 175 cities in the U.S., while CompuServe has about 1100 dial-up points around the world.

Once users determine that their provider offers a local access number in the place where they are headed, they are set. Users just need to make certain the phone number works with their modem speed. Once they know how to enter the new number, the long-distance problem is solved. Some access providers, however, do not offer a toll-free number for out-of-town access. These toll-free numbers often carry high surcharges, however. Users should check the cost before they go.

Users may also create a UNIX .forward (dot forward) file that will forward their E-mail to another online destination. The UNIX machine will then automatically forward their E-mail to the address contained in the forwarding file. If users do not wish to handle the influx of E-mail, they can eas-

# THE ELECTRONIC MESSAGING ENVIRONMENT

ily opt for a UNIX vacation program that messages everyone sending them mail that they are offline.

## Advertising on the Net

"Spam" is jargon that describes unsolicited messages of any kind. Although spam is usually some form of advertisement, most Internet users consider it to be the ultimate abuse of messaging.

Spammers who send their messages to members of a mailing list start an avalanche of complaints about wasting Net resources. Some users try to punish spammers by flooding them with multiple mail messages. This is not a good idea, however, because each message sent to the spammer may bounce back to the user's inbox multiple times if the spammer has left one Internet service provider for another.

Spam is not limited to advertising — it is any message that is inappropriate or unwelcome that is posted multiple times to many newsgroups or mailing lists without the benefit of cross-posting. Although spam may list a name, address, and telephone number, that may not mean it is accurate data. Revenge with a mail bomb or numerous responses is not an option because it risks bringing down the service provider's server as well as the files and accounts of several innocent subscribers.

Newsgroups (e.g., news.admin.net-abuse and alt.current-events.net-abuse) permit users to report spam and post the full spam, which makes system administrators aware of the problem and permits them to block it from their systems. By posting to these newsgroups, systems operators are alerted of the spam, giving them sufficient warning that allows them to block the offensive messages from their sites.

The best method in dealing with E-mail spam is to politely write the postmaster at the spammer's domain about the abusive actions of the poster. Several Internet service providers and commercial online services quickly cancel accounts when users abuse them.

## Telnetting to Home

If an Internet service provider (ISP) does not provide local access or toll-free number options, users can still pay local rates by employing a tool called Telnet, or a remote Internet login. Telnet, an Internet protocol, allows users to log in to remote computer systems and access their own account from another site.

To use Telnet, users need the address of the system they want to reach, a terminal where they are, and their user ID and password. The system will prompt the user once it accesses his or her provider. It then displays the login prompt just as if they were sitting at their home terminal. Users then can read E-mail, send messages, check on newsgroups, or access any information from their ISP.

## ONLINE FOR CORPORATE USERS

Online services' rush to market has turned their eyes toward corporate users. Although America Online focuses mainly on the consumer, CompuServe and The Microsoft Network deal with both kinds of users. The AT&T Business Network, however, is dedicated exclusively to business.

MSN appeals to businesses by proposing volume pricing and allowing corporate employees to sign on through their company networks with a shared TCP/IP connection, which can offer superior speed over dial-in connections. The key factor for online services is to maintain service reliability. MSN initial users have complained about the difficulty in logging on. As services grow, however, they periodically experience such problems. AOL experienced similar problems when its membership grew beyond system capacity, but has now shown itself to be stable.

Resellers find they must have online development expertise (e.g., graphics design) to survive in the online world. Businesses who have an online presence create an instant, long-standing job prospect between the company and the customer. Web pages can be instantly updated with new content; however, the fact that pleases clients the most is the resource the WWW site represents. The customer can access the site for updated information that provides a customer service department that never closes. Companies cannot only earn revenue from consulting, but can sell to a more specialized market through the Internet.

The Internet represents the least expensive and fastest means of establishing an online presence, yet most company benefits are reached by establishing an online audience through a commercial online service provider. These commercial services offer Internet access, but many users remain within the service's own boundaries. These people never see some Web pages. Therefore, most content providers find it beneficial to maintain content on a commercial service in addition to the Web to achieve the best exposure for their message.

### Secure Commerce

Although credit card transactions are conducted on private networks, most businesses fear commerce transactions through unsecured Internet connections. Most advertisers display a phone number or address for users who are interested in placing orders. Because Netscape's security was breached this past fall, recent developmental improvements have placed customers and businesses more at ease when doing business on the Internet.

## THE ELECTRONIC MESSAGING ENVIRONMENT

AT&T Corp. provides its own online Internet service meant especially for corporate and small business clients. The AT&T Business Network will use its 10 million business subscribers and then focus on providing AT&T-branded content services for businesses via the Internet. AT&T's venture competes directly with Internet service providers that extend their services into the Internet (e.g., CompuServe Inc. and America Online Inc.). AT&T is concentrating on the business community, however, whereas others mostly focus on the consumer market.

**SUMMARY**

Commercial online services offer an excellent means for corporate entities to reach their customers in an efficient and effective manner. The Internet will be the most effective means for communication in the next century. The minimal cost, in combination with speed and interactivity, of the Net will make the online world the most effective means that draws both consumers and business people alike.

# Chapter 17
# Selecting Internet Service Providers

*Stewart S. Miller*

The Internet continues to grow exponentially each year — it is now estimated at 30 million users! This has great implications for businesses, but what exactly does that mean? The Internet clearly offers businesses an instant and relatively inexpensive avenue for promotion, resource gathering, and communication. Business users can instantly contact branch offices throughout the world through electronic mail (E-mail), receive the latest information on a company's newest product and stock reports, or route through thousands of servers anywhere in the world for news on any number of issues within seconds of logging on. Business is built on timeliness and the ability to react quickly. The Internet exemplifies instant access to information.

The Commercial Internet Exchange (CIX) — a backbone created as a conduit for business-related packets — was established in 1991. Internet service connectivity was offered freely to the public, and then generally at quite a premium. Accessibility improved greatly in 1994 as more Internet service providers (ISPs) moved in. Today, however, numerous ISPs exist in almost every city. The Internet is the business tool of the 1990s. It will evolve, however, to be the information conduit that drives the future of business.

The Internet is not an online service; it is actually a collection of services comprised of a variety of protocols used to exchange data through a number of interconnected backbones and networks. The Internet was developed so that it does not rely on any one computer or subnetwork, and the packets of data that make up Internet traffic can be routed any number of ways to reach their destinations (packets are reassembled in the correct sequence at the receiving machine).

The Internet is composed of several thousand independent networks. The general direction of the Internet is organized by the Internet Society, which is a voluntary membership organization whose purpose is to promote global information exchange through Internet technology. The technical aspects (e.g., design and engineering) of the Internet are handled by the Internet Architecture Board (IAB) as well as its constituent subdivisions. They are responsible for the design and approval of new network protocol software and network applications that work in a large-scale capacity on the Internet.

## WHERE TO START

One or all of the following are the most common reasons behind purchasing an account with an Internet service provider: E-mail capabilities, information resources, and presence on the Internet.

First, users (businesses) must choose a service provider that matches their needs with the service being offered. If E-mail capabilities are all that is desired, there is no need to purchase an expensive dedicated line. In connecting to the Internet, users truly get what they pay for. The ingredients for establishing an Internet connection are discussed in the following sections.

### Managing the Internet

Just as there is no central help desk, there is no central governing body for the Internet. The Internet is best characterized as a self-governing anarchy with its own rules and codes. No authorities are there to manage the rules and standards that allow communications to work, and watch out for technical considerations to maintain and expand this growing network of networks. Primary among the open-ended organizations created to promote and improve technical standards for the Internet's continued existence, as well as maintain records of domains and IP addresses, is the Internet Society (ISOC), founded in 1992. ISOC, based in Reston, VA, actually contains several key suborganizations:

- *Internet Registry (IR)*. This organization records IP addresses in use.
- *InterNIC*. This organization provides technical advice, coordinates Internet services, and makes remote procedure calls (RPCs) available to the public.
- *Internet Engineering Task Force (IETF)*. This organization studies and creates technical standards designed to improve the infrastructure and interoperability of networks on the Internet. The Internet Architecture Board oversees the IETF.

### Internet Utilities

This section describes Internet utilities that, depending on an organization's desires, can effectively accomplish what it needs on the information superhighway.

**Electronic Mail.** This basic tool is the most widely used internetworking function. E-mail allows users to send anything from a short reminder to a full report through the Internet, arriving at its destination anywhere in the world in only seconds.

**Gopher Access.** The Gopher provides ASCII text-based searches of libraries, government institutions, and archives of corporate press releases, and much more. Although Gopher has lost some of its popularity to flashier colors and graphics of the World Wide Web, most resources still on the Gopher can be readily accessed through many WWW browsers.

**Usenet Newsgroups.** Newsgroups provide open forums for the exchange of news, opinions, and gossip. Business users find newsgroups have news feeds such as Clarinet, Reuters, and Newsbytes that are the most useful to retrieve news on any given topic before it appears in print.

**File Transfer Protocol (FTP).** This tool fosters the transfer of data from the server to a client. Data could take the form of freeware or shareware software, frequently asked questions (FAQs), or graphics. FTP could be helpful to a business in acquiring the latest version of popular software by downloading it from an FTP site.

## Defining the Purpose

The first thing to decide is why an organization wants an Internet connection and what it expects to gain from being on the Internet. Several Internet resources exist, and not all may be applicable to every business. One company may only need E-mail capabilities, while another company aims to be a presence on the Internet, open a storefront to advertise its products and services, or simply post a kind of billboard describing the company and what it can do for its customers.

## Doing Your Homework

Some important facts to research when choosing an ISP include: determining whether the ISP has too many subscribers and the number of T1 phone lines used to provide service.

Many ISPs are not big, sophisticated operations; therefore, it is important to determine that they neither oversell their available bandwidth nor provide insufficient underfunded Internet access through modems and routers. It is important to make certain the ISP is constantly connected to the Internet 7 days a week, 24 hours a day. The available bandwidth must be adequate to support the client's applications. The ISP must provide Internet users with corporate literature, file access, or E-mail by way of a common gateway interface (CGI) or applications supporting online transactions.

Evaluating providers is an important step in determining the best choice. If a company can find the least number of providers to service all

# THE ELECTRONIC MESSAGING ENVIRONMENT

of its needs, the better its intersite service will be. Ways of doing this include the following:

- Checking customer references
- Determining how long have they been in business
- Researching how long have they been offering the services the company needs
- Finding out if they serve all the locations of the company's sites
- Checking the distance of their POPs from the company's sites

A company must carefully examine the price list information for every item to determine what services it will and will not need for its business. It should eliminate the services it needs and make a price comparison to choose which value-added Internet services are provided against how important they are for specific business needs.

The future direction of an ISP must be determined regarding transmission technology additions, speed upgrades, and new POPs. Several providers offer network design, consulting, and systems integration services, but it is important to check with independent consultants who can offer an objective opinion and who are not trying to oversell Internet services beyond specific needs. When choosing an independent consultant, a company should search for people who have worked at Internet providers or at major corporations heavily implemented with the Internet.

It is important to create an Internet business plan with finite policies and procedures, adequate file security, comprehensive training, help, and technical support from an ISP. Patience is the rule, and the benefits of researching this medium will pay off in the long run.

**The Cost of a Connection**

Once an organization has decided on what it wants, it must consider the various costs involved, including installation costs, communications costs, and ongoing service fees.

Different service providers charge different rates, therefore shopping around will help find the best value for the money. Having a clear idea of what is needed and how much that service is worth is the best way to avoid paying for services an organization does not need or cannot use. The various costs for basic SLIP/PPP services can include the following:

- A setup fee for establishing the account
- Local or long-distance phone calls
- Hourly charges
- Monthly maintenance fee
- The purchase of software

*Selecting Internet Service Providers*

Clients may support a Web site either by a proxy server or a dedicated server. The proxy server is at an ISP or third-party location and does not require the installation of a leased line. It also offers the advantage of low maintenance from the client's perspective. Installation costs about $1000 in addition to the $100-per-month use fee. The disadvantage is that the client has little or no control of the operating system.

Establishing a dedicated server at the client's site requires a leased line at an additional cost of $150 per month in addition to the fee charged by the ISP linking the client to the Internet. The cost is $400 to $500 per month to purchase data rates of 56K bps. Higher speed requirements, however, increase the ISP fee to about $1,000 per month. Although building and maintaining a personal Internet server can cost between $40,000 and $50,000 a year, businesses also have the option to rent space through an Internet service provider's server for about $250 a month.

CompuServe's Internet Division offerings include Internet Office, which includes Internet access software that resides on a LAN and emulation products that allow connection to mainframe and legacy information. The Internet OfficeServer allows a company to set up a Web server. Depending on connectivity options, the price for the Internet Office ranges from $199 to $449.

The communications costs involved when establishing and maintaining an Internet server include circuit installation and frame relay charges. The circuit installation fees cost around $1,000 to a telephone company and the ISP setup fee is $1,000. The phone company's periodic frame relay charge can be $150 to $500 per month, while a similar price scale applies to the ISP for dedicated access. Network management, security, and other services may also be included. The cost, however, may not include a router, system installation and integration, and training. In addition, a communications link leased phone line, SMDS or ISDN service, metropolitan fiber, and microwave is needed to connect each site being supplied as well as the ISP's closest POP. If the Internet service is mission-critical, multiple links are desired. Such costs are often separate from and greater than the cost of the Internet service itself.

Once the breakdown of monthly and yearly costs has been determined, a clear picture of the total expense of an Internet account can be seen. It is critical to choose an ISP only after fully studying these costs and determining carefully your company's needs, because if the client is dissatisfied with the ISP, it is not an easy process to switch to another ISP.

**Finding the Best Connection**

When you decide on an Internet provider, there are four types of accounts to consider: shell (UNIX), dial-up SLIP/PPP, dedicated SLIP/PPP, and leased line. These four connections differ in both price and usability.

## THE ELECTRONIC MESSAGING ENVIRONMENT

A shell account is the most inexpensive of the four, but it requires knowledge of UNIX commands and does not provide a graphical user interface (GUI). A dedicated-line account requires the purchase of a server, router (for linking the Internet server to the company's network), and a direct high-speed line from the Internet service provider to the business's server. A leased-line account is the most expensive of the options. It provides access to the Internet at speeds from 19.2K to 768K bps. For this type of account, however, a company could spend anywhere from $20,000 to $40,000 a year. With a leased line, the server is actually on the Internet and available for users to log into: the most direct connection to the Internet.

Dial-up services are an affordable option from local ISPs — it allows a user to dial in and temporarily be placed on the Internet for access to services. Both serial line Internet protocol (SLIP) and point-to-point protocol (PPP) access employ local software on the user's own computer, allowing for fancy GUIs that make the Internet easier to use for novices. Many companies will find that it is easy and cost effective to purchase a SLIP or a PPP connection. Both SLIP and PPP are low-cost dial-up connections that use standard modems (14.4K bps is the lowest recommendation, with 28.8K bps being the preferred modem speed) and standard telephone lines to establish a link between the business and the service provider. PPP accounts are quickly becoming the industry standard and surpassing the SLIP connection.

There are some all-in-one packages that allow access to most Internet services from within the same application, but many SLIP/PPP users can choose a variety of tools, each one suited to a specific task or two (e.g., an Archie/FTP client, a World Wide Web browser that serves both Web and Gopher data, a Usenet news reader, and an E-mail program). Users can switch to tools they prefer for a given task, while background software (typically a TCP stack on a PC, or a Control Panel for a Macintosh) maintains the Internet connection.

Setting up a SLIP/PPP connection can be a difficult task, but once past the initial step of configuring the software, use is quick and easy. SLIP/PPP accounts can use either commercial software or a combination of freeware and shareware. These accounts include a TCP/IP client for establishing the Internet connection.

**Qualities of a Good ISP**

To find an ISP that provides consistently reliable service, a company must first look into the mechanics of the ISP's network. Much like the telephone service, an ISP's network should be have sufficient capacity to allow users to connect at an adequate speed to perform all the functions they need.

**Network Strength.** Network topology makes it possible to determine how vulnerable the network is to outages, including how much capacity is available when the network is loaded and whether it has heavier use than normal. Virtual backbone maps "virtually" tell nothing because data travels on a physical network, not a virtual one. Virtual network maps represent theoretical paths that could be implemented by the supplier's virtual circuit switching equipment. However, this is not a true representation of physical capability. Frame relay, ATM, or other technologies that use virtual circuits as part of the backbone are fine; however, a provider must understand the physical topology besides their virtual (logical) network. Much as in life, theory is not always representative of practical implementation.

**Network Speeds.** The speeds of the backbone links indicate that network connections can only be as fast as the slowest link in the path. If a user is connected to a T-3 node and there is a 56K-bps link between the user and his or her destination, their speed cannot be fully realized. ISPs can show links that are not operational listed as part of their backbone infrastructure. In some cases it is possible to see planned links labeled with solid lines and the operational links with dashed lines. It is important to check and see which links are operational and which are only planned.

**High-Speed Backbone.** If an ISP lists a high-speed backbone, users should find out if that is present or planned speed. ISPs may require users to buy the local loop segment from their base of operations to their closest POP. If speed is limited by the local loop because the price of a high-speed loop is not cost effective, the high-speed backbone is useless because users cannot take full advantage of its capabilities.

**External Network Links.** It is very important to determine if the ISP has a single connection to the rest of the world — this is a potential single point of failure.

An effective strategy is to find multiple, direct connections to other network providers. If they have a single connection to the outside world, however, companies can check the providers record to see how often it fails and how long they (and the user) are isolated.

**Technical Staff and Support.** A most important factor may be the quality of its technical support services and staff. Does the service provider have 24-hour, 7-day-a-week technical support? Or is support available only during regular business hours? (If this is the case, be sure to verify time zones — are their business hours Eastern time or Pacific time?). Most ISPs provide constant support via E-mail or provide a constant answering service that responds to inquiries in only a few hours. Research the experience of their staff in TCP/IP data networking (it should be close to 10 years). The provider should have adequate staff to handle daily situations.

**Organization.** Research how long the company has been an Internet provider. Their record can determine if they will be in business several months from now. Check to see if the company actually provides the network service, or if they are acting as a middle man for a large entity. If this is the case, check the status of the larger provider and determine if they will give accounts the same attention as their customers who have direct connects.

### The Big List of ISPs

Internet service providers have recently grown to service nearly every city. Most major cities have several ISPs competing for the same customers. A terrific site on the World Wide Web provides detailed information for most ISPs and enables the consumer to shop around for the best price in the area. THE LIST is the most comprehensive list of Internet service providers and allows users to comparison shop for the ISP with the best rate in their service area. Its URL is: http://www.thelist.com/.

The right ISP will provide at least 30+ hours of SLIP/PPP access for about $20 a month. It is also important to check to see if an ISP has local dial-up access numbers in a user's area so they can avoid paying long-distance charges. When choosing an ISP, the following guidelines should help:

- *Customer service.* Determine what customers think of the vendor's support. How quickly are E-mail inquiries answered? Is the staff familiar with the specific applications used by the company? Can they answer effectively or not at all?
- *Performance.* Is there adequate system capacity to handle user demand? Determine if the system response time is slow or shaky. Is the network connection up to speed? Did the ISP oversell bandwidth to the point that there is too much congestion to access the service? How often do users experience busy signals of degraded response time? Do ISP system managers install updated Internet applications?
- *News services.* What is the size of the ISPs newsfeed? Do these news services include premium news services or only free newsgroups (e.g., full Clarinet newsfeed)?
- *Reliability.* Is only high-quality equipment used by the vendor? What type of connection does the ISP have to the Internet backbone? Does the ISP have 24-hour monitoring of its network?

## THE ROLE OF COMMERCIAL ONLINE SERVICES

In choosing an Internet service provider, a company is going forth into cyberspace with little assistance. Commercial services, however, will offer constant help in every step of the process — whether they are wanted or not. Commercial sites are gateways through which nothing flows without their approval and under their regulations. They show users what they want them to see and nothing more. For a novice Internet explorer, this type of controlled exploration could be a good thing. The chance of getting lost is far less likely.

Persistent Net surfers, however, will find numerous hidden resources. Commercial service providers, unlike Internet service providers, follow a fairly standard pricing schedule, and most carry the same services. CompuServe, America Online, and Prodigy all offer E-mail, Gopher, WWW, and Usenet Newsgroups. Their standard pricing is comparable in that for $9.95 per month, each user is given 5 hours of time and additional time is $2.95 per hour.

## SUMMARY

The Internet continues to attract more and more business users and curiosity seekers alike. Internet service providers will help make the travel a little less difficult. The choice of a guide for any would-be Internet user, however, is as important as the tools he or she carries, for the wrong guide may bring the user to the edge of disaster, while a knowledgeable, forthright guide will show them the treasures they are seeking. The careful consumer can find the Internet to be an enriching experience for nearly every aspect of his or her professional and personal life. Shopping around will find the best value and services for the money.

# Chapter 18
# Using Internet Resources
*Stewart S. Miller*

The Internet is quickly becoming a tool for commerce that is both popular and profitable. The resources of this global network that now wires millions of people are staggering — it offers local access to most countries for virtually no long-distance charges and inexpensive connect fees.

It is all too easy to get lost in the jargon, misunderstand the conventions, or find oneself in an avalanche of flames (i.e., hate E-mail). In addition, misinformed users who do not know the boundaries of the Net may open a proverbial can of worms. This chapter defines the most common terms associated with this vast resource, offers tips for using the various Internet tools efficiently, and describes procedures for putting some of the most common tools to productive use.

## THE BASICS

Because the Internet does not rely on any single access method, users can perform some of the more basic Internet tasks (e.g., sending and receiving E-mail or downloading files by FTP), whether it be by personal computer or terminal that:

- Networks directly onto the Internet
- Connects by a terminal (UNIX) shell
- Implements serial line Internet protocol (SLIP) or point-to-point protocol (PPP) graphical user interface tools
- Is connected by way of a third-party gateway service (e.g., America Online, CompuServe, or Prodigy)

People still learning the basics of the Internet may find it confusing or difficult to master. The Internet is not a bulletin board system (BBS), but it can connect to BBSs and other remote systems (including some of the major online services) using a program called Telnet. This feature offers users the ability to control a remote computer and lets them connect anywhere on the network and use resources located physically at that host. Although

## THE ELECTRONIC MESSAGING ENVIRONMENT

this is not a graphical online service, it can resemble one when a graphical browser is used to visit the World Wide Web. The Internet offers E-mail that reaches a person at any point on the globe in only seconds, but this represents only one tool of many.

The Internet can be defined as a variety of mostly free resources readily accessed from a computer. A massive amount of data is available, but it is difficult to try to find something of value without knowing the right tool for the job. The Internet is also a widespread, varied community with its own rules. It should also be pointed out that breaking the rules can lead to severe consequences, because the Internet is mainly a self-policed medium.

The Internet offers many advantages, including convenience, depth of information, and speed. In contrast, the Internet may be slow, unwieldy, and more time-consuming than a simple, well-placed person-to-person telephone call. A little knowledge helps users pick the right tools so they can get what they need in only seconds. Circumstance often comes into play, however, because key computers may be unreachable or someone may have stored the material users seek in a way that makes it difficult or impossible to access when using the wrong tool. World Wide Web servers may not contain the same information at a company's Gopher site — all of which means that users can find what they need if only they know where to look for it!

The Internet depends mostly on its inhabitants or users, therefore finding information may depend on who reads a posted message or the popularity of a specific topic. Some topics that have been heavily addressed may have a wealth of archived data that may be downloaded or browsed. Several resources or topics may be new to the Internet and require a catalyst to get them started.

### TOOLS OF THE TRADE

The standard Internet tools for a user who has SLIP/PPP access include: E-mail, Gopher, World Wide Web, FTP, Usenet newsgroups, Internet relay chat (IRC), Finger, Talk, Telnet, WAIS, and translators or decompression utilities for files that are uuencoded, zipped, stuffed, or otherwise compressed or archived.

#### Electronic Mail

Computer techie or not, almost everyone knows of the existence of E-mail. Millions more use it daily, but most are not aware of its potential.

E-mail allows people with Internet accounts to communicate with each other quickly anywhere in the world. Because E-mail is an asynchronous communications medium, it will not interrupt any task the person at the receiving end is doing. The economics of this medium make E-mail the cheap-

## Using Internet Resources

est, most efficient communication for business and consumers around the world.

**E-Mail as an Interactive Resource.** E-mail is the easiest and most valuable information resource for any purpose. This resource accesses more than 20 million potential customers and is offered by all Internet service providers as well as all major commercial online services. This resource costs little and lets users personally interact with customers by responding to inquiries and comments about a company's products and services. Unsolicited E-mail (i.e., "spam") is hardly ever welcomed for marketing purposes.

Some systems automatically respond to incoming E-mail requests, eliminating any wait customers experience instead of a personal reply. LISTSERV is automated software (a discussion-list server broadcasts E-mail messages) that sends a message to hundreds or even thousands of subscribers. Participating in such specialized forums allows companies to keep tabs on their competition and establish their company's expertise, in addition to gaining exposure and name recognition. Some companies even set up their own list server to automatically send out company news and product information.

If users have questions, they can quickly E-mail the person who might be able to help them. If the person E-mailed cannot help but knows someone who can, he or she can forward or redirect the message to a third party. In any event, it is possible to receive a prompt reply in minutes or hours, often in the form of the original message quoted back with annotations that address each point. Quoted material most frequently has the ">" symbol to the left of each line, and sometimes includes a header citing the original author's name and the time and date of the original letter. This can be very useful and in some ways is preferable to a direct telephone communication, because the recipient gets questions in easy-to-follow written form, and the sender receives answers clearly written out, eliminating such mistakes as incorrectly spelled names.

E-mail communication allows the receiving party to answer messages at his or her convenience. An E-mail message is much more likely to achieve a prompt reply than would voicemail, a fax, or postal mail because responding is easy, convenient, and provides a written record. E-mail applications permit the sender to specify any desired number of carbon copies. (Note: the recipient can see the names of all cc: recipients, although some programs allow the sending of blind carbon copies to which the recipient is unaware.)

Some E-mail programs (e.g., Eudora) allow users to create nicknames, which are shortcuts or aliases for entire lists of E-mail addresses. This feature is extremely convenient for mailing duplicate messages regularly to the same group of people. Press releases or memos can be instantly routed without individually typing a long list of recipients. The blind carbon copy feature allows a message to be sent where none of the recipients are aware of the others. This is considered bad etiquette, because some E-mail applications indicate "recipient list suppressed" — but blind carbon copies are useful when it is important to conceal the identity of other recipients.

Several E-mail applications support signature (.sig) files, which are automatically appended to the end of all outgoing messages. It is important to keep signatures short; some recommend there be no more than four lines; some mailing lists reject messages with .sigs that exceed two lines. Some elaborate signatures that use ASCII graphics or designs may wish to choose a simpler alternate .sig when sending mail to online services (e.g., America Online or AOL) — these signatures sometimes are mangled when displayed using these systems' proprietary software because of the use of nonproportional fonts and look unprofessional. Users may wish to test signatures by sending them to addresses at online services to gauge the effect.

## ANATOMY OF AN ELECTRONIC MAIL ADDRESS

This section dissects the E-mail address, an exercise that will greatly assist users in searches. E-mail addresses generally are in the format of username@domain name.type, although there are many variations.

The address listed after the "@" sign is the domain. It is the actual address of the host computer for the account. There are never spaces, because the Internet does not support them, although some names contain a period (.) or an underscore symbol (_) in lieu of a space. When Internet addresses are spoken, the period (.) is pronounced "dot" and the "@" symbol is pronounced "at." Exclamation points (!) are pronounced "bang," but are used only in rare cases.

As a hypothetical example, john_doe@fish.widget.com might represent the E-mail address for John Doe at the Widget Company (note the "com" at the end). The suffix "com" is used for commercial locations, "gov" is for government sites, "mil" represents military installations, "edu" denotes educational institutions or universities, "net" is for network service providers, and "org" is a nonprofit organization.

Other suffixes users may encounter represent foreign countries; for instance, "jp" is Japan, "de" is Denmark, "uk" is the United Kingdom, and "can" is Canada. A suffix for the U.S. (us) is becoming common as well. In addition, some E-mail addresses may contain strange characters (e.g., an exclamation point (!), a percent sign (%), or a double colon (::), that are required for messages to pass through E-mail gateways connected to the Internet.

### Searching for a Specific E-Mail Address

Looking for the E-mail address of another person on the Internet is trickier than one would think. Because no single directory lists all Internet users, and many users prefer their privacy, this practice can be quite fruitless.

Because the Internet community is so widely distributed and continues to grow so quickly, no single definitive set of White Pages keeps track of everyone. Instead, several tools offer varying levels of abundance, authority, and ease of use. It is often necessary to use more than one of these tools to find an E-mail address.

Most Internet-savvy users recommend an old-fashioned method first — they call the person on the telephone and ask. Some oblique methods include using such protocols as WHOIS, X.500, and NetFind, which can scour Internet databases for user information, if it is available. However, some companies intentionally conceal this data from outsiders. These types of searches may be accomplished through search engines available through various Gopher servers, Web sites, and by using client programs designed for the task.

**NetFind.** NetFind is a powerful tool used to find Internet E-mail addresses. NetFind can assist users even if they do not know the domain of the person's address. Users only need to supply the person's last name and a basic description of where he or she works. NetFind then tries to find an Internet E-mail address that matches the search criteria.

### Domains

Examples of typical domain suffixes were described earlier. In addition to what was already listed, a fully qualified domain name (FQDN) is a site's domain name that indicates a site's organization or sponsoring agent. Internet numbers indicate a unique address for every machine on the Internet. The IP address, or commonly called a dotted quad, signifies a 32-bit number represented as four number groups joined by periods (.), as in 255.255.255.255.

A message is said to bounce when it fails to reach its destination. The message is returned to its sender with information on why it was not delivered. This can happen when the destination server is temporarily or permanently unavailable, the address does not exist, or can be the result of a typographical error.

All E-mail messages have headers that include routing information that is generally ignored by the user. Looking at the header of a bounced message, however, helps users determine the problem. If the destination serv-

## THE ELECTRONIC MESSAGING ENVIRONMENT

er was down, users can choose to wait a short time and try resending the message. If the message was an attempt to guess at a particular person's Internet address, a bounce usually indicates a failed guess.

**Mailing Lists**

Simple E-mail correspondence is only the beginning. Users can join a mailing list on a plethora of topics — many highly technical; others offer press releases that arrive periodically in the electronic mailbox.

Some lists are moderated, indicating that only pertinent information is transmitted. Lists used for delivery of press releases and electronic publications do not usually permit contributions from recipients. Mailing lists that are moderated weed out irrelevant postings, while unmoderated lists allow anyone to broadcast a message to everyone on the list. These lists often have a code of behavior, so it is worthwhile to "lurk" or only read without posting before contributing to the discussion.

Once a user joins a mailing list, his or her E-mail address is placed on a computerized routing slip. Whenever a message is sent to the server that maintains the list, that message is sent to everyone who participates. A thread (i.e., the message and all of its replies) can be followed simply by checking E-mail. Instruction by E-mail is a type of mailing list that is gaining popularity — although most are free, some require a paid subscription.

The procedure for subscribing or discontinuing mailing lists varies from list to list, but the basic protocol remains the same. The subscriber sends an E-mail message to either a person or a computer. The message must be sent to a specific address — should a computer process the request, the format must be followed exactly because the computer scans the message and processes it as if it were a script or computer program. Some lists require that the command "to subscribe" be written in the subject header of the message; most LISTSERVERS require all relevant commands to be typed in the body of the message.

The standard way to join a list is to enter: "subscribe <list name> <your name>" into the body of the message. The <list name> is the name of the mailing list, whereas <your name> is the user's full name, not his or her E-mail address. Most list request processors require users to give the name of the list they want to join, because the same processor may handle the subscriptions for many different mailing lists. Some lists may only require the subscriber's name be included in the command. A few lists, however, do need an E-mail address instead of the user's full name.

Some lists are offered in digest form daily or weekly for less active members who wish to keep up with the discussion. Some lists even provide methods for searching digests and other indexed data regarding a particular topic of interest. Users may even retrieve files or back issues of publications. The instructions on such advanced features as well as how to retrieve a list of commands are included in the welcome message sent to subscribers of lists that support them.

Warning: If the user's E-mail software automatically appends a signature file at the end of outgoing messages, he or she must disable this feature before sending a mailing list request. If the message is received by an automated system, the computer processing the request will attempt to interpret the lines in the user's signature, and he or she will receive a number of error messages in response.

It is also important to keep any instruction or auto-response messages received for each mailing list joined. These messages often contain the E-mail address that must be used to send an unsubscribe command, which is needed to sign off from a list. A different E-mail address is used to send contributions to a mailing list, although many users forget or become frustrated and send an unsubscribe request to everyone on the mailing list.

**LISTSERV Servers.** Interested in any given topic? There is a list of available mailing lists that can be received through FTP from various archives, or queried from any automated LISTSERVER. A listing of all known LISTSERV lists — a very large file — can be obtained by sending the "lists global" command to any available LISTSERV server address.

A mailing list that offers announcements of new lists and changes to old ones can be received by subscribing to NEW LIST by issuing a message addressed to "LISTSERV@vm1.nodak.edu" with the "subscribe NEW LIST <your name>" command in the body of the message. It is not as important to follow case-sensitive requests when issuing these commands or addressing messages on the Internet. Mailing lists often move and servers change frequently on the Internet, so the addresses listed in this chapter may change.

When messaging a LISTSERV, multiple commands may be sent in the same message — the only convention that must be followed is to place each command on a separate line. Some important commands include "help" to retrieve a help text file; "query <list name>" to find out how subscription options are set; and "unsubscribe *" (the universal command to cancel all LISTSERV subscriptions — however, some lists require that users unsubscribe to each list separately).

Individual LISTSERV servers handle different mailing lists apart from the few available "global" and "netwide" commands, which cannot all be accessed from any one LISTSERV address. Users must remember to always send such commands as "unsubscribe" and "help" to the server to which they subscribed to a particular mailing list. It is important that users al-

# THE ELECTRONIC MESSAGING ENVIRONMENT

ways send mail from the same E-mail address from which they subscribed, so that the LISTSERV can authenticate them.

Some lists offer additional features for which a password is assigned; details including how to change the password are included in the welcome message. Some of the more processor-intensive requests (e.g., global lists and database searches) may require time to process. Users should not send a request to another server unless they receive a bounce notification indicating that the server to which they originally sent mail is no longer in service. If the request was successfully executed by the LISTSERV server, users will receive a short verification message with use statistics. Automated subscription requests are usually processed in less than a minute; those processed by a human may take a day or longer, depending on his or her processing power.

Some mailing lists (e.g., LISTSERV lists) make their subscriber lists available to any user; however, they offer subscribers a method for hiding their names. To get a list of all subscribers' names and E-mail addresses that have not been hidden for a LISTSERV list, users can send the "review <list name>" command to the appropriate server. Users can choose to hide their names from such third-party access by sending the "set <list name> conceal" command to the server that holds his or her subscription.

Posts to mailing lists that are direct advertisements for a product or service are almost always viewed with extreme distaste on most lists and Usenet newsgroups that are not devoted exclusively to that purpose and may lead to "flaming," or heated, angry responses. Most users of the Internet resent being forced to read ads that they did not solicit, particularly if they arrive in E-mail as part of a mailing list to which they voluntarily subscribed. Discretion is key. Many users will gladly receive such information if it is relevant to the group's central topic and it is written as a legitimate contribution to the list's ongoing discussion. Some list members with no observable stake in a company will forward press releases and similar materials to a mailing list if they are relevant to a discussion; this is accepted more readily than if a person from the company that sells the product submits the message in multiple groups.

**File Transfer Protocol**

File transfer protocol (FTP) is the method used on the Internet to download files. Finding files can be accomplished using an Archie search, or, if the FTP site address where files are stored is known, users can directly access the server and scan the relevant directory.

Archives that are available on the Internet provide a variety of freeware and shareware programs for many different platforms, including IBM PC,

## Using Internet Resources

Apple Macintosh, and UNIX-based computers. Software companies post demos, bug fixes, and other software updates.

In addition, FAQs and requests for comments (RFCs) are widely available as sources of information on many different topics. Most are available for anonymous FTP, meaning the user logs into the host as "anonymous" to retrieve publicly available files. Some sites use the password "guest," while others require that users put in their E-mail address for the password. FTP software applications can handle this type of login transparently once the user's E-mail address is known.

**FTPmail.** An automated E-mail tool fosters the capability to transfer or download files by FTP. Much like automatic mailing list subscriptions, a carefully worded message sent to an automatic processor reads the contained lines in a computer program and executes it. It can FTP a file requested from a local server or an archive anywhere in the world. It then attaches the file to an E-mail message, which is sent to the requester's mailbox. This resource saves the user the trouble of signing onto the site and FTPing the file, which can be a two-step process when using a shell account. This is most popular, however, with Internet users who only have E-mail capability but wish to retrieve a file by E-mail.

Once a file's physical location is known, users must send a request to the server offering a system for retrieving the file. One such system is called FTPmail, named after the software used to accomplish the task. For use in the U.S., "ftpmail@ftp-gw-1.pa.dec.com" has been set up. As with all Internet services, choosing a server closest to the user's physical location reduces Internet traffic. For example, to retrieve a file called "test.txt" at "test.site.com" in the "/pub/information" directory, a user would send the following message to the FTPmail server:

>   connect test.site.net
>
>   chdir pub/test
>
>   get test.txt
>
>   quit

FTPmail accesses the server, changes to the correct directory (or path, in Internet parlance), FTPs the requested text file, and then quits. The text file is then returned to the user in the form of an E-mail message. Most applications, graphics, and other nontext files are called binary files, which require more commands. These files must be uuencoded, or translated to ASCII before transmission, where they are then decoded by the receiver. To get the file "program.exe" from "test.site.com" in the "/pub/binaries" directory, a user would execute the following commands:

```
connect test.site.com
chdir pub/binaries
binary
uuencode
chunksize 50000
get program.exe
quit
```

If users wish to specify an address other than the one they used to send the request, they must place the command "reply <e-mail address>" at the beginning of their message. The command "chunksize" can be omitted if the user's E-mail software or Internet gateway has no limits regarding message size. It can take hours or even days to receive files by E-mail, depending on the amount of traffic the FTPmail server receives. Once received, users need a uuencoding utility to reassemble its segments, which will enable them to translate the file from ASCII back into usable binary form.

**The Mechanics of FTP.** In an effort to handle the increasingly high traffic demands over the Internet, mirror site servers were established with identical archives that are maintained at various points on the Net. One particularly large FTP site sponsored by AOL — mirrors.aol.com — offers a vast, easy-to-access store of some of the most popular Macintosh and Windows files. Users are prompted to pick the site geographically closest to them. When users download files, they should do so at a time that is not going to interfere with the site's regular activity. In essence, users should not FTP during the standard daytime business hours for any given site.

Some sites, especially educational facilities, limit the number of off-site ports that are available during certain hours to keep traffic manageable. FTP site names usually start with "ftp." When searching for an unknown FTP server address at a specific domain, users should try "ftp.<domain name>" first (e.g., ftp.qualcomm.com). It is important to note that some sites do not allow anonymous logins, while others may limit the number of anonymous FTP ports. Most sites have a "pub" directory, where publicly accessible files are maintained.

Using a UNIX shell account to FTP a file is a two-step process. First, the file must be transferred to the user's remote host computer, which is often very fast because it can use high-speed Internet connections. Next, users can download the file to the their personal computer through such standard protocols as Kermit, XMODEM, or ZMODEM. If users are connected through a SLIP or PPP access, the file is transferred directly from the FTP

## Using Internet Resources

site to the user's computer. The transfer is then only as fast as the user's modem or network connection.

Files on the Internet are almost always compressed to reduce Internet traffic time users need to spend online. PC files usually are zipped, and have the extension ".zip" at the end of the filename. Macintosh files often have ".sit," "cpt," or "sea" (self-extracting archive.) These files are easily decompressed with freely available utilities. Macintosh files are usually binhexed ".hqx" for ASCII file transfer. UNIX files compressed with gzip have a "z" or "gz" extension. UNIX archives created with an application called tar have the "tar" extension.

Often, a file may have a combination of extensions, indicating that several translation methods have been implemented. Users need to begin by translating the protocol indicated by the extension at the far right of the file name and work their way left until the file is in usable form. For example: application.sit.hqx indicates a program that is binhexed and compressed.

Files retrieved from public FTP archives can be possibly infected with viruses. Users should always scan any downloaded programs with an up-to-date virus program before launching them. It is usually the case that only executable files have viruses; text files, E-mail messages, and image files at worst may be corrupted, but they cannot transmit viruses or Trojan horses. The only exception is PostScript documents, which can contain executable instructions and can be used to infect a system.

**Internet Shorthand**

Written communication lacks the humor and emotion of verbal communication. Visual shorthand, called "smileys" or "emoticons," represent several variations on this theme. To read the smiley :-), look at it sideways.

Internet conventions describe ALL CAPITAL LETTERS as a practice considered rude and referred to as shouting. The asterisk symbol (*) before and after a word or phrase takes the place of *italics,* a font-based convention that cannot as yet be transmitted by the ASCII-only TCP/IP networks.

In addition, Internet users employ many abbreviations and symbols to get their message across quickly and succinctly. Emoticons indicate symbols that look like a face when viewed on their sides. These symbols indicate sarcasm or inflection that otherwise might not be understood in a plain-text message. Some of the most commonly used acronyms and emoticons include the following (the first three are used in live interactions known as "relays" or "chats"):

|     |     |
| --- | --- |
| AFK | Away from keyboard |
| BAK | Back at keyboard |

# THE ELECTRONIC MESSAGING ENVIRONMENT

| | |
|---|---|
| BRB | Be right back |
| BTW | By the way |
| FWIW | For what it's worth |
| FYI | For your information |
| IMHO | In my humble opinion |
| LOL | Laughing out loud |
| NRN | No reply necessary |
| OTOH | On the other hand |
| RTM | Read the manual |
| TPTB | The powers that be |
| YMMV | Your mileage may vary |
| <g> | Short for <grin> |
| : ) or :-) | Basic smiley |
| ; ) or ;-) | Sly wink |
| : ( or :-( | Unhappy |
| :> | Sarcasm |
| O :) | Halo |

## USENET NEWS

The Usenet is an extensive compilation of bulletin boards that cover practically any subject for discussion. More than 20,000 newsgroups are now in existence, and most Internet service providers (ISPs) carry an average of 10,000 topics. Internet users find group discussions on current events, science, society, computers, and alternative views on anything they could ever possibly imagine.

### Reaching Communities through Usenet

Usenet newsgroups, like LISTSERVs and mailing lists, serve as a forum for discussion on any given topic of interest. Newsgroups vary greatly in traffic. Many are organized in a hierarchy used to name the groups, where the main classifications include the following:

- alt — alternative
- bit — gatewayed
- BITNET — mailing lists
- biz — business

- comp — computers
- rec — recreation
- sci — science
- soc — society
- talk — discussion

Usenet newsgroups are much like conferences in the typical BBS, except that the messages posted to Usenet are echoed to servers across the globe so postings are seen by a vast number of people. New newsgroups are added daily. Newsgroups are often added within hours of a globally reported tragedy; for example, two newsgroups appeared the day of the April 1995 bombing in Oklahoma City to discuss this traumatic topic.

Users often choose newsgroups that tailor to their interests or needs. Messages are threaded, meaning there is an initial post and then a series of responses. The groups are categorized under several prefixes. The most common group categories include the following:

- *Alt.* A catchall category that generally offers the most frivolous groups, but also contains a balance of technical topics (e.g., television, computer security issues, communications, and software); however, because of the explicit nature of this category, not all sites carry them.
- *Bit.* Devoted to reprinting the contents of various LISTSERV mailing lists.
- *Clari.* Groups offered by ClariNet, a licensed gateway news service that requires payment for use, although some ISPs offer access to this feed as part of their basic fee or for an added charge.
- *Comp.* Computer-related topics and issues.
- *Misc.* Groups that do not fit in elsewhere, including announcements, job offerings, and for-sale items.
- *News.* Topics related to Usenet itself and the software that runs it; mainly of interest to Usenet administrators.
- *Rec.* Recreational topics.
- *Sci.* Scientific research.

Usenet newsgroup messages sometimes contain binary files that are uuencoded using a UNIX ASCII translation scheme and often have a ".uu" or ".uue" filename extension. Users who do not have high-speed connections usually find it faster to locate an FTP site that contains the same file.

## FAQs

Users should always first read the frequently asked questions (FAQs) file — if it exists — before posting to any newsgroup. These files are created and maintained because the newsgroup has already had many basic and complex questions that have been dealt with in an effective and fairly conclusive manner. To ask a question that has been dealt with in this way is considered rude, and often will get users flamed, either publicly or directly to their E-mail. Newcomers (newbies) to the Internet often hit this pitfall; however, several users escape such labels by lurking through the group without immediately joining in until they are reasonably familiar with a newsgroup's flow.

A few newsgroups regularly post their FAQs in the form of a message in the group. Some groups require new visitors to FTP the file from an archive. The main archive for FAQs is stored on a server at the Massachusetts Institute of Technology (ftp://rtfm.mit.edu/pub/usenet).

FAQs are useful for questions that should have an answer readily available. Reading these FAQs is an easy way to make certain a particular newsgroup is the appropriate place for queries. If the quest for information becomes too time-consuming, however, users can often turn to a human resource that is often much faster. If too much time is wasted searching FTP sites or waiting for an FAQ to appear, the purpose of using the Internet is defeated. After users study the procedures, they are ready to post. Some questions might appear to be insignificant, but they may be unanswered and could spark a new debate or discussion that is of interest to the entire group.

FAQs are an excellent resource for the beginner learning a new field. Browsing the FAQ list may uncover a plethora of valuable information on several topics that were collected simply because people became tired of answering the same questions over and over again. Although some FAQs are highly technical, many focus on the problems of a user of a particular product or technology. Newsgroups that are more structured request that users post a message only after all other available options for finding an answer have been exhausted. These groups are often archived, which means all or edited selections of past messages are available for FTP. They also may be searchable by WAIS servers.

**Usenet Netiquette**

The following advice is useful when users participate in Usenet discussions or mailing lists on the Internet. Whenever a question is answered on the Internet, even if it is not a useful answer, it is Internet protocol to send a short thank-you by E-mail. If newsgroup answers to a newsgroup query are sent only to a user's private E-mail, it is common net courtesy to compile and post the responses to the newsgroup so others will also benefit.

Usenet posts are often replicated repeatedly to reach a large number of servers across the globe; therefore, each line of text adds to the Internet traffic. Quoting back entire messages can sometimes provoke flames in some corners of Usenet; so users should cut quoted material down to the minimum needed to convey its meaning effectively.

## Using Internet Resources

It is also useful to use a subject line that elucidates the meaning of a message so readers uninterested in the topic can skip it. If the subject is changed in a thread, the subject line should match so readers that filter the newsgroup or are otherwise ignoring a thread can note the change in focus.

Long paragraphs should be segmented into sections to make the topic more digestible. Cross-posting messages to a number of newsgroups should only be done when the user is certain the message relates to a number of groups. Mass cross-posting of any commercial message or spam only invites flaming. If a user wishes to test their newsreader's ability to post, they should send a test message to a group devoted to the task. "Alt.test" and "misc.test" are two examples. Autoresponders at some sites that carry Usenet groups may send an E-mail message that will allow users to determine how far around the world their message reached. Automatic responses can be prevented by typing "no reply" in the test message's subject line.

It is a good idea to check replies to Usenet posts to make certain they are not being inappropriately cross-posted to other newsgroups. A common Internet hoax is to post an inflammatory message that is cross-posted to one of the Usenet test groups. Unaware respondents find they are flooded with automatic responses from Usenet servers all over the world.

Users might also notice that some Internet users resent users who post to newsgroups from gateway Internet services such as America Online (AOL), Prodigy, or CompuServe. They tend to hold these users in contempt because many users rush to post into newsgroups without the benefit of learning the ropes when the initial gateway was opened. Since then, AOL users among others have been considered intrusive to the Internet culture. In response, some newsgroups have been created simply to flame AOL users.

Usenet newsgroups permit businesses to keep current with related corporate news and products. In addition, it also broadens a company's exposure to potentially millions of Internet users. Companies should take care not to make the error a few businesses have made regarding Usenet. Broadcasting unsolicited advertisements (spam) will only bring hate mail from many angry users. Newsgroups are intended for specialized discussions, but marketing indiscriminately is looked upon with disdain.

**Usenet Filters**

Usenet provides such a wealth of content that many users implement methods for filtering out what they do not need. Configurable kill files, offered by some newsreader programs, are one method. They can eliminate posts from specific users or sites, or using any number of other criteria, including words mentioned in the text of posts. Usenet news messages can even be filtered according to user-defined criteria.

## THE ELECTRONIC MESSAGING ENVIRONMENT

**TELNET**

Telnet, and an IBM variant called tn3270, are programs that allow users to log on to a remote computer. Telnet emulates a dumb terminal where the programs being executed are located entirely on the host computer. Telnet is much like being logged onto a BBS. The user generally is prompted for a name and, in some cases, a password to gain entry. He or she then works through the text-only command line interface by selecting menu items.

Some Telnet addresses require a port number, which usually is separated by a space after the main server address. For example, port 23 is the standard connect. To Telnet to the main CompuServe service, users can type: Telnet CompuServe.com 23.

Many companies offer Telnet, though generally for their employees' use only. The U.S. Government offers many free Telnet sites, including FedWorld, offering access to a large number of government databases. It can be accessed by Telnetting to: fedworld.gov.

Several Telnet sites, including FedWorld, are actually dial-up BBSs that have been linked to the Internet. The advantage to accessing them through the Internet is that users avoid paying long-distance phone call charges that could be across the U.S. or the world. Some BBSs are free, yet others require a subscription.

**Telnetting to Online Services**

After AOL released its revised client software for Macintosh and Windows in late 1994, it began offering TCP/IP access as an alternative to its regular SprintNet and Tymnet local access nodes in North America. This access method has the following advantages: faster downloads for users with SLIP/PPP high-speed modem access or direct Internet connections, a local connection method for users lacking a local dial-up node from one of AOL's providers, and an alternative method for accessing the service when local nodes are busy.

The only drawback is that users must pay an hourly connect rate for Internet access, which will be charged by both AOL as well as the Internet provider. In addition, usernames and passwords are sent out over the Internet unencrypted; therefore, users who access AOL by TCP/IP are at risk of having their passwords intercepted and stolen. In addition, when connections are accidentally dropped by the AOL server or disconnected improperly, they are not immediately timed out by AOL. This makes it not only difficult to reconnect, but can lead to excess charges by AOL for time not actually used. Such online services as CompuServe and Delphi support Telnet connections without the elaborate GUI — users can reduce connect charges when they have no local access nodes, but do have a low-cost Internet connection.

## GOPHER

Gopher, once considered the most high-powered tool for navigating through the Internet, now lies deep in the backdrop of the Web. Gopher's main advantage is that its gopherspace is readily searchable using such tools as Veronica and is easily linked to other information sources (e.g., WAIS search engines, X.500 directories, and WHOIS servers).

Gopher is list driven, often leading from the general to more specific content. Although sometimes organized as erratically as the Web, Gopher generally is quick and easy to use. To use Gopher, users need only navigate by picking an entry from a list. The University of Minnesota, where the Gopher protocol was created, maintains the most heavily used home site, so directions to other Gopher sites are frequently given from this starting point (Gopher://gopher.tc.umn.edu, port 70). At times, a Gopher server's address is given for direct access. Gopher client software usually allows the user to maintain and edit a list of bookmarks that offer quick access to useful servers that contain pointers to other servers.

Gopher server addresses look much like other types of site addresses (e.g., FTP), with the exception that they often start with the word gopher. When searching for an unknown Gopher server address at a known domain, users can try "gopher.<domain name>" first (i.e., gopher.microsoft.com). Gopher servers usually offer access through port 70, by default. In a case where a different port is assigned, users should make certain to input the correct number.

The only drawback to Gopher is that the client software does not always offer a method for directly entering selector strings, which are essentially subdirectories on the Gopher tree at a particular server. The user must often access the top-level Gopher directory (i.e., peg.cwis.uci.edu in the above example) and then search for menu items that delve into the subdirectories of the tree. Although this is not always successful, many Gopher clients are imitating Web-style addresses called uniform resource locators (URLs), which follow a standardized format and allow convenient resource access. Modern World Wide Web browsers can interpret Gopher URLs among others.

### Veronica and Her Gopher

Veronica permits keyword searches of several Gopher server menus. The search results can connect users directly to the source of the data. Veronica is accessed through the standard Gopher client and yields access to several different data types.

Veronica search results are automatically produced in a Gopher menu according to the user's keyword specification. Menu items are obtained from many Gopher servers. These items are immediately accessible

THE ELECTRONIC MESSAGING ENVIRONMENT

through the Gopher client. GUI clients allow users to read files, perform other searches, or access many Gopher sites.

**WORLD WIDE WEB**

The best resource for interactivity on the Net is the World Wide Web (WWW). This resource arranges and catalogs linked resources through graphical user interface (GUI) software, the most popular of which is Netscape Navigator, which fully formats information, references, and archived resources with extended text (fonts and markup embellishments), icons, and graphics. Users can navigate through a simple point-and-click environment. The WWW transparently sends users to any sites regarding related information and resources. The WWW is growing phenomenally as a visually attractive resource for potential customers.

Users no longer require full interactive access to the Net to access the WWW. All the major commercial online service providers have a Web browser that lets users navigate the Internet. In addition, several new tools are on the market to make writing hypertext markup language (HTML) easy for businesses wishing to control their own content on the Web.

**Links Everywhere.** The most popular method these days for gathering information on the Internet is through the WWW. The Web is not only well suited to information gathering, but it is the most graphical way available to disseminate information in an easy and effective manner. Browser applications (e.g., Mosaic, Netscape Navigator, Microsoft Explorer) allow users to click on hypertext links, sending them anywhere on the Net to find their desired information. These are generally depicted as colored and underlined words and phrases used to gain access to other pages or collections of text or multimedia elements that may be located on the same server or in another location.

It is called the Web because pages at different sites are interconnected by hypertext links, while access to different sites can be anywhere in the world. Both Netscape and Mosaic browsers offer a hot list feature, which can be used to store bookmarks of frequently visited sites in the same manner as Gopher clients. Web pages, written in HTML, include embedded graphics, which are esthetically pleasing and make using the WWW more like reading a magazine or newspaper.

Different types of links enable a user to download files or programs, play sounds, or view movies as long as the user's client software is configured to accept it. Examples include a verbal welcome from a university or company president, or an MPEG or QuickTime animation that demonstrates a product. The main page that users see when they sign on to a particular service is known as the home page, which usually has links to other documents within its site or other similar ones.

In some cases, links can be within graphic image maps, although the more considerate sites offer a text-only version for those users whose browsers do not support graphics or who lack sufficient bandwidth. Browsers often employ helper programs — freeware or shareware applications that separately display graphics, play sounds, show movies, and extract binary files from ASCII-encoded messages.

**Download and Search Performance.** The major drawback to using a Web browser is low bandwidth. Users with a dedicated network connection through a T1 or other high-bandwidth delivery system may be able to receive very large graphic or movie files in seconds or fractions of a second. Those who access by modem — even at speeds as high as 28.8K bps — however, may have to wait minutes or hours to retrieve the large images. Small graphics can also offer a significant delay at 14.4K bps, although many of the new Web browsers load images after the text has finished loading, which allows the user to continue work. In addition, such features as auto load of images can be disabled. Some sites list the size of movies and sounds that are attached to links so a user can estimate the download time before choosing them.

The Web does not always have items linked in an orderly fashion. Browsers are aptly named because users must try out different links, often leading to few sites with useful data. It is very easy to waste significant time searching endless gigabytes of useless information. Knowing the site address where a particular piece of data is stored does not always mean it can be rapidly found and extracted. Browsing, however, can often lead (sometimes accidentally) to useless results.

Some people find it useful to try both Gopher and WWW tools if a site offers both. Data is sometimes easier to access with one resource, or may only be offered by one of the services. Gopher, FTP, Telnet, and even Usenet newsgroup servers are accessible through the Web. E-mail can be sent using HTML "mailto" links with most browsers. The WWW has quickly earned a reputation as the most flexible of the Internet access methods, allowing access to the most services and features with a single application.

**Uniform Resource Locators.** Web sites can be accessed directly by inputting the URL. The URL often contains a Web server address followed by subdirectory information and, in some cases, personal user login information (e.g., FTP passwords) for direct access to a particular page or resource.

URLs for Web addresses always begin with the letters "http" for hypertext transfer protocol, followed by a colon and two backslashes to signify the beginning of an address. It is then followed by the server address, which often starts with "www." Port numbers, when required, are preceded by a colon. Some URLs may have a number of odd characters in them (e.g., tildes, ~), and many end with the "html" filename extension.

For example, if a user wants to read *Time* magazine's technology articles on the Web, their browser might list the following URL: http://pathfinder.com/@@LJPyhaGf5gAAQOdI/pathfinder/pulse/news/time/timetech .html/.

When seeking an unknown WWW server address at a known domain, users should attempt to type in: http://www.<domain name> to try and access the site. There is a UNIX convention that is true for all Internet services, including FTP and Gopher: slashes in URLs are the opposite of slashes used for directory paths on PCs.

Some browsers will default to http:// if no protocol is given; others require users to type in the complete URL each time. URLs that do not end in .html, the PC three-character equivalent of .htm, or some other filename extension (e.g., .txt, .gif, or .jpg,) generally require a trailing slash. Many browsers are forgiving and will allow this trailing slash to be left off, but some will not connect properly unless it is included. The distinction here is that URLs with filename extensions appoint HTML filenames, while URLs with the trailing slash represent a directory structure similar to FTP or Gopher. It is also important to note that URLs are case sensitive; the convention is for most characters to be lower case, although upper/lower case exceptions (e.g., in the Time Pathfinder URL example) must be entered exactly.

URLs that indicate access to another type of service begin with the name of the protocol. FTP servers accessed through the Web have URLs that start with ftp://, and Gopher URLs start with the gopher:// designation. Telnet URLs begin with telnet:// and typically require a client-side Telnet application to be configured as a helper program.

If a WWW server address is not known, many search engines and indexes are available to help a user find these resources, even though no centralized catalog of all Web offerings exists. Some features require a Web browser that supports forms, a method for inputting data into onscreen fields so that it can be transmitted to the host and processed by a script on the remote server. Some sites (e.g., Lycos, found at http:/lycos.cs.cmu.edu/) run programs, known as spiders or 'bots, to automatically and continually map and index the Web.

Many effective WWW resources are Web pages that are made available by other users — who often provide their services for free — and contain carefully organized directories of related links. Most will offer an E-mail address or allow a user with forms support to add or suggest new entries. Some of these sites are not well maintained, so users may discover a large number of invalid or broken links.

## WIDE AREA INFORMATION SYSTEM

The wide area information system (WAIS) is an excellent means of finding information that can be accessed through the Web. If companies have a large database, WAIS lets users search their product offerings and allows customers to identify exactly what products/services they want.

Most business information is provided on the Net in a passive manner that allows customers to come to them. Some companies may subtly advertise by offering their expertise through discussion forums and document archives. Although these restrictions may sound confining, it prevents them from being flamed or receiving numerous angry messages from the majority of users who are highly appalled at Internet advertising. WAIS servers permit users to get information from a plethora of hosts. The user inputs the request into a client application (usually a WWW client), which then searches several WAIS servers. The user then is given documents that answer the user's search criteria.

## OTHER HELPFUL TOOLS

Numerous other tools are available for using the Internet. Some are platform-specific, while others (e.g., WAIS) are available from within clients (e.g., Gopher) or separately with a specific client application. WAIS is particularly useful because it rates the quality of information that matches a user's selected search pattern that has the highest degree of match closest to 100%.

Many standards are instituted on the Internet, and many tools support those standards. Users should start by using the ones they feel are the easiest to use, then try a variety of tools that accomplish the same job to find the best one.

For example, users should not try using only one Usenet newsreader — assuming they have access (e.g., SLIP or PPP), they can check several freeware or shareware options. They can then choose the one with the features that suit them best. Some newsreaders have built-in binary file extraction, while others allow users to conveniently save multiple subcategories of groups to address users with a variety of interests. Some applications allow offline reading of newsgroups — a convenience for users with high Internet connect-time fees. If the product a user ends up using is shareware, they are obligated to remit whatever fee the author requires or delete the software from their collection.

### Videoconferencing

Another method for finding information through the Internet is person-to-person contact in real time. Although videoconferencing through the Internet is available in a test capacity in some quarters, it consumes large amounts of bandwidth and is still unavailable to most users — however, shareware options do exist (e.g., CUSeeMe). Alternate methods are available for online person-to-person contact that include the use of Talk for one-to-one contact with an individual, although it is not recommended that users contact a stranger in this way. Most users ignore or turn off Talk request capabilities, so it is unlikely to achieve a response, although it can work if users plan in advance with someone else.

## IRC

Internet Relay Chat (IRC) is a popular method for finding quick answers to questions on a variety of topics. IRC chat areas, called channels, are designated with a "#" symbol before their name. Enlisting patience will make it is possible to drop in on an IRC session devoted to a particular computer platform (e.g., #Macintosh) and receive an immediate response to a query. This method does not have as far a reach as posting to a newsgroup. If an in-depth discussion of a topic involving the maximum number of people is sought, IRC will not satisfy those needs. For an immediate response in real-time, its appeal cannot be disputed.

Long-distance voice communications are increasingly possible over the Internet with shareware offered for Macintosh and PC by commercial companies. The quality of connections is questionable, and the software is usually limited to minute-long discussions until the shareware fee is paid. In addition, the various offerings are not compatible with each other at this time, thereby limiting users who can contact each other. It can cut down overseas telephone charges, however.

### Finger

Finger is an excellent location resource that can be used if a user knows another person's domain address. They need only type: finger user@domain, or finger @domain, to see a list of users who are currently logged on. If the partial user name is typed, the host will list all the names that match the inquiry. It is a good idea to start with a broad search, then narrow criteria as appropriate. Users should type as little information as possible, then use the all-encompassing tool: guessing.

Users can improve their chances of finding what they want. Some users store and update useful or interesting data in their plan files (i.e., daily news), and they usually advertise this in their E-mail or Usenet signatures. Most commercial sites decline Finger requests, however, because it is a known hacker tool often exploited to discover weaknesses in a private network.

## SUMMARY

The Internet is an immense information resource that is limited only by the total population of its users and their ability to contribute. The Internet itself is founded on donations of human labor and resources (e.g., servers, networks, software, publications, and databases). Commercial ventures have constructed sites that offer free information, and some of these services are moving from a free to a paid subscription service.

As the Internet makes the partial transition to a commercial environment, the standard of volunteerism and the knowledgeable users helping the newcomer remains. The unwritten code is that one should give back as much as one takes from the Internet. Because of the enormous open archive facilities on the Internet, users can find information in numerous sources if they persist. But mailing lists and newsgroups, for instance, work only because their members give out answers or opinions nearly as often as they ask for them.

Finding useful business information on the Internet, like anything else, becomes easier over time. The best rule is to observe and practice using the wide variety of available access tools to become comfortable with the Internet's conventions before actively joining in and contributing. All in all, the Internet was designed as an open environment that encourages the free flow of information. The tools described in this chapter should give users a comprehensive overview of what is out in cyberspace and how to best obtain it. Happy net surfing!

# Chapter 19
# Netting Web Customers

*Stewart S. Miller*

The Internet personifies the ultimate in communication. Its primary purpose is to transmit data. Internet users will find new and increasingly efficient methods to exchange data, be it business transactions, ideas, or social commentary. The World Wide Web is now widely accepted as a powerful business tool for the dissemination of information to prospective clients. As a means of advertisement, it offers the ability to reach thousands of customers at minimal expense. The Web is much like a 24-hour infomercial — all a business has to do is promote its site so people can shop there.

It is imperative to develop an Internet business plan, finite policies and procedures, adequate file security, comprehensive training, and help and technical support from an Internet service provider (ISP). Patience is the rule. The benefits of researching this medium will provide an excellent education in the future of communication and in understanding and reaching a vast online customer base.

## DOING BUSINESS ON THE INTERNET

Business and marketing activities are gradually gaining increased acceptance on the Internet, considering the switch of funding from the National Science Foundation to private enterprise. An increasing number of businesses now advertise and sell products on the Internet, and there are certain acceptable commercial practices and approaches that explain how to do this without making any virtual enemies along the way.

The best and most accepted way to advertise on the Internet is through the World Wide Web (WWW), which allows a seller to organize and index resources. This effective means of promotion uses an easy-to-use graphical user interface (GUI) software. Most commonly available browsers include Netscape Navigator, Mosaic, and Microsoft Explorer. Most sites, however, are configured to take advantage of the enhanced GUI features Netscape has to offer.

Customers can access esthetically pleasing pages of information that both reference and archive resources. WWW pages embody extended text, fonts, and enhanced formatting. Navigation tools include icons, graphics, and mouse-driven point-and-click links for customers to readily research and explore product information. The Web allows sellers to transparently send users to other sites on the Internet in search of related information and resources. It is an effective virtual marketplace that attracts attention and draws potential customers.

At first, businesses faced significant limitations on the Web because users needed full interactive access to the WWW resources, which limited the market to less than half of all users. Because all major commercial online service providers now offer WWW access, however, the market share for Internet users has grown dramatically. This expansion is an excellent indicator for marking the advantages of doing business over the Internet.

For those businesses that must prepare and maintain WWW documents and databases (time- and resource-intensive tasks), several user tools exist to greatly simplify the act of formatting documents with required HTML. And whereas businesses were once required to maintain servers connected to the Net over an expensive, very high-speed connection to eliminate performance bottlenecks, today numerous ISPs offer to host and maintain a company's Web site on their servers at a reasonable cost, depending on the amount of information and traffic the business handles. Overall, the cost has decreased while the potential for generating revenue has increased.

The alternative to building Internet and WWW services for business purposes from the ground up is to contract with service companies (e.g., the Internet Company of Cambridge, Massachussetts, and NovX of Seattle, Washington), which build and maintain a company's presence on the Internet. The expense of contracting with these services offsets the cost of staff networking and systems experts. An excellent starting point for comparison shopping is to go online and point your browser to http://www.thelist.com. This site systematically indexes ISPs by location and area code.

## INVESTING IN COMMERCIAL CYBERSPACE

Businesses are attracted to the Internet for one simple reason — to achieve a good potential return on investment. For an investment as small as $30 a month, anyone with a personal computer and a modem can set up a base of operations on the World Wide Web. Web pages have the potential of being seen by millions of users on the Internet.

The potential number of consumers the Internet can reach has companies rushing to get onto the Web. Most businesses know the Internet exists and instantly know they want to be part of this phenomena and for good reason — executives realize that this electronic medium has the potential of being the main marketplace of the 21st century.

At present there are greater than 100,000 Web sites, a number that is forever increasing. World Wide Web sites such as the Yahoo guide (http://www.yahoo.com) list more than 23,540 companies, while a site directory called The Shopper lists over 370 shopping malls on the Web (http://www.hummsoft.com/hummsoft/shopper.html). The Web defines a new commercial era in which a single electronic medium combines text, voice, video, and graphics. Businesses that will be the first to embark into this virtual community include publishing, banking, retailing, healthcare suppliers, insurance, and legal services.

Nearly two thirds of Web users have a college education, while half are in professional and managerial occupations. Most users spend about five hours a week using the Internet for E-mail and Web browsing. A growing number of users are female, whereas this was a male-dominated field only a year or two ago.

Business opportunities on the Web will require a new outlook regarding advertising and marketing. Traditional broadcasting and print sent a one-way message to a passive consumer. Alternatively, the WWW has enormous potential as an interactive two-way medium. Interactivity draws people back for repeat visits to Web sites. The key is to make documents both appealing and rich in useful information. The more often the content is updated, the more appeal it has for users to visit the site again. Consumers who use the Web are independent and do not tolerate standard or direct advertisements.

Businesses that use the WWW as a passive advertising medium do not risk alienating large groups of "netizens" (customers on the Internet) or run the risk of violating the unspoken law of the Internet. However, in allowing customers to come to a company's site, the business generates a positive virtual storefront that promotes sales, hits (accesses), and repeat business.

The Internet is an information-based community reaching thousands of new customers. When planning a Web site, it is important to decide what unique content can be offered, how much can be expended for connectivity, and price check different access providers to achieve the best return on investment. Adding information and expertise that enriches the Internet builds a positive reputation and also improve sales revenue.

## NETTING WEB USERS

Web use has grown phenomenally since its inception only a couple of years ago, mainly as a result of its user-friendly interface, flexibility, and ease of authorship. Many individual users as well as major corporations, newspapers, and government bodies have home pages on the World Wide Web. Web traffic is second only to E-mail, with an estimated 2 to 5 million users accessing the 15,000 or more Web servers packed with data, images, sounds, and links to additional sites.

The fastest-growing segment of the Internet, the graphics-based World Wide Web (WWW or Web) has become the present battleground for corporate America. Storefronts, virtual malls, and corporate home pages have emerged all over the Web, changing its landscape from a relatively unknown curiosity used almost exclusively by academics and government employees to a virtual Las Vegas with all its beauty and glitter.

Businesses that establish only a basic Web presence with a few pages of stationary information charge anywhere from $5000 to $10,000 a year. However, users are attracted more to sites that constantly update their information with fresh ideas. Companies that update the site frequently with several links charge about $60,000 to $120,000 a year. Some companies pay more to instantly link their site to the world by signing with such companies as Time Warner's Pathfinder. For a price, a site that receives numerous hits a day is automatically linked to another — instantly generating substantial traffic.

**USING THE WEB AS A MARKETING TOOL**

Creating a presence on the Internet fosters constant contact between the business and its customers while giving a company an edge over its competition. Many businesses are adapting the Internet as a marketing tool. In the near future, businesses will find the Internet is a great asset that finds new customers. Successful Web pages are linked with several others, increasing the reach of a company's content to many more customers on the Web.

Finding success on the Internet is much more involved than creating a home page and placing it on the World Wide Web. When building a site, a company should try to tailor it with its customers' interests in mind. The site should be used to create a resource of information that attracts customers, in much the same way as a magazine intermixes content with advertisement. Attract users with content — once there, it becomes possible to sell them services or products.

Many businesses only wish to simply establish a Web presence at first, rather than use it to promote the company's sales and marketing efforts. The Web is much like a marketing executive who gives prospective clients company information 24 hours a day. Companies now integrate their WWW marketing into the overall marketing effort. As a rule, all print advertising should always include an E-mail address and Web site address, giving customers an easy path to use their Web browser to look at the Web page.

## TELECOM REACHES THE NET

Communications operators are building out their networks in an effort to provide businesses with Internet access in many major cities. In addition to traditional Internet access providers, major online service providers also plan Internet offers that present strong competition to communications ventures in this area. America Online (AOL), the largest commercial online service, has acquired O'Reilly & Associates' Global Network Navigator (GNN). This site is one of the most highly visited sites on the Web that offers a prevalent Internet directory. In the future, AOL will use GNN as the formation of a new exclusive Internet online service.

All users have the option of creating their own Web sites through the AOL service. Each user is afforded 2MB of space for each of the five screen names each user can have. This leads to a total of 10MB of available space for each standard user account. In comparison, Web providers usually offer one quarter to one half of a megabyte for a standard Web presence.

The decision a company makes in selecting an access provider to use for a client's project is extremely important. Access providers, much like telephone companies, sell access to the Internet. Some local companies only sell access in a specific region, while national ones have points of presence (POPs) in many major metropolitan areas in the U.S. Serious business platforms will find the Internet to be a valuable virtual medium that presents a business tool of the next century.

## WEB PIONEER DEVELOPERS

Many software developers are producing applications that help businesses save time and money in search of a professional developer. Because most companies do not have in-house developers who use HTML, the language that instructs the browser to implement the graphics and text on another browser, software products that write HTML are in demand. These programs automatically create HTML and allow a cost-effective solution to produce and control the material on a home page.

### Earthlink's Web Solution

Earthlink Network Inc., Southern California's largest Internet access provider, is establishing itself as a comprehensive provider for World Wide Web transactions. It offers businesses Web sites and high-speed Internet access at reasonable rates as part of their new services. Earthlink's setup fees range from $1121 for 25 megabytes of disk space to $5606 for 400 megabytes of space, whereas monthly maintenance fees start at $109. Earthlink offers shared addresses for businesses with a tight budget. The setup fees start at $224, while monthly maintenance fees are between $30 and $60 for less than 10 megabytes of space.

Although businesses who pay a premium are seamlessly integrated into Earthlink with their own domain, shared sites usually indicate the Earthlink's site address with the company name indexed as a subdirectory. Sharing an address on Earthlink's server is an excellent way to develop a small company's Internet efforts. In comparison, companies who set up their own Web sites using their own software spend an estimated average of $30,000 to $100,000.

### BBN Planet's Web Advantage

BBN Planet Corporation, a subsidiary of Bolt Beranek and Newman Inc. (BBN), offers the Web Advantage service. Customers using Web Advantage are given the ability to develop a corporate presence on the World Wide Web. BBN Planet provides its customers with a dedicated Web server connected to a router (part of the high-speed Internet backbone network.) These servers are capable of transmitting high-volume publishing traffic directly onto the Internet when the customer's server is colocated at one of BBN Planet's network centers.

Web Advantage service provides businesses and organizations with the ability to produce large amounts of data on the Internet, alleviating the customer's responsibility to install, operate, and maintain the required computing and networking resources. Web Advantage offers electronic publishing for commercial, educational, and government organizations. The servers route heavy-volume publishing traffic through the backbone of the Internet without increasing the load on the customer's normal Internet connection. BBN Planet's colocation service offers Web Advantage users full 24-hour, 7-day monitoring of their server's electronic information delivery.

BBN Planet's third-party partnership program for the Web Advantage service produces substantial content for World Wide Web sites. BBN Planet, in combination with its partner organizations (e.g., Media Circus), promises to take full advantage of this new medium. Web Advantage services connect to the Internet using 10M- to 45M-bps links, which enable the data to flow to its destination without experiencing bandwidth bottlenecks. Web Advantage service pricing starts at $2000 per month for 10M-bps service.

### O'Reilly & Associates' WebSite

O'Reilly & Associates, Inc. has its WebSite product, which is a 32-bit World Wide Web server for the Microsoft Windows NT 3.5 and Windows 95 operating systems, that lists for $499. The WebSite software is designed for businesses, organizations, educational institutions, and home users who publish and do business over the Web. This type of software makes it possible for anyone to create and control their own publishing and business presence on the World Wide Web.

Establishing a Web server is an expensive operation to set up and maintain. Most businesses are using low-cost, user-friendly NT server-based systems that enable users to publish their own information on the Web.

**THE INTERNET POTENTIAL**

Estimated Web revenues will grow exponentially in the next few years. Revenue will be generated from spending on the Web's infrastructure, charges for access and online services, advertisers paying to reach the high-income consumers who browse the Web, and direct sales. The Web may prove profitable for news organizations, publishers, and other content providers in an information medium. Web sites will find that simply providing information is only the beginning. Events that allow users to congregate may be the best inducement to generate traffic (e.g., providing specialized newsgroups and chat features.)

Magazine publishers merge content, readers, graphics, brand names, and advertisers when establishing an online presence. These assets yield the ability to create successful sites of interest. Venturing into the Internet may generate a high degree of customer interest for companies with a World Wide Web home site. Although the Internet is still young, companies find they receive about 1000 to 2000 hits a day, while popular companies get more than 100,000 hits a day.

Many of these sites attract a substantial number of hits; however, products sold using the Internet are the exception and not the rule. The main reasons that discourage commerce are the lack of security and Internet breaches and lack of customers with the right purchasing power. These reasons, however, are becoming out of date.

With the advent of Netscape 1.12, security has been enhanced for purchases. The Internet was designed to be an open system; however companies such as Netscape are making strides to secure and encrypt data to deter and stop information theft. In addition, a growing number of female users with buying power are on the Net.

Companies that have gained popularity have placed their full electronic catalogs on the Web and several others place order forms online, which enables customers to order directly. One site, called CDNow (http://www.cdnow.com), has its full line of music listed through a powerful search engine, as well as full ordering information complete with prices and shipping. Businesses that take the time to make their sites rich with information and ease of use will dominate the online market that will pervade purchasing to a greater extent in the near future. Heavily linking a company's site to others is the best way to attract prospective customers. To generate repeat traffic, however, Web pages need be interesting and changeable enough to educate, entertain, and inform users.

**SUMMARY**

When is the best time to enter into the Internet? The answer is: yesterday. There is no better time than the present, because the future will be inherited by those who take advantage of this technology today.

The Internet is growing faster than anyone could have anticipated. An increasing number of businesses see the future as a growing electronic village where most transactions and comparison shopping will be done at home through the Internet. The Web has grown out of its infancy and into adolescence; an awkward time for companies, to be sure. Those who do not get on the Internet will find themselves severely underpowered on the information superhighway. Because of the quick pace of technology evolution, it is of paramount importance to check out the WWW and determine what kind of presence will benefit your company in the growing electronic village.

# Section II
# Leveraging the Electronic Messaging Infrastructure

Once the messaging system is implemented and functioning at the required levels of availability and reliability, the organization needs to develop ways to add value by leveraging the investment in the system.

Mail-enabled applications that ride the messaging infrastructure — such as calendaring and scheduling, workflow, task and document management, and electronic commerce — enhance productivity and better position the organization to compete in the global electronic marketplace. In this section we take a look at how Lotus Notes enables organizations to exploit and extend their messaging environments in groupware applications and executive information systems. A chapter on Notes versus the Internet (Chapter 24) points out the advantages and disadvantages of deploying these technologies within an enterprise.

Electronic data interchange has now been engulfed by electronic commerce, which includes messaging, transaction processing, and just recently, Internet communication. Terms, concepts, and issues relating to electronic commerce are explored as these environments are constructed in real-life scenarios.

For electronic commerce to reach critical mass, a trusted infrastructure must be developed to permit the exchange of confidential information. In this section we discuss the requirements for creating an electronic commerce environment that rests on messaging.

Most electronic messaging systems are becoming the premier repository for messages of all kinds. E-mail, voicemail, fax, images, pagers, all are coming together in the most logical place — the universal inbox. From this mailbox, an end-user can receive and act upon messages sent there by people as well as devices. And providing access to the mailbox from a telephone gives users easy remote access.

## LEVERAGING THE ELECTRONIC MESSAGING INFRASTRUCTURE

Telecommuting is definitely on the rise. Workers need to have fast access to all the services and resources that they commonly used while in the office. Messaging is critical to the telecommuter and to the mobile worker. And as messaging systems offer advanced routing services, a user can receive a fax and have it forwarded as an E-mail message to a recipient who would listen to it from a pay phone.

# Chapter 20
# Exploiting and Extending E-Mail
*Audrey Augun and Eric E. Faunce*

Enterprisewide messaging requires an infrastructure that is solid, reliable, and secure. In addition, this infrastructure must be flexible enough to accommodate rapid changes in the enterprise itself and be able to accommodate new technology (e.g., the Internet and the World Wide Web).

In many organizations, decisions regarding E-mail are often made to meet the needs of individual groups, departments, or business units. Often originating with a departmental focus, growing an E-mail implementation into a corporate messaging infrastructure can be an enormous challenge. The system must constantly respond to a growing demand for high-fidelity electronic communication across departmental (and even organizational) boundaries. This enterprise and inter-enterprise expansion can quickly exceed the limitations of disparate E-mail systems and the gateways connecting them.

## GOALS OF ENTERPRISEWIDE MESSAGING

Information technology executives are charged with consolidating heterogeneous E-mail systems, and building a robust messaging infrastructure that meets the needs not only of individual departments, but also of the enterprise as a whole. The ultimate goal is to provide a reliable and flexible messaging infrastructure that meets services of the entire enterprise while keeping users both satisfied and productive. Attaining the goal of enterprisewide messaging requires the following:

- *Reliability*, resulting from field-proven technology, combined with a predictable, manageable, industrial-strength client/server architecture that scales to the largest enterprise.
- *Architecture*, enabling users to lookup information they need on their timetable. It must go beyond traditional store-and-forward messaging, which puts the burden of selecting who needs to receive what information on the sender and can result in lower user productivity. Instead, a push-and-pull paradigm must be implemented that eliminates

the need to send an object multiple times to multiple users, yet allows users to collaborate on a single topic using the most recent information. All this should be protected by an integrated public key security system that authenticates users and protects their information.
- *A low cost of ownership* that saves money because it decreases the number of messaging system components that need to be managed, and provides superior tools to support messaging administration.
- *Enhancement of the current infrastructure with an extensible messaging system* that leverages the environment already in place while offering a broad range of connectivity and application options. This includes the ability to fully support both custom communications-based groupware applications with complete Internet and World Wide Web integration on the same infrastructure.

Lotus Notes, with its premier messaging, groupware, and Internet/intranet capabilities, is uniquely suited to fulfill the IT manager's needs for a corporate messaging infrastructure. Reliability, superior architecture, a low cost of ownership, and infrastructure enhancements, along with compatibility and connectivity with other environments, are the hallmarks of Lotus Notes from an extensible messaging services standpoint. The following sections discuss each of these areas in greater detail.

**WHERE MESSAGING, GROUPWARE, AND THE INTERNET MEET**

Today, the Internet and the World Wide Web provide an international resource for both communication and information. Users expect to not only send simple E-mail by way of the Internet, but also to send attachments and convey graphics and photographs. They have also come to expect that they can search, browse, and access information repositories interactively. Lotus Notes was designed to incorporate these different means of communications, collaboration, and coordination beyond simple messaging. All of these capabilities are now extended to the ubiquitous access and reach of the Internet.

For example, some messaging applications available today allow users to click on a URL in a received mail message, launching a browser that lets the user view the Web page in question. Essentially, the user ends up in another application — not a particularly straightforward operation from the user perspective. Notes takes this a step further, enabling the user to remain in his or her original application to view, read, and manipulate the Web page from there, effectively placing the Internet in the user's inbox.

The growing number of browsers and Web-enabled applications has created legions of Internet-savvy users. Often, these applications are offered as stand-alone tools or are not completely integrated into current company platforms and services. Experienced users immediately see the value in integrating important tools (e.g., Web access, E-mail, and other desktop ap-

## Exploiting and Extending E-Mail

plications). For them, integration means quicker access, greater ease-of-use, and higher productivity.

For IS organizations, integration means a huge cost savings derived from a less complex infrastructure, along with a lower cost in user training. With such immediate benefits to both organizational and individual users, the demand for complete and transparent integration of E-mail, groupware, and the Internet will continue to increase. Today, with Lotus Notes, corporations can easily meet the user demand for convergence. Why? Because Notes was designed with the vision of integrating diverse elements rather than a single element (e.g., E-mail).

## RELIABILITY AND ARCHITECTURE

The following sections review the elements provided within Lotus Notes, enabling enterprise information sharing and collaboration. The architecture implementing these features forms the foundation for bringing together the leading groupware and communications platform with the universal connectivity of the Internet.

### Superior Scalability

When consolidating departmental E-mail systems into an enterprise messaging infrastructure, managers should carefully consider three major requirements: scalability, universal information access, and robust application development. An enterprise server should easily support the following:

- *Scalable server platforms* (e.g., UNIX, OS/2, and Windows NT), where a single server can support thousands of simultaneous mail users and provide extremely fast message routing speeds
- *Universal information access* that includes tight-knit Internet integration, enterprise fax support, voicemail, document imaging systems, plus optional MTA support for diverse messaging systems
- *Application development of custom communications-based applications* with both industry-standard APIs (e.g., MAPI, VIM, and XAPIA), and other robust development tools, including standard and de facto standard programming access to core services
- *A wide-ranging selection of compatible third-party products and services,* particularly for vertical market use

Lotus Notes meets these requirements because it can support organizations of any size, from to an entire enterprise system, spanning tens of thousands of users, suppliers, and customers. Notes addresses the requirements for scalability in large enterprises with an architecture that includes the following:

- 32-bit multithreaded design to support concurrent routing and replication on all major server platforms

# LEVERAGING THE ELECTRONIC MESSAGING INFRASTRUCTURE

- Symmetrical multiprocessor (SMP) support for up to eight processors so that core server functions can be spread over multiple processors or multiple machines (e.g., mail server, hub router, or application server)
- Clustering capabilities to support high availability and fail-over operations
- Open interfaces and standards-based protocols required for current and future commercial and customized applications
- A wide-ranging selection of both horizontal and vertical products and services from a worldwide business partner base of over 11,000 companies

**Multilevel Security**

Lotus Notes uses a security model widely regarded for its flexibility and robustness. Notes security is based on RSA public key encryption technology, which eliminates the need to transfer passwords across the network. Used along with each user's private key, it makes possible digital signatures and end-to-end encryption of messages. Notes provides four levels of security:

1. *Authentication,* which reliably verifies that users seeking access to network resources are who they claim to be
2. *Digital signatures,* where Notes verifies the authenticity of the sender, and that information received from another Notes user was not modified during transmission
3. *Access control,* for specifying who can use a resource and what they can do with it; access control is applicable to servers, individual databases, documents (including those referenced via links) and fields within documents
4. *Encryption,* for secure communication of information between individual users

Encryption can be applied to: databases, including documents and fields within those databases, whether they are located on servers and/or clients; and data in transmission channels, including both bulk data transmission between servers and client-to-server transmission. This permits secure transmission across non-secure media, including the Internet.

Features built for these basic security capabilities include the following:

- Separately configurable security administration roles (a form of access control) that permit safe delegation of administrative responsibility.
- Certificate revocation, to instantly deny all access to users whose authorization is revoked.
- Encryption of local databases means that Notes uses the private key to perform the encryption, providing secure password protection. Pri-

## Exploiting and Extending E-Mail

vate key encryption functions also enable administrators to enforce local security (including the enforcement of access control levels), ensuring that data is secure when end-users or third parties replicate a protected database.

### Fault Tolerance with Store-and-Forward and Least-Cost Routing

The Notes Server implements store-and-forward routing with support for dynamic routing paths, which can be altered by factors like the cost associated with a particular route, the urgency of a message, or the availability of required links.

Because Notes routing is also multithreaded, a Notes Server can initiate multiple message transfers simultaneously by multiple ports, by spawning separate threads. Administrators can define two types of message routing configurations — mesh routing and hub routing.

**Mesh Routing.** Servers on the same high-speed network (typically T1 or faster) will initiate the immediate transmission of messages without the use of connection records or other administrative overhead. This innovative concept shifts the responsibility for routing messages from the messaging system to the network itself — taking advantage of the robust, managed infrastructure that is already in place, rather than creating another routing infrastructure for mail on top. With mesh routing, users gain three major benefits:

1. Faster delivery and enhanced fault tolerance by eliminating most hub routers, which act as points of potential congestion or failure
2. Enhanced message delivery performance, because messages are sent from the server as soon as they reach the outgoing queue. They need not wait for a scheduled delivery time
3. Lower administration costs, because no connection records are required for the messaging infrastructure. Messaging relies on the network infrastructure already in place.

**Hub Routing.** Typically used for connections over slow links (e.g., WANs running at less than T1 speeds). Notes supports scheduled, prioritized connections that can be easily and centrally configured in the Notes directory. Fault-tolerant routing is also supported — if a link is unavailable, a backup link is automatically used. The implementation of hub routing on a high-speed LAN is possible to schedule the movement of mail (e.g., to confine some mail traffic on the network backbone to off-hours only).

### Object Store Flexibility

Notes provides for a distributed object store, supporting location-independent references. Highly flexible in design, the container does not distinguish between messages and other objects. Links automatically provide access to an object in the most conveniently located replica of the database in which it resides, whether on a server or the user's desktop. Any form of multimedia information can be easily stored, managed, and retrieved, whether by way of links or through manipulation by users, agents, or other programs. Notes databases are self-contained and self-defining. Database design elements are also part of the object store.

From a messaging standpoint, the Lotus Notes distributed object store differentiates Notes from traditional point-to-point, or point-to-multipoint messaging systems. With Lotus Notes, users are part of a push-and-pull paradigm that improves their productivity. For example, instead of attaching or embedding an object in a mail message and transporting it to recipients, Notes users have the option to embed a document link to an object in a message. The object can reside in a Notes database, a page on the World Wide Web, or even a Windows OLE link. Recipients merely click on the link to pull the document into view. Groups of users can work on the most recent information without having to send the same object over the network multiple times.

For messaging, the Notes object store can be configured to use a single copy object store (SCOS) — saving significant amounts of disk space and greatly reducing message traffic. When a message is destined for multiple recipients on the same server, only one copy of the message is stored on the server, and each recipient receives a pointer that provides access to the message in the SCOS. For example, if a 1MB message is sent to ten people, the result would be one 1MB instance of the message in the object store, and 10 associated pointers in various mail databases; a storage savings of nearly 9 MB for that one message alone.

The Notes SCOS is unique in that it supports a message store of unlimited size that can be scaled to comfortably accommodate growing numbers of users. An administrator can authorize a maximum SCOS size in 4-gigabyte segments, with automatic cascading between segments. The SCOS can even span physical storage system boundaries to meet the needs of the largest customers.

**Broad Cross-Platform Communication**

The Notes component-based architecture design supports a broad range of messaging clients, server platforms, transport protocols, and applications programming interfaces, making it easy to leverage the existing environment and also interoperate with a customer's or partner's infrastructure (see Exhibit 20-1). The resulting flexibility makes any move to a client/server architecture easier and cheaper because the infrastructure can be upgraded without changing the clients on users' desktops. A resultant savings on software purchases, training, and related costs is experienced.

**Exhibit 20-1. Notes Component-Based Architecture Support.**

| | |
|---|---|
| Client platform support | Windows95 |
| | Windows NT |
| | Windows 3.1 |
| | UNIX (Sun Solaris, HP-UX, IBM AIX) |
| | Macintosh (Power PC, 680x0) |
| | OS/2 Warp |
| Server platform support | OS/2 Warp |
| | NetWare NLM |
| | Windows95 |
| | Windows NT (Intel and DEC Alpha) |
| | UNIX (Sun Solaris, HP-UX, IBM AIX) |
| Transport protocol support | TCP/IP/MIME |
| | LAN Manager |
| | DEC Pathworks |
| | AppleTalk |
| | NetBEUI/NetBIOS |
| | Banyan VINES |
| | LAN Server |
| | SPX |
| | X.25 |
| | SNA |
| Applications program interface (API) support | Notes API for C and C++ Access |
| | LotusScript for cross-platform BASIC-compatible, object-oriented scripting |
| | Visual Basic |
| | VIM (Vendor Independent Messaging) |
| | XAPIA Common Mail Calls (CMC) |
| | MAPI (Messaging Applications Programming Interface) 1.0 |

## A LOWER COST OF OWNERSHIP

### An Extensible, Replicated Directory

The Notes directory (known as the Notes Name and Address Book, or NAB) is a another scalable and secure infrastructure component. The NAB is completely extensible — enabling customers to completely tailor and add information and fields required by their implementation. The NAB can be partitioned and distributed in logical subdirectories.

A master directory maintains information about the directory knowledge tree to direct queries to the appropriate piece of the enterprise directory. In every instance, changes made to one part of the NAB can be propagated by Notes' replication technology, which guarantees synchronization of all copies of the NAB, throughout the organization.

# LEVERAGING THE ELECTRONIC MESSAGING INFRASTRUCTURE

A cornerstone of the Notes security model, the Notes NAB holds both the certificates used to authenticate all users when they log in, and the public keys used for signatures and encryption. The NAB is central to Notes network management and configuration. It maintains users, groups, connections, roles, and other means of access control that allow centralized (even offline) management of the entire network infrastructure.

The NAB supports delegated network management by way of access controls over fields in the NAB itself. Any field or group of fields can be restricted to a given user, group, or role. For instance, this allows management of enterprise mail server connections to be restricted to the mail administrators, while delegating the maintenance of user office locations to human resources. As directories become rich and useful stores of information beyond mail addresses, the Notes Name and Address Book provides a flexible model that allows an appropriate mix of centralized and decentralized administration.

Notes supports two levels of directories:

- Personal (desktop-based).
- Domain (server-based).

This division gives users a smaller, more manageable directory without compromising enterprise directory requirements. When resolving an address, Notes automatically cascades from the personal directory to the domain directory. It provides type-ahead addressing — as the user starts typing a name, Notes fills in the rest — allowing faster message addressing and fewer errors or bounced messages.

The NAB directory also supports configuration of a personal mobile directory, which automatically replicates a copy of the domain directory containing only name and address information to the desktop. This enables mobile users to carry a dramatically smaller image of the domain directory on their laptops.

The Notes NAB features an X.502-compliant format and X.400 hierarchical naming structure. In the future, as organizations move to an X.500 native Notes directory, redefinition of the address space will not be necessary. The Notes NAB will support LDAP.

## Replication and Synchronization Capabilities

The replication and synchronization capabilities of Lotus Notes let users access the same information. Mobile users can maintain local replicas of Notes databases, including their mail database, and work offline. Once they reconnect to the server, Notes replication synchronizes the changes across the enterprise in real-time. Only those fields that have changed must be replicated, which minimizes communications costs. The powerful attributes of replication include the following:

- *Bidirectional.* Users in various cities, or any site to which a database is replicated, can make changes to the database — adding new documents, modifying or deleting others — and bidirectional replication will synchronize all changes made at all sites.
- *Efficient.* When synchronizing databases, only those fields within documents that are new or that have been changed on either side of the replication process need be replicated. Field-level replication ensures the optimal use of resources and the shortest synchronization cycles.
- *Client replication.* Occasionally, connected users (e.g., mobile users working off-site) require the same level of access to information as connected users. With Notes replication, these users have the same replication capabilities as their on-site colleagues.
- *Selective replication.* Notes users can replicate a subset of information contained in a Notes database. Users can define the profile of documents that they must replicate to their client workstations and replicate only those documents that have changed within a set time period (e.g., 30 days), or those authored by a particular workgroup member.
- *Background replication.* Here, replication runs in background mode, allowing users to continue working on other tasks.

**User Agents**

Agent support in Notes provides users with state-of-the-art automation very similar to rules or macros; however, they are much more powerful and easier to use. Both system administrators and end-users can take advantage of agent technology. Agents within Notes can be:

- Either time- or event-triggered, and can run on either the Notes client or the Notes Server
- Created using simple actions, or written in the LotusScript language or in the Notes macro language
- Used to perform a wide range of simple and complex tasks, becoming a powerful and flexible end-user tool to help organize and customize the mail environment

Agents are built by an intuitive interface that prompts users step-by-step to fill in fields, check boxes, and specify actions (e.g., "Copy to Folder") or conditions (e.g., "If Documents have been created or modified"). As the agent is built, the interface lets users see the flow of actions to be performed.

Users can easily design agents that save time and ensure quick access to critical information. For example, a user can create an agent that looks through all incoming mail for the words *urgent* or *important* and copies those messages to a folder called "hot issues." This process ensures that if the user chooses to work remotely, but does not have time to peruse and gather all documents, critical information can be obtained simply by replicating the hot issues folder.

Agents can play an important role with regard to frequently performed processes. For example, in minutes a user can create a mail agent to move high-priority mail to an "urgent mail" folder and route phone messages to a "call back" folder. From a messaging aspect, one of the template agents included with Notes lets people know if recipients, to whom they have sent E-mail, are out of the office and unable to reply. Users can customize the agent's behavior to send differing messages to different groups of people. The agent can even track the names of individuals who have received the out-of-office message, so that no one gets it more than once.

Agents set to run on Notes servers can be scheduled and controlled by Notes administrators. The administrator can control which users are allowed to run agents on a given server and how much of a server's resources are devoted to running agent tasks. For example, a system administrator with the correct permissions can control such things as disk space with the use of agents in Lotus Notes. Agents can be set to automatically seek free disk space on servers, or can notify those users using more than 200MB of the object store to clean up their areas.

All agents make full use of the Notes security model, which will authenticate the requester, guarantee the request to run the agent by way of digital signatures, and verify access privileges to the target information.

**Integrated Management and Administration: Distributed or Centralized**

**Flexible Administration Features.** Lotus Notes offers manageability with a lower cost of ownership not readily found in other enterprise mail systems. Policy-making and monitoring functions can be centralized, yet other administrative functions can be distributed to the lowest functional level in the organization.

From an administration standpoint, the Notes' administration control panel provides a graphical interface for the administration of the Notes Messaging infrastructure. Administration activities can be initiated from a single location (e.g., maintaining user and group records in the Notes Name and Address Book, managing keys and certificates, or performing a mail probe and starting a remote console session). The administration control panel facilitates the centralized management of remote sites.

Mail tracing, another aspect of the administration control panel, allows the performance monitoring and tuning of the Notes network. Tracing determines the fastest way to send mail between servers, identifying problems along the routing path and specifying the precise path that a message must take to reach its destination.

Administration agents can automate tasks such as log analysis. Custom views of logs can be created to display information to suit the organization's specific needs. Notes' replication can deliver updated directories, routing tables, logs, and connection records throughout the enterprise. Agents based on Notes formulas can be used to check for errors when creating directory entries.

**Management.** Because Notes allows for centralized configuration management, the organization can reduce skill level requirements at remote sites. For example, all server connections can be defined in the Notes Name and Address Book and replicated to remote sites. The Notes Administrator can choose between delegating or centralizing responsibility for maintaining the Name and Address Book. Forms-based directory maintenance allows for addition or deletion by filling out a form. Many aspects of directory management can be automated (e.g., using agents to synchronize the NAB with other databases maintained by other departments like human resources or payroll).

To simplify enterprisewide administrative hierarchy, it is possible to delegate responsibilities to a role rather than an individual user. Multiple levels of control over user access, down to the level of fields within documents, allow the delegation of responsibility for maintaining specific fields in the NAB or specific network configuration parameters, for example.

**Remote Monitoring.** Using the Notes' Event Manager, the administrator can survey a Notes network with several predefined statistical monitors, including: communications/network, security, mail, replication, resources, and servers. Each monitor's notification threshold and warning severity levels can be separately configured.

Distributed logging provides data on how much information is logged by each server, including a broad range of statistics (e.g., peak and average number of users, peak and average message delivery time, peak and average transactions per minute, and mail queue depth). Notes can then replicate the statistics to a central database where views, agents, and other Notes tools can be used to analyze the log data automatically. For example, an agent could monitor the log file for server disk usage information and then E-mail a local administrator when disk capacity at his or her location runs low.

Notes supports four methods of event notification: log entries, event relay to another server, mail or pager message to a predefined user or group, and posting an SNMP trap, which notifies any properly configured SNMP management station. Certain events can be deemed trouble tickets. They can then be tracked automatically to ensure timely problem resolution.

**Policy Management.** The Notes policy enforcement features provide for the definition of and monitoring for compliance with policies that standardize user expectation and improve the network's reliability. For instance, policies can specify a standard server hardware and operating system platform — a standard Notes Server version and an acceptable user load. The Notes administration control panel can create reports on all network servers and notify administrators of policy violations. Unlike other such systems, enforcement of administrative policies is far easier with Notes because its policies apply across all the sites in the Notes Domain (one or more Notes networks).

**Server Maintenance.** The Notes architecture features 24-hour availability every day of the week. There is no need to take a Notes Server out of service for routine maintenance. Multinational organizations and round-the-clock operations can always send and receive mail.

**NotesView Integration.** NotesView graphical management software provides control over an enterprisewide Notes environment. Users can easily integrate the messaging management of the Notes Messaging infrastructure with the rest of their distributed environment by the Simple Network Management Protocol (SNMP). From a single management station, NotesView allows monitoring and control over Notes Servers with real-time access to Notes network information. For example, NotesView can be used to restart a server, even if Notes is not running.

NotesView enhances network reliability by providing a powerful level of control over Notes servers, Notes databases, mail routing, and replication. Because NotesView integrates with existing network management stations to provide a complete view of the enterprise from both the perspective of mail traffic and of the network itself, mail problems be can quickly differentiated from network problems. For example, NotesView makes it easy to determine that a failed Ethernet segment is the source of a mail problem. Network-wide trends can be charted for capacity planning purposes and proactive problem resolution. The following views of the Notes network topology can be maintained:

- All networks to which a server is connected, and the status of each connection
- The network's entire mail routing topology, including routes to and from selected servers
- The network's database replication topology, including replication maps of specific databases

## Exploiting and Extending E-Mail

NotesView provides real-time monitoring for collection and analysis of statistics from across the network with the following:

- Server monitor, providing such statistics as server up-time, LAN and COM port status, disk space usage, mailbox sizes, and users logged on.
- Mail monitor, providing such statistics as mail volume and number of pending and dead messages, running utilities to force delivery, or restart a router.
- Replication monitor, providing such statistics as replication time and status, running utilities to force replication, or restart a replicator.
- Network monitor, providing statistics on server protocol configurations and packet traffic.
- Database monitor, providing remote views of database information and running database utilities remotely.
- Mail Probe, allowing for continuous assessment of the speed and stability of the network's routing functions by measuring the performance of probe messages sent across the network.

NotesView's SNMP support and central administration capabilities mean that fewer administrators are required to support a Notes user base. Remote servers can be controlled without on-site staff, significantly reducing the cost of owning a Notes infrastructure.

## ENHANCEMENTS TO THE INFRASTRUCTURE

### Seamless Internet Integration

Today's organizations view the Internet and the World Wide Web as an entity upon which they can build a single architecture for the deployment of client/server applications for both internal use or to reach customers, business partners, and suppliers. Lotus Notes is based on the concepts of a rich document-oriented database, fielded forms and document linking — concepts that are very much part of the World Wide Web. By supporting native Web protocols within Lotus Notes, the years of Notes deployment in real business situations can be leveraged to fulfill an obvious need on the part of many companies who have come to see the Internet and the Web as an integral part of their inter-enterprise.

Notes directly supports the native protocols of the Internet and the World Wide Web. Notes clients continue to provide value-added capabilities in such areas as mobile user support and a rich, active user interface, while Web browsers enable any user with net access the ability to reach Notes capabilities.

In a practical sense, the Notes user interacts with the Internet as simply as with any local Notes-based information. When viewing a Web page from

## LEVERAGING THE ELECTRONIC MESSAGING INFRASTRUCTURE

a Notes client, all the elements remain active. Because the user views it as a Notes document, however, they can use all the Notes tools: the Notes Editor to annotate the document and simply forward the document without worrying whether the recipient has Web access. It is no longer necessary to copy down the URL in a mail message, forward the message, and hope the recipient can correctly copy it carefully into their browser and read the Web page in question.

**Standards Support**

Within the enterprise, users of diverse E-mail systems must be able to communicate quickly and easily. To ensure this capability for Notes users, Notes Servers can optionally integrate MTAs (message transfer agents) for SMTP/MIME and X.400 and cc:Mail messaging environments. All MTAs provide three basic capabilities:

- Scalable, high-performance routing and relaying of messages in their native format.
- Encapsulation, enabling customers (whose networks run on an enterprise messaging backbone connecting multiple dissimilar messaging systems) to use the SMTP/MIME and X.400 MTAs for Notes-to-Notes communication without loss of message fidelity or security. For example, the Notes SMTP/MIME MTA can route full-function Notes groupware messages across the Internet with complete security and integrity.
- Protocol conversion from the Notes environment to standards-based networks, while maintaining fidelity with the rich content of the native Notes environment.

Integrated multiprotocol MTAs reduce the administrative overhead and additional costs associated with configuring and maintaining add-on mail gateways or connectors, while providing industry-standard protocols for enterprise backbone or inter-enterprise communication over value-added networks and the Internet. Unlike gateways, MTAs are part of the network infrastructure. This means they support the following:

- Automatic, high-performance mesh topology routing over high-speed LAN connections
- Configurations for fault-tolerance: if one path fails, another path is automatically used
- Higher throughput as a result of multithreading
- Installation and configuration with the Notes directory
- Integrated console control and reporting of error and statistical information with Notes Messaging
- Support of SNMP for standards-based management

*Exploiting and Extending E-Mail*

For cc:Mail integration, the cc:Mail MTA provides full-function connectivity to the cc:Mail network. The cc:Mail MTA fully integrates cc:Mail and Notes, including messaging interoperability, directory interoperability, and integrated network management. Large cc:Mail customers making use of the Notes cc:Mail MTA can take advantage of Notes message routing and management technology and gain access to SMTP/MIME and X.400 networks without having to change cc:Mail clients or Post Offices.

Because Lotus Notes provides support for the messaging application programming interface (MAPI) 1.0 specification, POP3 and HTTP, it is possible to front-end a variety of mail clients to Notes Servers. Examples include the cc:Mail Release 7 client or the Exchange client in Windows 95 or any POP3 mail client. Also, developers of messaging-reliant applications can create MAPI-compliant applications that work not only with Notes clients, but also cc:Mail, Exchange, or other MAPI systems in a heterogeneous environment.

**A Powerful and Consistent User Interface**

Notes clients deliver a myriad of capabilities for every level of user by way of a powerful user interface that enables automation and customization of messaging tasks. Among the many features included, the Notes rich text editor lets users embed a broad spectrum of objects in mail messages, like graphics, fonts, color, and OLE 2.0 attachments.

Notes Mail is a world-class ActiveX container, where a user can embed a 1-2-3 or Microsoft Excel spreadsheet in a document. When the recipient opens the message, the spreadsheet is immediately displayed and can be directly edited.

Notes further extends the capabilities of rich text to include such personality features as mood stamps to help convey confidentiality, thanks, or sentiment. Predefined letterhead options can be customized by individuals or by organizations to fit a corporate image.

**Integration with Workflow, Authoring Tools, and Document Libraries**

Notes' ability to manage document-based workflow is well known, particularly because it can simplify document management processes, specifically the following:

- Routing documents for review and publication
- Tracking the status of routed documents
- Enterprise and inter-enterprise document distribution
- Providing robust, flexible security for documents or components of documents down to the field level
- Making documents accessible, either widely or selectively

- Organizing and linking documents
- Document maintenance and archiving

Notes uses full text search and query capabilities along with customizable views and hierarchical folders. Users do not have to remember file names, authors, or dates to retrieve information. They can control the varying parts of the document life cycle while the Notes architecture effortlessly handles security, version control, and storage and retrieval issues.

The strength of the groupware features of Lotus Notes is apparent in an example of an interactive Web application effort that streamlines the corporate hiring process. In this example, a hiring manager lists her personnel requirements in a Notes form and, with the push of a button, routes the form to upper management for approval and authorization. On approval, those requirements are then published on the Web site. Potential applicants can review job requirements and respond with background and salary requirements. Although it is important to be able to publish static information (e.g., the agenda for a trade show), a greater feat is to publish dynamic information that results in the recruitment of new, high-caliber employees.

**Calendaring and Scheduling**

Demand is growing for calendaring and scheduling functions to be integrated with every user's workgroup and messaging functions. This integration assures customers of high performance and high fidelity of scheduling processes across any size environment, from small workgroups to worldwide implementations.

Notes calendaring and scheduling offers a graphical, feature-rich, and intuitive user interface. Users are able to see busy and free time information in real-time, send and accept meeting notices via E-mail, and manage their calendars offline just as they do with E-mail. An intuitive extension of its messaging and groupware environment, Lotus Notes calendaring and scheduling features are fully interoperable with Lotus Organizer users working outside of Notes and with IBM OfficeVision users.

Users are provided with real-time views (security privileges permitting) to the calendars maintained by other users on the network. Summary information that reflects the status of a group scheduling activity is also provided in real-time. A user who has sent meeting invitations to ten individuals, for example, can instantly track who has accepted or declined the meeting.

From an administration standpoint, there is no administrative overhead associated with native Notes calendaring and scheduling or Notes-based Organizer clients. The same tasks executed by the administrator (e.g., adding or removing users from a Notes server or granting users mail privileges) can also be used to grant users calendaring and scheduling functions.

*Exploiting and Extending E-Mail*

**Complete Remote Connectivity**

To maximize offline, any Notes client can be configured for dial-up access. Notes does not require remote users to carry the complete Notes Name and Address Book on their systems. Notes gives them full access to their Notes messages by supporting replication of mail files, discussions, and other databases directly to remote systems.

Remote users can opt to download mailbox subsets to save space in their laptops; for instance, their inbox and drafts but none of their folders. With Notes, they can also automatically limit messages kept locally to a specified time period (e.g., the last 90 days).

Notes supports selective downloading of specified documents (e.g., message headers only or the full message minus attachments) for additional convenience and time savings. Mobile users can even download part of a message and request the rest later. Notes' ability to replicate only the parts of a Name and Address Book that have changed, down to the field level, further reduces both replication time and the message traffic associated with replication of mobile directories. Notes provides constant user feedback during replication regarding the number of documents and estimated time remaining.

Location profiles let users set up a wide range of criteria about any location from which mobile access may take place, including communication methods and related parameters, connection intervals, and scope of replication (e.g., send only, receive only, and receive E-mail headers only). Once a location is defined, all a user need do to access the Notes object store from that location is select an item from a pop-up menu. Notes clients adjust communication parameters automatically.

**Global Service and Support**

Lotus Notes is backed by global IBM/Lotus premium service and support people. This worldwide organization provides on-time installation and support services, which range from hotline support to on-site consulting services. Even global organizations can be assured that the most remote sites will receive the support they need to fully participate in making the business successful.

**SUMMARY**

Messaging, groupware, and Internet/intranet technologies must converge because of the corporate user demand for a single, reliable, and manageable infrastructure that is understandable and usable by corporate users and at the same time sports a low cost of ownership. Lotus Notes is the first product to deliver on this convergence. It has a powerful and consistent architecture that is platform- and communications-independent, enabling construction of a reliable and flexible communications infrastructure.

# Chapter 21
# Electronic Commerce
*Rik Drummond*

Electronic commerce — the way businesses will do business in the future. Why use paper when things can be done easier and better without it? E-mail, EDI, electronic funds transfers, credit card transactions, electronic banking of all kinds, and all the thrills of electronic shopping are becoming available from the comfort of home, office, or car. Finance and banking are now available to even the novice home computer user. Where will it stop and what is in store? Will the Internet change the way we live or just the way we shop? This chapter looks at where and how electronic commerce is used, and even though many types of electronic commerce payment vehicles exist, it will focus on the credit card because it is currently the best documented.

## THE BASICS OF ELECTRONIC COMMERCE

The question of how electronic commerce (EC) fits with E-mail and EDI frequently comes up in EC-related conversations. It is best defined in the following vignette. Joe, in front of a large audience, was leading a discussion that kept attempting to define what electronic commerce is and is not. Holding up a piece of blank paper, Joe said, "The best way to define electronic commerce is to picture doing business without this," as he crumpled the paper and tossed it into the waste can.

The definition of electronic commerce is doing business without using paper. Electronic commerce is not EDI, it is not E-mail, it is not Web browsers — it is all of these and more. It is electronic mail, Web servers, workflow systems, process changes and reengineering, EDI, Internet, video on demand, and voice messaging — electronic commerce is not using paper. Not using paper reduces errors in communications, speeds information flow, reduces mailing costs, increases opportunity costs, and helps make business processes more effective.

Because EC affects so many areas — all the ways companies do business — a broad set of disciplines must be harnessed to cover EC. This includes such disciplines as process reengineering, communications, databases, and applications (e.g., Web and E-mail clients).

## LEVERAGING THE ELECTRONIC MESSAGING INFRASTRUCTURE

This chapter covers several of these areas: EC interchange participants, EC interchange categories, EC human interface requirements, EC bandwidth and connectivity requirements, and electronic payment technologies.

### ELECTRONIC COMMERCE INTERCHANGE PARTICIPANTS

Different participants in the EC product chain have different requirements and roles. Each industry has its own slightly different model. The following model is sufficient for these purposes. From the creation of a product to its purchase, five participants are involved in the process. They are:

Supplier → Manufacturer → Distributor → Merchant → Consumer

**Supplier.** The supplier produces raw material components used by the manufacturer to make the end product. Examples include a fuel injection system supplier for an auto manufacturer, or a paper supplier for a daily newspaper. The supplier-manufacturer relationship is recursive in that a supplier may also be a manufacturer for another supplier. Whether they are considered a supplier or a manufacturer depends on whether they produce the end product or not. The fuel injection company is a supplier if the fuel injection system is sold as part of the car, and a manufacturer if it is sold directly to the consumer.

**Manufacturer.** The manufacturer is an entity that produces an end product (e.g., cars, trucks, Monopoly games, auto fuel, paper, or Barbie dolls). Each offers a final consumer-ready product. They are often categorized as being part of the textile, automotive, transportation, grocery, petrochemical, or steel industries.

**Distributor.** The distributor takes the product made by the manufacturer and delivers it to the merchant or consumer. Mobile Oil produces automotive fuel and distributes it through outlets it owns or franchises called gas stations or marts. Ford manufactures cars and distributes them through Ford dealerships. At times, the distributor is also the end merchant, as in the case of the gas stations. At other times, it may be a middleman who moves the product from the manufacturer to the merchant, and has nothing to do with the final sale.

Making things even more complex, distributors use shippers, which may be partial or full load shippers, rail, sea, air, truck, pipelines, or others to get the product to the appropriate destination point. The area of transportation is so paper-intensive and complex that this is the area where the need for reduction of paper was first recognized. Transportation is where EDI first started, with an organization called Transportation Data Coordination Committee (TDCC) in the late 1960s. Even though EC is not EDI, EDI was the first to address the problem, and is arguably the grandfather of EC.

**Merchant.** The merchant is the organization that sells the product to the end consumer. The product is usually a physical entity (e.g., jeans, or computer software on a CD-ROM). For these purposes, it is not a service like telephone long-distance or CompuServe. The merchant who has contact with the consumer is the one who helps the consumer choose the product, explains product limitations, and sets consumer expectations. Merchants and the manufacturer are the ones the consumer blames if there is a problem, or goes to for help.

**Consumer.** The consumer is the one who consumes and uses the product — the end of the product chain and the reason the product is even manufactured in the first place.

In most industries over the last 25 years, EDI has helped reduce the paper used to negotiate, deliver, buy, and sell the product between the first parties in the supply chain. EDI was used to reduce paper between the supplier and the manufacturer, between the manufacturer and the distributor, and, in large retail or merchandising organizations (e.g., JCPenny or Wal-Mart), between the distributor and the merchant. What has been missing is the reduction in paper between the consumer and the merchant. Technologies that support EC between the merchant and the consumer have just fallen into place with the advent of Web technology.

## SIMPLIFIED ELECTRONIC COMMERCE PARTICIPANTS

To keep things as simple as possible, it is helpful to assume that only three participants are in the supply chain by combining the manufacturer, the distributor, and the supplier. From a consumer viewpoint, the automotive industry fits this model. Thousands of suppliers send subcomponents to the manufacturer (e.g., General Motors), who assembles it and then sells it to the consumer.

Each of these manufacturers, merchants, and consumers interrelate and exchange at least four types of information in the relationship. Two types of relationships exist, each with different needs and requirements: manufacturer-merchant and merchant-consumer. These relationships exist to support the buying and selling of a product or service. Four basic process categories involved in these relationships will be described, as shown in Exhibit 21-1. This chapter will then go into further detail on the exchange of electronic value. The four categories of exchanges between participants are:

1. Negotiating the product and relationships
2. Order fulfillment, shipping, and delivery of the product, physically or electronically
3. Paying — exchanging the value

# LEVERAGING THE ELECTRONIC MESSAGING INFRASTRUCTURE

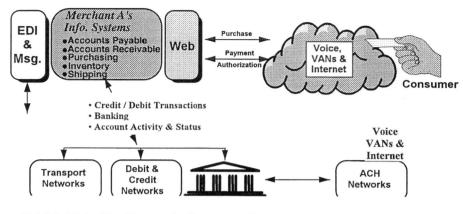

Exhibit 21-1. The Electronic Commerce Process.

4. The ordering process — exchanging the paperwork (e.g., PO, invoices, and shipping notices)

## Category One: Negotiating the Product and Relationships

This category encompasses processes necessary to establish the business relationship — in the EDI world that would be the trading partner agreement. This category would normally not cover the merchant-consumer exchange because the relationship is not an ongoing, frequent, or consistent one. The processes in this category are used to establish initial and ongoing relationships. These processes are not used for sporadic, unpredictable, unplanned purchases and exchanges like those often found in the consumer-merchant exchange.

This category will often use electronic messaging, paper mail, and voice to establish the relationship, negotiate the deals, and confirm the purchase price and terms for the relationship. Predominantly, this area would use electronic mail and not Web servers in the manufacturer-merchant relationship.

## Category Two: Order Fulfillment, Shipping, and Delivery of the Product (Physically or Electronically)

Delivering the product electronically has to do with actually shipping the product electronically. In lieu of the transporters used in Star Trek, category two is only workable for products that start as electronic in nature. Examples include online magazines, software, music, video, books, and online services (e.g., AOL). Translation services — physical items (e.g., cars, clothes, and groceries) are also in this category. The delivery of these items would be supported by electronic paperwork and electronic purchase, but themselves are not electronic in nature. A new example of electronic products is Encyclopedia Britannica. For a yearly fee, it is no longer necessary to buy the set of books. The Encyclopedia Britannica may be accessed online through a Web server.

This category, the delivery of products electronically, uses both E-mail and Web clients and servers. Electronic products will probably be retrieved primarily through Web servers. For example, a subscriber orders a video by accessing the merchant on the Web and starts the download (i.e., delivery) of the movie *Red Tide* online.

## Category Three: Paying (Exchanging the Value)

Delivering the value electronically covers the exchange of credit, debit, E-cash, and other payment types electronically. This is the area holding back consumer-merchant EC — it is the mechanism where the consumer purchases electronically.

Different value transfer needs and technologies exist. The oldest is electronic funds transfer between banks — the automated clearinghouse. Another example is the use of voice or the digital network to convey the consumer's credit card or debit card information to the merchant. Automated teller machines (ATMs) are changing the way banks do business, and overtaking ATMs are smart cards. New standards are being developed (e.g., secure electronic transactions, or SET by MasterCard and Visa), which will directly support consumer-merchant exchanges though existing monetary exchange vehicles (e.g., the credit card).

Category three will be electronic messaging and Web servers. Merchant-consumer exchanges will be using Web servers to exchange value and pay for products; the networks run by banks and credit card clearinghouses will, however, still exchange bulk reconciliation data and value exchanges (i.e., EFT) and EDI by the messaging structure. At the heart of the merchant-consumer relationship problem is an unsecure and bandwidth-limited network.

## Category Four: The Ordering Process (Exchanging the Paperwork)

Delivering supporting documentation electronically is the historic EDI. This is the transmittal and receipt of purchase orders, invoices, shipping notices, credit information, and college enrollment information between participating parties electronically. This has historically been supported by the large value-added networks that often, but not always, handle the transmission between companies. This is where EC started 30 years ago. Much of what was learned during these manufacturer-to-merchant implementations is transferable to the merchant-to-consumer portions of EC.

The ordering process is currently messaging, but it is already moving to a more interactive focus. EDI is not going away any time soon, and will remain the primary means to conduct business-to-business exchanges, in the same way that electronic funds transfer will remain the way the banks reconcile value exchanges. Both of these are ultimately electronic messaging.

EDI is not the only technology needed for EC, but it is a primary one. Several infrastructure issues are holding back the full-scale implementation of EC. In very general terms, they are: the end-user device, the bandwidth and connectivity availability, and security and protection issues.

## EC TECHNOLOGY ISSUES

### Connecting the Consumer's Device

One of the major issues in EC is how to connect the consumer's device to the EC network. Glitz attracts users, and graphical user interfaces, or GUIs — visual graphic-based exchanges (e.g., the Web or multimedia-based E-mail) — deliver that glitz. Frequent use of these technologies requires devices at higher speeds than the current 28.8K-bps modems. ISDN is probably not a valid option over the long term. It is an old technology filling a short-term gap until other lower-cost, higher-bandwidth options come into production. There will probably be some sort of a hybrid option for the foreseeable future for most consumers.

Two options are available for wiring the consumer's device with sufficient bandwidth: create new ways to use the existing media (e.g., twisted pair or coax entering homes), or run new transmission media to the user's home. The latter is much too expensive and manpower-intensive, in that it requires that $300 to $500 per connection be invested up front just to install new wire or fiber from an existing junction box to the home. Some industry experts estimate it would take $120 billion to rewire U.S. homes to support the high-bandwidth needs. The price does not include other infrastructure costs that are attributable to upgrading the backbone infrastructure.

This may not slow the expansion of EC in the manufacturer-merchant area, however, where major savings will be realized just on the reduction of paper and information cycle times. The further automation of paperwork exchange between organizations is normally not a high-bandwidth requirement, as it is the case for exchange of graphical, video, and other types of products online.

### Bandwidth and Connectivity Requirements for EC

Three infrastructure requirements must be addressed for EC to be implemented widely for both consumer and business-to-business communications: security, the ease of reaching others (i.e., interconnectivity), and the size-volume of the exchange. Negotiations, product delivery, value exchange, and paperwork exchange have different security, connectivity, and

### Exhibit 21-2. Manufacturer-Merchant Exchange.

**Manufacturer-Merchant Exchange**

| Exchange type | Security requirement | Exchanges size | Interconnectivity |
|---|---|---|---|
| Negotiating the product and relationships | High | Small to medium | Low |
| Product delivery | Low | Low | Low |
| Payment — exchanging the value | Low to high | Low | Low |
| Paperwork exchange | Low to high | Low | Low |

exchange size needs. This is referred to as the *manufacturer-merchant exchange* (see Exhibit 21-2).

Interconnectivity needs pertain to the number of parties who are capable of intercommunicating, and are relatively low in the manufacturer-merchant infrastructure. A limited number of trading partners are typically in this arena and the relationships are often planned. This is not always the case and will be less so in the future, but it is true now.

Exchange size is not as important in the manufacturer-merchant relationship as it is in the merchant-consumer relationship, because the amount of information exchanged per dialog is often low. These objects are dense, and are often textual and filled with large amounts of data required to support the exchange — not deliver the product. They are often in the hundreds or thousands of bytes — not millions. With the case of the product data and the product delivery, the size can be very large (millions of bytes) if the exchange is an engineering or integrated circuit drawing. This is different than the *merchant-consumer relationship* needs for EC, as shown in Exhibit 21-3.

### Exhibit 21-3. Merchant-Consumer Exchange.

**Merchant-Consumer Exchange**

| Exchange type | Security requirement | Exchanges size | Connectivity requirement |
|---|---|---|---|
| Negotiating the product | Low | Low | High |
| Product delivery | Low | Low to high | High |
| Payment — exchanging the value | High | Low | High |
| Paperwork exchange | Low to high | Low | High |

## LEVERAGING THE ELECTRONIC MESSAGING INFRASTRUCTURE

In the manufacturer-merchant exchange, the connectivity requirements were relatively low, while in the merchant-consumer exchange, the connectivity requirements are high. For EC to work on the merchant-consumer side, many consumers must be able to get to many merchants in an ad-hoc manner to conduct shopping exchanges. In the supplier-merchant exchange, the communication is less ad hoc and more planned. Pervasive supplier-merchant interconnect needs are not as high as in the merchant-consumer area.

In the case of a manufacturer-merchant exchange, high bandwidth among a few players in a planned, non-ad hoc loop will often solve the needs of the exchange, while in the case of merchant-consumer exchange, high bandwidth among the entire consumer and merchant universe would solve the needs. The supplier-merchant exchange requires a more limited connectivity, while the merchant-consumer exchange by definition requires broad connectivity across a large heterogeneous population. In the future, both will require high connectivity and high bandwidth; but for now, the merchant-consumer interchange, in all but a few cases, does not require high bandwidth. The exchange requires high connectivity penetration.

The Internet currently offers high connectivity, at low to medium bandwidth for the consumer. Most consumers will be operating at speeds of 28.8 kilobits or less for some time, so they can only support EC applications that require low- to medium-speed networks. In Exhibit 21-3, the only place in the exchange size that requires high speed is that of product delivery. The delivery of things like movies, digital voice, and complex multimedia documentation are important, but are not precluded by bandwidth limitations.

**Human Interface Requirements and Needs**

What is so exciting about EC in the advent of the World Wide Web? Because EDI is already making major contributions between companies, what will the Web offer?

Corporations buy and sell in bulk, which fits the electronic messaging communications methodology well. Most consumers do not buy in bulk, however, and do not make a large amount of repeat purchases of the same thing. Consumers often window shop, which is exactly what is offered by the Web. The user decides when to shop, where to shop, and how often to shop — frequently without notice — just like going to the mall.

Electronic messaging is often called an example of a *push technology,* and the Web or database queries examples of *pull technology.* In a push technology, the user receive things when the originator wishes, as in the case of broadcast TV, postal mail, or radio. A pull technology is one in which the recipient decides when to retrieve the information, as in the case of the Web, video rentals, audio tapes, or video movies-on-demand.

## Electronic Commerce

**Exhibit 21-4. Electronic Commerce Status.**

| Description | Status | Who/What |
|---|---|---|
| Ubiquitous communications infrastructure | In place | Internet/Vans |
| Uniform address space | In place | Internet IP |
| Protected exchange of electronic value | Partially in place | Banks, Credit cards |
| Reduced cost and complexity of interface devices | Partially in place | TV, Internet terminals |
| Ubiquitous medium-speed wire to the consumer's device | In place | 28.8 K-bps modem |
| Ubiquitous high-speed wire to the consumer's device | 2000+ | Coax, twisted pair, satellite |
| User-friendly push-and-pull technologies | Partially in place | E-mail, Web |
| Reduced cost to participate in consumer EC | Ever? | Not clear |

People use both methods to effectively work and accomplish tasks. An office worker may be notified of the need to approve a purchase requisition by E-mail, at which time the worker accesses a database to actually read the supporting data and approve the requisition. Or a consumer is notified of a sale by broadcast radio, and then purchases the product by means of the pull technology of the Web. In implementing EC programs, it is important to remember to implement both technologies, and not make the user attempt to work with only a pull technology when they also need a push.

## ELECTRONIC COMMERCE TECHNOLOGIES

Several items affect EC implementation, either as attractors or detractors. These items, their status, and who or what is needed are depicted in Exhibit 21-4. The items affect EC implementation in the following ways:

- Ubiquitous communications infrastructure is required to support the consumer access to the merchant. Every household must be able to access the merchant by way of the communications backbone.
- Uniform address space is the reason that the Web has taken hold so quickly. It is not the user interfaces on the Web clients; it is not the links between servers (hyperlinks); it is not the graphics — it is the uniform address space offered by the IP network that allows functions to be easily accessible (IP addresses).
- The protected exchange of value, or money, over the Internet is not hard for a limited group of participants. They are able to manage and coordinate the encryption keys. The banks were doing this without encryption keys for years. As the consumer becomes active, it is no longer a limited number of participants — it is hundreds of millions. How are the encryption keys that form the basis for signature and encryption for the value transfers managed? This area is just now being solved.

# LEVERAGING THE ELECTRONIC MESSAGING INFRASTRUCTURE

- Reduced cost and complexity of the interface device is very important. Currently, the large majority of those using the Internet are college educated and have PCs, UNIX workstations, or Macintoshes. These are not low cost for most people, and are not easy to support. The interface device must become simpler and lower in cost. As in the cellular phone industry, it must be given away as part of the purchase if hundreds of millions of people will be online.
- The wire to the consumer's device heavily affects the amount of functional ability, ease of use, and the type of product the end-user can receive over the network. This wire is the last 50 to 100 feet — the run to the residence. Until satellite or cable television is cost-effective to the residence, this will be a major limiting factor.
- The high-speed wire solution is the same as what is required for medium-speed wire.
- User-friendly push-pull technologies are essential. The Web is a fairly user-friendly technology for those who are computer literate. It is difficult for those who are not; just as Microsoft Windows is not always user-friendly for those who do not have a computer background. The Web is a start; however, it is a long way from being simple enough for the hundreds of millions to use.
- A reduced cost to participate is necessary to facilitate the merchant-to-consumer EC interaction. This means it should be around the cost of a telephone (e.g., $10 per month, not $30 to $40 per month).

## MERCHANT-CONSUMER ELECTRONIC PURCHASE PAYMENT PROCESS

As discussed earlier, the exchange of payment information in a secure manner in the merchant-consumer relationship is the last requirement for full-scale EC. A consumer can browse and window shop the Web servers for products using current technology; however, they cannot safely pay for it on the Internet without encryption technology. Protection requirements, each depending on encryption, have been identified as follows:

- Provide ease in identifying valid credit, debit, check, and cash type accounts
- Preclude interception
- Ensure that payment information is not alterable without being identifiable
- Ensure that the transaction has as much anonymity as possible
- Detect and prevent fake storefronts from masquerading and collecting credit and debit card account information.

Both MasterCard and Visa are leading the effort to implement the standard payment information exchange of EC. They are both working with others, including Microsoft, Checkfree, IBM, Netscape, GTE, and CyberCash. The merchant-consumer electronic payment exchange process is com-

*Electronic Commerce*

posed of two areas that must be addressed in concert before things become workable: processes and technologies. This includes: payment processes, encryption used in EC, electronic signatures, public/private key encryption, and certificates.

## Payment Processes

Payment processes fall into several categories: establishing means to readily identify merchants, consumers, and banks, issuing certificates necessary for validating electronic signatures, and establishing the actual payment process. The payment process will be discussed later, and the first two categories will be addressed as the discussion weaves through the technologies.

The MasterCard and Visa consortium published the Secure Electronic Transactions (SET) specification in February 1996. They have spent much time defining the processes and technology, and their document will be used as a basis for the following discussion. Because the SET document is several hundred pages, many areas will be abbreviated here to simplify and consolidate.

**Merchant-Consumer Purchase and Payment Process.** Two types of processes must be supported for merchant-consumer EC: one based on E-mail and the other using the Web's capabilities, both offline and online. E-mail-based ordering will support offline shopping; this shopping uses catalogs or CD-ROMs to browse for products and then initiates an order by sending an E-mail message. The interaction between the consumer (who has a credit card) and the merchant, supported by a third-party credit card verification entity, and the acquirer bank is shown in Exhibit 21-5.

The consumer issues an E-mail request to the merchant; the E-mail is a purchase order (PO). The merchant responds with a PO response. Embedded in the PO from the consumer are credit card and digital signature information, in addition to the product and quantity requested. The credit card and digital signature information is passed on to the acquirer or bank for verification of the credit card numbers, account availability, and digital signature. The merchant never sees the actual credit card information. They are only told whether it is valid or not. The merchant sends the encrypted user's card information on to the acquirer for verification in an authorization request message, and is returned an authorization response that completes the transaction. The user may at some later time query the merchant for status of the order by issuing a PO inquiry message. The credit card and digital signature information are contained in a special package that will be discussed in the technical basis for this exchange.

Online, Web processing may be somewhat of a different process than E-mail because the user is already attached to the Web site in an interactive

# LEVERAGING THE ELECTRONIC MESSAGING INFRASTRUCTURE

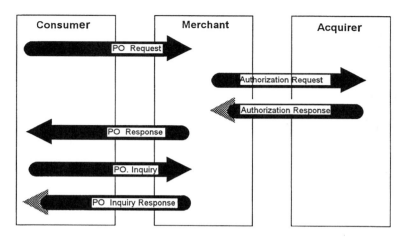

Exhibit 21-5. Merchant-Consumer Purchase and Payment Process.

---

session. Because of this, the request sequence is somewhat different from the E-mail offline transactions. For example, because the consumer is already attached to the server and is assumed to have identified the product online, a PO request is not necessary.

It can be assumed that the price and terms of the purchase have already been agreed on. The consumer, or cardholder, starts by issuing an initial message to the merchant. The message contains the consumer's credit card information. The merchant responds with an invoice message that contains, among other things, the electronic signature identifying the merchant to the consumer and what is being ordered or purchased. The consumer then responds with a purchase order that includes the consumer's signature, certificate, and payment instructions. The consumer's payment information is passed on to the acquiring bank for verification. (The payment information is not revealed to the merchant. The merchant does not know the card number of the consumer transaction.) After the acquiring bank verifies the consumer's information, the merchant receives a purchase order response with the authorization code, the acquirer's signature, and authorization.

## Encryption's Use in Electronic Commerce;

Often, two types of encryption schemes are used in an attempt to maintain the privacy of the data being transmitted. The first is a symmetric scheme that uses only one key — a key both parties must possess to encrypt and decrypt the data. The one most often used is the Data Encryption Standard (DES), which has been used by the U.S. Government for years. The other encryption scheme is the public/private key scheme, whose best known example is RSA encryption. Each has its positives and negatives and because of these, each adds a unique capability for information protection.

**Data Encryption Standard.** DES, the *symmetric* scheme, is used to encrypt large amounts of data. It is very effective and efficient at encrypting and decrypting data. Both parties, however, must know the same key — which means it is only usable between two parties that possess the key. The key must remain secure or other parties can decrypt the data sequence.

The DES scheme works very well with online encryption where two cryptographic devices, one on each end, exchange protected data. It does not work well for more spurious uses unless there is some way to distribute the key for this transmission, along with the transmission, in a protected manner so that only the intended destination is able to get the key and decrypt the transmission. This is where the public/private key encryption scheme comes into play.

Data can be encrypted with the DES algorithm. Then, the DES key, with the public key of the key-pair, is encrypted and sent to the destination, which recovers the key and decrypts the DES data stream. More on how the public/private key scheme works is included in the following sections.

## Electronic Signatures

Credit card transactions must be signed to identify and confirm the entry initiating the purchase and resulting transaction. What is a signature? When a customer signs a check, it means the check is from the customer and the amount on the check is correct. Signing indicates that the piece of paper (i.e., the check) is from the customer and should be honored as from the customer. The same is true in the electronic world. However, there is a problem. Unlike the paper check, which is not easy to tamper with (e.g., change amounts or names without being discovered), electronic documents are easily modified without it being evident. If a customer signs a general electronic document, how is it known that the document is not subsequently changed, and how is it known what the signature applies to? Where does the document start and end?

To take care of this unique electronic document problem, the hash algorithm was introduced, which indicates to which document the signature applies. The document — the electronic check — is fed into an algorithm, which takes all the bytes of the document and computes a unique bit sequence from those document bytes. The bit sequence is designed so that if any part of the original document is changed, the bit sequence is changed. By applying the same algorithm at the receiving end, the receiver can tell whether the signed document is valid or not valid. The hash-produced bit sequence is then encrypted using the source's private key.

# LEVERAGING THE ELECTRONIC MESSAGING INFRASTRUCTURE

The private key uniquely identifies the source, and the hash verifies that the document received is the same as the one signed. Both the hash algorithm and the public/private key pair are required to produce an electronic signature.

## Public/Private Key Encryption

The encryption scheme these electronic transactions depend on is of a category of schemes called *nonsymmetric* or public/private key schemes. What is unique about these schemes is that each user is identified with one to many sets of unique key pairs. One pair can be for encryption and one pair for signature — one pair for can be personal and one for business. A key pair is very unique in that one half of the pair, the public, may be widely published without giving away the contents of the private half. What this means is that if some trusted third party (e.g., government, post office, MasterCard, Visa, or American Express) generates the keys for a known, verified entity, and publishes the public key widely, the entity may be easily identified. Before describing how they are identified, it is important to understand how the keys work together.

## Certificates

Certificates are used to tie the user's identity to the public key in a way that does not allow tampering and is protected. It is the method used to distribute the public part of a key-pair with the user identities. X.509 certificates are the predominant means for doing this at this time. The latest version is from the 1993 X.500 standards and is the basis for the MasterCard signature methodology. In the X.500 world, these will be kept in the directory. In the initial phases of the MasterCard implementation, it is assumed that a directory may not always be available, and that the communicating entity must include the certificate with the signature so that the signature may be verified. This may happen in a completely secure manner and without loss of protection. This will become evident when the certificate creation and structure is discussed.

A certificate is composed of at least the elements in the following list. MasterCard has added a few additional fields to the certificate, as others will also do over time. The X.509 certificate contains:

$$CA(A) = CA\{V,AI,CA,UCA,A,UA,AP,TA\}$$

where:

- CA(A) is the certificate or user A produced by certificate authority CA
- V is the version of the certificate
- AI is the identifier of the algorithm used to sign the certificate — the CA's algorithm
- UCA is the X.500 type distinguished name of the CA

- A is user A distinguished identifier in X.500 format (in the case of the MasterCard use, it is an alias known only to the issuing bank)
- UA is a list of algorithms for A's public keys
- AP is a list of A's public keys
- TA is beginning and ending dates for which the certificate is valid

The public/private key pair is generated, and the identity of the user confirmed to the appropriate level for the use of the certificate. Is some cases, the user may have to show birth certificates and fingerprints, while in others much less identification will be required. In this case, for credit card use, the identity will probably be just an address, name, and signature. These will be sufficient to generate the public/private key pair, and tie the name to the published key.

The certificate is constructed in the following manner: A customer is sent the private key on disk or in a smart card. The public key may or may not be sent to the customer — either way is fine and it makes little difference. The customer's X.500 distinguished name is issued by the credit card bank, assigned a certificate serial number, and given a valid time frame for the certificate. All of these are appended together with the distinguished name of the CA. These are run through a hashing algorithm so that they cannot be changed, and the hash is appended to the rest of the data after being encrypted by the private key of the CA. The certificate is built and is then either given to the issuee or put in a directory for worldwide access.

Because of the construction of the certificate, the issuer is known without a doubt. They used their private key to sign it. It was not tampered with, because the entire string was hashed and the consumer's public key is known to be authentic because it is in an unaltered certificate from a verifiable certificate authority. In this case, the information signed by public key A for consumer A is valid for a credit card transaction.

**SUMMARY**

Electronic commerce is not just one process — by definition, it must cover all business relationships, those between consumers and business, and those between businesses. The large spread of EC indicates different needs and requirements. Some areas (e.g., electronic product exchange) may require large amounts of bandwidth, while others (e.g., the exchange of paperwork) may require much less bandwidth to be effective. Other applications (e.g., consumer shopping) require the ability to connect to a large number of sites and businesses over and above the need for sufficient bandwidth. This was solved when the Internet became available for wide-scale use.

The last item that must be addressed is the security of information exchanges, which the payment processes and technologies being driven by

## LEVERAGING THE ELECTRONIC MESSAGING INFRASTRUCTURE

MasterCard and Visa will address. They are both based on public key cryptographic systems. These systems allow uniquely identified entities to ensure privacy on information exchanges. Two things will continue to reduce the speed of the EC expansion: high costs of the end-user device (i.e., currently the PC) and the inability to solve the ubiquitous higher bandwidth requirements.

# Chapter 22
# Developing a Trusted Infrastructure for Electronic Commerce Services

*David Litwack*

The use of internetworking applications for electronic commerce (EC) has been limited by issues of security and trust and by the lack of universality of products and services supporting robust and trustworthy EC services. Specific service attributes must be addressed to overcome the hesitation of users and business owners to exploit open systems — such as the Internet — for commercial exchanges. These service attributes include:

- *Confirmation of identity (nonrepudiation).* This indicates proof that only intended participants (i.e., creators and recipients) are party to communications.
- *Confidentiality and content security.* Documents can be neither read nor modified by an uninvited third party.
- *Time certainty.* Proof of date and time of communication is provided through time stamps and return receipts.
- *Legal protection.* Electronic documents should be legally binding and protected by tort law and fraud statutes.

## SERVICE ATTRIBUTE AUTHORITY

To support these service attributes, an organization or entity would need to provide:

- Certificate authority services, including the registration and issuance of certificates for public keys as well as the distribution of certificate revocation and compromised key lists to participating individuals and organizations.

- A repository for public key certificates that can provide such keys and certificates to authorized requesters on demand.
- Electronic postmarking for date and time stamps, and for providing the digital signature of the issuer for added assurance.
- Return receipts that provide service confirmation.
- Storage and retrieval services, including a transaction archive log and an archive of bonded documents.

These service attributes could be offered singly or in various combinations. The service attribute provider would have to be recognized as a certificate and postmark authority. The following sections describe how a service attribute provider should work.

**Certificate Authority**

Although public key encryption technology provides confidentiality and confirmation of identity, a true trusted infrastructure requires that a trusted authority certify a person or organization as the owner of the key pair. Certificates are special data structures used to register and protectively encapsulate the public key users and prevent their forgery. A certificate contains the name of a user and its public key. An electronic certificate binds the identity of the person or organization to the key pair.

Certificates also contain the name of the issuer — a certificate authority (CA) — that vouches that the public key in a certificate belongs to the named user. This data, along with a time interval specifying the certificate's validity, is cryptographically signed by the issuer using the issuer's private key. The subject and issuer names in certificates are distinguished names (DNs), as defined in the International Telecommunications Union-Telecommunications Standards Sector (ITU-TSS) recommendation X.500 directory services. Such certificates are also called X.509 certificates after the ITU-TSS recommendation in which they were defined.

The key certificate acts like a kind of electronic identity card. When a recipient uses a sender's public key to authenticate the sender's signature (or when the originator uses the recipient's public key to encrypt a message or document), the recipient wants to be sure that the sender is who he or she claims to be. The certificate provides that assurance.

A certificate could be tied to one individual or represent an organizational authority that in turn represents the entire organization. Also, certificates could represent various levels of assurance — from those dispensed by a machine to those registered with a personally signed application. Additional assurance could be provided by the personal presentation of a signed application along with proof of identity, or by the verification of a biometric test (e.g., fingerprint or retina scan) for each use of the private key.

*Developing a Trusted Infrastructure for Electronic Commerce Services*

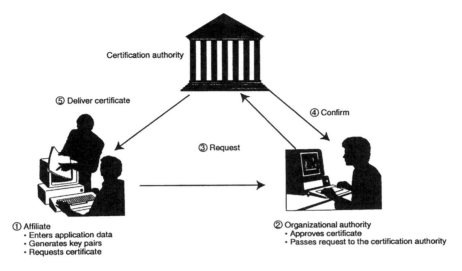

Exhibit 22-1. The Registration Process.

---

Exhibit 22-1 shows a possible scenario for obtaining a certificate. The registration process might work as follows:

1. The affiliate (i.e., candidate for certificate) fills out the application, generates private/public key pairs, and sends for the certificate, enclosing his or her public key.
2. The organizational authority approves the application.
3. The organizational authority passes the certificate application to the certification authority.
4. The certification authority sends back a message confirming receipt of the application.
5. After proper proofing, the certification authority sends the certificate to the applicant-affiliate.
6. The applicant-affiliate then loads the certificate to his or her workstation, verifies the certificate authority's digital signature, and saves a copy of the certificate.

**Digital Signatures.** Exhibit 22-2 illustrates how a digital signature ensures the identity of the message originator. It shows how a message recipient would use an originator's digital signature to authenticate that originator.

On the Web, authentication could work as follows:

1. The originator creates a message and the software performs a hash on the document.

# LEVERAGING THE ELECTRONIC MESSAGING INFRASTRUCTURE

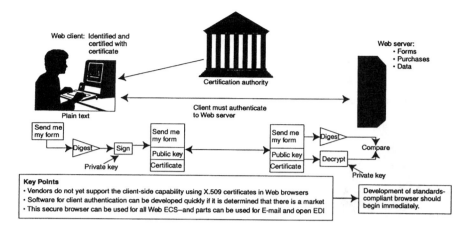

Exhibit 22-2. Client Authentication.

2. The originator's software then signs the message by encrypting it with the originator's private key.
3. The originator sends the message to the server, attaching his or her public key and certificate to the message if necessary.
4. The server either requests the originator's public key from a certificate/key repository or extracts the certification from the originator's message.

With this service, the authentication authority could either attach an authentication message verifying the digital signature's authenticity to the originator's message or provide that authentication to the recipient via a publicly accessible database. Upon receipt, the recipient would either acknowledge the originator's authenticity via the attached authentication message or access the public key and certificate from the publicly accessible database to read the signature.

To provide such levels of assurance, the certification authority must establish proofing stations where individuals and organizations can present themselves with appropriate identification and apply for certificates. The authority must also maintain or be part of a legal framework of protection and be in a position to mount an enforcement process to protect customers against fraud.

## Certificate Repository

The certificate authority (CA) also provides the vehicle for the distribution of public keys. Thus, the CA would have to maintain the public key cer-

## Developing a Trusted Infrastructure for Electronic Commerce Services

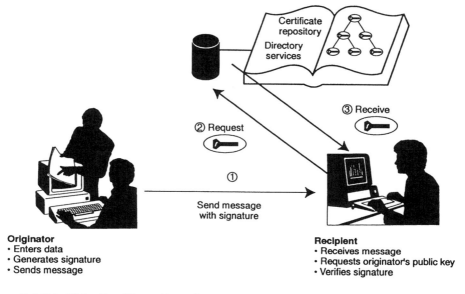

**Originator**
- Enters data
- Generates signature
- Sends message

**Recipient**
- Receives message
- Requests originator's public key
- Verifies signature

**Exhibit 22-3. Certificate Repository.**

---

tificates in a directory server that can be accessed by authorized persons and computers.

Exhibit 22-3 shows how subscribers might use such a repository. Certificates could be retrieved on demand along with their current status. Additional information, such as E-mail addresses or fax numbers, could also be available on demand.

The repository would work as follows:

1. The message originator creates a message, generates a digital signature, and sends the message.
2. The recipient sends a signed message requesting the originator's public key from the certificate repository.
3. The certificate repository verifies the requester's signature and returns the public key to the recipient.

The CA could also use the certificate repository to maintain a certificate revocation list (CRL), which provides notification of certificates that are revoked pursuant to a suspected compromise of the private key. This service could also require that the authority report such compromises via a compromised key list to special customers — possibly those enrolled in a subscribed service — and that such notifications be made available to all customers.

365

# LEVERAGING THE ELECTRONIC MESSAGING INFRASTRUCTURE

Finally, transactions involving certificates issued by other certificate authorities require that a cross-certification record be maintained and made publicly available in the certificate repository.

## Electronic Postmark

A service providing an electronic date and time postmark establishes the existence of a message at a specific point in time. By digitally signing the postmark, the postmarking authority assures the communicating parties that the message was sent, was in transit, or received at the indicated time.

This service is most useful when the recipient requires the originator to send a message by a specified deadline. The originator would request the postmark authority to postmark the message. The authority would receive a digest of the message, add a date and time token to it, digitally sign the package, and send it back to the originator, who would forward the complete package (i.e., signed digest, time stamp, and original message) to the recipient, as shown in Exhibit 22-4.

Electronic postmarking functions as follows:

1. The originator sends a request to the postmark authority to postmark a message or document (i.e., a digital digest of the message or document).

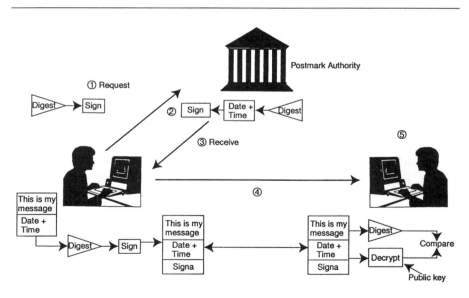

**Exhibit 22-4. Electronic Postmark.**

## Developing a Trusted Infrastructure for Electronic Commerce Services

2. The postmark authority adds date and time to the message received and affixes its digital signature to the entire package.
3. The postmark authority sends the package back to the originator.
4. The originator sends the original message or document plus the postmarked package to the recipient.
5. The recipient verifies the postmark authority signature with the authority's public key and reads the message or document.

### Return Receipts

This service reports one of three events: that a message has transited the network, that it has been received at the recipient's mailbox, or that the recipient has actually decoded and opened the message at a specific date and time. In the latter instance, the transaction delivered to the recipient that has been encrypted might be set up only to be decrypted with a special one-time key, as shown in Exhibit 22-5. This one-time key could be provided by the postmark authority upon receipt of an acknowledgment from the recipient accompanied by the recipient's digital signature.

Here is how return receipt might work:

1. The originator sends a message digest to the return receipt and postmark authority (the authority) with a request for a postmark and return receipt.

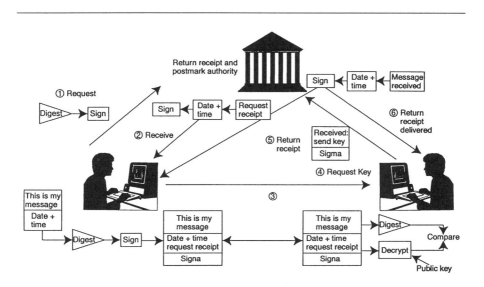

Exhibit 22-5. Return Receipt.

2. The authority receives the message digest, adds date and time, encrypts the result, attaches a message to the recipient to request the decryption key from the authority upon receipt of the message, and affixes its digital signature to the package.
3. The authority returns the postmarked, receipted package to the originator, who sends it to the recipient.
4. The recipient receives the message package and makes a signed request for the decryption key from the authority.
5. The authority receives the recipient's request, verifies the recipient's digital signature, and sends the decryption key to the recipient, who then decrypts and reads the message.
6. The authority simultaneously forwards the return receipt to the originator.

**Storage and Retrieval Services**

These services include transaction archiving where copies of transactions are held for specified periods of time, as illustrated in Exhibit 22-6. The service might also include information (i.e., documents, videos, or business transactions) that can be sealed, postmarked, and held in public storage to be retrieved via any authorized access. Likewise, encrypted information (i.e., documents, videos, or business transactions) can be sealed, postmarked, and further encrypted and held in sealed storage for indefinite periods of time. Each of these storage and retrieval capabilities must carry legal standing and the stamp of authenticity required for electronic correspondents.

Storage and retrieval works as follows:

1. The originator sends a request to the archive in order to archive a document or message for a specified period of time and designates this information as publicly retrievable.
2. The archive adds date and time to the message, verifies the identity of the originator, affixes a digital signature to the package, and archives the package.
3. A customer requests the document from the archive.
4. The archive retrieves the document, adds a date and time stamp to the package, affixes another digital signature to the new package, and sends it to the recipient.
5. The recipient verifies the first and second archive signatures and reads the message.

## USE OF COMMERCIAL EXCHANGE SERVICES

Electronic Commerce services (ECS) may be used in one of three ways:

*Developing a Trusted Infrastructure for Electronic Commerce Services*

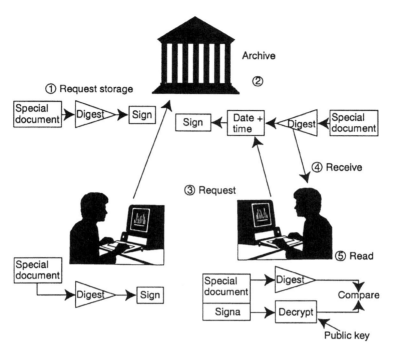

**Exhibit 22-6. Storage and Retrieval.**

1. The originator sends a message to the authority with a request for service; the authority provides the service and returns the message to the originator; and the originator then forwards the message to the recipient.
2. The originator sends a message to a value-added network (VAN), which then forwards the message to the authority with a request for services. The authority provides the service and returns the message to the VAN, which then forwards the message to the recipient.
3. The originator sends a message to the authority with a request for service and the address of the recipient. The authority then forwards the message directly to the recipient.

All these services could be provided by a single authority, by a hierarchy of authorities, or by a network of authorities, each specializing in one or more of these services.

## AVAILABLE TECHNOLOGIES FOR ELECTRONIC COMMERCE

Currently, three major technologies are capable of providing electronic commerce services: E-mail, the World Wide Web, and open electronic data interchange (EDI). Typical of advanced technologies, security elements are the last to be developed and yet are essential if these technologies are to be deemed trustworthy for electronic commerce.

The issues of confidentiality, confirmation of identity, time certainty, and legal protection apply to all these technologies. The solutions — certification, key repositories, postmarking, return receipts, and storage and retrieval — are equally applicable to each of these technologies. Although the state of universality and interoperability varies among these technologies, they are all in a relative state of immaturity.

### Secure E-mail

Electronic messaging's most classic manifestation is E-mail. Because of its capacity for handling attachments, E-mail can be used to transfer official business, financial, technical, and a variety of multimedia forms.

**DMS and PEM.** Both the Department of Defense standard for E-mail, which is based on the ITU's X.400 standard for E-mail (called the Defense Message System or DMS), and the Internet E-mail standard, the simple mail transfer protocol (SMTP), have made provisions for security. The DMS uses encapsulation techniques at several security levels to encrypt and sign E-mail messages. The security standard for the Internet is called Privacy Enhanced Mail (PEM). Both methods rely on a certificate hierarchy and known and trusted infrastructure. Neither method is fully developed.

### Secure World Wide Web

The phenomenal growth of the Web makes it a prime candidate for the dissemination of forms and documents. Organizations see the Web as a prime tool for services such as delivery of applications and requests for information. However, Web technology has two competing types of security: one at the application layer that secures hypertext transfer protocol (HTTP) formatted data (known as SHTTP), and one at the socket layer that encrypts data in the format in which it is transported across the network.

In addition, vendors do not yet support either client-side authentication or the use of X.509 certificates. Although software for such activities as client authentication can be developed relatively quickly, vendors have to be convinced that there is a real market for such products. This technology is about to emerge and, although it will emerge first to support Web applications, it will also speed the development of E-mail and EDI security services.

### Secure Open EDI

Until now, EDI has been used in closed VANs where security and integrity can be closely controlled. Signing and encryption have been proprietary to the EDI product in use or to the value-added EDI network provider.

By contrast, open EDI, running across open networks, requires adherence to the standards that are still being developed and a yet-to-be developed infrastructure that can ensure trusted keys. To date, the various schemes to accelerate the use of open systems for EDI have not captured the imagination of EDI users and providers.

## THE OVERRIDING ISSUE: A PUBLIC KEY CERTIFICATE INFRASTRUCTURE

The suite of services and technologies described in this chapter depend on trusted public keys and their bindings to users. Users could be completely assured of the integrity of keys and their bindings if they were exchanged manually. Because business is conducted on a national and international scale, users have to be assured of the integrity of the registration authority and the key repository in an inevitably complex, electronic way.

One as-yet-unresolved issue is whether such an authority or authorities should be centralized and hierarchical, or distributed. The centralized, hierarchical scheme would mean that certification authorities (and purveyors of the accompanying services) would be certified by a higher authority that, in turn, might be certified by yet a higher authority — and so on to the root authority. This kind certification would create a known chain of trust from the highest to the closest certification authority. This scheme is often referred to as the Public Key Infrastructure (PKI).

The alternative assumes that the market will foster the creation of a variety of specialized certification authorities to serve communities of interest. A complicated method of cross-referencing and maintaining those cross-references in the certificate repository for each community of interest would then develop.

The outcome of this debate is likely to result in a combination of both methods, such as several hierarchies with some kind of managed cross-referencing to enable public key exchanges between disparate communities of interest when required. Following are some of the issues yet to be resolved:

- Agreement on the exact contents of certificates
- Definition of the size of prime numbers used in key generation
- Establishment of the qualifications required for obtaining a certificate
- Definition of the identification and authentication requirements for certificate registration
- Ruling on the frequency with which certificates are renewed
- Agreement on the legal standing and precedence for such technology

## SUMMARY

Groups such as the Internet Engineering Task Force (IETF), the Federal Government Public Key Infrastructure (PKI) users group, and even the American Bar Association are tackling these knotty issues.

## LEVERAGING THE ELECTRONIC MESSAGING INFRASTRUCTURE

In fact, with toolkits now available that allow the user to become his or her own certificate authority, everyone can get into the act. Private companies such as VeriSign are establishing themselves as certification authorities so that users will give their public keys and certificates credence. The National Security Agency wants to become the certificate authority for the U.S. federal government. The U.S. Postal Service is intent on offering electronic commerce services to businesses and residences by acting as the certificate authority and provider.

An infrastructure will emerge, and it will probably work for users in a very similar manner to the way that it has been described in this chapter.

# Chapter 23
# The World of Electronic Commerce

*David A. Zimmer*

The world of business is changing at a rapid rate. Some say that within five to ten years, the way we do business will not resemble our current methods. Electronic data interchange (EDI) and electronic commerce (EC) will prevail, and paper methods of buying and selling goods will be obsolete. In several vertical markets, this prediction has become reality.

The advent of electronic communication throughout all sectors of society and the proliferation of the Internet and World Wide Web would seem to support the move to EC. Understanding the current capabilities of the EC technology, business propositions, and electronic markets is key to evolving our businesses into thriving competitors in the future instead of hulking dinosaurs.

This chapter describes the current landscape and the future horizons of electronic commerce. Understanding how it ties into current electronic communications systems is important. Knowing the evolutionary process from systems today to the future world of commerce will help better position costly networks to support tomorrow's transactions.

## A BIT OF HISTORY

EDI dates back to the 1960s, with its greatest growth during the 1980s when several large industries (e.g., trucking and automotive, separately) decided that interconnecting computers for transferring information would provide a cost- and time-savings benefit to business. EDI is used by companies of all sizes, but its greatest use is by companies whose revenues exceed $500 million. With the advent of the World Wide Web and the rapid adoption of PCs into the homes (the network computer will also have an effect), electronic commerce will begin making in-roads into the consumer market.

# LEVERAGING THE ELECTRONIC MESSAGING INFRASTRUCTURE

**Paper-Pushing Scenario**

Companies wanted to develop a machine-readable data stream format that could be used to replace the multipart paper forms currently in use. Typically, to order supplies from a vendor, a customer would submit a paper order form to the vendor, the vendor would circulate the form to the appropriate internal organizations, the product would be shipped, and the vendor would issue a paper invoice. The customer would circulate the invoice to its internal organizations for the proper approvals and issue a paper check to the vendor. The vendor would deposit the paper check into the bank and get a paper receipt that had to be circulated to the proper organization for crediting. The customer, receiving the canceled check, would have to balance the books.

Although convoluted, paper-based systems worked for a majority of companies. Unfortunately, it inhibited their ability to expand — the time delays causing extra costs in the products and slowing their businesses. In addition, because many had implemented computer-based systems to facilitate tracking, each stop in the paper's journey required reentry of the data. It is estimated that an entry mistake occurred every 300 keystrokes. In the life cycle of one order, many mistakes could occur.

To compound the problem, each company had its own forms and formats. If two large companies needed to conduct business between themselves, they had to agree on whose form to use, knowing that they were not the same format. Having a foreign form floating throughout the corporation could increase the entry error rate.

**Moving to Electrons**

In the 1970s, no standards existed between trading partners. As a result, each partner had to agree on the format of the transaction data. Because no standards existed, trading with a third partner required a new set of agreements and customized development. After aligning with several trading partners, this ad-hoc approach became expensive.

In 1975, the transportation industry, under the auspices of Transportation Data Coordinating Committee (TDCC), developed and published a standard for transmitting data. In 1979, the American National Standards Institute (ANSI) approved TDCC's standard as ANSI X12 (pronounced X-12, not X-dot-12. If you say X-dot-12, the other person will know you are a neophyte in the discussion).

X12 defines a standard of field-value pairs (a field name defines what the information following is, and the value is the actual data), for example:

> PER IC*J. Smith*TE*4125551212*FX*4125551213
>
> IC = Information Contact

TE = Telephone Number

FX = Fax Number

The field-value pairs were separated by a delimiter — in this case, an asterisk (*). Transmission was held to 7 bits, meaning that they could only use less than 127 characters, resulting in an almost human readable format. All the characters are readable, but the format is horrendous for casual reading. The published standard allowed companies to buy third-party EDI products and services.

Unfortunately, each industry has its own terms and needs. X12 has become a large collection of specifications, some specific to an industry while others are used for all industries (e.g., the trucking industry has bills of lading). The automotive industry has it forms. The medical industry has claim forms. They all have purchase orders and invoices.

X12 is not without its problems. The data can vary in length, requiring a delimiter so that software can determine where one data value stops and another begins. X12 does not define a consistent delimiter. Some companies may use "*" while others use ":" for delimiters. Therefore, formatting messages to a trading partner — a company that you trade with electronically — can become a complicated process of negotiation. Whose delimiter is used? Who supports the cost of developing the converter to convert one format to the other format? Who pays for the communication line?

Fortunately, over time, many of these issues have been resolved. Many large companies are trading electronically now. Medium and small companies are getting linked because of the proliferation of powerful PCs and inexpensive (e.g., $800 to $1200) EDI software.

To move the electronic data interchange into the global market, the United Nations defined EDIFACT — a globally accepted electronic representation of business. EDIFACT is used primarily in European businesses; X12 is the predominant standard used in the U.S. Both standards describe only the message format, not the transmission protocol. Therefore, companies can hook to other companies' computers by synchronous or asynchronous connections. X.435 defines EDI over X.400, a messaging standard. And now, there is much discussion about EDI on the Internet/WWW (i.e., EDI over TCP/IP transport).

## A CASE STUDY: A BUSINESS TRANSFORMED

Arrow was a company that built custom cabinets for sewing machines. The sewing machine industry has suffered a decline in sales for a long time, resulting in less need for cabinets. Arrow needed to transform its business or risk going out of business. As a result of the company president's friendship with an executive at Sears, Arrow embarked on a new line of business.

# LEVERAGING THE ELECTRONIC MESSAGING INFRASTRUCTURE

Sears wanted to service its customers by shipping vacuum cleaner bags to its customers. Unfortunately, Sears could not handle the low-volume shipments required.

Arrow established EDI links to Sears and some of its suppliers. Arrow stocked the necessary vacuum cleaner bags and received shipment requests from Sears electronically. As the inventory dwindled, Arrow would replenish the stock by electronically ordering replacements. The electronic transactions were extended to cover all aspects of the stocking and shipping systems of a transaction. Invoices, shipment requests, and shipment acknowledgments were all done electronically.

Arrow has expanded its business to include more than just vacuum cleaner bags. Today, they handle many large companies' customer fulfillment needs. Arrow drives all business electronically. To use Arrow's services, large companies must comply with Arrow's electronic systems. (An interesting twist: a small company forcing a large company to do business its way!) Although Arrow has not gone into officially helping other companies implement EDI, it works closely with its customers and suppliers to ensure a smooth operation of EDI transactions.

Arrow ships hundreds of thousands of items, handles tens of thousands of orders, and generates tens of millions of dollars in revenue, all with only 35 employees. Although the employees work very hard, without implementing EDI transactions, Arrow would not be the fulfillment house it is today.

## DEFINITION OF ELECTRONIC COMMERCE

What is electronic commerce? Is it simply the electronic exchange of purchase orders and shipment notices, or does it go beyond that? Electronic commerce can be categorized into three areas:

- Business-to-business
- Business-to-consumer
- Consumer-to-consumer

In each category, the needs are different. As a result, the solutions are different. For example, business-to-business has higher-volume traffic, larger transaction values, automatic data entry into various databases, stringent security, and high reliability. Consumer-to-consumer requires simple interfaces, no database input, and smaller transaction amounts.

Each category has different electronic connection needs as well. Business-to-business needs secure networks safe from hackers trying to change transaction values. Consumers can use a simple dial-up connection to place items on bulletin boards or place an order.

## The World of Electronic Commerce

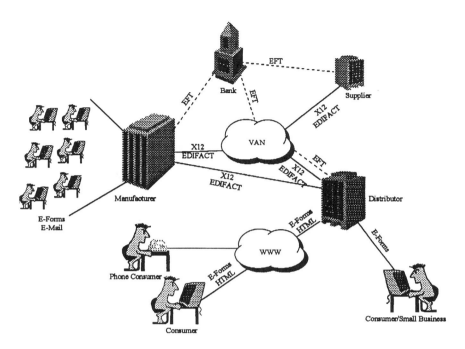

Exhibit 23-1. The World of Electronic Commerce.

Exhibit 23-1 depicts all three categories, the types of transactions between parties, and the various connection types used by each party. Each section must be analyzed to understand its perspective. As time evolves, however, parts of this picture will drop out as connections from manufacturer to consumer tighten and intermediate parties are squeezed from the process.

Electronic commerce has several components. Traditional EDI contains the electronic transactions between suppliers' and buyers' computers. EDI is a computer-to-computer data format protocol. Internal software translates the EDI data into internal business applications (e.g., the accounting database, order entry, manufacturing, and shipping and receiving).

Electronic funds transfers (EFT), a computer-to-computer protocol, deals with the transfer of money or payments electronically.

The third component involves electronic forms that generate the transactions. It provides the interface between humans and the automated systems.

# LEVERAGING THE ELECTRONIC MESSAGING INFRASTRUCTURE

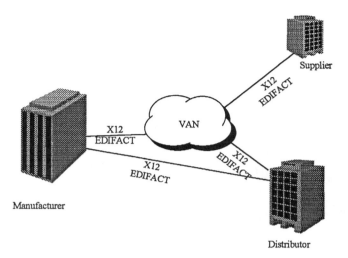

Exhibit 23-2. Business-to-Business Electronic Commerce.

## Business-to-Business Electronic Commerce

Business-to-business electronic commerce involves translating internal business information into EDI transaction sets. Mapping software extracts data from various business applications and formats the data into the appropriate EDI format for trading with partners. The translation may be specific to a particular trading partner or it may be more general in nature, depending on which connection is used with the trading partner.

Businesses can use two methods of trading transactions: point-to-point connections with trading partners or value-added networks (VANs) (see Exhibit 23-2).

Point-to-point connections require leased telephone lines between the partners. In some large companies, they choose to manage all their communications with their trading partners by providing the leased-line connections. In this case, the large company is considered a hub. Companies they trade with are considered spokes (see Exhibit 23-3). This configuration gives the greatest control over the transactions to the large company. If the company must transact business with a number of other large companies, also acting as hubs, the number of leased lines may become unmanageable. At this point, the company may opt to use a VAN to lower leased-line costs.

Businesses implement EDI for a variety of reasons. The top reason by far is customer-supplier requests. Other reasons usually cited are (in descending order of importance):

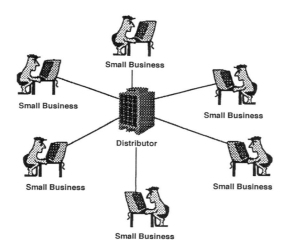

Exhibit 23-3. Hub and Spoke EDI Topology.

- Improving customer service
- Gaining competitive advantage
- Increasing data accuracy
- Cost savings
- Increasing speed of information access/flow

This interconnection between companies helps to speed the process of acquiring needed materials and supplies, shipping products, and managing the information flow necessary to transact business. The increased speed of the information flow means faster turnaround times for procuring products, manufacturing the final goods, and quicker time to market. Product evolution cycles decrease so that competitive advantages are gained. In addition, manual reentry errors decrease immensely — eliminating waste and lost time.

EFT. Electronic funds transfers provide electronic transmission of funds. The process of electronic ordering, tracking, and delivery receipt would not be complete if the final leg of the process required a paper check to be issued for payment. The EDIBANX standard provides a standardized mechanism for financial EDI. EDIBANX uses X12 formatting so that existing systems require only simple upgrades to support electronic transfers of funds.

The electronic funds transfer supplies several benefits to companies: quicker access to funds, lower processing costs for the funds, and no lost or misdirected paper checks (see Exhibit 23-4). EDIBANX permits compa-

## LEVERAGING THE ELECTRONIC MESSAGING INFRASTRUCTURE

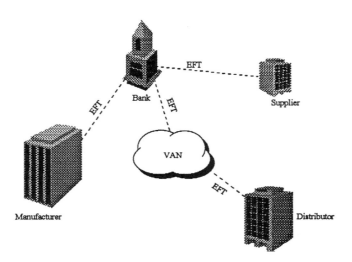

**Exhibit 23-4. Electronic Funds Transfers.**

nies to tie the money transfer directly into their business applications (e.g., accounts payables or account receivables). For financial transactions to be feasible, networks must provide security and reliability. In some cases, companies have chosen to implement point-to-point connections directly with their banks or they have chosen VANs. For most, the Internet is still too risky for sensitive transactions.

**E-Mail Backbone.** Integrating EDI with the electronic mail backbone within large companies is the next logical step. Employees already use E-mail for communication, and workflow for other business processes is commonplace. Integrating EDI translation into the backbone provides employees with a direct link to the EDI process. For example, accountants could put stop-shipment orders on bad accounts, shipping clerks could authorize shipment orders, and employees responsible for ordering parts can do so by filing an electronic form. A workflow could be developed so that a parts order could be validated against an open purchase order or sent on to the authorizing agent for approval.

Using the electronic mail backbone, the process of ordering products can be pushed further into the organization, eliminating the need for middlemen (see Exhibit 23-5). By pushing the task further into the organization, errors and miscommunications are lessened because the person most affected by the order would be responsible for order placement. The rest of the process is automated so other business applications (e.g., accounts payables, receiving, and manufacturing schedules) are notified accordingly. Notices of purchase acknowledgments, shipments, and backlogs are re-

*The World of Electronic Commerce*

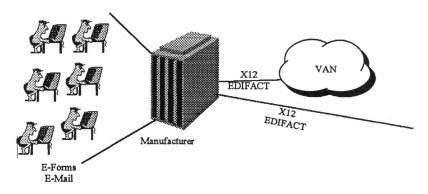

Exhibit 23-5. Electronic Commerce Integrated with Electronic Mail Backbone.

turned electronically. Following the reverse path to the originator, the notice can affect the business applications, as well. For example, back-order parts may affect the manufacturing cycle. The cycle manager could address the effect of the back-order and request that a different supplier be used.

Today, unfortunately, most electronic mail backbones and EDI networks are separate. A few companies (e.g., Control Data Systems and Sterling Software) provide mapping software that pulls messages from the E-mail store and maps the information into EDI transactions. Incoming EDI transactions are mapped back into E-mail format and forwarded to the originator. A tighter integration between E-mail and EDI must occur for full EDI integration into corporate workflow and everyday communications.

**Business-to-Consumer Electronic Commerce**

For years, businesses have used telephone mailorder to eliminate the paper chase with consumers. The next step is to develop ordering by PC or MAC using the World Wide Web. The Web browser is a software program that runs on virtually any machine. The company can provide product information to any degree the consumer needs in a very simple and compelling way. Using graphics, sounds, and motion, the company can grab the consumer's attention. Because the Web browser provides easy navigational tools through hot-links (i.e., hypertext links), the company can provide a brief summary of the product with links leading to additional details. Once the customer is ready to purchase, a link will lead them to a secure page where sensitive information can be passed using the secure socket layer (SSL) protocol. The SSL encrypts the data being transmitted for protection.

# LEVERAGING THE ELECTRONIC MESSAGING INFRASTRUCTURE

Several companies have started to use this method for product sales. The volume of sales is still low because of the perceived security risks of sending credit card information over the Internet. Interestingly, the security of the secure links is more secure than the telephone mailorder that generates billions of dollars of sales per year. It is only a matter of time before the misperception is gone and commerce to consumers soars.

Any size business can use the Web to sell its products. The low-entry cost of creating Web pages, retrieving orders, and processing them by way of the Web is enticing to small companies. Unlike traditional channels, small companies have worldwide presence by using the Web. They can easily load search engines to advertise their sites and draw attention to them. Big companies are not excluded from this medium, either. In fact, in this channel, the consumer may never be aware of the company size backing a particular product.

Telephone-based browsers are being developed. Those who do not own a computer to surf the Web can use a browser operated over a conventional telephone. Using interactive voice response (IVR), listeners can listen to a Web page (text only, graphics are hard to describe) and press a button when they want to travel to another page. Using this method, a company can describe a product, and if customers wants more detail, they simply travel to the more detailed pages. Ordering can be done by phone as well (see Exhibit 23-6). Voice recognition is sufficient for credit card input. The products would be shipped to the billing address of the credit card. Of course, this type of information would have to be coordinated with the appropriate credit card company.

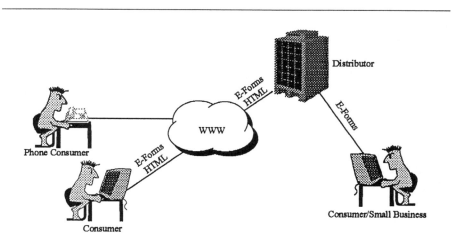

**Exhibit 23-6. Business-to-Consumer Electronic Commerce.**

*The World of Electronic Commerce*

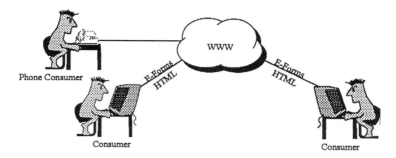

Exhibit 23-7. Consumer-to-Consumer Electronic Commerce.

---

**E-Forms.** Large businesses might make electronic forms available to small businesses for input of orders and fulfillment information. The link might go directly into the larger companies' machines. This type of interface provides a faster, easier method of ordering multiple or large quantities of items with direct feedback access of inventory levels, and part assemblies with a greater sense of security because access would be limited to selected partners. The company could use an intranet to support such interactions so that the interface is a browser supporting forms.

## Consumer-to-Consumer Electronic Commerce

Consumer-to-consumer electronic commerce will soon begin to emerge (see Exhibit 23-7). Currently, consumers have electronic banking and electronic bill payment. These could be considered the first forms of consumer electronic commerce. As more banks open their electronic capabilities, we will see consumers providing payments for purchases electronically. Consumers can create Web pages to advertise various items for sale (e.g., cars, boats, lawn tractors, and other items commonly found in newspaper want ads). Digital cameras can be used for placing pictures on the pages. Buyers interested in buying the item can contact the seller to make the purchase.

Using some mechanism, the seller and buyer enter into an agreement and the transaction is completed when the buyer transfers money from his or her account to the seller's account. This type of service, although a bit futuristic, may be provided by a third party or by a bank. The transaction is logged and validated by the transacting entity and sealed with the money transfer.

Most likely, consumer-to-consumer commerce transactions will be done over the Web rather than by a specialized service. The ubiquitous nature of the Web, common protocols, easy interface, and rapid acceptance makes the use of private networks obsolete. Using telephone browsers, anyone can take part in the buying network. The same transactions that are completed by PCs can be accomplished by the telephone. Transactions and money transfer authorizations can be completed by telephone interfaces.

## THE EVOLVING ELECTRONIC COMMERCE LANDSCAPE

As more companies implement electronic commerce technologies and streamline their business strategies, evolution in supplier-buyer interaction will continue. In his book *Digital Economy*, Don Tapscott's overriding theme was that everything is going electronic. He outlines the following 12 themes of the new economy:

1. *Knowledge.* The new economy is a knowledge economy — things consumers buy will be smart.
2. *Digitization.* All information will be stored digitally.
3. *Virtualization.* As information shifts from analog to digital, physical things can become virtual.
4. *Molecularization.* Old corporations are disaggregated, replaced by clusters of individuals that form the basis of economic activity.
5. *Integration-internetworking.* The economy becomes a network of molecules re-aligning for the creation of wealth.
6. *Disintermediation.* Middleman functions between producers and consumers are being replaced by digital networks.
7. *Convergence.* Computing, communications, and content are becoming one.
8. *Innovation.* Make products obsolete — otherwise, someone else will.
9. *Prosumption.* The gap between consumer and producer blurs.
10. *Immediacy.* The key driver and variable in economic activity and business success.
11. *Globalization.* Customer bases are becoming global because of electronic networks.
12. *Discordance.* Unprecedented social issues are beginning to arise, potentially causing massive trauma and conflict.

Each theme has profound effects on the way people do business. Using electronic commerce, manufacturers will be able to work more closely with the end customer. Knowing their buying habits and consumption rates, manufacturers can more finely tune production runs. This fine-tuning could result in smaller lots of a particular item being made, requiring less inventory and warehouse space before being directly shipped to the consumer's outlet. Wal-Mart has created the model for such activities. Its suppliers work with Wal-Mart to replenish stock only when needed. As a result, Wal-Mart has very little inventory sitting in warehouses costing the company money. Consumers benefit through lower prices and more convenience (one-stop shopping for many items). The manufacturer benefits because they know exactly the buying trends of the consumer.

Manufacturing lines will change. Rather than producing single items, lines will be retooled to produce smaller lots of several items. Orders will be fed directly into the computers that schedule the lines. Workers may produce widgets one day and do-hickies the next. Using the incoming EDI orders and the production line capabilities, computers can schedule the optimal production line to lower inventory and warehousing needs.

Computers will receive data from several sources: consumers buying the finished products, suppliers supplying the required parts, and managing back-orders so that the process can run smoothly. The result of this automation is less people between raw goods and buying consumers. Middlemen will no longer be needed. Consumers and manufacturers will deal directly with each other. Key components the manufacturer needs to track the transactions and consumers are directories. Consumers can find manufacturers of desired products easily through directories.

## DIRECTORIES: KEEPING TRACK OF ADDRESSES, TRANSACTIONS, AND INFORMATION

Directories enable users to do business electronically without complicated addressing schemes. The propagation and synchronization of the directory information between companies is a very important success factor. Without accurate addresses, purchase orders and payments will be misdirected.

Synchronizing changes within a single vendor's product environment is simple — the vendor supplies the synchronization tools. Directories used for electronic commerce will not share the single-vendor luxury. Companies implementing directories for use in electronic commerce buy from different vendors. Synchronization becomes complex in this multivendor environment. This complexity has already been experienced within enterprise E-mail networks. Directories open to the public for the purpose of electronic commerce will be even more complex.

### A Model Demonstrating Directory Synchronization Complexity

A few years ago, two major West Coast banks announced their merger. Within weeks, an E-mail connection was established between the two banks permitting the exchange of electronic mail between key personnel planning the merger. Directory entries, however, had to be propagated manually by the respective mail system administrators. More than two years passed before an automated process was developed to distribute and synchronize directory updates among the merged bank's 35,000 PROFS, cc:Mail, MS Exchange, and Lotus Notes users.

The lengthy process occurred because no standard existed and each mail system had its own characteristics for naming and addressing. Understanding the differences between naming and addressing became clear in the process. Naming is the tag by which a user is known to other users on the system, while addressing is the tag by which the computer knows the user.

The name contained in a directory is not necessarily a user's name; it may be a server name, an internal organization name, an application, or the name of a mail domain. The corresponding address information is used by the computer for routing purposes. For example, on the Internet, the name of the mail domain known as ameagle.com corresponds to the IP address of the computer that serves as the gateway to that domain. When a user addresses a message to Accounts.Payable@ameagle.com, the messaging system refers to the corresponding IP address to route the message through various computers and gateways until it eventually ends up in the mailbox.

Understanding these concepts is key to developing directories that are useful in the interenterprise realm of electronic commerce. Just as each vendor of directories and E-mail systems has its own ideas on address and name construction, so do the companies that use these products. Each incarnation reflects the personality and culture of the company. This mixture of naming and addressing convention makes it difficult to easily create a network between trading partners.

**The State of EC-Related Standards**

To facilitate continued growth of electronic commerce, standard protocols must be instituted. Fortunately, the vendor companies have not been idle, but have foreseen the need of such products.

X.500. In 1988, the first version of X.500 was introduced. Products based on the standard have been available for some time. In 1993, a second version of X.500 was introduced. The second version fixed some of the weaknesses of the 1988 version. Products now on the market support the 1993 standard.

It is important to understand that X.500 is best suited for large enterprise networks or inter-enterprise environments. It is not suitable for smaller-scale operations (e.g., desktop or small LAN environments). In fact, proprietary solutions are better fitted for smaller operations if interaction with the directory comes from within the organization.

LDAP. Another protocol becoming popular from the Internet RFC process is the lightweight directory access protocol (LDAP). As the name implies, it is less resource-intensive (lightweight) than the X.500 DAP standard. Therefore, it can run at the desktop level. In addition, LDAP ac-

## The World of Electronic Commerce

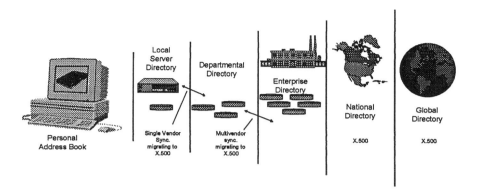

Exhibit 23-8. Six Levels of Directories.

cesses more than just X.500 directories — it can access other directory structures that may exist at all levels of the organizations. LDAP unifies all directory types into a single access method, making it extremely powerful and useful from the end-user perspective.

Rapport Communications, in their report "The Rapport Messaging Review: X.500 Directory Services," predicts that X.500 will not be positioned at the personal address book level and will be slow to emerge at the local server and departmental level. They feel that typically these environments have limited processing power and, given their local nature, will not fully gain from the benefits of the global nature of X.500. X.500 is best suited for complex environments, unlike these localized networks.

In Exhibit 23-8, each band is a summation of the information of the directory entries in the band to its left, starting with the local server band. Typically, the personal address book contains entries specific to a user that would not be of value to the rest of the organization. Starting with the local server directory, entries describe users located on that server. The departmental directory contains a summation of local server entries within its domain. Synchronization is done within that domain. The enterprise directory is a summation of the department directory entries. The pattern continues all the way to the global directory.

Companies restrict the access to their directories at the enterprise directory level. Because of privacy issues, data security, and other reasons, enterprise directories are closed to outsiders. The enterprise directory may contain entries of people in another organization by an automatic registration process offered by some vendor's products, but access to that entry is only from within the organization. Accordingly, outsider entries are not synchronized with the originating organization. Invariably, the entry becomes stale and is no longer accurate. Inaccuracy in entry values would break routing of information and would impede electronic commerce.

**Three Types of Directory Products**

In preparing the directory infrastructure for electronic commerce, it is important to understand the different types of directory products: proprietary, standards based (X.500), and X.500-like.

Many of the directory products in use today are proprietary based. Standards and products based on standards did not exist when companies were implementing their enterprise solutions. Vendors (e.g., Soft*Switch — now a part of Lotus Development Corp., IBM Corp., and Digital Equipment Corp.) offered directory products in conjunction with their E-mail systems. As a result, support of the embedded base of products is a must in any transition or implementation of another electronic commerce directory structure.

Standards-based products means they support the X.500 specifications. Hewlett-Packard, Digital, and OSIware offer X.500 directory products. Other vendors offer X.500-like or X.500-architected products. Simply put, the information is organized in a manner similar to X.500, but it does not support the access protocols as specified by the X.500 standard.

**A Trio of Concerns: Security, Authentication, and Repudiation**

Electronic commerce has three concerns: security, authentication, and repudiation. Transactions must be secure. They must be impervious to outside tampering or misdirection. Values inside purchase orders must be accurate.

For example, a buyer places an order for 10 widgets. During the transmission of the order, the 10 is increased to 20. The seller receives the order with an incorrect value. Unfortunately, the wrong number of widgets will be shipped, requiring a return, extra processing, and expense on both sides of the transaction. Using directories with public key information, buyers can secure their transactions from tampering. The buyer would need to know that the public key information was available and be able to access it. Without using standard protocols for obtaining the information, the transaction between buyer and seller may not have occurred.

Buyer authentication is very important. The seller needs to be sure the persons or organizations placing the order are really who they say they are. With directory entries, the seller can authenticate the buyer through the use of passcodes or other mechanisms (see Exhibit 23-9). The buyer can change those passcodes regularly to avoid impostors obtaining access to their accounts. When the buyer changes the passcode, the process inside the seller's company is not affected. Using the directory, each party is insulated from the inner workings of the other company.

## The World of Electronic Commerce

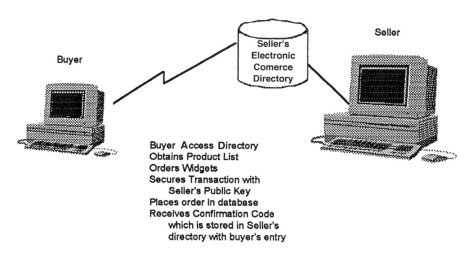

Exhibit 23-9. Secure Orders, Tracking with Directory, Accessible by Buyer.

In the case of differing opinions on orders (e.g., buyer placed an order for 10 widgets and seller shipped 20), the directory record can be used to settle disputes. The directory can track confirmation numbers that track back to the order and shipment logs. The directory is used by the automated tracking system to obtain confirmation information and buyer information. Additional information may include credit limits, current balance, and last payment received. The directory is a natural repository for this information because all information for a trading partner would be stored in one place and accessible by both parties. Software would check balances and limits so that orders would be placed only if values were within the limits.

### VALUE-ADDED NETWORKS VS. INTERNET AND WEB

Currently, many companies are discussing which networks they should use for transporting electronic commerce data. The two main considerations are value-added networks (VANs) or the Internet. VANs charge for transporting data across their networks and provide mailboxes for those customers not requiring full-time connection to the network. The Internet provides a low transport cost option, but lacks in several features critical to data transmission integrity.

### DataTransmission Integrity

Companies have five areas of concern regarding data transmission integrity: security, authentication, nonrepudiation, reliability, and customer service.

VANs are universes in which members are required to subscribe to the service prior to the transmission of data. When accessing the service, the members are required to log onto the system identifying themselves. Only certain types of transmissions are permitted (e.g., sending data destined for others and receiving data sent to them). They are not able to surf to another member's mailbox and peruse its contents. The data sent to a recipient is secure from other members. The VANs maintain their service to meet high standards of up-time, data transmission integrity, and data delivery. They track all transmission through their networks, giving customers the ability to determine where a transmission may have gone astray.

VANs satisfy all five concern areas, as well as provide consulting and other value-added services that ensure the security and reliability of transactions over their networks. The Internet has provided increased access to all levels of trading partners, but it is not a single source of service. As luck would have it, a trading partner may use the same Internet service provider (ISP) as another, but most likely that will not be the case. Therefore, the transmission of data may pass through one or several other ISPs before reaching its final destination. Each ISP maintains different levels of service standards. Typically, no tracking of data transmission is done, no single source of customer service exists, authentication of users may be lacking, and other security breaches may exist.

Recently, products have been coming to market to help overcome the Internet inadequacies for electronic commerce. Several ISPs have implemented secure networks within themselves. It is only a matter of time before the technical issues are resolved and electronic commerce over the Internet is reliable. Customer service is the remaining issue to be solved. AT&T WorldNet has advertised that they can provide all five concern areas and are promoting electronic commerce over the Internet. Others (e.g., PSI, MCI, and UUNET) will follow suit shortly.

The Web provides an interesting method of electronic commerce. It provides a cross-platform interface. The plug-ins provide powerful mechanisms to take the data and either translate it into EDI transactions or place it directly into company databases. Within companies, intranets help process orders from manufacturing to shipping to customer delivery. Inventory levels can be monitored and new supplies ordered from suppliers. Agents and Web crawlers can be used to find the best supplier for the need.

The question that remains to be answered for customers is, which system is best — VANs, the Internet, or the Web? The answer remains the same — companies must use the system that best fits their needs and level of quality required. For the foreseeable future, all methods will be used. They will be integrated together so that they work seamlessly and transactions flow reliably from buyer to supplier and back.

## SUMMARY

Business is changing rapidly. Product cycles that used to be measured in years are now being measured in quarters. Major corporations have announced that they will update their product lines every three to six months. Time is no longer measured in solar months, but in Web weeks. Cutting-edge products are obsolete shortly after they hit the store shelves.

What does all this imply? First, companies must understand the buying habits of the buyers. Second, the time interval from purchase to delivery must be shortened. Third, inventory needs must be calculated accurately. Fourth, companies must ensure that suppliers provide raw materials when they are needed. Fifth, slow, antiquated paper processes must be replaced with automated systems that update various business applications so that better decisions can be made faster. In a nutshell, this whirlwind of faster and faster will continue. Therefore, tools must keep up.

EDI and EC, a technology and a method of doing business, will help companies meet the challenges. Refusing to break the old working methods of today to shape them for tomorrow will cause current thriving businesses to become dinosaurs. The U.S. Government has mandated the use of EDI if a company wants to do business with them. Large manufacturers (e.g., Ford and Chrysler) require that all parts houses use EDI. Other businesses must follow suit.

Electronic commerce, the methodology of using EDI and other technologies to enhance business processes, encompasses the full breadth of the food chain from raw materials to final consumption by consumers. Each intermediary stop must align with the electronic flow of information or else risk planned extinction. It spans business-to-business transactions, business-to-consumer, and eventually, consumer-to-consumer transactions. Beginning with mainframe-to-mainframe transactions, it has spread to the World Wide Web. Anyone with a PC or a telephone can transact commerce with anyone else in the world. A technology born in the 1960s is ready for prime time today and tomorrow.

# Chapter 24
# Intranets: Notes vs. the Internet

*Brett Molotsky*

A buyer is looking for some new transportation and a friend recommends a dealer who specializes in many different types of vehicles. The dealer suggests either a car or a tractor, based on the logic that the vehicles are essentially the same; both have engines, drivetrains, steering and braking mechanisms, and both will transport the driver from one location to another.

Although the similarities between the two vehicles may be strong, there are also strong differences. The vehicles are designed and built for different purposes and different environments, and will work differently on different types of terrain. The reasons for choosing one over the other depend entirely on what the driver's transportation needs are and how those needs are affected by the environment in which the vehicle will be operated.

This metaphor can be applied to a current trend in corporate IS strategy. Companies are starting to question how Lotus Notes as a groupware product may fit into a corporate technological infrastructure that seeks to exploit the current rush toward the corporate intranet. Many managers are wondering why they should implement or continue to use Lotus Notes when a World Wide Web browser and a Web server can do the job.

But like the differences between the car and the tractor, there are significant differences between Lotus Notes as a corporate technology and the Web as a corporate technology. They are not the same vehicles, and they each have strengths and weaknesses in different environments.

This chapter points out those differences and raises some of the key issues related to using either or both of these tools in a company's Internet and intranet initiatives.

Some people believe that Web technology will supplant the need for products like Notes in the very near future, and others feel that Notes and products like it offer features that Web technology will never be able to provide on its own. The reality, however, may fall somewhere in between.

LEVERAGING THE ELECTRONIC MESSAGING INFRASTRUCTURE

## THE CONCEPT OF THE INTRANET

The basic premise of the whole Internet is founded on the idea that the computers that are connected to each other need not share the same operating system, platform, or software. All that they need in order to communicate is a common networking protocol, such as Transmission Control Protocol and Internet Protocol (TCP/IP), physical connections, and software capable of displaying and retrieving files in specific formats. By providing each node on the network with the proper tool to access information, the source and/or location of the information becomes secondary.

### Managing Distributed Information

As more companies move toward TCP/IP as their standard networking protocol, they have the opportunity to begin exploiting this concept of distributed information within their own organizations. Like the Internet, information will be placed on distributed servers within the company in a standard format (in this case, hypertext markup language [HTML]), and users across the organization will be able to view, print, and give feedback regardless of their hardware/software configuration. As long as they have a Web browser on their workstation, corporate information will be accessible.

Extending this model through a large corporation has tremendous appeal, mainly because browser software is very inexpensive (if not free), and no other client-side tool is required to access data. Coupled with new products emerging for the back-end of the equation, including products that allow HTML publishing of database data, companies see an opportunity to provide client/server computing solutions to all users at a fraction of their originally estimated costs.

The intranet is the collection of servers and workstations connected to each other using the same paradigm as the Internet: the workstations are connected via TCP/IP and browser software to various information stores that may be simple HTML documents or complex database applications using powerful back-end database tools. All of this functionality comes at a low cost per workstation, is generally platform independent, and does not have to disrupt current development projects using standard database development tools.

## LOTUS NOTES IN THE ENTERPRISE

Many companies have already found all of this functionality and more by installing Lotus Notes within their organization. Notes, with server and workstation software for all of the major operating systems and platforms, provides exactly the same basic functionality as an intranet: the ability for low-cost client software to access a wide variety of information and database applications using the existing corporate network. Notes also allows users to provide feedback and input into databases, as well as develop sophisticated applications within its own development environment.

Lotus Notes is almost as pure a client/server environment as exists today for the PC platforms. A server task runs on a file server in the network and the Notes client software makes requests for information from that server task. The information is fed to the client, where all the processing is done. Any changes made to the information on the client are submitted back to the server on request.

Companies have implemented Lotus Notes for a variety of reasons. Notes provides a secure, monitored environment for the management of information, has built-in messaging and E-mail features, and allows groups of users to share and exchange information in a controlled environment. Coupled with the development tools built into the product, Notes provides a stable environment for developing and distributing applications in which groups of users need to move, share, or store information.

**Costly Implementation**

All of these benefits come at a cost, however. There are costs associated with the software, but that is only the beginning. Notes generally requires close administration and management on the server side, and database management on the application side. As the size of the Notes network grows, the administration tasks become more critical to the overall performance of the network.

In the past, Notes has been criticized for the lack of flexibility of its development tools and the overall cost of implementation. Lotus has repositioned Notes with the release of version 4.0. The client software now retails for as little as $70 per workstation (which is not much more than the license for a standard browser tool like Netscape) and $750 per server. Steep discounts exist for volume licensing and members of Lotus' special purchasing programs.

This pricing policy has put Notes right into the mix among companies looking to either leverage their existing Notes infrastructure into a company-wide intranet or extend Notes' influence to build new standards for computing within the organization.

**NOTES VS. INTRANETS**

There is a pervasive feeling among IS managers that Notes and intranets are somehow mutually exclusive. According to popular opinion, the rise of the intranet means the demise of Lotus Notes as a useful information platform, simply because Web technology coupled with an intranet provides all the functionality of a Notes installation at a fraction of the cost.

## LEVERAGING THE ELECTRONIC MESSAGING INFRASTRUCTURE

For many IS managers, supporters of Notes technology fail to see the bigger picture and the advantages of Web technology that reaches across database platforms and client workstation issues. For managers in organizations using Notes, the reverse is often the case; the benefits they have realized by using Notes cannot be replaced simply by installing a browser on every desk. IS professionals at all levels may wonder why they must make an "either/or" choice with respect to intranetworking and the tools that support it.

If the two approaches are considered opposite technologies as diverse as the car and the tractor, with Web technology at one pole and Notes at the other, the true area of opportunity and possibility lies in the middle. The questions that IS managers should be asking when considering the implementation of an intranet in their organization is not whether Web technology or Notes technology should be used.

Instead, an organization needs to look beyond the technical limits of both technologies into how each might function within the organization. Instead of either/or, the question may become: where are the areas of synergy among these technologies and how can that synergy be used to build the most effective intranet solution for the organization?

**Browser-Based Intranets**

Netscape, Microsoft, and others already ship or have announced cross-platform versions of their browser tools. Netscape has even stopped calling its tool a browser, preferring instead to call it "an environment." The implication is that the tool used to view HTML content on the network is not just a viewing tool, but with the use of add-in programs, it allows users to execute sophisticated client/server applications from within the tool itself.

The browser, in Netscape's view, becomes the front-end to everything — content on the Internet, content on the corporate network, client/server applications, E-mail, and groupware applications. Because the interface is standard across all platforms and the basic functionality never changes across applications or information types, the focus returns to the information being presented to the user, and not the effort typically required to design and build a comfortable user interface for every application. Because this approach requires only that the user have a compatible browser installed, there is no inherent reason to choose a single browser from a single manufacturer.

Currently, several companies offer low-cost, commercially available browser tools that run on a variety of platforms. The key feature they share is the ability to retrieve and display information from any hypertext transfer protocol (HTTP) server. Minor differences among platforms are accounted for by the browser without the user or the programmer having to deal with them. Font differences, screen resolution differences, even interface object rendering differences (i.e., the way buttons and form fields are drawn by the operating system) are removed from the responsibilities of the programmer.

IS managers may welcome a degree of standardization that is possible by using a browser tool in the context of a corporate client/server computing strategy, especially when coupled with the types of HTTP server software now available. The HTTP server is the backbone of the corporate intranet. Many of the most respected enterprise database software companies have announced products or plans to incorporate HTML document publishing into their database architecture. What this means for the corporate IS strategy depends on which tools are used in the company.

All of the major database vendors have either released or intend to release products that will incorporate the information stored in their databases into Web-compatible content. The features of these new products vary, but most will allow the retrieval and display of information via live queries into the database, as well as allow dynamic updates of information in the relational database via the browser.

### Notes-Based Intranets

Notes performs the same basic functions as an intranet installation — the ability to access information in a friendly, comfortable format tied with the ability to generate and store large amounts of data in a database format. The Notes client software is cross-platform in nature, allowing users across the organization to share and update information.

Notes' built-in database development tools set it apart from the browser market; any user with an appropriate license may become an applications developer in Notes simply by invoking the Design menu on the toolbar. The development environment is simple enough for novices to grasp quickly and robust enough for some sophisticated development to take place.

However, it is the inclusion of InterNotes tools, released by Lotus in 1995 and to be upgraded this year for the new release of Notes, that places Notes squarely in the realm of a practical intranet solution. Using InterNotes, a standard Notes workstation can become a browser tool.

### THE HYBRID ENVIRONMENT

A hybrid environment can be created in which Notes may function for certain users as their primary interface to both workgroup solutions as well as the corporate intranet. For other users who may not need the applications or information currently stored in Notes, a standard browsing tool may be all that is required.

# LEVERAGING THE ELECTRONIC MESSAGING INFRASTRUCTURE

Using InterNotes, users of a standard browser are not kept out of Notes-based information. The InterNotes Web Publisher, a component of the InterNotes tools, allows Notes databases to be published, searched, and updated via Web technology. Regardless of the front-end tool being used, the Notes-based data is just as accessible as other types of information.

## Administrative Benefits

Using the standard Notes administration and database development tools, an organization can implement and manage Web-based information. The Web Publisher can publish any Notes database into HTML documents and refresh those documents on a scheduled basis.

This means that users do not have to learn HTML codes in order to generate or update Web information. Much of the cost related to managing Web information comes with maintenance and upkeep of the documents themselves. By using the Notes Web Publisher, that administrative task can be reduced significantly. Because all of the changes are made using Notes, editing and publishing of the HTML information can be greatly simplified.

## Drawbacks

The biggest disadvantage to the hybrid approach is that it requires an investment in Lotus Notes. For organizations that have already made this investment for other reasons, there is a logical extension to the Notes environment. Leveraging existing experience with the Notes interface and environment reduces the learning curve and provides seamless integration with corporate information on an intranet. However, the desire to build an intranet or to extend the company's reach to the Internet is not justification for building a Notes infrastructure.

## INTRANET IMPLEMENTATION ISSUES

Regardless of the approach a company decides to take — browser-based or Notes-based — there are issues relating to the implementation of the technology that must be addressed.

## Fit with Existing Technology

Decisions already made for workstations, servers, and the network play a key role in deciding which directions the organization will take in the future. Platform decisions, standardization of desktop applications, and decisions related to workgroup computing environments all play a part. IS must consider how the new environment will take advantage of the best features of the existing one, and determine which tools will be the most useful when the intranet is in place. IS planners should also determine internal and external workgroup computing needs. As the new infrastructure takes shape, IS may find an opportunity to reshape how the corporate computing environment should be put together.

### How Is It Connected?

The current state of the corporate network must be factored into the plans. If an organization is already using TCP/IP, then it already has in place the communication layer necessary to extend an existing Notes network to an intranet or onto the Internet. Otherwise, a network protocol upgrade will need to be planned for the entire corporate network. If the network needs to be upgraded, or if the standard protocol needs to be changed, then the company will have an opportunity to change the fundamental paradigm related to how users interact with the information on the network.

### Learning and Training

There is a significant learning curve involved in training someone to develop and manage Internet content. This education extends well beyond the IS staff to the people who will be charged with building individual Web pages. These are the people who, if they are not using a tool like Notes to manage content, are going to have the longest learning curve.

Once the intranet is in place, training will have to be conducted for each end-user on browser techniques, basic Internet protocols, and the use of the intranet itself. With Notes, the standard training on the workgroup computing environment is simply mapped onto the corporate intranet. The tool is already in place from the users' perspective.

### Management

If, for example, it is 4:00 p.m. on Friday and a ten-page press release needs to go online immediately, someone must be responsible for putting this content on the company's Web page — be it an IS employee, a Webmaster, or a public relations agent. Once a company Web site is running, there is a daily need to watch and administer the activity. Allowing end-users to manage this content through a standard interface like Notes can reduce production and development time and allow IS to monitor the activity in the site.

### SUMMARY

IS staff and managers can use the following questions as general guidelines when setting a corporate strategy for information management and the development of workgroup applications:
- What are the company's strategic information goals?
- Do the goals, in general, include the use of workgroup computing tools such as Notes or Microsoft Exchange?

# LEVERAGING THE ELECTRONIC MESSAGING INFRASTRUCTURE

- Is the organization interested in gaining specific expertise in the management and implementation of the intranet?
- Is the company interested in making the producers of the intranet content experts in HTML and Web technologies, or should those technologies be as hidden and seamless as possible for these users?
- Who is going to administer and manage intranet Web sites?
- Where is the Web site expertise going to come from?

**Existing Investments**

The following questions can help IS evaluate the existing corporate technology investment:

- Does the company have an existing investment in Notes?
- What are the strategic directions for workgroup computing in the organization?
- Have standards been set for the use of browsing tools or client/server database applications to be used on the corporate intranet?
- Given the direction in the industry toward a more distributed form of information dissemination, how do the company's existing database investments and database systems (i.e., Notes, DB2, Sybase, or Oracle) fit into an overall corporate strategic plan for client/server and workgroup computing?

**Long-Term Thinking**

IS may want to consider the following questions to plan more efficiently for the future:

- What are the company's long-term plans for integration of all database and non-database content on a corporate network?
- Given the integration of workgroup computing tools into the basic structure of the corporate intranet, are there standards and guidelines in place for the construction, implementation, and maintenance of the enterprise workgroup computing environment?

# Chapter 25
# Using Groupware to Enhance Team Decision-Making

*Naomi S. Leventhal*

Few managers today would suggest that a single individual should have responsibility for making all of the important decisions required within an organization. The team-based, problem-solving model has been endorsed throughout government and private industry. Despite the prevalence of this model, it is also widely acknowledged that teams suffer from several common ailments, among which are tendencies to defer decision-making, to lapse into political posturing over objective analysis, and to produce voluminous reports with little content.

This chapter discusses the way in which groupware tools can minimize the inherent weaknesses of teams and maximize their strengths. Groupware tools enhance idea generation and analysis and expedite the decision-making process. Some of the management objectives to which groupware products can be most readily applied are discussed, along with some critical success factors for introducing groupware tools into an organization.

## GROUP COMMUNICATION

The term "groupware" encompasses many different types of automated tools. The two categories of groupware tools with which most people are familiar are: meeting management tools and information exchange tools.

### Meeting Management Tools

These tools support real-time meetings that occur at one time and, generally, in one place. Although there is growing interest in the use of teleconferencing and networking technology to include individuals not present at a designated meeting site, meetings involving remotely located participants are still the exception rather than the rule.

LEVERAGING THE ELECTRONIC MESSAGING INFRASTRUCTURE

By providing an automated means of analyzing issues and reaching consensus, meeting management tools help to ensure that teams do more than communicate, but that they take action as well. These types of tools have the highest impact on the achievement of team objectives.

**Information Exchange Tools**

These tools enhance different time, different place communication among a group of individuals. Tools can be as simple as electronic mail or corporate bulletin boards, or as complex as database tools designed to support group access and update. Although such tools may provide, in some sense, a virtual meeting capability, they are not meeting management tools.

**THE GROUPWARE SESSION**

To the casual observer, a groupware session conducted in an electronic meeting system (EMS) facility looks similar to any other team meeting led by a facilitator. In an EMS room, a group of about a dozen participants is seated around a U-shaped table. Participants are there to discuss focused issues, such as how high to set sales objectives for the following year, how to centralize (or decentralize) computer systems support, or how to implement the new work team organization model. During the process, participants will agree or disagree, and amplify, refine, define, and, in general, wrestle with questions to which there are no easy answers.

There are, however, several significant differences between this meeting and the nonautomated meetings with which most of us are familiar. When the facilitator requests that the group generate ideas or evaluate alternatives, participants perform this task using a computer, keyboard, and screen. In a state-of-the-art facility, the equipment is hidden from view in the individual desks that together make up the U-shaped seating arrangement. This provides for greater privacy in the use of the individual workstation and prevents the computer from becoming the focus of attention or interfering with the free flow of ideas during discussion.

An electronic meeting room has none of the usual clutter of easel paper, yellow adhesive notes, and paper forms. Instead, there are two large projection screens on which are displayed the ideas being reviewed by the team members. One screen shows the ideas generated by the use of the groupware; on the other screen is the analysis of those ideas presented in a graphical model.

Compared with meetings conducted without the use of interactive automated tools, this meeting accomplishes its objectives more quickly, is likely to stay on track more reliably, and can develop consensus more effectively. The team also produces more accurate documentation more quickly than would be possible in a nonautomated meeting. These outcomes increase the sense of accomplishment among team members and reduce the frustration many people feel when attending unproductive meetings. The benefits of meetings with groupware are summarized in the following table:

| Meetings without Groupware | Meetings with Groupware |
| --- | --- |
| Long meetings without clear focus | Shorter meetings with clear focus |
| Unbalanced participation among team members; some members dominate | Participation balanced among team members; all voices heard equally |
| Difficulty translating different team-member language and jargon | Common vocabulary easily documented and referenced during and after session |
| Difficult to track discussion in progress | Discussion structured; objectives clear |
| Uncertain documentation quality | Accurate documentation quickly produced |

## GROUPWARE TOOLS AND THEIR CAPABILITIES

Groupware tools designed to support real-time meetings work by managing the interaction of team members. They provide a focus and structure for that interaction that, when properly implemented, define the boundaries and rules of group communication. In a sense, whereas the team provides the content and vocabulary of the discussion, the groupware provides the structure and boundaries of the discussion — in effect, its grammar.

Groupware tool capabilities vary, depending on the particular software tool selected. However, there is a great deal of similarity in some of the capabilities offered. The following description of tool capabilities is not intended to be a profile of any one product; rather, the list describes capabilities that are commonly available.

Common capabilities provided by groupware tools include:

- *Idea generation.* The tool supports the simultaneous input of ideas by multiple participants, and promotes creativity by allowing team members to generate ideas at the same time, to read the contributions of others, and to respond to those ideas with new ideas of their own. Without the input, routing, display, and storage capabilities provided by the groupware, this ability to support simultaneous interaction would be impossible to replicate.
- *Idea analysis.* The tool supports the examination and refinement of ideas generated by the team, as well as the organization of those ideas, allowing the team to determine the factors to be used to organize or evaluate ideas. This capability enables the team to work jointly to

group-related ideas, to examine the relationship of parts of an issue to the whole, and to rate and rank ideas.
- *Matrix analysis.* The tool supports the development of a two-dimensional matrix to examine relationships among ideas. Team members can evaluate current business processes, for example, in terms of two factors: efficiency and effectiveness.
- *Voting.* The tool makes it possible to test the level of agreement among team members. Some tools provide several different types of voting capabilities, including the ability to prioritize choices.
- *Group dictionary.* The tool ensures that the common vocabulary of the team is recorded to ensure consistency and clarity in analysis and to support the sharing of results of the session with others.

Some groupware tools offer the capability to create an agenda within the tool itself that tracks the exercises to be conducted and the specific tool capabilities to be used to support each exercise. All offer some type of report production capability as well, although many tool users will not be satisfied with the appearance of the reports generated. If high-quality report production is an important requirement, it is necessary to translate tool output into final form using a word processing package.

Tool capabilities vary somewhat from vendor to vendor. Some packages are more flexible than others in allowing the team to define a unique approach to each meeting. In some cases, for example, information generated in one software function can easily be transferred to another function for further work. The team may want to develop a list of proposed action steps in one function, organize those steps into categories in a second function, and vote on step priorities in a third function. Better tools allow this to be done easily.

These capabilities are valuable because they make it possible for all team members to speak at the same time. In a nonautomated meeting, if everyone is talking at the same time, no one has the chance to learn from anyone else; thus, valuable ideas are lost and focus is impossible to control. Groupware turns the potential liability of multiple views into an asset. No one team member can hijack the meeting and shut out less assertive participants. The emphasis is on the joint creativity of the group rather than the opinions of an individual team member.

**Current Limitations**

The sophistication of the available software is increasing; however, the number of groupware packages that qualify as meeting management tools remain limited. GroupSystems V (Ventana Corp., Tucson, Arizona), VisionQuest (Collaborative Technologies Corp., Dallas), and Meeting Room (Eden Systems Corp., Indianapolis) are the most widely used packages. All are available in both DOS and Windows versions. All support the creation of meeting agendas, brainstorming, idea analysis and organization, voting, and report generation. Their common strength is the ability to support simultaneous idea generation, which maximizes participant interaction. Their common weakness is their limited ability to support detailed, structured analysis of complex technical issues.

Although any of these tools alone can adequately support a session devoted to high-level planning, policy development, or key issue identification, none is entirely adequate to support such tasks as redesigning a critical business process (such as order fulfillment) or developing a systems requirement definition for a new accounting system. For these technically complex tasks, structured modeling tools are currently required to supplement groupware capabilities. Advanced analysis capabilities are likely to be added to groupware tools in the near future; in addition, links between computer-aided software engineering (CASE) tools and groupware packages might soon be available.

**GROUPWARE APPLICATIONS**

Although groupware can be used to support any type of meeting — from a 1-hour meeting on how to structure assignments for the following week, to a 5-day meeting on reengineering the organization's records management function — it is most helpful in addressing complex tasks that require several days of intensive effort. Typical groupware applications include:

- The development of strategic plans
- The design of new organization structures
- The reengineering of key business processes
- The development of comprehensive plans to address information systems, personnel, or financial problems

In a strategic planning session, for example, the issues addressed could include a definition of the organization's mission and objectives, an analysis of customers and competitors, and the development of a plan to develop internal resources to increase the organization's competitive position. The discussion of these issues would be supported by different groupware capabilities, including:

- Idea generation to define the corporate mission and brainstorm key mission objectives
- Idea analysis to rate and rank objectives
- Matrix analysis to evaluate objectives in relation to customer needs
- Voting to evaluate current resources
- The group dictionary to maintain consistency in definitions of the terminology used

For a business process reengineering (BPR) session, the tool's capabilities can be applied, though somewhat differently from how they were used in the planning session. For BPR, the idea generation capability might be used to support team members in brainstorming ideas about process improvement opportunities. The idea organization capability might be used to create a hierarchical analysis of a specific process area, such as inventory management. The matrix capability might be used to evaluate both the relative importance of processes and the effectiveness of their current performance.

Groupware capabilities are most effectively used to address issues of a less complex nature. Somewhat narrower applications might include a meeting held to write new job descriptions to support a departmental reorganization plan, a session devoted to brainstorming ideas to launch a new product or service, or a session held to evaluate the effectiveness of an ongoing program or initiative (e.g., a cultural diversity program or a new compensation program).

The idea generation and organization capabilities of the groupware tool are invaluable in maximizing the input of team members and keeping the group focused for these tasks. The meeting documentation provided by the groupware tool ensures that even a 2-hour session devoted to a simple problem-solving assignment will be accurately documented, thereby reducing misunderstandings about what was agreed to in a session and providing an audit trail when the plans created through a groupware session are implemented.

## CRITICAL SUCCESS FACTORS FOR USING GROUPWARE

Five key factors must be in place for a groupware session to be successful. These factors are related to the environment of the session, the facilitator leading the session, the makeup of the group itself, the objectives of the session, and the capabilities of the groupware tools used.

### Designing a Groupware Session Environment

Although a groupware session can be set up anywhere that a temporary local area network can be installed, several factors greatly enhance the chances for a successful session. An appropriate groupware environment should have adequate space for as many as 18 participants in the session. Within that space, it must provide each participant with easy access to a keyboard and easy viewing of a screen. Specially designed desks are available that place the screen below eye level or even entirely beneath a see-through desk surface. This arrangement facilitates the direct verbal exchange of ideas among participants and helps to emphasize the role of the groupware tool as supportive, not primary.

In addition to allowing participants to view their own input and that of the group at an individual workstation, the well-designed groupware center provides participants with a clear view of one or (preferably) two public screens. (Because most groupware capabilities are currently text-based, a second screen allows for the display of additional graphics that can be used to amplify the information presented on the groupware screen.) The ability to project the ongoing group interaction using the groupware software is essential to maintaining a group-focused environment. Good lighting designed not to interfere with a clear view of either the participant's screen or the public screens is also a necessity.

**How a Groupware Session Operates.** To support this type of environment, a typical EMS room configuration might include up to 20 workstations driven by a 486/33 server operating with a 200M hard drive and 8M of RAM, controlling a LAN (e.g., Novell's Netware, Version 3.11), and running a groupware software package (e.g., Ventana's GroupSystems V). Color monitors are needed at each workstation, as well as two active overhead projectors to support the two public-viewing screens. A laserjet printer should be available to produce hardcopy documentation of the group's decisions if needed.

**Software Management.** A groupware operator is responsible for controlling participant access to the tool's capabilities. The operator needs a 486 workstation, and if graphics or modeling software is used to supplement the groupware, a second workstation and software operator are required. These software management workstations are generally set apart from the participants' workstations, often in the back of the room and out of the direct view of the participants so that they do not become a distraction. From this workstation, the groupware operator sends participants a screen or set of screens that support an assigned task. As an example, participants might receive a single screen for brainstorming alternatives to an existing travel voucher policy, a set of color-coded screens for identifying reporting requirements associated with a predefined set of government regulations, or a score sheet for ranking items on a previously defined list of strategic corporate objectives.

**Portable Possibilities.** The EMS environment can be either portable or fixed, depending on the type of workstations and projection equipment used. If notebook computers are used as workstations and an overhead projection pad is available, an EMS site can be set up almost anywhere. However, some sacrifice in environmental quality has to be anticipated; not all conference rooms offer adequate space, ventilation, and lighting for an intensive 6-hour meeting with more than a dozen people. Nonetheless, the ability to bring the groupware to the group can sometimes be beneficial.

## Choosing a Groupware Session Facilitator

Whether a facilitator is working in an automated environment or a nonautomated one, this individual needs skills in such areas as group dynamics, organization values and culture, communication strategies, and leadership style. The best facilitator is someone who knows how to allow the group to achieve its objectives and yet not be seen as the author of the decisions reached. Facilitation requires the practitioner to steer the ship (a directive act) while assisting the group (a supportive act) in determining the ship's course, speed, and ultimate destination.

All of these skills are required of the groupware session facilitator. In addition, the groupware session facilitator must be thoroughly conversant with the software being used and able to define the best application of its capabilities to the objectives of the session. The groupware session facilitator must be able to structure a groupware session in such a way that the session participants forget about the tools and focus on the tasks at hand.

**Acquiring General and Package-Specific Skills.** Relatively few training programs exist that are targeted at producing skilled groupware facilitators. Those training programs that are available are offered either by the vendors of the different groupware packages or by consultants acting in their behalf.

This training can be valuable because it is keyed to the capabilities of a specific software package and is designed to produce facilitators who know how to get the best results out of that package. Package-based training has its limitations, though. Chief among these is that it assumes that training participants are already skilled facilitators.

Before attempting to get training on a specific groupware package, the prospective groupware facilitator should have attended a program in generic facilitation skills; that is, a program that focuses on how to define group meeting objectives, how to structure an agenda, how to design group exercises, and how to deal with uncooperative participants. If possible, the facilitator chosen for groupware training should have several years' experience in a nonautomated facilitation environment. Without this type of background, the new groupware facilitator may have difficulty in performing at a level of expertise that will justify the expense and effort of maintaining a groupware facility.

Not everything a facilitator needs to know is taught in either a package-based or general skills program. Although a facilitator may have an academic background in organization development or human resources, some of the best facilitators have backgrounds in sociology, psychology, business management, general systems theory, and/or linguistics.

Academic experience in the characteristics of group communication strategies, traditional and nontraditional approaches to problem-solving, and the development and deconstruction of heuristic paradigms can greatly add to the development of an expert facilitator. Because the use of groupware multiplies the complexity of the problem-solving process, this type of advanced education is, if anything, even more valuable for groupware facilitators than for those who work with nonautomated methods. Although it is not practical to require every prospective groupware facilitator to have this type of advanced education, it is possible to insist that the selection process be rigorous and that a broad view be taken of what constitutes an appropriate background for this important work.

**Software Expertise and Session Design.** Any effective facilitator must have substantial knowledge of the groupware tools being used in a session. It is also essential, however, for the facilitator to be supported by a groupware software expert, sometimes called a technographer. This individual should assist the facilitator in creating a detailed session design, a design specifying the tool capabilities to be used during the session, the techniques to be used to integrate the results of the different tools, and the strategy for translating tool output into session documentation. In addition, the session designer should serve as groupware operator during the meeting.

## GROUP COMPOSITION

The type of group that benefits most from the type of interactive communication found in a groupware environment is a group characterized by diversity of background, responsibility, attitude, and interests. A groupware session team works best if it is composed of people who are at different levels of expertise or status, or who represent different functional areas within the organization.

### Diversity as an Asset

Whereas it is difficult to run a successful nonautomated work session if the session participants include, for example, the vice-president of marketing, a programmer from the IS department, and a group of engineers from the product development team, the use of groupware enables this group of people to communicate despite the different language and jargon they may speak. Because groupware allows for the analysis and synthesis of large amounts of information, any result produced by use of the tools is enhanced by the quality and variety of the input. In other words, the greater the initial disagreement within the group, the higher quality the resulting outcome is likely to be, and the stronger the consensus that supports that outcome. In effect, groupware can help transform a group into a team.

### Boundaries for Economy and Manageability

The typical groupware session includes from 6 to 16 participants. Groups with fewer than six participants may benefit from the software capabilities of a groupware tool, but the intimacy of such a small group may be compromised by the tool. In addition, it is not often worth either the effort or expense to arrange for small meetings to be supported by groupware. Groups with more than 18 members tend to be unwieldy, and the free-flowing generation of ideas that characterizes groupware supported discussions may turn into a liability if the boundaries of the meeting are not carefully defined.

### Session Objectives

Too often, meetings fail because their purpose and focus is unclear. Groupware sessions can support any general meeting objective, but it is critical to the success of a session that all objectives be clearly specified before an agenda is created.

This is not to suggest that a groupware session cannot be adapted to unknown situations that arise. It is quite possible for a good facilitator to change course in midstream if necessary. If participants are confused, if a discussion is not leading to a clear decision, if the groupware capability selected turns out not to be appropriate to the analysis required, then changes can easily be made. A skilled groupware facilitator will understand that the session objectives should drive the use of groupware capabilities, not the other way around.

## SUMMARY

The opportunities exist to extend an organization's reach beyond the 6th-floor conference room and beyond the boundaries of the organization itself. Although groupware tools are not yet widely used for team decision-making, experimentation with these tools, as well as the development of electronic meeting facilities, is rapidly escalating among those organizations committed to team-based problem-solving. Groupware involves a different way of communicating, analyzing issues, and solving problems. Those organizations that make this new technology work for them demonstrate that broad-based employee involvement, supported by a focused communication medium, strengthens the ability of teams to achieve their maximum potential.

Organizations that are early adopters of groupware have a distinct competitive advantage in the marketplace. Using groupware tools for same-time, same-place communication forces the organization to structure and focus the communication process. In turn, global organizations can move effectively to the same-time, different-place communication capability provided by videoconferencing and the different-time, different-place communication capability offered by the Internet. The structural norms for exploiting these opportunities are only beginning to be developed.

One of the leaders in using groupware tools for general problem-solving is IBM Corp., which has dozens of sites around the country in which it conducts groupware sessions to meet internal needs. The Department of Defense applies groupware to a wide variety of requirements. In one project run jointly by the Air Force, Army, Navy, and DOD's Office of the Inspector General, a suite of groupware and modeling tools is being used to support the enhancement and standardization of criminal investigations. These tools are expediting an ambitious project life cycle that includes strategic planning, business process and data reengineering, systems migration, and information systems design.

An organization's potential for introducing groupware with any degree of success may rest with its:

- *Readiness to accept teams.* An organization that already accepts and uses teams to address important issues is likely to succeed with groupware. If the organization has pushed responsibility and authority down into the ranks of middle- and lower-level managers, it will embrace a tool that facilitates the communication that is vital to team action.
- *Approach to risk-taking.* Those organizations in which risk-taking is rewarded are likely to do well with these new tools. Because implementing any new tool usually results in a few misfires, the organization that can initially accept a less-than-perfect performance has a greater chance of graduating to the ranks of experienced groupware user than one that is not.

# Chapter 26
# Using Lotus Notes in Executive Information Systems

*Barbara J. Haley and Hugh J. Watson*

The primary purpose of an executive information system (EIS) is to provide executives with the internal and external information required to effectively perform their jobs. Executives have always had systems to supply needed information. Traditionally, executives have relied on printed reports, subordinates, meetings, networks of people inside and outside the organization, telephone calls, newspapers, and industry newsletters as information sources. Contemporary EISs use computer and communications technology to deliver much the same kind of information as before, but in a better, more timely, accurate, and relevant manner. EISs have unique characteristics. For example, they:

- Are custom tailored to individual executives
- Extract, filter, compress, and track critical data
- Provide status information, trend analysis, and exception reports
- Access and integrate a broad range of internal and external data
- Are user-friendly and require little or no training
- Are used directly by executives without intermediaries
- Present graphical, tabular, and textual information
- Provide support for electronic communications
- Provide data analysis capabilities
- Provide organizing tools

## EIS EVOLUTION

The first EISs were implemented in the late 1970s using custom-built software. Although there were a few notable successes, there were more failures. These specially tailored systems were often too difficult to build, maintain, and use. In the mid-1980s, Comshare and Pilot Software introduced their Commander EIS and Command Center products, respectively.

These mainframe-based offerings provided a set of tools that greatly facilitated EIS development and use.

The emergence of appropriate software, combined with executives' demand for information, fueled a rapid growth in the number of EISs throughout the decade. The EIS software market has changed seriously, however, as firms have been moving from mainframe to client/server-based applications. Consequently, the EIS software vendors have moved away from their more expensive mainframe offerings to less expensive client/server products.

EIS products were evolving concurrently with general-purpose software. In many ways, general-purpose software began to be more like EIS software in terms of the capabilities they provided. For example, most database and spreadsheet software evolved to include a graphical user interface. Many companies began to question what they should pay for EIS software when they could get nearly the same capabilities in Microsoft Excel, Powersoft's PowerBuilder, or Visual Basic, and at a lower cost. Although these general-purpose alternatives still do not match the functionality provided by specialized EIS software, the gap is narrowing.

The nature of EISs has also changed. Originally developed for a handful of senior executives, they evolved to support the top management team, and now in many organizations have spread to serve hundreds or even thousands of users. For this reason, EIS now informally stands for everybody's information system.

## LOTUS NOTES

Lotus Notes entered the EIS software scene in 1988. Notes first appeared in the marketplace targeting groupware and workflow applications. It has been widely recognized as the first popular commercial product to serve this market. Today, Notes is used in more than 2000 companies and on the desks of more than 1 million employees.

Notes serves a variety of purposes, ranging from basic E-mail to complex workflow applications that are closely interwoven with critical business processes. Many companies have capitalized on the product's excellent data storage and sharing mechanisms. Others have been overwhelmed with Notes' extremely diverse capabilities and lack direction or an overarching strategy in their internal Notes development projects. Companies that learn how to harness the power of Notes to create well-planned, business-driven applications are much more satisfied with the results. The benefits from Notes surface when all of Notes capabilities are integrated to create robust solutions.

### What Notes Can Do

**E-Mail.** An effective communications product, Notes provides the foundation for electronic mail exchange. Users can send and receive messages using standard mail forms that can be adapted to corporate standards or individual needs. Messages can include simple text or data that is saved in a variety of formats such as a Microsoft Excel spreadsheet or Lotus Ami Pro word-processed report. E-mail allows a company to reach beyond organizational boundaries through communications with customers and vendors.

**Standard Templates.** Notes also provides an applications development environment. Users can create applications from scratch or through templates. Notes users can select from several boilerplate applications for sales tracking or internal discussions. Developers can enhance or change these template applications to meet specialized needs. The result is an attractive, personalized form in which users can input data and, with proper access rights, later change or view the information.

**Replication.** After entering information into a repository through forms, workers are able to share the information that is spread over a variety of departments or locations. Notes' unique replicating ability facilitates information sharing among the most distributed of corporate structures. In a Notes network structure, distributed databases periodically synchronize their information to create mirror images of data located throughout the company. Users can then access information locally, regardless of the location of the original data source. A Notes administrator determines how often data needs to be refreshed and how to best meet user needs.

**Searching and Viewing.** A powerful full-text index search engine allows executives to filter out pertinent information once it has been distributed to users. In addition, flexible views can be created and manipulated to display the information.

## Notes Applications

**Broadcast.** These applications often resemble electronic bulletin boards that display timely information that managers can check regularly for updates and posted messages. They also are popular repositories for news. Packages such as Newsedge and Hoover compile the day's news from numerous sources and store it in Notes databases. The news can be filtered with user-defined criteria, such as a client name or a particular industry. The information is delivered to executives as conveniently as the morning paper, only the Notes solution supplies articles from news sources all over the world.

**Reference.** These applications are similar to broadcast applications and serve as libraries for robust, mostly static data. Meeting minutes, management reports, and policy manuals can be stored and categorized for users to access later. This application usually saves companies in the cost of duplication and dissemination of documents that are best stored and updated centrally and accessed on an as-needed basis.

**Tracking.** These applications contain information that is valuable to a number of employees. Users can record an event and its current status. This event is then monitored, passed along to another user for action, or stored for future access. This dynamic manipulation of documents provides great benefits for processes that need to be automated.

**Discussion.** These applications provide forums for users to pose questions and dialog through hierarchical replies. These saved discussions serve as an important part of organizational memory because they illustrate the thread of a decision-making process that can later be accessed when similar problems or situations arise.

## THE TEN-COMPANY STUDY

A study was conducted to learn about the potential role of Notes in EISs. The study was exploratory in nature because of the relatively limited experience firms have, to date, had with the product.

First, firms using Notes for EIS purposes were identified. This information was compiled from a list of companies in the University of Georgia's EIS database and references from vendors of EIS software. Ten firms agreed to participate in telephone interviews. The industries represented include gas distribution, natural gas, banking, consumer products, insurance, pharmaceuticals, consulting, and manufacturing. The companies are located throughout the U.S. and Canada.

In each company, the most knowledgeable person about the use of Notes for EISs was interviewed. The interviews were scheduled in advance and took between 20 and 30 minutes to complete. The focus was on current and planned uses of Notes in the organization, specifically with regard to EISs. The interviews also explored the strengths and weaknesses of the use of Notes for EIS.

### Why Use Notes for EIS?

Interviewees were first asked how Notes was chosen for EIS use. Answers revealed that the origins were either opportunistic or strategic. In the majority of cases, Notes was brought into the company for non-EIS applications. Its EIS potential was recognized quickly, however, as developers saw the opportunity to enhance their systems by using Notes. A few firms made the strategic decision to acquire Notes specifically for use with the EIS.

Already having Notes in-house increases the likelihood that it will be used with an EIS, especially if the licensing agreement covers the EIS user base. If not, the incremental cost of obtaining Notes may be prohibitive.

## How Notes Is Used

The interviews revealed three different ways that Notes is being used for EIS: as part of the EIS software; as a separate, complementary EIS; or as the primary EIS software.

**Using Notes as Part of EIS Software.** Notes is often used in combination with other EIS software because users appreciate its capabilities for entering, maintaining, updating, and retrieving textual information. One of the banks profiled in the ten-company study is considering using Notes this way. Its EIS was originally developed using Lightship and Lightship Server from Pilot Software. This software provided an effective solution for handling numerical data but was inadequate for textual information.

Notes, however, provides a management topics application that allows users to initiate topics for discussion and to enter and review comments made by others. All the information is textual and is used as a kind of electronic bulletin board for sharing soft information. The bank's EIS is designed to allow data suppliers and users to add textual commentaries to screens. For example, a comment entered by a data supplier might explain why performance as measured by a key service indicator has dropped. A user might add, for example, that certain actions should be taken to improve performance.

**Using Notes as a Separate EIS.** Lotus can also serve as a stand-alone system used to complement another EIS. The Notes-based system handles the kinds of applications for which it is inherently well suited. For example, one manufacturing firm in the study operates an EIS based on Commander EIS, but uses Notes for document management applications. The EIS manager would like to integrate the two systems, but has encountered compatibility problems. Both systems operate separately to serve the firms' executives.

**Notes as the Primary EIS Software.** Notes can also be used as the primary EIS software. This is the case at one company where a Notes-based EIS has replaced a system using more conventional EIS software. Several factors drove the decision to rebuild using Notes. First, there were bugs in the commercial EIS software, and the system's response time was too slow. In addition, the company recently decided on Lotus SmartSuite as an organizationwide desktop strategy, and Notes was compatible with this strategy. Finally, the company's EIS focuses on the display of information rather than on data analysis; this need could be well served by Notes. The company's

# LEVERAGING THE ELECTRONIC MESSAGING INFRASTRUCTURE

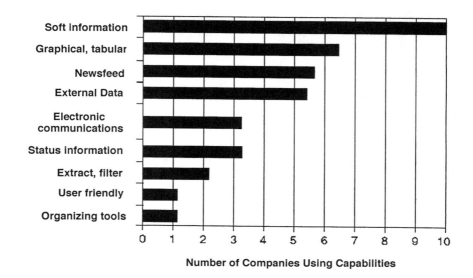

Exhibit 26-1. How Businesses Use Notes.

new EIS was developed in 6 weeks using Notes, and the reactions of users have been positive.

## EIS CAPABILITIES PROVIDED BY NOTES

Notes provides many capabilities that are associated with EISs. Exhibit 26-1 shows the number of companies from the study that use Notes for these capabilities.

### Support for Soft Information

Successful EISs often include soft information such as predictions, opinions, news, ideas, and even rumors. Most senior managers recognize that hard data is not always sufficient for decision-making.

In the document management system for a consumer-products manufacturer, executives are able to add comments to forms that track standard reports at a corporate consolidated level. These comments provide a valuable exchange of reactions, explanations, and issues among the many executives who regularly use the application. This function, in fact, has become so popular that the company has added the ability for commentary as a standard for all of its internal Notes forms.

As shown in Exhibit 26-1, every company in this study uses Notes to exchange soft information. In addition, each person interviewed noted the ease of incorporating soft information into Notes applications.

## Presentation of Graphical, Tabular, and Textual Information

Executives often examine documents that contain graphics, tables, and formatted text. A rich presentation supports analysis and decision-making. Notes manipulates robust documents much better than standard EIS packages. Graphics and tables created in Excel and PowerPoint can be embedded easily into Notes documents. Tools such as Lotus F/X can be used to link the graphics dynamically with their data sources, so that data changes are reflected in the graphical presentation.

For example, at BC Gas Utility Ltd., graphs and analyses are prepared by various functional area groups and deposited in Notes databases. Executives can examine these graphs and tables instead of summary reports that are often more difficult to interpret. Furthermore, the company has used this process to implement management report standards. Users who once constructed numerous paper reports in various graphics packages now deposit reports into Notes documents that have a uniform look.

## Support for Electronic Communications

Notes provides the primary E-mail system in some companies. Arthur Andersen & Co. first brought in Notes as a replacement for its Wang E-mail system. Notes E-mail unites its 35,000 consultants worldwide, offering companywide E-mail standards and customized forms. Other businesses integrate Notes E-mail capabilities with existing systems. For example, a consumer-products manufacturer uses Notes E-mail in parallel with MS Exchange, PROFS, and CompuServe's E-mail. Many companies, however, have invested in an E-mail infrastructure other than Notes and do not want to reinvest in an additional E-mail technology.

## Presentation of Status and Trend Information

Many executives need to closely monitor events or processes to identify status, trends, or exceptions. Notes displays categorized documents so that users can easily identify the status of a process or event. At Toronto Dominion Bank, a Notes tracking system monitors and maintains relationships of customers for managers and executives. These relationships are vital to the bank's success, and their development is tracked carefully. A bank vice-president can access statistics on how many meetings have been conducted and by whom, presentations that have been given, and the number of prospects each person currently has.

## Extracting, Filtering, Compressing, and Tracking Critical Data

Most EIS data resides on mainframe systems or in non-Notes applications, so companies need to download this data directly into EIS applications for their executives to access. Third-party products for Notes, such as InfoPump, offer developers the tools to set up links between various data stores so that data that was not created in Notes can still be used in Notes applications.

Arthur Andersen & Co., for example, is developing a system that can extract data about its consulting engagements from a mainframe using a medium-dependent interface (MDI) gateway. The mainframe collects detailed information about client billing and employee time reporting. Periodically, this mainframe data will be summarized and exported into the Notes engagement information system where the data is categorized and reformatted. Executives can then view this information along with documents such as action items and status reports that are created within the Notes applications themselves.

## COLLABORATIVE WORK SUPPORT AND OTHER BENEFITS

Notes provides a few capabilities not typically associated with EIS software. Especially noteworthy is support for workflow applications where the work task requires collaborative efforts.

For example, Toronto Dominion Bank plans to use Notes in its EIS for processing commercial loan applications. A loan officer performs the initial processing on a loan application and enters the information on a Notes form. This information is reviewed by higher-level officers, who add comments and conditions for the loan. The entire history of the loan application can thus be maintained in Notes.

In another example, an international bank has an EIS that provides considerable support for the personnel function. The bank plans to use Notes for processing employee requests for exceptions to personnel policies and procedures. A personnel manager can enter the request on a Notes form. Higher-level managers can review the request and enter comments as the request works its way through the approval chain. A powerful feature of the Notes-enhanced system is the ability to retrieve information about similar requests and the decisions that were made.

In both cases, multiple users are involved in the work and have the need to share information — an important benefit in a corporate environment.

Notes is a relatively inexpensive product to use with an EIS. If Notes has already been licensed for use within a firm, there may be no costs involved in integrating it with EIS software. Even if Notes requires additional expense, its per-user cost is typically less than that of traditional EIS software. Recently, Lotus released an inexpensive runtime version of Notes for use on computers that do not need to access development tools.

IS departments tend to choose generic rather than specialized software because of concerns about compatibility and technical support responsibilities. For this reason, Notes may be preferred over more specialized EIS software.

## LIMITATIONS OF NOTES

Notes offers so many functions that it is difficult for some companies to find direction as to how to best leverage its capabilities. In addition, many of the Notes tools and third-party products are just now starting to mature. Although it is possible, it still is not simple to access mainframe-resident data or build applications that run on multiple platforms. The embedded data analysis and display capabilities are not as advanced as in traditional EIS software.

As for the user interface, users access documents through a nontraditional navigational system. They do not maneuver through a hierarchical menu structure, but must learn how to manipulate views and forms to find the appropriate document or piece of information. All reporting must be created through views as well. In addition, building a graphical presentation of data is not inherent to Notes and must be created offline in other packages. The graphics are then pasted into Notes.

## THE FUTURE OF NOTES FOR EIS

Future product upgrades and more enhanced toolsets are easing current limitations, and further improvements should make Notes more suitable for EISs. For example, a more robust user interface is expected in the new release of Notes. In the past, developers have been restricted to objects that follow the rules of Notes forms and views. Menus and complex macros required complicated workarounds. The increasing popularity of packages such as VIP and PowerBuilder Library for Notes hint at the growth of more powerful user interface development resources.

Integration between Notes and other databases exists, but capabilities are expanding for companies that want to take advantage of data located in existing applications in a variety of formats. Toronto Dominion Bank is investigating how to integrate its credit processes in Notes with related mainframe information. Once the mainframe systems download customer information into Notes, and the credit is approved, the information will flow back into the financial systems. DataLens for Windows, for example, is a set of drivers that can connect Notes with leading relational database management systems.

Overall, the consensus among the companies participating in this study is that there is an enormous amount of potential in using Notes with executive information systems in ways that reflect both traditional and nontraditional EIS characteristics. As more companies become familiar with Notes and realize its potential, its use for EISs should grow.

# Chapter 27
# Universal Message Services
*David A. Zimmer*

Communications is the very heart of our personal and business lives. The amount of communications that any one person must perform daily requires multiple methods and many people. Just 20 years ago, the amount of communication necessary was much less. A few people had secretaries to take messages if they were not available, while the rest of us simply hung up the phone, only to try later. Today, telephones, fax machines, electronic mail accounts, pagers, cellular phones, answering machines, and voice mail are all used to maintain our communications. Rather than lessening the need for one communication medium, we have information coming at us from all angles. As a result, most people have more information than they can handle and more places to check for messages. The prospect of the "less frantic life" is fading quickly.

## NOWHERE TO HIDE

There is a proliferation of "mailboxes" to receive messages. Many working people have voice mail at work and at home (or at least an answering machine), a fax machine (many more are appearing at home as well), an electronic mail account, an online service such as CompuServe or America Online, a pager, a cellular phone, and the ubiquitous paper mailbox. Some people have many "addresses" that someone trying to reach them might use. Recipients spend as much time collecting all the messages as they do replying to them. Fortunately, messaging permits recipients to shift the processing of the messages to a time convenient to them.

With all the advances of communicating avenues, the originator (i.e., sender) of messages has gained many advantages and services. The recipient, on the other hand, has little control over the influx of messages. The recipient must wade through the messages, determining which are important, which should be viewed casually, and which should be ignored entirely. The recipient receives messages as sent by the originator. Voice messages are voice messages, electronic mail is electronic mail, and fax remain as faxes. At various times, one particular medium is more convenient to the recipient than others. Unfortunately, the recipient cannot edict that all messages at a particular time be received in a particular manner — at least not yet.

## Message Storage

The industry has seen the need for a unified messaging platform and has been developing products that meet some of the needs of both message recipients and originators. Vendors are building products that provide a visual interface to voice mail, fax, and E-mail from a PC desktop. Typically, these systems exist within corporations using local area networks to transport the messages to the user's desktop. Some products permit remote users to use two phone lines to interact with their unified inbox — one for voice playback and one for message control via the visual interface on the PC.

These products focus on the message storage area or "mailbox." Some call it the *universal inbox* or *universal mailbox*. These products are the first incarnations of *universal message services*. However, they have a long way to evolve before they match the overall concept of the universal message service.

## QUICK START DEFINITION

What are universal message services? Universal message services (UMS) refers to a collection of products and services that convey information in a more natural way. The services provide added functionality to both the originator (i.e., caller or sender) and the recipient (i.e., the called party or recipient) of a message. Either party could be a person or an application, such as information services.

UMS gives the recipient intelligent controls over the reception and processing of the incoming messages. Because of its multimedia nature, UMS is the single place that recipients need to access in order to manage all their communications.

## Breaking Geographical Chains

Today, telephone numbers are associated with a physical location, although we can place extension phones on the same line so that we do not have to run from room to room in order to answer the telephone.

A fax machine sits in one location; so in order to retrieve a fax, someone must walk to its location (if the fax machine is across the country from where you might be currently, you are out of luck). E-mail is a tad more portable with the use of a notebook computer, so users can usually grab it as long as they can find a phone jack or have a strong enough wireless connection.

## Universal Message Services

For each geographical location, the individual has a separate address. To reach other people while they are working, a caller must dial the work phone number, not the home phone. When addressing messages or calls, people typically address them to a physical location, not the actual person. Universal message services let people assign an address to a person, not a location. In this way, the message can be routed to the person regardless of location. Accordingly, the recipient of a message can process incoming messages more easily since they will be coming through one path.

Other features of UMS let the recipient manage the incoming messages more quickly. Recipients can "tune" the UMS to meet their needs at the moment. If the person wants all calls routed immediately, that can be accomplished. Other times, the person may want throttle incoming messages so that only important messages are routed directly.

The address is assigned to the person and travels with him or her. The same address is used by originators regardless of the communication medium. Whether the message is voice, fax, E-mail, or video, the originator uses the same address.

## FOUR LEVELS OF INTEGRATION

The central component to UMS is the message box. (For this chapter, the term *message box* is preferred over *mailbox* because it connotes that multiple media types may be stored in the storage area. The term *mailbox* is used when only one type of medium is stored in the storage area.)

The message box may be one physical location or several storage areas in different systems, such as voice mail platforms, fax servers, and electronic mail post offices. Regardless of the physical implementation, the user sees the message box as one unit with all messages accessible to him or her via the access device in use, whether it is a telephone or PC.

### Category 1 — Existing Services, Switch Box

Category 1 products provide a unifying front-end to existing services. The voice mail, fax server, and electronic mail post office are separate, distinct services. Each has its own native interface.

The switch box permits the user to access voice mail, then faxes, and then electronic mail (see Exhibit 27-1). To the user, each media type appears to be separate, although the user has a single interface to view each message.

This type of solution saves the user time by not having to log into separate services to retrieve and process messages. It does not, however, provide the user with any type of control over incoming messages; there is no

## LEVERAGING THE ELECTRONIC MESSAGING INFRASTRUCTURE

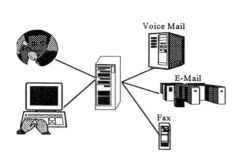

- Uses existing voice mail, E-mail, and fax
- Interface through PC or touch-tone phone
- Separate, incompatible systems
- No inter-operability

Exhibit 27-1. Existing Services Using Switch Box.

integration of services and no ability to reply in a different medium than the original message.

For example, a user retrieves a voice message. If the user wants to reply by voice, he presses the proper button and records the reply. If the user prefers to reply by E-mail, he would have to use a separate method.

### Category 2 — Single Logical Storage, Disjointed Parts

The second category of integration uses the same logical store, but the messages from differing mediums are not integrated together (see Exhibit 27-2). The user may use either a telephone or a PC for accessing the messages.

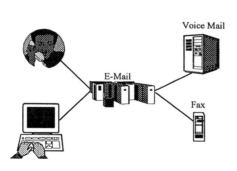

- Messages collected into single inbox
- Access through PC or touch-tone phone
- Replies via same medium as original message
- E-mail message may contain body parts of another type

Exhibit 27-2. Single Logical Store Using Disjointed Parts.

The user processes the messages after listening/viewing the message by either saving, deleting, or responding to the message. The reply message returns in the same format (voice, E-mail, or fax) as the original message. Addressing of the reply message takes the same form as the original. For voice messages, the address remains as a telephone number; for faxes, the user would have to address the message with a phone number; and E-mail retains the originator's E-mail address.

The messages may be multiple media in nature with "attachments" or body parts appended to the end of the message. For example, an E-mail message may contain body parts of fax, E-mail, or voice. In the case of a reply, the E-mail addressing would be used. Each part is an integral part of the message and cannot be broken out separately without explicit user actions.

Phone interfaces for these types of messages provide the user with basic controls simply because of the nature of the 12-button interface of the telephone. The PC provides a richer set of interface options, letting the user perform more advanced operations. The visual interface of the PC lets the user order the messages by date/time message received, message type (e.g., all voice messages, faxes, then E-mails), and message priority (e.g., urgent messages first, normal priority second). The PC interface permits the user to break messages into their components for storage or forwarding to others.

### Category 3 — Single Logical Storage, Integrated Parts

The third category furnishes integration between message types. Messages are stored in one logical place, with interfaces via the telephone or PC (see Exhibit 27-3).

Message replies may take the form of the original but may easily take another form. By the user simply requesting the desired format, the message address is translated to the appropriate format. The proper interface is presented to the user for reply creation.

If the user selected a voice reply, a "voice record box" is presented on the PC screen. If the user selected an E-mail reply, the "editor box" is presented. The original message may or may not be attached to the reply message.

In essence, the system begins to act the way the user would expect and in the manner in which the user works. For instance, frequently a user will want to call the originator shortly after reading an E-mail message. The user should be able to react the same way while responding. The voice recorder should be activated to accept the person's voice response. Just as E-mail messages are annotated with interlaced comments during a reply, voice comments can be interlaced into the original E-mail message before being sent back to the originator. Although the response may be imple-

## LEVERAGING THE ELECTRONIC MESSAGING INFRASTRUCTURE

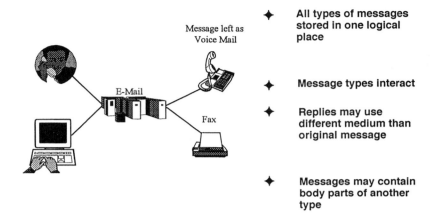

Exhibit 27-3. Single Logical Store with Integrated Parts.

mented and physically stored as independent body parts, the recipient receives the integrated message in the same manner in which the reply was created — with interspersed verbal comments.

The integration of message media is more difficult than the disjointed parts method described previously. Standards and definitions are being written to provide such functionality.

### Category 4 — Single Logical Storage, Translated Parts

The fourth category permits the most flexibility possible to the recipient of the message. In previous categories, messages were translated or rendered into an acceptable format for the interface. For example, E-mail messages were converted from text to speech for users accessing them by telephone. Voice messages required connections to a telephone or speakers while being "viewed" via the PC interface.

In category 4, full rendering to the recipient's desires is permitted. For example, voice messages will be translated (or rendered) to text through the use of speech recognition. Faxes will be recognized using OCR technology for reading to the user of a telephone handset. Admittedly, the full functionality of these services is a ways off for general-purpose use, but early examples are emerging.

These translation services are useful for several populations of users, including:

*Universal Message Services*

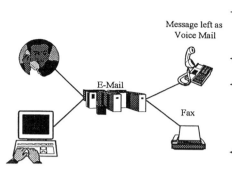

- All types of messages stored in one logical place
- Message type interact
- Replies may use different medium than original message
- Message may contain body parts of another type
- Message types may be translated to another type

**Exhibit 27-4. Single Logical Store with Translation of Message Parts.**

---

- Mobile workers who may not always have the proper equipment to handle messages
- Disadvantaged individuals who may not be able to use a particular device
- Adopters who simply want to choose their medium "du jour"
- People that learn by extensions of current services rather than evolution of technology

The fourth category of integration (see Exhibit 27-4) provides the user with the foundation necessary to reach the ultimate universal message services. The central component to UMS is the message box. It is the interface for the user for all interaction with UMS. From within the message box, the user configures and controls incoming and outgoing communications.

The integration and translation of message parts is necessary for the user to reap full benefits of the universal message services. Information gathered by UMS will not be useful if users cannot obtain it in a form understandable by them. If weather information is obtained in text-only format, a person who cannot read would be disadvantaged. The message box could translate that weather information into speech so the illiterate person is still able to receive information.

## FOUR KEY COMPONENTS OF UMS

Universal message services go beyond the integration of message media within the message box. The message box is only one of four key components to universal message services.

## LEVERAGING THE ELECTRONIC MESSAGING INFRASTRUCTURE

UMS provides the recipient with control over message receipt, information gathering, and message routing. However, to support the necessary functionality for the recipient's control, new messaging infrastructure must be created and deployed. Services must be provided in the message box, the post office or routing services, and the communications network.

Many of the necessary functions exist today for supporting the universal message services, but the integration of those services is not fully understood. Understanding the roles of each part and its functionality is crucial to the creation of the UMS.

Universal message services combine four key components:

- The universal message box
- The intelligent post office
- The intelligent network
- The intelligent services

### KEY COMPONENT 1 — UNIVERSAL MESSAGE BOX

The *universal message box* permits the recipient to receive all messages into one logical location. This single-location approach allows the user to process messages from a single interface. The interface device may vary (e.g., telephone, fax machine, PC, interactive TV). The single-access interface provides the user with additional functionality in creating new messages.

For example, while listening to a voice mail using today's technology, the recipient cannot easily respond to it via E-mail unless he turns to his computer first. If the recipient was not near the computer containing the file, he would have to save the voice message and act on it later.

The universal message box (UMB) permits the user to respond to the voice message by simply attaching the file and replying to the caller. Because the UMB can be accessed from a variety of terminals — telephones, PCs, PDAs, fax machines, and soon pagers — the recipient could respond to the caller upon first hearing the message, regardless of location to the computer containing the file. Thus, the user's productivity goes up, the list of "to-dos" goes down, and the recipient can dispense with the voice message immediately after responding, rather than saving it — as required today.

The UMB provides additional capability. As the recipient of many messages, a user needs to prioritize messages. Today, voice mail only gives a serial interface to messages. The recipient listens to the first message, then the second, the third, and so on until he either listens to them all or tires of listening. As the recipient listens, he jots notes as to the names and numbers of the callers, actions to be taken, and so forth. Electronic mail, on the other hand, gives the ability to randomly process messages, responding to the most important first and processing others later. The recipient can easily respond to information requests by attaching files or typing the response. Faxes are a hybrid between voice and E-mail in that faxes can be scanned randomly, but the recipient has to make notes for the responses.

**UMB Functionality**

Using rules and agents, the UMB will process some of the messages before they even reach the recipient. For example, if the originator is requesting the latest ad copy for the Jones Camera ad, the UMB could retrieve the file and notify the recipient of the request. If the user sanctions the response, the UMB would forward the ad to the originator. This is an example of a person-to-person request that resulted in an application-to-person authorization fulfilled by an application-to-person message.

Very easily, this same functionality can be extended to application-to-person or application-to-application requests. For example, the accounting software requests the latest sales figures from the salespeople. The response could be returned in the form of a voice message from a salesperson at an airport listening to voice mail, a fax message from a person at a remote site, an E-mail from a user in a hotel with a laptop computer, or a spreadsheet from another person who has established rules to fulfill the request automatically.

The UMB user could set other rules as well. For example, she might choose that all messages should be prioritized so that important calls from the boss, spouse, or daycare center appear first. The messages could be prioritized based on content, originator, or media type. Messages may be converted from E-mail to voice so that they can be retrieved easily from the airport. Information requests could be handled automatically. Electronic forms containing requests for products, purchase orders, shipment requests, workflow items, or authorizations could be handled automatically, be brought to the recipient's attention, or sent to a pager. With the use of two-way paging, the response to the page could initiate a workflow process internal to the company approving expenditures, digitally signing work items, or rejecting the latest request.

The industry is very active in defining and building the UMB. Companies such as Digital Sound in California and Boston Technology in Massachusetts are manufacturing products for local telephone companies. Telephone companies such as PacBell and GTE are implementing early versions of the UMB in the public phone network for small-business use. AT&T and Lotus Development Corp. have joined forces to produce offerings for customer premises. AT&T Intuity and Lotus cc:Mail will be combined to provide an early version of the UMB. Octel has articulated a unified messaging strategy and developed a working unified system. Centigram has developed unified messaging products as well. We will see more offerings in the next year from other vendors.

# LEVERAGING THE ELECTRONIC MESSAGING INFRASTRUCTURE

## Three "Rooms" of the Universal Message Box

The UMB provides the user with a visible interface to universal message services. From the UMB, the user configures all aspects of the UMS operations on her behalf, manages all incoming and outgoing messages, and retrieves information and electronic sale items. The UMB can be broken into three conceptual components or three "rooms" for ease of use and understanding: the message room, the control room, and the activity room.

**The Message Room.** The message room handles all the messaging needs of the user. A user processes incoming messages by listening/viewing, replying or forwarding, then deleting or saving the message. She creates and sends messages from the message room. The message may be sent to one or several people at the same time. To help address messages, the user has a directory that contains names and addresses (e.g., personal and business telephone numbers, fax numbers, E-mail IDs) available to her.

The message room replaces the user's need for an answering machine. It stores the incoming messages and notifies the recipient of new messages. Outgoing messages are chronicled in a "Sent" log for later use. The message room may contain other folders for user-initiated filing of messages.

From within the message room, the user can create a message and send it to one or many recipients. The recipient addresses are extracted from the available directory. The directory is a multilevel service. The user creates and maintains a personal directory that stores often-used addresses. Accessing the public directory helps the user locate any recipient, but the user might incur a charge similar to the directory services for the telephone system today. Once the address is obtained, the user may place the directory entry into the personal directory for future use without charge. Emergency numbers/addresses may be provided by default for all personal directories (some people may prefer the term "address book" when referring to the personal directory).

The message room handles all forms of communications: voice, fax, electronic mail, video, and sound. Messages are left by the originator in one of two ways:

- The originator attempts a real-time connection such as a voice call or video call, but the recipient is not available, so the sender leaves a "message."
- The originator creates a message and addresses it to the recipient. The message is deposited into the UMB.

The message room may be multidimensional. For example, a family might subscribe to UMS, with each family member having his or her own private message room and a general message room for the entire family. The private message rooms would be password- or PIN-protected, so only the family member who owns that room can manage the messages within that room. The general message room can be used for receiving messages not targeted for a particular person. A message room could be partitioned into business and personal correspondence as well, at the user's discretion. Additional message rooms could be created for other areas of interests, such as clubs and associations. Messages are filtered into the proper room by use of filters.

In other words, the message room can be configured to suit the personality of its user. For some, creating business and personal folders would be sufficient and the use of filters would organize the messages accordingly. Others might prefer separate rooms for sorting incoming and outgoing messages. Because a message room is protected by a password, the user can establish a clear separation of messaging communities (i.e., business correspondence, personal messaging, and extracurricular activities).

**The Control Room.** The control room contains the controls that help the user in managing messages. Using filters and rules, a user can direct incoming messages to another number/address or select other messages for automatic deletion or refiling. The control room permits the user to add (subscribe) or delete (unsubscribe) to services (e.g., message waiting indicator selection, news/stock/weather services, or other information services).

*Announcements and Template Messages.* Outbound announcements are configured from within the control room. The outbound announcements can be personalized to the originator of the message. If the originator is the user's business associate, the service uses the "business" announcement. If the originator is a personal friend, the service uses the "personal" announcement. The distinction between categories of originators is controlled by the user from within the control room using rules and other parameters.

A third level of outbound announcements can be established that responds based upon the originator's identity. For example, a personal friend calls. Instead of receiving the generic "personal" announcement, the announcement might be customized (or individualized) to speak the person's name and play a specific message, such as inviting the friend to an upcoming party.

To facilitate individualized announcements and other types of messaging, the subscriber would create and store "templates" in the control room. Templates are messages (e.g., voice, fax, or E-mail) that have embedded fields that will be filled in at a later time. For example, the subscriber might create a generic personal greeting that contains a Name field that is replaced with the caller's name. A Little League coach could create a template message that announces the next game where the date, time, location, and opposing team will be filled at a later time.

## LEVERAGING THE ELECTRONIC MESSAGING INFRASTRUCTURE

The templates could interact with the personal directory to select names during message sending. During incoming-call reception, the template would use the caller's ID to retrieve the person's name from the personal directory.

Announcements can become active agents for the message box owner. For example, say that Bill is the message box owner and he works with Sue, the account representative for the Jones account. Bill could personalize the announcement so that when Sue calls, she is greeted with, "Sue, this is Bill. We need to meet at one o'clock, Tuesday, to go over the Jones account status. Can you meet me at my office at that time?" If Sue answers yes, both Sue's and Bill's calendars are updated with the meeting. If Sue answers no, the calendar program can search for another available time on Bill's calendar (within the parameters set by Bill), and the meeting can be established.

In short, the announcement becomes an assistant to the message box owner, not simply a passive message. Similar activities can transpire using announcements for workflow items and approvals.

*Personal Directories.* From the control room, the subscriber would build and maintain a personal directory (address book) of numbers and addresses. For ease of use, the personal directory may be populated from within respective message rooms, during message creation or message retrieval. Adding entries to the directory should not be limited to one specific location within the UMB; otherwise, the user will become frustrated by the limited access.

A network-based personal directory provides several benefits to the user:

- It is easy to track information.
- Information is accessible from anywhere (i.e., other rooms in the house, remotely from hotel rooms or public pay phones, electronically from a PC).
- Speed-dial codes can be established by user and activated from any phone extension, public phone, or the PC.
- Electronic addresses and telephone numbers can be updated by a UMS network (although it may be an additional charge for the public network to monitor and update a personal directory).
- Directory entries could be used to populate personalized greetings.

The personal directory is multidimensional, just like the message rooms. A general directory can be established so that those entries are available from any message room. Other entries can be added but marked so that they are specific to either a particular user of the message room or specific to a message room. Entries may have a designation such as business, personal, or social so that it is available from any message room with that attribute. Of course, with all this functionality comes the challenge for ease of use.

*Privacy and Security Parameters.* The control room lets subscribers control the privacy and security of their cyberhome. Just as we place security systems and window shades on our physical homes to protect us and block out curious onlookers, we need to have the same safeguards when living and communicating in the electronic world.

Subscribers can designate privacy parameters on the information to be distributed. For example, the subscriber could choose not to have the telephone number known to the public. Unlisted telephone numbers provide privacy, but the number still "rings" when dialed. The control room provides a method of unlisting the addresses. The subscriber may choose to publish the general number, but not the other message room extensions. Or, the subscriber may publish the general mailbox and some message rooms supporting home business while leaving others private.

The subscriber's message room must be secure from others trying to access it illegally and from others placing unwanted material in it. The control room is the place the subscriber uses to establish and change passwords (PINs) for the various message rooms.

In addition, the subscriber may place "blocks" on certain types of messages or messages from specific origins. Objectionable material, such as solicitations from pornography houses, religious groups, or credit card solicitors, could be blocked from entering the mailbox. The block can be done by categorical specifications, categorically with exceptions (e.g., "block all charities except Red Cross, United Way, and Salvation Army"), or by specifying a particular origination address. The subscriber has complete control over these blocks.

*Message Rendering.* While in the control room, the user can establish message-rendering rules for converting text messages into speech and speech into text. The render rule will be performed in advance of the user processing the message. It can be applied to all or specific messages based on additional criteria. The user will have the option of rendering a message on a per-message basis from within the message room. Other options may include language translation as well.

*Enhancements to Services Offered Today.* Many services offered today in the telephone network, such as call waiting and call forwarding, could be configured by the user. For example, currently, a subscriber must contact a company representative to subscribe to a service such as call waiting and set its values. The control room may provide options for the user to select the service and set its operating settings.

In addition, these basic services can be enhanced to provide more functionality than is currently possible today. Today's call forwarding service forwards all calls to the forwarding number. The newer service might forward only specific calls or calls meeting a certain criteria. The criteria could be based on the personal directory, time of day, or originator's ID.

Suppose that the user runs a home-based business and maintains working hours between 9 a.m. and 5 p.m., Monday through Friday. During nonworking hours, all business calls are forwarded directly to the universal message box that plays a specific announcement for that time of day. Personal calls are patched through to the user's phone. If the user does not answer, the call reverts to the UMB with the personal greeting played. The user can segment which calls get forwarded by simply adjusting the values in either the personal directory (entry-level forwarding) or call-forwarding parameters (group forwarding).

**The Activity Room.** The activity room contains services that are more interactive in nature. The user would use the activity room for accessing the public directory (also accessible when creating messages), multi-user services such as voice party lines and online games, real-time chat services (interaction via PC), and online magazines, newsgroups, and special interest boards. The activity room could be used for Internet/Web access and for paying bills.

Teleconferencing and videoconferencing are conducted from the activity room. Attendees are selected from the personal directory, the public directory, or entered at call setup time by the originator.

The information accessible through the activity room resides in several places. To the user, it will appear to be stored locally. In fact, some information such as bill payment logs may be stored locally since the information is specific to the user. Other information — especially "streaming information" such as news, stocks, and weather reports — will constantly be refreshed when accessed but never really stored, producing a ticker-tape effect. Information with broader appeal may be downloaded to the person's viewing device at the time of access, similar to World Wide Web technology using hot-links to the data.

## KEY COMPONENT 2 — INTELLIGENT POST OFFICE

*Intelligent post offices* provide routing capabilities for messages. The electronic intelligent post office directs messages in accordance with recipients' wishes. For example, important messages may be routed to an assistant while the boss is on vacation, while other messages will be routed to the boss's UMB. Some messages may be sent back because of lack of interest in the subject. The post office may route messages of interests such as news or stock quotes directly to the message box owner regardless. The recipient has complete control over the disposition of messages at all times.

The post office could support other types of services, such as bulk mailings by generating the mailing list dynamically from an electronic list. The list could be based on interest areas of subscribers. These topics may be ongoing, long-term interest areas or transient.

Transient interest areas come and go at various times in a person's life. For example, people are not always interested in the latest car models or prices. But, when it is time to buy a new car, related newspaper ads grab a prospect's attention, radio spots land on interested ears, and TV commercials catch the person's eyes. Suppose the prospective customer is not looking for a certain brand of car but is interested in midsize cars, luxury sedans, or minivans. The post office would "watch" bulk mailings and forward the ads that fit the prospect's established criteria.

The criteria is set in the activity room. The prospect would fill a form that lists the category of cars, colors, options, and price ranges. In addition, if a specific geographic area for car dealerships is important, the prospect could note that as well. The prospect could also eliminate some car makes or models or options from the search.

This type of service benefits the originators of messages as well. Rather than paying for mail going to people in general, they only pay for those messages sent to people specifying an interest area. The recipients are targeted, the rate of return would be higher, and the quality of the mailing would be better. The message does not have to rely on just color to catch a person's eyes. The message could contain animation, sound, and text.

Going back to the car-buying example ... once the prospect has purchased a vehicle of choice, she can eliminate the interest criteria. Immediately, she stops receiving new car ads. Long-term interests can be served in a similar manner.

For example, many people access online services to get the latest stock quotes and weather. This same information could be sent to them as well. The information may be delivered in one of two methods: it either could be deposited in the message room for a snapshot of current information or it could be streamed to a ticker tape. The user controls the method of presentation.

The message box owner could set limits on her stock portfolio so that she would be warned of major price swings on investments. If a limit is hit, the post office forwards a message to the message box owner based on the person's routing criteria (e.g., it could be placed in the message room, passed to an assistant, or sent to the owner's pager).

The post office could have access to a user's electronic calendar, and after noting travel plans, supply weather information, entertainment information, hotel accommodations, restaurant locations, and other amenities. The post office uses personal interest areas and previous habits to filter information interesting to the traveler.

Much of this functionality exists today. How, then, is the intelligent post office different? Several differences come to mind.

**Coordinating and Distributing Information**

First, today's systems require that a local PC or other device be running and that the person be present to get the information. When traveling, the information sitting on the device does the person no good. Some information could be sent to a pager, so the person is notified immediately, but he has no avenue for immediate response unless he either carries a cell phone at all times or is near some other phone. Hopefully, the phone will be an adequate response vehicle.

Using the intelligent post office, the person is not limited to the local PC. The post office helps select the interest items that are immediately important and stores others that can be processed later. If an item (e.g., stock quote) usually handled during leisure time hits a critical point, the post office can decide to forward it to the recipient. Or, it could forward it to the UMB, where another rule is set to act on it in accordance to preset instructions.

To further broaden the scope of the information routing concept, consider that an ad agency is working with a client on a new campaign. The deadline for final print is tomorrow and the client decides to make some last-minute changes. Today, they would have to fax those changes to the agency and follow up with a voice call. The agency must assemble a team that may be working on other projects. A great deal of manual coordination and distribution of information is required.

The intelligent post office could be used for the coordination and distribution of information. The client sends the changes to the agency. Using the post office's knowledge of the team members' calendars and work items, it notifies them of the changes and distributes the information. Each team member can respond positively or negatively to the schedule impact. The leader of the agency is quickly informed of the possibility of making the changes and meeting the deadline. A workflow process can be created so the changes are incorporated appropriately.

Transient interests are handled more efficiently in the intelligent post office. Most online information exists under two basic types of subscriptions: subscription to a general service with many areas of interests, such as CompuServe or American Online; or a subscription to a single-purpose feed, such as stock quotes or news.

For the general type of service, a person must search for the information. It requires an active participation on the researcher's part. The single-purpose subscription usually streams the information to the recipient. For transient interest areas, a single-purpose subscription for a short period of time may not be cost-effective.

Using the features of the post office, a person can set up an interest area and tear it down all in the same day. It is very dynamic in nature. The recipient only uses it for as long as she wants, only pays for the information received, and has control over it. If the recipient receives too much information, she can restrict the area criteria immediately. If the user is not receiving enough information or the wrong information, a simple change to the interest area criteria provides better results.

## A Central Clearinghouse for Information

The intelligent post office becomes a central clearinghouse for all types of information. This clearing house gives the user a greater range of capability.

For example, suppose a traveler has scheduled a trip to Chicago. While there, he would like to find a local club with live jazz entertainment, a modestly priced Italian restaurant, a hotel within a particular price range and quality that includes a health spa either colocated or nearby, and all of these within walking distance of the meeting center. The traveler would like to see the review of the hotel, the jazz club, and the restaurant. Today, the traveler would have to access three or four sources of information. The post office, acting as the clearinghouse, already has access to the information. Using the "transient interest area" concept, it could ferret the information for the traveler, who could pick his choices from the list of selections and automatically make reservations at each (such applications are discussed further in the section on "Intelligent Services" later in this chapter).

## Buying and Selling Goods and Services

This same concept can be used in the buying and selling of services and goods. A company needs to purchase 10,000 widgets. The widgets must be blue or green and weigh less than 1 pound each. An acceptable price range may be between $1.00 and $2.00, with quantity discounts available. Sellers have published their product sheets in the post office or in a public area. The post office helps buyers locate sellers offering widgets that meet those criteria.

A buyer could float requirements specifying product features and functionality, expenditure ranges, and delivery options. Companies supplying products that meet the needs could respond by issuing a general response containing company information and product literature. They could follow up with a more specific response to the request. The buyer's request may

# LEVERAGING THE ELECTRONIC MESSAGING INFRASTRUCTURE

contain voice, graphics, sound, images, text, and video to describe his or her needs. The response could contain the same media formats.

## KEY COMPONENT 3 — INTELLIGENT NETWORKS

*Intelligent networks* help the post office locate people as they move around. Similar to the cellular networks that track a person's movements as she drives, intelligent networks understand her current position. The ability to locate a person when necessary will complete the services of the universal message box and the intelligent post office. For the UMB to notify the user that a message arrived from her spouse or an emergency call came from the daycare center, the UMB must have a path to send the notification. Today, we simulate that path in a crude way through paging.

The intelligent network provides a single address to subscribers for all communications. Currently, we have work telephone numbers, work fax numbers, home phone numbers, home fax numbers, cellular phone numbers, and electronic mail addresses.

Whether the recipient is at the office, at home, or in his car, the intelligent network would route an originator's call to him; the originator needs only to dial the recipient's personal address. By the same token, if the originator wanted to send a fax, she could use the recipient's personal address and the network would locate him in order to deliver the fax. E-mail would be sent to the personal address as well.

Working in conjunction with the intelligent post office and the UMB, the message may be passed directly to the recipient or placed in his UMB. The recipient controls the incoming messages via the features of the intelligent post office and the UMB while still providing the capability of actually reaching him.

The message box owner can easily adjust the messaging rules from his portable device. These parameters, stored in the control room, can be changed to reflect the current desires of the person. For example, if he does not want to be disturbed with any messages, the owner sets the parameters so that all messages are routed to the message room. He could set a parameter that all emergencies are routed to another person, unless the message comes from a particular person such as a spouse or daycare center. An hour later, the person can reset the parameters so that important messages are routed to him directly while all others route to the message room. Important messages that arrived during the previous hour are forwarded automatically to the owner.

### Sales Aid Networks

The intelligent network will aid business as well. Information returning queries or potential sales leads may be directed to the appropriate salesperson or placed in the UMB for later processing. Salespeople spend a lot of time traveling. The intelligent network knows the current location of the salesperson. If a sales lead arrives from a company in the same vicinity as the salesperson, the network can notify the UMB so that the lead is passed directly to her. The salesperson can possibly contact the prospect immediately for a visit. Today, the salesperson must process voice mail in order to collect the leads.

In another example, salespeople typically have contracts faxed to them, even while they are out of the office. Contracts could arrive that require immediate attention. The salesperson could configure the UMB to pass those types of items along to the next person in the contract-processing chain rather than let them sit on the fax machine. Now, the inbox (i.e., the UMB) becomes an active tool instead of a passive repository of communication.

## KEY COMPONENT 4 — INTELLIGENT SERVICES

*Intelligent services* will play an important role in universal message services. Without intelligent services, UMS will experience slow growth, with people using it simply as a convenience item and time-saver. To fully leverage the power of UMS, enhanced services or value-adding messaging must be introduced. These services go beyond some of those already mentioned, such as news and weather reports. These services must work in conjunction with the rest of the UMS.

First-generation services are being built for business-to-business activities as well as consumer services. Several companies are working on products that will permit a customized morning newspaper listing only those topics of interest. Others are working on electronic shopping with graphical images, personal interest areas, community interest areas, and more.

Although we are bombarded with information daily, we seem to have a need to find different information. The Internet is a testament to that point. People use the Internet to search for information on topics, communicate through newsgroups, and chat in chat sessions. But the vast resources of the Internet are the biggest dilemma as well. Locating quality information is difficult; and locating a complex scenario is even more difficult.

Information services pass their information through intelligent filters so that only information matching the subscribers' desires are funneled into the message box. Services may include news services, financial data, weather, or items for sale. In our previous example where the traveler was looking for a jazz club, hotel, and restaurant, the Internet would be a start, but it does not actively aid the searcher. Intelligent services could provide the needed assistance.

# LEVERAGING THE ELECTRONIC MESSAGING INFRASTRUCTURE

Launching intelligent agents, or programmatic search tools, the services could locate the desired items. These agents interact with databases of information to combine the needed materials to satisfy the request.

Information repositories are being designed to interact with these agents. News services, weather, stock quotes, and publication companies are some of the early adopters. Additional services are fast emerging as this technology continues to expand.

## OVERCOMING THE COMPLEXITY OF UNIVERSAL MESSAGE SERVICES

The infrastructure and underworkings necessary to support universal message services are complex. UMS requires integration between different media types, translations, easy-to-use interfaces, rules and agents, networking, and portable devices. If the UMS is provided in a public network where users pay for its use, a price may be established for each element.

Universal message services are so complex, no one company or vendor can provide the entire end-to-end solution. Standards already exist for some parts of the equations, and new standards will need to be defined for the other parts. As all standards go, they are open to interpretation by the various players; thus, an extraordinary amount of testing and interworking between companies must occur.

Another factor must be considered. Should the UMS be implemented as a public network service run by various telephone companies or other providers? Should it be a customer premises equipment product? Is it a combination of the two? Clearly, as the products evolve, the market will help decide.

### Integration and Directory Synchronization Issues

The first difficulty involves the interworking within media of the same type. For over 10 years, the electronic mail industry has been working to make transmission of E-mail messages between various platforms seamless, and the problems are not yet resolved. Try sending a message with an attachment through the Internet. It is a crapshoot if the other person can read the attachment easily. In other cases, messages created by systems rich with features lose a lot by going to older systems with less features. Synchronizing directory entries within a company is still a difficult task.

Voice mail integration also remains problematic. Companies have typically settled on one solution, which poses less need for integration. But what happens when larger companies possessing different voice mail systems merge? How do they integrate those systems? Voice mail switches and gateways do not exist as they do with E-mail; directory synchronization is totally nonexistent. Besides, how would you synchronize the directories? Using a four- or five-digit extension number to identify a particular person

## Universal Message Services

would certainly conflict with another person's extension in the newly merged company. Do we synchronize the full telephone number? Is that ten digits for North America, but less for other countries? We could use human-friendly names instead of telephone numbers. Unfortunately, on a phone handset, three letters are bundled onto one number, not all telephones are touchtone (which makes "addressing" from a rotary phone cumbersome), and there are no standards for supporting the Q and Z letters.

Faxes are an interesting anomaly. Typically, a fax machine is located in a business, usually supporting six to ten people. Faxes are transported by sneaker-net technology. Although PC-based faxing is growing in use, there is still a tremendous amount of paper that gets faxed. Consequently, businesses need to understand how to tie fax capabilities into the network.

The "edge" between E-mail and voice needs to be defined. Will a communication simply be an E-mail message with a voice message attached? How about a voice message with an E-mail communication attached? Or E-mail-annotated voice messages, or voice-annotated E-mail messages? Throw in some text-to-speech and speech-recognition capabilities so that users can easily go from one medium to the next. Make all this work with both private and public services with many levels of features and functionality. Combine the directories so that a person using a public telephone can dictate an E-mail message to a co-worker. Have that same person create a broadcast message that is sent to several people on different platforms. When received on those platforms, the message should follow the recipient rules so that it is forwarded to another person, translated into fax, or simply returned because of no interest. If intelligent services are integrated into this picture, someone may subsidize this complex network enough so that the average user or company can afford to use it.

## SUMMARY

Clearly, the job of building universal message services requires overcoming many obstacles and challenges. At present, many vendors are working on building a universal message box that, at a minimum, can hold the various media types. In some cases, the products permit some integration between media types, although the focus has been to make these services work within closed user groups such as corporations. Full-blown integration of messaging services is slow because of the lack of knowledge of how to make some of the infrastructure work, such as network management, directory synchronization, and media translation. Part of the slow evolution depends on advancements in related technology such as speech recognition and optical character recognition.

Greater functionality has been added to products and services on the message originator's side of the communication. The originator of a message determines the type of message, the contents, the time at which it is sent, and the urgency or prioritization of the message. The recipient, on the other hand, has not gained much in the way of control over the communications. He simply has gained more mailboxes in which to capture more messages, resulting in information and communication glut. The concept of universal message services helps eliminate a number of mailboxes and provides the recipient with control over the incoming messages.

UMS provides the highest level of integration between message types, permitting the user to translate messages from one format to another. Using the access device at hand, the user can process incoming messages anywhere. The powerful rules established by the user permit processing of messages without intervention from the user. Messages and their responses may trigger workflow processes within and between corporations. Using newer technology, such as wireless communications and two-way paging, a user might approve transactions and other work items with digital signatures by simply pressing a button on the pager.

Using intelligent messaging, the message box can pass messages through the intelligent post office to the message box owner or subordinates. The intelligent network follows the user so that important and emergency messages can be passed directly to the owner. Intelligent services provide information within the message box owner's interest areas.

Universal message services work with other aspects of the user's life as well, by interacting with the person's calendar. UMS can help coordinate several people's calendars, intelligently offering alternative times and keeping the user informed of changes.

The goal of universal message services is to simplify people's lives. Rather than having to check several message boxes, manually respond to each message, locate information, or arrange meetings and work items, a UMS subscriber can establish working criteria for the services to accomplish. One message box handles all communications for the owner. In a way, the UMS is similar to the butlers of the aristocrats — an intelligent assistant that knows the employer's wants and desires, completing tasks so that his or her employer can do other more enjoyable things.

# Chapter 28
# Client/Server Messaging and the Mobile Worker
*David A. Zimmer*

Have you ever noticed how the mobile mode of messaging tests the mettle of any network? The mobile environment has so many variables to overcome that vendors who do well in the mobile market usually have excellent networked products as well.

Mobile workers must be able to manage their communications while sitting in a restaurant, driving down a highway, squatting in an airport, huddled in a hotel room, telecommuting from home, and working in the office. Mobile workers use cellular phones, public phones, home phone, office phones, fax machines, and E-mail user agents on notebooks and desktop machines. Keeping all the information synchronized and on the proper device has been a challenge.

Experienced mobile workers have worked diligently to narrow the number of devices used to manage the influx of messages. Typically, they settle on using notebook computers for the computing needs outfitted with a fax modem, fax mailboxes for fax receipt, and a company-sponsored cellular phone (although lines have not diminished appreciably at public phones during conferences). Pagers are used to keep the mobile workers instantly updated as to new message arrival and other incoming messages from coworkers.

Electronic calendaring and scheduling of appointments helps others coordinate time with the mobile person. Typically, calendaring software requires some type of static database of information either on the notebook, making it inaccessible to others; or on a file server back at the office, making it inconvenient for the mobile person to use easily.

Mobile workers are required to be productive members of teams by participating in collaborative projects and completing tasks within workflows.

## LEVERAGING THE ELECTRONIC MESSAGING INFRASTRUCTURE

Collaborative projects and workflows are complex uses of technologies and networks within the confines of an office environment. Client machines are continuously attached in a known manner with plenty of transport bandwidth. The unknown environment of a mobile worker with lower transport bandwidth, lower quality connections, and the need to work in a disconnected mode can make for an interesting and challenging evening while huddled in a hotel room.

To support the advanced communications needs of today's businesses, and especially the mobile worker, a messaging architecture that supports new demands for communicating is needed. Business models are changing rapidly. Communication infrastructures must keep pace with the change.

## EVOLUTION OF COMMUNICATION: FROM HORSE AND BUGGY TO THE FAST LANE

### The 1800s

Prior to the late 1800s, we were mainly an agrarian society. Businesses consisted mainly of family units working their plot of land. Daily schedules revolved around the rising and setting sun, with long-range planning depending on the seasons. Communications lines were simple and real-time. Most communications remained within the family unit, with more general news coming from a visit to the general store, barber shop, or blacksmith. The need for quick decisions or turnarounds of communication was nonexistent. Virtual teams consisted of neighbors during special events (e.g., house or barn raisings). After several days and a few hearty meals, the team was disbanded. The only means of noninteractive communication would be by a paper letter. The need for rapid information transfer was low. Life and communication traveled at a much slower rate.

### The 1900s

In the 1900s, after the industrial revolution, our communications structure changed. Companies began to grow large with many workers. They formed supplier-buyer relationships. The communications circle of influence grew to encompass several surrounding communities. The invention of the telephone let people located in geographically dispersed areas communicate freely. The advent of the automobile made it easy for people to travel long distances to report to work. As a result, worker coordination had to be managed by calendars and schedules. Deadlines and other milestones became important to ship a product to market on time. Teams were assembled by departments, with workers sitting near each other. Extended teams might exist, but most interaction was within a team that lasted a long time. The outputs of teams were products or services.

The centralized workplace increased communication lines between workers. Rather than working with only a handful of people, many workers needed to interface with hundreds of other workers. Real-time communications slowed the progress. With the advent of answering machines, fax machines, and E-mail, communications could be shifted to another time. As a result, the transfer of information shifted to a higher level of speed and volume.

The pace of life and information flow in the 1970s was appreciably slower than the pace we must keep now. It almost seems a remote idea to live without fax machines, voice mail, E-mail, and cellular phones. Because information and communications flies around at tremendous speed, the need to quickly process them increases as well. We understand that if we do not deal with issues and other decisions quickly, our competition will. The tools that we use to process the information must keep pace with the speed at which we live.

**Approaching 2000 and Beyond**

As we approach the next century, the manner of work continues to change. More and more of the team's effort centers on information and information flow. Teams are dynamic and can be formed by simply creating an electronic mailing list or constructing a workflow process. Teams are no longer exclusive to a single company. Teams often consist of internal employees and personnel from another company. Partnering with other companies for information exchange is the norm. Teams extend across time zones, countries, and expertise areas.

The distributed nature of partnerships causes work to occur 24 hours a day. Many times, the mobile worker will complete a task very late one night, only to wake up early the next day to see that a colleague has commented and needs an immediate response. The morning newspaper with coffee may be replaced with a computer and espresso.

The flow of information is very complex. Scheduling of people is difficult because of the many meetings already engulfing people's time. As a result, most of the communications must be done in a nonreal time or time-shifted mode. Groupware facilitates the exchange of documents and responses. Calendaring programs help to coordinate people's schedules. E-mail, fax, and voice mail keep team members connected and in the information loop. Mobile workers require these tools as they travel so that they can remain productive. Client/server technology should help vendors produce products that aid the mobile worker to communicate more effectively and remain informed, even while moving around.

LEVERAGING THE ELECTRONIC MESSAGING INFRASTRUCTURE

## HOW MANY MOBILE WORKERS ARE THERE?

Mobile workers comprise a sizable portion of the market. They have been divided into various segments consistent with their travel characteristics. Approximately 55% of all mobile workers are situated in buildings: 18% are in metropolitan areas; 14% on campuses; 12% travel domestically, and 1% travel outside the country.

Messaging for these various segments is similar in many respects, but very different in others. For example, in all cases, mobile workers labor in a disconnected mode most of the time. At the appropriate time, they connect to the network at the main office and upload/download new messages, call in for voice mail, and check for faxes. Depending on the location, connection speed and quality can differ. Certainly, those traveling abroad experience more complex connection scenarios if they must use international dialing codes and lower quality transmission lines.

So, what is the effect of client/server messaging on mobile workers versus the file-sharing systems of today?

## FILE SHARING VS. CLIENT/SERVER MESSAGING

Two messaging architectures are prevalent today. They are file sharing and client/server. The evolution of these architectures has followed the computing architecture prevalent for the day.

File sharing has been used since the days of the mainframe era. Client/server messaging has been around since the time that client/server computing hit the scene in the mid-1980s. To better understand the difference between the two architectures, it is helpful to look at the evolution of the architectures in layman's terms.

### Architectural Evolution

Computing, and thus messaging, has gone through three stages so far. They are:

- Mainframe-server
- PC client
- Client/server

A new form of computing is beginning to evolve, called *client/network*. It will be described later, but it is so new that its impact is largely unknown. Each method is determined by the location of the computing power and execution of programs.

**Mainframe-Server Model.** When computers were first invented, the main machine would contain the computing power and resources. This machine — the mainframe and eventually the minicomputer — performed all com-

## Client/Server Messaging and the Mobile Worker

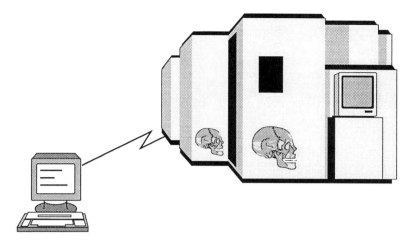

**Exhibit 28-1. Mainframe-Server Model.** *Note:* Processing intelligence performed on central server; client device has little or no processing capabilities.

---

puting tasks and stored all information. Users interfaced with the machine through a terminal that had no or very little processing power (see Exhibit 28-1). Through the terminal, users would request various actions to be performed. One of those actions would be electronic mail.

The E-mail post office typically would be one large database of messages with user indices pointing to the appropriate database record that represented a message for a particular user. The record could be shared by several users representing a message sent to several recipients. As recipients read the message, their individual pointer would be updated to reflect that they had read the message. As users deleted the message from their view, the message record would be maintained in the database store until the last user deleted the message, causing the record to be freed.

All processing of the message was done by the mainframe computer. It was difficult to enhance the messaging infrastructure because of the architecture. The database and terminal device would support only certain types of data. Such message types as fax or voice could not be handled because of the inflexibility.

Message content, being viewed by the user on a terminal attached to the mainframe, was sent to the terminal when the user opened the message. If the user was remote (or mobile), the message file was never stored on the remote device. Only the information was sent for viewing. Consequently, the user had to be connected to the mainframe to view messages. No offline message processing could occur.

# LEVERAGING THE ELECTRONIC MESSAGING INFRASTRUCTURE

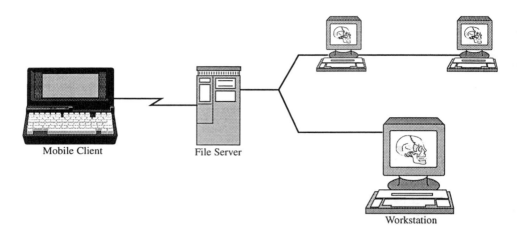

**Exhibit 28-2. PC Client Model.** *Note:* File server provides no processing of information; client provides all processing of information.

A primary benefit of mainframe computing is centralized maintenance and administration of computing resources. A primary disadvantage is that all computing is performed on one machine managing many users, resulting in possible poor performance and long wait times.

**PC Client Model.** The PC client shifted the computing power away from the mainframe or central server to a client. The low-powered terminal was replaced by a self-sufficient computer that was able to perform many tasks. Only one person at a time would access this computer. Therefore, all of its resources could be employed to perform the user's request without having to share with any other user. Because all the processing resided on the client machine, the server became a file server, where information could be stored and shunted to the client for processing (see Exhibit 28-2).

The evolution of computing technology permitted a more flexible architecture of messaging, but it still followed the model described earlier. Messages were stored in one central post office database, with user index files pointing to the appropriate message record. By now, the message structure supported message attachments consisting of word processing documents and spreadsheets.

Mobile users connect their client computers to the file server, typically through dial-up modems, and download message files to a post office residing on their notebook. By downloading the messages, they can work in a disconnected mode, lowering phone line costs. In the disconnected mode,

*Client/Server Messaging and the Mobile Worker*

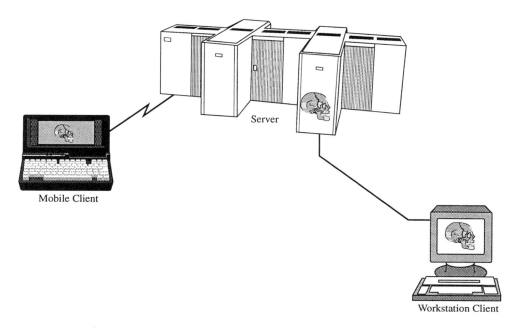

**Exhibit 28-3. Client/Server Model.** *Note:* Server processes information; client processes information; division of labor.

---

they can detach message attachments, create replies, and forward messages. They can delete messages from their local store and move messages into various folders.

Unfortunately, the processing is performed on the local store with no reflection of these actions in the main store on the file server at the office. Therefore, if the mobile user uses a desktop machine connected to the file server and main post office while in the office, the main post office will appear as if the user never processed any messages. As a result, mobile users opt to continue using their notebook computers and downloading messages to their local message store while in the office. A primary benefit of PC client computing is better performance of many user-oriented functions. A disadvantage is the distributed nature of data, resulting in more complex maintenance and administration.

**Client/Server Model.** Client/server distributes the processing across both the client and server platforms (see Exhibit 28-3). Certain functions are performed better in one place than another. For example, data presentation and computation of numbers is best performed on the client machine. Data storage and movement of shared information is best done on a server.

## LEVERAGING THE ELECTRONIC MESSAGING INFRASTRUCTURE

*Comprare and Contrast.* A very simple example will contrast the three architectures of computing. A user wants to copy a file that resides on the server (central store) to another location on the server.

In the mainframe environment, the user issues a copy command and the mainframe creates a copy in the desired location.

In a PC client environment, the user issues a copy command. Because the processing functions are performed on the client, the file (stored on the server) must be opened, and the data must be transferred across the LAN to the client, which sends it back across the LAN to the appropriate place on the server.

In the client/server model, the copy command is sent to the server, which performs the copy without transporting the data across the LAN. The result is a faster copy and more efficient use of the LAN. In this manner of computing, the client/server provides benefits in some major areas, overcoming the disadvantages of the two preceding forms of computing:

- *Division of labor.* Tasks are split across computing platforms to provide the most efficient use of computing platforms and LANs.
- *Specialization of tasks.* The computing platform can focus on specific tasks, doing those items well versus being a jack-of-all-trades. The net result should be better performing software, smaller memory footprints, and more flexible architectures.
- *Centralized and distributed maintenance and administration.* Critical information can be stored on a centralized site for ease of backup and tuning. Administration can be performed easily.
- *Client flexibility.* Other architectures force the messaging product to be a monolithic process where the user agent and the back-end processing are combined into one memory footprint. By necessity, the user agent is specific to the back-end implementation. Client/server permits the user agent to be separated from the back-end processing. Therefore, the user agent can be developed to handle the capabilities of the device the user would be using. For example, instead of requiring the user to always use a notebook computer, the user agent could be optimized to work through a telephone interface. The back-end store and processing can remain the same.

**Client/Network Model.** A new computing architecture is emerging with the advent of the Internet (or intranet) and popularity of the World Wide Web. Instead of the back-end store being a monolithic storage area, it may be spread across several machines distributed around the world (see Exhibit 28-4).

Several messaging systems permit embedding live hyperlinks within a message body. The recipient references the link that causes the information to be displayed. The client is responsible for rendering the informa-

*Client/Server Messaging and the Mobile Worker*

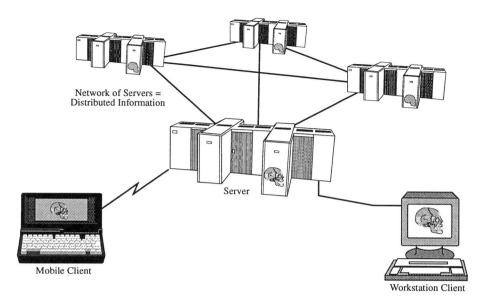

**Exhibit 28-4. Client/Network Model.** *Note:* Distributed networked servers process information; client processes information; information remains at original location or replicated.

---

tion into an acceptable form for the user's device, just as in the client/server mode. The back-end is responsible for linking the information and transporting it to the client. The effect of this type of messaging is not fully understood — the effect on the mobile professional is even more awesome.

### EFFECT ON MESSAGING COMPONENTS

Electronic messaging has evolved from routing simple text messages to complex structures that contain multiple types of data, active components, and other business communication types. Dr. Michael Zisman of Lotus Development Corp. describes the next generation of electronic messaging in his paper, "Fourth Generation Messaging" (http://www.lotus.com/corpcom/219e.htm).

First-generation messaging permitted short, static plain-text messages. Second-generation messaging added the capability of including binary attachments. Third-generation (or current) messaging provides enhanced support for rich text, including graphics attachments.

Fourth-generation products broaden messaging to fully support enterprise and mobile messaging with object stores, security, mobility support, manageability, and directories. The fourth-generation messaging infrastructure more fully supports the mobile worker in four key areas: media, transport, command-control, and applications.

**Media Modules**

Client/server messaging systems take advantage of newer development technology. Object orientation isolates information processing into a small, well-defined module. This module can focus on supporting only tasks it needs to manage, display, and manipulate. As a result, the server can provide an interface to the client portion of the process that is consistent, while the foundation modules handle the data. Consequently, more data types can be supported.

For example, file-sharing systems were retrofitted to support fax and voice messages in addition to the original text messaging. The messaging format typically resulted in an E-mail message with voice or fax attachments. The entire program was required to change in order to support the new types. The new version forced users to upgrade their software to take advantage of the new message types. For large corporations, the upgrade would result in a large investment in money, installation, administration, and training.

Client/server permits the server to support new media types without affecting the client portion. Exhibit 28-5 is a graphical representation. Upgrading to newer media types will be less costly. Some changes may need to occur in the client software, but their effect should remain minimal.

**Transport**

File-sharing systems require mobile users to download entire messages to their notebook before viewing/processing them. Some systems have

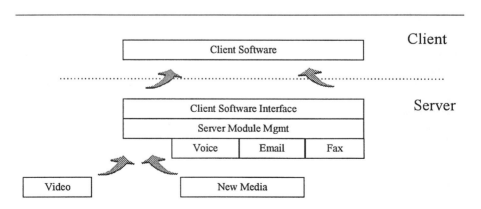

Exhibit 28-5. Client/Server Interaction with New Media Types.

tried to aid the users by permitting a selective download of messages while traveling. The system offers a summary of the unread messages (usually the subject, originator, size, and priority), allowing the users to select only those to be downloaded at that time. Using rules and filters, the users can semiautomate the selection process so that messages below a certain priority level or larger than a particular size are not accepted. When the users return to the office and connect to the post office with high-speed connections, they can download those messages.

Some of the newer media types require larger than normal message sizes (average E-mail message size is 1 to 2K bytes). For example, voice messages require approximately 3K bytes per 1 second of speech. A 10-second message (caller leaves name and telephone number only) requires 30K bytes. One page of a fax averages around 50K bytes. Both of these types of messages may be very important and would demand longer download time before the recipient can begin viewing them.

In addition, even though notebook machines support larger disk drive capacities, the capacity seems to fill more quickly than desktop machines. Large messages are an unwelcome sight. If the user must refer to the message several times, the message must be retained rather than deleted.

Client/server divides the labor of message processing across the client and the server. In the case of voice messages, the server would not download the voice message to the client, but would pass a tag to the client. To the user, it appears as a regular message that had been downloaded to the client. When the user views the message, the client could request connection to the server (or do it automatically, depending on configuration options). The media module would then stream the voice message to the client so the user could listen to it. The voice message would not be stored on the notebook computer, but on the server. This division of labor eliminates the user needing to delete the message to preserve disk space.

The client software is responsible for rendering the message contents into a form appropriate to the client device. For example, if the message is an E-mail message and the user is currently using a notebook computer, the message is displayed on the screen. If the user accesses messages from a public telephone, the client software would render the message to speech so the user could listen (the user could configure the client software so that only portions of the message are played and opt to listen to the entire message if necessary). File-sharing systems require client software to be tied to the message store — making flexibility very difficult.

To summarize, client/server messaging architectures will permit a variety of client devices to access the message store. The client software will decide if it is appropriate to download the message content to the client device or simply process it while residing on the server. Meanwhile, the server media module remains unaffected and simply provides the message content. As new client devices develop (e.g., PDAs, two-way pagers, and cellular phones), the system can easily add support for them. As new media types develop, server modules can manage the information. Client software plugs can be downloaded to the client device to support the new media type. User training is minimized because the device interface remains essentially the same and user actions are familiar.

Client/server architectures efficiently use the transport medium as well. Mobile workers may use high-speed, high-bandwidth channels in the office, modems at home or in a hotel, or cellular and packet radio wireless services in the field. The system understands the connection type and manages the message transport appropriately. The transport criteria may be user-tunable for cost savings or by the administrator for security and connection facilities optimization.

Client/server messaging can support replication of information. Rather than the information residing on one server, the information can be copied from one server to another in a synchronized manner. If a mobile user frequents particular field offices, his or her messages can be replicated to servers located within those offices. Rather than connecting to his or her main server, the user can attach to the local server by a high-speed, high-bandwidth connection for messaging needs.

**Command and Control**

Under the current file-sharing architecture, a user's actions on messages in the local store are not reflected in the main post office (e.g., if the user downloads messages to a notebook computer while traveling, reads the messages, replies to some, and deletes others). The main message store does not record or reflect those actions. If the user were to use a desktop machine when in the office, the messages already processed would appear as unread messages.

Client/server messaging systems replicate the actions of the user in the main post office. Messages that have been read are recorded as being read. Messages that were deleted are deleted. If the user were to process messages while at a remote office (a home could be considered a remote office), actions taken at the remote office would be reflected in the main office's post office.

Regardless of the access device (e.g., PDA, notebook computer, two–way pager, or public telephone), message processing actions would be reflected appropriately. In other words, the messaging system acts the way the user thinks. Delete messages are deleted. Read messages are read. Replies are recorded accordingly. This replication of actions is important in many areas beyond simple messaging. Calendar functions must be recorded appropriately for the calendar information to be valid. Workflow actions need to be reproduced faithfully; otherwise, downstream actions may be affected.

## Applications

Application support is critical to keeping mobile workers in the information loop and productive team members. While in the office, mobile workers can participate in workflow processes, messaging, calendaring, and other collaborative work. As soon as they step out the door, mobile workers are severed from most of these applications except for messaging. Keeping calendars synchronized becomes a nightmare, receiving workflow tasks is difficult, and collaborating requires long transfer times when sharing large documents. Aside from messaging, client/server messaging can more easily support applications that are related and can use the same backbone infrastructure as messaging. These applications leverage the investments made by companies.

**Workflow.** Often times, mobile workers are key personnel in business processes. The mobile worker may be a manager with responsibility for authorizing purchases, employee benefits, and work projects. He or she may be the knowledge worker responsible for the next task in the process. If the mobile worker is out of the office during a critical period of the process, downstream tasks may be delayed. Especially in today's environment where one day could mean an advantage over a competitor, mobile workers must be tied into the information stream so that decisions can be made, tasks accomplished, and projects finished.

Fortunately, many workflow systems use the messaging backbone to transport work tasks to the mobile worker. The worker can accomplish the task and send it along to its next stop. Using the advantages of the client/server architecture, workers can access the task flow from a variety of clients and have their actions reflected in the main message store.

**Calendaring.** Calendaring and scheduling has become a critical need for many information workers today. A current adage jokes that meetings are a practical alternative to work. It seems that people are running from one meeting to the next — some are local, some are remote. Being at the right place at the right time with the right equipment is increasingly difficult. Arranging a meeting with more than one co-worker is difficult and almost impossible if several are mobile workers. Fortunately, calendaring software helps.

The replication facilities of the client/server model and flexible interface options help the mobile worker keep electronic copies of schedules synchronized. A big disadvantage of many calendaring programs is the interface. They usually require a computer. The client/server versions could permit additional interfaces such as a telephone or PDA. The "instant on" of a PDA or telephones makes it easier to enter new meetings or changes than waiting 5 minutes for the notebook computer to boot.

Another advantage to the client/server model permits synchronization of calendars with other colleagues. Rather than having a monolithic calendar repository, several repositories might reflect a user's schedule through replication. Free-time searches can be accomplished more easily from several locations.

## CLIENT/SERVER FLEXIBILITY MEANS COMPLEXITY FOR ADMINISTRATORS

Client/server messaging provides the user with tremendous flexibility in access device, transmission medium, and processing capabilities. Providing the user with this added flexibility complicates the administrator's tasks.

The administrator must administer replication rules, replication sites, and additional transmission modes. Additional gateways and rendering engines must be configured so users can use a variety of devices and locations for message processing. Workstation and notebook computers may need new hardware to accommodate newer uses of messaging.

Understanding the ramifications of adding new media types, access methods, and applications falls on the administrator's shoulders. Client/server, as many companies have already experienced, is much more complicated than anything else we have known. Effects on resources are large. Careful planning and meticulous study is required for successful implementation.

### Mobile User Support

**Helpdesk Activity.** Successful mobile workers need support from those back at the office. A helpdesk staffed with knowledgeable people aids the mobile worker in overcoming many obstacles encountered under the worst of conditions. Notebook computers get dropped, modems die, phone cords get yanked, connections are lost by dismantling a hotel phone, and other problems regularly occur. The helpdesk staff must thoroughly understand the software and its configuration options in case parameters are destroyed. The mobile worker should not be the expert for the messaging software used.

Client/server messaging may cause additional trouble for the helpdesk staff. First, they need to understand how client/server messaging functions in the desktop world, how it differs for the mobile worker (possibly a separate message store, replication rules, and filters and agent activities), and how messages flow to the client and back to the server. The helpdesk staff must be trained specifically in client/server messaging and also remote and mobile worker issues.

**Training.** For the most part, the mobile worker needs the same software training as the desktop user. The additional issues that should be illustrated are the needs for replication, multiple location support, software configuring for a new location, and some of the concerns listed earlier to help offload the need for hotline support. By training workers about mobile issues, they will be able to deal with situations outside normal work hours (e.g., traveling abroad in drastically different time zones).

**Software Additions and Plug-Ins.** The client/server messaging infrastructure will support many more types of communications media than the file-sharing systems. As a result, as the server begins to support more types, the client must as well. Problems arise concerning upgrading the mobile worker's machine to support the new types. In several cases, upgrading software requires the worker to either bring his or her machine to the office for upgrading, mail it back to the office, or upgrade the software him/herself.

A recent method has been employed on the World Wide Web using plug-ins. These programs are viewers for the new type of communication. For example, plug-ins exist for presentations, audio messages, and graphics. When the browser encounters a data type it does not know how to display, it requests the user to download some plug-in applet.

In the case of client/server messaging, the server should know if the client needs the additional software and download it before the message so that that portion of the message can be viewed easily. The applet download capability does not currently exist, but some functional ability along those lines must exist to fully support the mobile worker.

**Data Backups.** A key concern the mobile worker has is data backup ability. In client/server messaging, a copy of the message is stored on the server until the user deletes the message. In most cases, the user will delete the local copy of the message and the deletion action will cause the server copy to be deleted on the next replication. This two-stage approach to a message storage is beneficial to the mobile worker. In cases where the local copy is corrupted, the worker could delete the local store completely and replicate it from the server. If the server version of the message store is destroyed, the worker could restore it from the local copy.

## Security

Client/server messaging can provide additional security over the file-share model. A company can institute a policy that says that all messages are stored on the server until the user decides to view them. The user would have to be online to view the messages. Although inconvenient to the user (not to mention potentially expensive), the data is safely stored. The major security problem is the fact that assets are located outside the corporation. If a notebook computer is lost or stolen, valuable information could be lost.

The information need not reside in the message store only. External access to corporate information is necessary for the mobile worker. Typically, access by mobile workers is protected by passwords. In some cases, for the convenience of the worker, the passwords are configured into the software. This storage of passwords in the software rather than requiring the worker to enter it when logging in is dangerous. Anyone with access to the notebook computer (e.g., if it is left behind in a hotel room while the worker is at a meeting) can easily enter corporate systems without much difficulty. By forcing the mobile worker to enter the password at logon time, the corporation has greater protection for its data. In addition, the software on the notebook computer should be protected by passwords so that sensitive information cannot be accessed illegally.

**SUMMARY**

Client/server messaging is the next evolution in electronic communications. The need for a flexible messaging architecture is apparent in the mobile worker's work style. Working from many locations using several access devices demands that the message server be disconnected from the client. Increased productivity while traveling is the goal of the new architecture. Mobile workers must stay informed and connected with other colleagues regardless of their location. Their ideas, opinions, and decisions are critical to the continued success of their businesses.

Client/server technology aids the mobile worker by making the messaging infrastructure more flexible. The infrastructure can handle new forms of communication and advanced uses of communications paths (e.g., calendaring, workflow, and collaborative work). The separation of client and server permits programs to perform specific functions. The division of labor makes more efficient use of processing resources and transport mediums. The flexible and powerful features of client/server messaging permits new uses of the system. Such applications as workflow processing and calendaring are only the beginning. As more is understood about client/server messaging, additional applications will be added.

# Chapter 29
# Telecommuting
*David A. Zimmer*

To many, telecommuting represents the newest state of the art for working environments. To others, it represents a method of working that they have been doing for years. Both perceptions are correct. Telecommuting is a combination of old techniques aided by advances in technology, work philosophies, and the growth of the Information Age.

About 100 years ago during the Industrial Revolution, the style of work changed from an agrarian culture to an industrial society where jobs were located at a central site. In the agrarian culture, each family was a self-contained business; their labor was done just outside their windows. The Industrial Revolution created a consolidation of effort to one location; workers were required to go to "work."

Today's information-based businesses are able to bring the work to the worker. Information-based businesses do not require a centralized location for work to be accomplished. Work is something that people do, not a place to which they go.

Advances in computer and telecommunication technologies permit people to carry large quantities of information and tremendous computer power with them. Merely 20 years ago, we relegated the same data and processing power to large, raised-floor, glass rooms. The telecommunication advances permit transmission of other information from remote computers halfway around the world in a matter of minutes; 10 years ago, the same task required dedicated lines or extremely long wait times.

Other, nontechnological factors favor telecommuting. Government mandates for higher-quality air have required companies to reduce employee travel. Employee retention and recruitment at minimal costs nudge companies to consider alternate forms of employment. Employees' desire for better family relations and higher-quality lifestyles push them toward telecommuting.

# LEVERAGING THE ELECTRONIC MESSAGING INFRASTRUCTURE

## TELECOMMUTING DEFINED

Telecommuting is defined as the practice of an employee performing his or her normal duties from a remote location, typically home, on a full- or part-time basis.

This definition distinguishes between the person who takes work home for an evening or opts to work from home on a particular day, and a person who follows a schedule of working in the main office certain times and from home or remote locations on other days. The formalization of the work location schedule becomes an agreement between employee and employer. Company policies cover insurance, working relationships, employee and employer responsibilities, and office setup.

Although the term *telecommute* is beginning to be used to describe the occasional work-from-home session, telecommuting requires a formal agreement between employer and employee. For example, company policy statements must consider which party supplies the equipment to be used (e.g., computer, fax machines, copiers, extra phone lines), who will pay the recurring expenses (e.g., additional electric bills, higher or additional phone bills, special requirements such as ISDN or other broadband lines), equipment protection (e.g., company equipment policy or a rider on the homeowner policy if located in a home office), and other issues such as injury protection, work hours and schedules, and job responsibilities and milestones.

In some cases, the employer has determined that the employee will be saving enough in commuting expenses to more than offset the additional costs of such items. In other cases, the employers have split the expenses or will pay for "business-only" expenses. Other companies have a more liberal policy that pays for all additional expenses. Exhibit 29-1 is a list of alternative office strategies, and their advantages and disadvantages, that provide flexibility to accommodate the many forms of telecommuting and the needs of the telecommuter.

## TELECOMMUTING DEMOGRAPHICS

Because of regulation, employee requests, and need for reduced office real estate, many companies are turning to telecommuting. According to a Gartner Group study, more that 80% of organizations will have 50% of their staff engaged in telecommuting by 1999. Company size is not a factor in telecommuting. Other notable statistics on telecommuting are that:

- Productivity increases by an average of 20 to 30%
- Average work time increases by 2 hours a day
- Annual facility savings, per telecommuter, are $3000 to $5000

### Exhibit 29-1. Alternative Officing Strategies.

| Strategy | Benefits and advantages | Disadvantages and pitfalls |
|---|---|---|
| Reengineered space Flexible schedules | Takes advantage of segment of labor pool interested in part-time work<br>Work is accomplished when worker is most productive | Problems may arise that require attention when employee is not in office<br>Requires employee to structure time and work differently |
| Modified offices (refined standards to improve productivity and efficiency) | Reduced space required, better utilization<br>Less hierarchical space distribution<br>Increased flexibility for employee moves | Organization cannot use office/workstation size to indicate seniority<br>Inhibits customization of office |
| On-premise options<br>Shared space (two or more employees sharing single, assigned space) | Better utilization of space<br>Increases headcount without increasing space required | Employees may be reluctant to give up own space<br>Requires employees to work closely with one another |
| Group address (designated group or team space for specified period of time) | Team orientation of users ensures high, ongoing utilization rate<br>Encourages interaction between team members | Team size may create space shortage<br>Difficult to manage turnover of spaces among users<br>Requires accurate projection of team size |
| Activity settings (variety of work settings to fit diverse activities) | Provides users with choice of setting that best responds to tasks<br>Fosters team interaction | Requires advanced technological equipment<br>File retrieval can be problematic |
| Free address (workspace shared on a first-come, first-serve basis) | Maximizes use of unassigned space<br>Minimizes real estate overhead<br>Suitable for sales and consultant practices | Access to files can be problematic<br>Probable scheduling conflicts<br>Substantial investment in equipment |
| Hoteling (reserved workspace) | Accommodates staff increases without increase in facility and leasing costs<br>Can result in upgraded office amenities | Storage can be problematic<br>Probability high for scheduling conflicts |
| Off-premise options<br>Satellite office[a] (office centers used full-time by employees closest to them) | Lowers rentable cost/sq. foot | Remote management a challenge<br>Employees may feel disconnected from organization |

## LEVERAGING THE ELECTRONIC MESSAGING INFRASTRUCTURE

**Exhibit 29-1. Alternative Officing Strategies.** *(Continued)*

| Strategy | Benefits and advantages | Disadvantages and pitfalls |
|---|---|---|
| Telecommuting (combination of home-based and office workspace) | Transportation and real estate costs reduced<br>Improved quality of personal life<br>Potential increases in productivity | Inadequate home office equipment<br>Staff interaction is reduced<br>Home office must be quiet |
| Remote telecenter[a] (office center located away from main office, closer to clients) | Achieves Clean Air Act mandates<br>Fosters productivity and employee loyalty through improved family life | Clear guidelines for and support of supervisory staff are required |
| Virtual office (freedom to office anywhere, supported by technology) | Increases employee productivity<br>Potential increase of time with clients due to reduced commute time<br>Reduced space and occupancy costs | May impact employee connection to organization<br>New criteria for evaluating performance required |

[a] Telework centers and satellite offices are very similar. They make take the form of noncompany-owned office space shared by several companies. The space provides shared services, such as copy centers, rest rooms, and reception areas. Satellite offices may be company-owned and not shared with other companies, but are based closer to employees' homes.

Source: Compiled by consulting firm Hellmut, Obata & Kassabaum, Inc. (415) 243-0555.

---

- Average annual recurring costs are $1,000 to $1,500
- Average one-time startup costs are $2,000 to $4,000

## TELECOMMUTING BENEFITS AND WORRIES

Telecommuting affects several parties: the telecommuter, the manager, and the company. Properly planned telecommuting programs can maximize the benefits for each party while minimizing any disadvantages.

In postmortem studies conducted concerning successful programs versus those that failed, the key ingredient was sufficient planning before instituting telecommuting. Although employees may have been working from home on occasion, formalizing the process protects each party and sets the proper expectations.

The first step of the planning process is understanding the benefits and disadvantages of telecommuting. Identifying each will facilitate proper guidelines and procedures.

Telecommuting involves individual employees and companies. Thus, each telecommuting program will be unique because of company culture, type of business, work tasks to be performed, and many other factors. The benefits and disadvantages discussed here are based on general findings from other telecommuting programs. Specific programs may or may not experience all advantages or disadvantages.

**Impact on the Individual Telecommuter**

Telecommuters benefit from a telecommuting program in many ways. These benefits are frequently stated as reasons for telecommuting. The benefits are not always quantifiable, however, although emotional "feel better" reasoning has been shown to produce positive results in employee morale, work productivity, and overall quality.

**Less Commute Time.** Less commute time is an obvious benefit. Studies have shown that the average worker spends at least 1 hour commuting one way each day for an average total of 2 hours a day. If there are approximately 250 workdays per year, then the average worker spends 500 hours, or 12.5 workweeks, in the car annually. If half of those hours were eliminated by telecommuting, the employee would work an additional 6 weeks per year.

Studies have shown that telecommuters typically work the hours at home that would have been spent commuting. Is there any wonder why telecommuters are more productive?

Besides the benefit of additional work time, the telecommuter benefits from less wear and tear on the automobile, lower monthly gasoline costs, lower insurance costs, less chance of accidents, and cleaner air. The telecommuter misses the daily stress-filled grind of rush-hour travel and therefore can enter the workday more relaxed. A more relaxed person is more productive and works better.

The benefits to the individual telecommuter will vary depending on the amount of time telecommuting and the distance traveled. The measurements for these benefits are both tangible and intangible: lower commuting expenses, greater available work time, and an overall feeling of less stress.

**Family Focus.** For many commuters, the average 8-hour workday typically turns into 12 hours away from home and family. Between the commute time, sitting in traffic, and spending extra hours at the office, the workers cannot spend as much time with their families. Also, the workday is scheduled during the "activity hours" of the family so that by the time the commuter arrives home, the children are off to bed and the spouse is too tired to appreciate any quality time.

Exhibit 29-2. Survey Results: Resolving Work/Home Conflict Issues Through Flextime.

| | Employees on flextime (%) | Employees on rigid schedules (%) | Difference (%) |
|---|---|---|---|
| I find carpooling arrangements: | | | |
| a. very difficult to arrange | 6.6 | 8.3 | −1.7 |
| b. difficult to arrange | 6.6 | 36.7 | −30.1 |
| c. easy to arrange | 60.0 | 43.3 | −16.7 |
| d. very easy to arrange | 26.8 | 11.7 | 15.1 |
| Making arrangements for childcare during the workweek is: | | | |
| a. very difficult | 0.0 | 0.0 | 0.0 |
| b. difficult | 3.8 | 20.0 | −16.2 |
| c. easy | 46.2 | 33.3 | 12.9 |
| d. very easy | 50.0 | 46.7 | 3.3 |
| My opportunities to spend time with my family during the workweek are: | | | |
| a. very inadequate | 2.0 | 3.2 | −1.2 |
| b. inadequate | 2.0 | 24.2 | −22.2 |
| c. adequate | 69.4 | 67.7 | 1.7 |
| d. very adequate | 26.5 | 4.8 | 21.7 |
| My opportunities for taking care of my personal business during the workweek are: | | | |
| a. very inadequate | 2.0 | 9.7 | −7.7 |
| b. inadequate | 2.0 | 27.4 | −25.4 |
| c. adequate | 57.1 | 51.6 | 5.5 |
| d. very adequate | 38.8 | 11.3 | 27.5 |
| My opportunities for off-the-job recreation during the workweek are: | | | |
| a. very inadequate | 4.6 | 1.7 | 2.9 |
| b. inadequate | 2.3 | 18.3 | −16.0 |
| c. adequate | 53.5 | 56.7 | −3.2 |
| d. very adequate | 39.5 | 23.3 | 16.2 |

Flexible work schedules help people manage their family and work lives. Exhibit 29-2 shows the differences between telecommuting and commuting employees and their ability to balance work and home conflicts.

Telecommuting permits the person to eliminate the extra hours commuting, shift work schedule around "family time" such as when the children return from school or the other spouse returns from work, and enjoy quality time with the family when it is most appropriate. The increased family cohesiveness and morale is the measurable factor of this benefit.

**Sense of Control and Freedom.** Many telecommuters report the feeling of a sense of control over their time. With fewer distractions, such as impromptu meetings and interruptions, the telecommuter controls his work schedule.

The telecommuter determines the start and stop times of work, if the work period is one contiguous time block or split across various blocks of the day, and whether starting in the early morning or working well into the night is more effective.

Scientists have studied the circadian rhythm, the natural activity cycle lasting 24 hours, of individuals and have found that not all humans work most effectively between the hours of 9 a.m. and 5 p.m. Energy levels rise and fall according to this rhythm. For some, their best or highest level of energy is during the morning hours. Others are at their optimum energy level during the afternoon; still others have the most energy during evening hours.

The telecommuter can take advantage of the natural rhythm to be most productive, using times of highest energy to tackle the toughest part of their jobs and using the lower-energy times for more mundane tasks.

Telecommuters obtain a greater sense of freedom because they are not under the scrutiny of the manager. While at work, the 9-to-5 office worker is less likely to run an errand in the middle of the day. If that is the only time it is possible to do the errand, the worker may feel urgency and stress to return to the workplace as quickly as possible. The telecommuter's schedule is more flexible and can accommodate the errand run without the added stress. For example, the bank run can be done other than at the noon time rush of other workers going to the bank.

The sense of control and freedom provides the telecommuter with a sense of being trusted and a valued employee. No longer is the manager watching over the employee to make sure the work is accomplished. The work program is switched from a time-based measurement (e.g., worker has spent 40 hours in the office this week, therefore the work requirement has been met) to a project/task-based measurement (e.g., project A was finished within the deadline necessary).

The telecommuter has an additional advantage because of the project-based work schedule. The project has been broken into milestones to be met with the requisite reviews. As each milestone is met, the telecommuter sees the progression to the ultimate goal: the completed project. Meeting the milestones provides the telecommuter with a feeling of satisfaction and accomplishment. It is puzzling that management has not switched to this style of management, even for those who work in the office.

**Control over Communications.** Telecommuters gain control over their communications. Because of modern technology, telecommuters select the time and style of communicating.

Voice mail and E-mail communication can be done during their working schedule, which may be different than other workers. The ability to retrieve and process E-mail from anyplace at any time lets the telecommuter maintain productivity. By attaching spreadsheets, word processing documents, and reports to their E-mail, they can be as effective as office-bound colleagues.

The flexibility of today's LAN-based E-mail systems lets the telecommuter work in a remote (or mobile) mode. Essentially, remote mode means the person need not be connected to the LAN server continuously, but can work offline preparing, replying to, and otherwise processing messages. Once new messages are to be sent, the telecommuter instructs the computer to connect and upload the messages.

Voice mail is very similar to E-mail. The remote worker can either access the voice mail system at the office via phone or have all calls forwarded to the home line. If the message is stored on the office voice mail system, the person can reply to the messages, send new ones, and forward others as if he/she were in the office. Because the voice system is employing messaging, the recipient need not be there to receive the message and can process it at his or her convenience rather than always being "on call."

**Less Interruptions.** Telecommuters suffer less interruptions to their workflow than commuters. By nature, there are less people around to cause interruptions, such as casual conversations in halls and offices, phones ringing, other people's discussions, and ambient noise. Although the telecommuter can be interrupted by the occasional telephone ringing at home, the number of calls is usually diminished. The telecommuter gains greater spans of time that can be devoted to concentration. As a result, critical idea development is not hindered and can be thought through to completion.

**Out of Sight, but Not out of Touch.** The telecommuter may be out of colleagues' sight, but they need not be out of their reach. Workflow and other groupware products permit the remote worker to participate. Using the E-mail infrastructure for information transport, the workflow software passes tasks to the telecommuter. After completing the tasks, the worker presses the "send" button directing the software to wrap the results and place it into the E-mail stream.

Calendars and schedules can be synchronized in similar fashions. Changes to calendars, requests for meeting rooms and other equipment, and searching for free time can result in a message sent via E-mail. As a result, the telecommuter can be scheduled for meetings just as any office-bound colleague. Ad hoc or spur-of-the-moment meetings can accommodate telecommuters via audio- or videoconferencing.

Because the telecommuter can be so well connected via the various communication technologies, he or she remains in the communication loop that is necessary to stay productive.

**Impact on the Manager/Employer**

Just as the telecommuter benefits from telecommuting, the manager gains by having clearly defined projects, established milestones, and set work expectations.

**Employee Recruiting.** Employees have changed their focus from a "whatever is good for the company is good for me" mentality to more of one that states, "whatever is good for me is good for the company." Employees with outstanding reputations and excellent skill sets may be recruited by a company, but the employee may not want to move to the company's location. Rather than lose the recruit's skills, companies have begun to offer telecommuting as a benefit. The employee works from home with regularly planned visits to the company site. Companies have seen this arrangement work successfully with consultants and so have instituted it with employees as well.

**Employee Retention.** As the adage goes, "Good workers are hard to come by." Managers find it hard to assemble teams of quality workers to meet project objectives. Several factors make it hard to keep such a team together once it has been developed.

The dual-income family causes a team member to move because the spouse's job changed. The introduction of a child into a family causes shifts in work patterns to either part-time or flextime. Company location moves may require employees moving, which could result in some employees staying behind. Regardless of the reason, the manager may continue to keep the group together by instituting telecommuting.

**Increased Documented Communications.** Because the opportunity for meeting face-to-face with a telecommuter is less, managers must use other forms of communications. In many cases, the manager will use E-mail. Its use provides a history trail of directives, discussions, and other information flow. The documented nature helps in clarifying ideas, lowers miscommunications, and aids in settling any disputes.

**Productivity.** Managers shift from time-based management to project-based management. Rather than counting the hours employees are present, the manager can more easily see the progress toward completion dates by seeing the milestones being (or not being) met. Adjustments to work projects can be done sooner, keeping the project more on track. As a result, more projects are completed on time, more planning and handling of issues are done proactively, and less "firefighting" or crisis management needs to be performed. Overall, the group's productivity increases.

**Lower Absenteeism.** Managers have reported that sick days taken by telecommuters are lower than the number taken by commuting employees. Several hypotheses have been proposed to explain this phenomenon. Some thoughts include less stress, less contact with infectious people, and work time that is broken into short periods during rehabilitative times. Another reason is that some people will work at home even when not feeling their best, but would not have gone to work that day because of the illness.

## Impact on the Company

**Clean Air and Environmental Regulations.** Companies are being forced to meet certain regulatory guidelines in order to positively affect the environment. The automobile, especially the idling car stuck in traffic, has become the number-one cause of pollution today. By reducing the amount of traffic on the roads, pollution should go down.

Several methods eliminate car usage, such as mass transit, car pooling, or telecommuting. All are necessary to overcome the pollution problem. According to calculations performed by Jack M. Niles, a noted telecommuting expert, an organization with 16,000 telecommuters working from home an average of 1.4 days per week reduces annual pollution on the order of:

- 6,150,000 pounds of carbon monoxide
- 380,000 pounds of nitrogen oxides
- 1,150,000 pounds of unburned hydrocarbons
- 26,000 pounds of particulates

Exhibit 29-3 illustrates the annual levels of reduced car miles for the Los Angeles CMSA (Combined Metropolitan Statistical Area), an area in southern California that includes almost half of the state's population. Half of these miles represent automobile cold starts — the most polluting phase of car use. To correlate this information with air pollution reduction, using data from California's South Coast Air Quality Management District, telecommuting expert Jack M. Niles further examined the effects of telecommuting. In 1991, the pollution contributed by cars was measured to be 1,580,000 tons of carbon monoxide; 221,000 tons of hydrocarbons; 243,000

*Telecommuting*

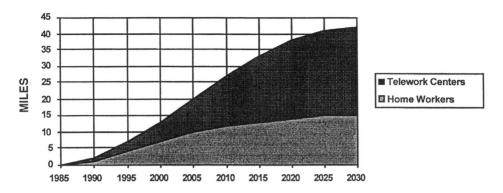

Exhibit 29-3. Annual Levels of Reduced Car Mileage.

tons of nitrogen oxide; and 20,000 tons of particulates. Following the projected telecommuting trends, Niles estimates pollutant reduction by the year 2000 of 19, 23, 8, and 4%, respectively. Exhibit 29-4 shows annual tons of pollutants eliminated.

**Office Space Reduction.** Companies are lowering their need for office space by instituting telecommuting. The telecommuting employee may work from home either full-time or part-time. If the employee works from home full-time, obviously the company would no longer need to supply office space for that employee. In those cases, the company might reserve some office space for those times when a telecommuting employee reports to the office for meetings or whatever.

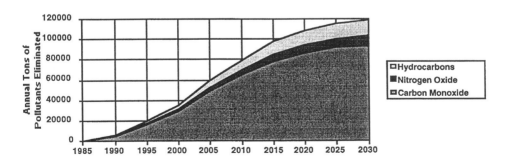

Exhibit 29-4. Pollutant Reduction Trends.

Other companies employ "hoteling," or a program of sharing office space and desks between employees for part-time telecommuters. Two part-time telecommuters might share the same desk on opposite days. For example, one telecommuter might use the desk Mondays and Wednesdays while the other telecommuter uses the desk Tuesdays and Thursdays. Friday is left open for the times when one of them needs to report to the office for an extra day. Each telecommuter is responsible for clearing the office of personal items at the end of the workday.

## SKILL SETS FOR TELECOMMUTERS

A telecommuter must possess certain skills to be successful. Managers can usually determine which employees might best suit telecommuting. Typically, these same people are those that are already performing well within the company. The same traits that make them successful inside the company will make them successful as telecommuters. Successful telecommuters are:

- Self-starters
- Motivated
- Organized
- Goal-oriented
- Disciplined
- Knowledgeable
- Trustworthy

A more in-depth discussion of personality types and determining those that might make good telecommuters is described in *The Telecommuting Handbook* (Electronic Messaging Association, David A. Zimmer, 1995). The handbook goes into great detail about personality typing using the Myers-Briggs Type Indicator battery.

Who would not make a good telecommuter? Studies have shown that employees new to the company or who are younger, shy away from telecommuting. They reason that they need to "establish" themselves in the minds of others, especially management. They are sensitive to career management issues and want to ensure they do not miss an opportunity because they were working from home. In addition, younger employees usually do not have the pressure of family life and do not relish the idea of working in an environment where there are no people. They get enough time by themselves since they may live alone. As a result, they typically turn down the opportunity of telecommuting.

## TELECOMMUTER TRAINING

Regardless of their skill set, telecommuters need to be trained. Areas to consider in the training sessions would be organization skills, time management, care and troubleshooting of office equipment (maintaining PCs and fax machines, upgrading software, reloading paper, and attaching wires), company policies, zoning laws, and insurance considerations. Some of the items to be covered are general to the telecommuting program, while others are specific to the telecommuter, such as zoning laws and insurance concerns.

Sociological and psychological areas should be addressed as well. The telecommuter going from the social environment of the office to the isolated arena of working from home needs to know the signs of isolation-related stresses, the workaholic conditions, and other maladies. These concerns hinder the telecommuter's effectiveness and productivity, as well as the physical and psychological well-being. Periodic reviews of these areas are highly recommended.

The telecommuter should be trained in basic hardware and software troubleshooting. More important, the telecommuter should be trained to communicate most effectively by telephone and electronic mail. Phones, faxes, and E-mail are the telecommuter's lifeblood for working effectively. Without proper skills, the telecommuter will suffer from information deprivation leading to isolation, which lowers the person's productivity.

## SKILL SETS FOR MANAGERS OF TELECOMMUTERS

The manager of telecommuting employees must possess certain skills as well. Unlike supervising people they can see, managers must work differently with the telecommuting employee. The manager must facilitate the telecommuter's continued involvement within the organization. The manager must set an example for commuting employees in interacting with telecommuters. The manager must determine that no special allowances or favoritism are shown toward the telecommuter because they are telecommuting. In short, the manager must have the following attributes:

- Leadership — knows how to extract the most from each employee regardless of the employee's commuting status.
- Trusting — believes the best of people and that the people will meet the deadlines imposed.
- Respectful — does not require the telecommuter to work 24 hours per day just because the work is always nearby; respects the timetable established with the telecommuter.
- Communicative — keeps everyone in the group informed of information regardless of their location.

## TRAINING FOR MANAGERS

The manager must be trained in managing telecommuters. The skills listed previously may be freshened with techniques to facilitate proper support of the telecommuter. Also, the manager must be trained to manage by objectives and tasks rather than the time-based manner of managing. It forces the manager to think differently and to see the value of being less concerned over the amount of time spent on the project and see that meeting the milestones are more important.

Just as the telecommuter needs to be trained in effective communications by telephone, fax, and E-mail, the manager should receive similar training. Keeping people informed and coordinating projects encompass most of a manager's time. Effective use of the communication infrastructure is vital, especially in a virtual group situation involving telecommuters.

## ESTABLISHING A POLICY FOR TELECOMMUTING

### The Company Policy

The company policy is an important document that clearly states the guidelines of the telecommuting program. It lists such things as program functioning, party responsibilities, insurance coverage for incidences, and equipment and information security. It lists any additional measures the telecommuter must perform, such as purchasing additional house insurance or car insurance or making home improvements to create a suitable work environment.

### Telecommuter Contract

The Telecommuter Contract describes the duties of the telecommuter, the work schedules, other provisions, and the timelines that must be met. The contract serves as an understanding point between telecommuter and manager. Typically not done between manager and employee, the contract clarifies the person's duties in order to minimize misunderstandings and support more productive work. (Several sample telecommuting agreement forms and rules are presented later in this chapter.)

## HOME OFFICE INFRASTRUCTURE

Telecommuting may be implemented by a company in several different manners: employees working from home (telecommuting), employees reporting to community sites (telework centers), and others (satellite offices). If the company chooses to implement community sites, the company typically outfits the office with equipment equivalent to the main office. Rather than address the office infrastructure for community sites, however, the discussion here focuses on the equipment and concerns necessary to outfit a productive home office.

Unfortunately, many employers consider the home office as a temporary or "second-hand" place of work. Therefore, the equipment "donated" for use in the home is antiquated by computer standards. While an employee can produce work using older equipment, faster machines with up-to-date functionality can make the employee more productive at a relatively small cost. With that in mind, outfitting the home office should be the same as if outfitting the employee for the office.

The best way to outfit the home office for productive use is to consider the equipment the employee uses in the office on a day-to-day basis. If the employee uses a computer in the office for generating reports, processing electronic mail, programming spreadsheets, or entering data, the employee needs a computer at home. Similarly, the employee will need other equipment such as fax machines, copiers, and modems.

While planning their home office needs, employees should consider three main areas:

- Communications
- Computing resources
- Home office environment

**Communications Technology**

The proper inflow and outflow of information is the lifeblood of telecommuters; without it, the telecommuter becomes isolated from the real world. As someone once said, "Home office plus communications to the main office equals telecommuting."

Communications come in three primary forms today, with several others emerging very quickly. The primary communication methods are voice, fax, and electronic mail. The emerging mediums that should be considered are videoconferencing and online connections.

**Voice Communications.** Voice communications account for the majority of interactions with others. As a result, the telephone has become a ubiquitous tool available to all.

The adoption of voice mail inside companies almost rivals the adoption of the telephone. Because the telephone has become such an important business tool, the employee should have an extra telephone line installed at home specifically for business use, especially if the employee's job or personality requires extensive use of the telephone.

The type of phone equipment can be left to the discretion of the employee. Many home office workers prefer speakerphones if they spend hours tuned into audioconferences. The hands-free mode of speakerphones lets the employee move paper, take notes, and even do other work while the conference call is in progress. If a speakerphone is going to be used, the employee is wise to buy a high-quality phone that provides clarity in speaker mode.

# LEVERAGING THE ELECTRONIC MESSAGING INFRASTRUCTURE

**Voice Mail or Call-Answering Services.** If the person uses the phone heavily for the job, voice mail purchased from the local phone company is another must. Unlike a simple answering machine, the voice mail provides coverage while the phone line is busy and presents a more professional image.

**Answering Machines.** Answering machines are a good substitute to voice mail if the added expense of the voice mail is not justifiable. Answering machines come in two formats today: digital and cassette tape.

Digital machines record the announcement message and incoming messages in computer memory, so the recordings are clear. The machines permit random saving and deleting of messages as opposed to the sequential recording of tape-based machines. If purchasing a digital answering machine, the employee should be aware of the maximum message time the machine will store. If more time will be needed, the employee must purchase a machine that supports the additional time.

Tape-based machines are the least costly and most trouble-prone. The tapes have a tendency to stretch and wear from usage, causing messages, announcements, or incoming messages to sound muffled or distorted. Messages cannot be randomly eliminated — either all the messages are saved or they are deleted.

**Call Waiting.** If the employee requires extensive phone usage, call waiting is another feature that should be considered. Call waiting is a service purchased from the local phone company. It "announces" another call on the same line that is currently being used. Call waiting permits the employee to "flash" the telephone hook to put one call on hold while answering the other. This feature is useful, especially when the employee is expecting important calls that should not be routed to the voice mail or terminate in a busy signal.

**Three-Way Calling.** Three-way calling is another feature that is useful. A service purchased from the local phone company, three-way calling permits the telecommuter to hold a phone conversation with two other parties in separate locations. If the telecommuter coordinates activities of others, three-way calling is a must.

**Call Forwarding.** Others may find locating a telecommuter by phone frustrating if they are not used to the telecommuter's schedule. The caller may have to dial a second number in order to reach the telecommuter. To eliminate the need to call additional numbers, a telecommuter can employ call forwarding, which directs all calls placed to one location to automatically forward to another location. By using this service, a telecommuter need only provide one number and the caller need not know the telecommuter's location.

For example, the telecommuter may need to run an errand during normally scheduled hours. Forwarding the phone line to a cellular phone, the person will not miss any calls. In cases such as telemarketing or customer service personnel, reaching the person is most important, while the person's location is not. Call forwarding permits location independence.

**Automatic Number Identification or Caller-ID.** Caller-ID (automatic number information), a service offered by the local telephone company, lets the recipient of a call see the telephone number of the calling party. This feature is particularly useful during normally scheduled work hours. The telecommuter can determine if the call is of a business nature or not. If it is not and it is not an emergency, the home office worker may decide not to answer the call. The opposite is true as well. The person may chose not to answer business calls except within the agreed-upon time schedule.

**Follow-Me Services.** New services permit calls to travel along several hops in order to locate a person. These services are called follow-me services. The telecommuter programs the telephone network to direct calls through a series of numbers (office phone, home office phone, and cellular phone) until either the telecommuter answers or the network returns the caller to a voice mail system. These services are beneficial to mobile telecommuters who are constantly moving from one location to another and when maintaining a preset schedule is difficult.

**Message Waiting Indicator.** Regardless of how simple or how fancy the telecommuter makes the call routing, a very important feature needed is the message waiting indicator (MWI). If a message is left but never retrieved, the systems become useless. MWI can take several forms, which should be chosen to suit the telecommuter's work patterns. The forms are stutter dialtone (an intermittent tone when first lifting the phone handset), LEDs on the phone or answering machines, and pages sent to pagers.

## Fax Technology

The fax machine has become quite pervasive within businesses. Even the smallest businesses have fax machines, and many fax machines are starting to penetrate the home/home-office space.

Fax machines are very important tools. Even with all the electronic advances made, much business is conducted by pen and paper. The fax machine lets telecommuters move information from one place to another quickly and cheaply. The home office will require fax capabilities. Faxing can be done in two ways: through fax machines and directly from a PC.

**Fax Machine.** The fax machine is great for the times information resides on a piece of paper and cannot be easily transcribed electronically. In addition, receiving faxes is a simple process.

## LEVERAGING THE ELECTRONIC MESSAGING INFRASTRUCTURE

**PC Fax.** The PC fax does not run out of paper while receiving a large fax. The fax can be viewed, rotated, and cleaned before printing. In some cases, the employee can use optical character recognition (OCR) to translate the fax image into text that can be imported into a word processor.

Regardless of which fax receptacle is chosen, an extra telephone line should be considered. The extra line will permit sending and receiving faxes while the employee is on the other line. If the extra line is not installed, the employee should use a line detector that automatically routes the call to the proper receiving device — telephone, answering device, or fax machine.

**Fax Mailboxes.** Fax mailboxes permit retrieval of faxes from anywhere. The fax mailbox acts like a receiving fax machine from the perspective of the fax sender. The fax images are stored on computer disk until the recipient decides to retrieve them. Using voice prompts (also known as interactive voice response, or IVR), the fax mailbox reports the number of received faxes and directs the recipient in downloading the faxes to a fax machine. Fax mailboxes come in two flavors: one-call systems and two-call systems.

*One-Call Systems.* One-call fax mailboxes require the person retrieving faxes to place the call to the mailbox from a fax machine. When instructed, the person presses the "Start" or "Send" button on the fax machine and the fax mailbox downloads all the faxes (or selected faxes, if the mailbox provides selective functions) to the fax machine.

One-call fax mailboxes provide simple call-charge processing and security. For example, the person retrieving the faxes pays for the call to check for messages and for downloading. If an 800-number is used to retrieve the faxes, then all call charges are billed to the 800-number for easy tracking by the company. This is especially nice if the fax mailbox service is outsourced to an outside service bureau where additional costs might be charged if they were required to dial a number, as in the case of the two-call systems. Retrieving sensitive-information faxes is more secure since the recipient is present as the faxes are downloaded.

*Two-Call Systems.* Two-call systems overcome the major disadvantage of one-call systems — accessibility to a fax machine. While staying at a hotel, a guest may not have immediate access to the hotel's fax machine. A two-call system permits the guest to call the fax mailbox from the room, arrange to have the faxes sent to the hotel's fax machine, and pick them up at a later time. Secondly, accessing a one-call fax mailbox and downloading faxes to a notebook computer is very difficult because of the combination of IVR prompts and receiving modem commands. The two-call system permits the person to handle all the voice prompts, hang up, and let the computer answer the incoming call.

## Electronic Mail

Electronic mail (E-mail) has become a mission-critical tool in business today. Because of demands on people's time, harried travel schedules, and desires to improve family life, coordinating schedules is very difficult. E-mail lets the originator communicate messages at a time most convenient and the recipient to respond at a different time. Furthermore, E-mail has become a transport agent for all types of information, such as word processing documents, spreadsheets, and calendar requests.

E-mail becomes increasingly important to telecommuters who must juggle their work schedule around family issues. For example, the telecommuter may agree to work very early morning hours and later evening hours so the middle of the day is open for family issues, whereas colleagues work the normal 9-to-5 hours. Real-time communications such as telephone calls or audioconferencing would be difficult, at best. E-mail lets the person communicate during working hours and the recipients communicate during their working hours.

**Enabler of Virtual Workplaces.** If the telecommuter is working with offices spanning the world, managing time zones becomes tricky. E-mail eliminates the time boundaries, permits transmission of data and information easily, and ties the remote offices into a single virtual office. Work projects can progress because of the worldwide nature of E-mail.

Newer mail systems permit greater ranges of data-type attachments. While older systems permitted text-only contents, newer systems support advanced types such as audio and video. Rather than receive a text description of complicated projects, the sender can record a sound bite that more fully describes the situation. Video may help the recipient understand the communication better because of the visual representation. Three-dimensional graphics can aid in the design of new components or products.

Many companies use electronic mail for exchanging information with business partners outside their organizations. Protocols such as X.400 and the Internet let the companies exchange information from disparate systems. Gateways and switches translate one system's format into another so that sender and recipient can exchange ideas.

**Internet Mail.** The Internet with its pervasive network lets people communicate around the world as easily as two neighbors talking over the backyard fence. The information repositories make researching topics easy and fruitful. Information can be sent back to the researcher or other recipients effortlessly via E-mail. Browsers on the World Wide Web (WWW) make moving from one computer on the network to another simpler than logging into the company computers. Information is displayed graphically.

## LEVERAGING THE ELECTRONIC MESSAGING INFRASTRUCTURE

**Emerging Universal Mailbox.** In the not-too-distant future, E-mail systems will support the same type of look and feel as they integrate with the Internet and WWW. Mail messages will no longer be a monolithic document, but a series of hyperlinks to other supporting documents interspersed with the originator's text. This free flow of ideas and information will further support the concept of remote workers and telecommuters.

Eventually, the telecommuter will be collecting all messages from one, unified mailbox. All voice, fax, E-mail, and other communications will reside in one logical store, eliminating the need to access several places in order to gather communications from others. (Chapter 27 explores in detail the concepts of "Universal Message Services.")

### Data Communications

The telecommuter connects to the office by both voice (already covered) and data. Just as consideration for voice communication is important, the telecommuter must plan the most effective way to communicate using data networks. Depending on the need for data, the amount of data, and the speed necessary for the computer linkage, the telecommuter may need an additional telephone line, an ISDN line, or even a dedicated line.

**POTS Line with Modem.** If data traffic consists of mainly electronic mail transfer, a simple telephone line may suffice. The computer must be outfitted with a modem to do the data transfer. In today's market, the fastest modem available runs at 28.8K bps. The fast speed will let the telecommuter quickly gather data, cruise online services and the Internet without time-consuming wait periods, and lower connect-time charges. The money and time saved by using the fastest modem quickly returns the extra expense of the modem.

**ISDN Line.** If the telecommuter will be using videoconferencing or surfing the Internet and World Wide Web extensively, an ISDN line is a must. Videoconferencing is not tolerable on regular telephone lines. The Internet and World Wide Web are vast resources of information. The line speed directly impacts the time required to obtain any information.

The ISDN line can be used for connecting the telecommuter's PC as a remote node on the office LAN. In this type of a connection, the telecommuter has all the network services (e.g., printers, fax servers, electronic mail, files, and databases) available as if he or she were in the office. Although the ISDN line is much slower than the office cabling, the impact is not as noticeable as a regular telephone line.

**High-Speed Dedicated Line.** In cases where heavy data traffic is necessary, the telecommuter may need a high-speed dedicated line. Some companies may be hard-pressed to support a telecommuter in this manner, but the option is nonetheless available.

To install such a line, a telecommunications expert should be contacted so that all equipment and connections are made properly. Also, local zoning laws need to be checked to ensure no code or other ordinances are being violated.

Data communications bandwidth (i.e., the speed of the connection) should be matched with the task at hand. Just as it would be unwise to outfit a CAD/CAM programmer with an old IBM PC, underpowering a telecommuter's communication path wastes more money in idle time and frustration than it saves in costs. The telecommuter and employer must plan jointly the most cost-effective means of connecting the telecommuter.

**Videoconferencing**

Videoconferencing is gaining acceptance by businesses today as a way of "meeting" with someone without having to be physically present in the same location. Videoconferencing provides the ability not only to speak to another person over a telephone line, but also to see the person (or drawing or sketch, if necessary). Videoconferencing equipment comes in a variety of sizes ranging from a personal desktop version to large, auditorium setups.

Because of its emerging state, videoconferencing still requires some professional help in establishing a working system. Until now, most systems were based on proprietary technology that precluded mixing equipment and conferencing networks of different vendors. Standards have been established to eliminate this barrier. H.320 is an industry standard that describes video transmission. T.120 describes computer "desktop" information transmission. Any new installation of video equipment should support these standards.

To have a satisfactory videoconferencing experience, the conferencing network must support enough bandwidth to provide smooth picture motion, typically 30 frames per second. The industry has standardized around 384K-bps transmission speed between large sites. For home office use, 128K bps is acceptable and not too costly. To reach that speed, the telecommuter must install an ISDN basic rate interface (BRI) line in the home. ISDN lines are broken into a voice channel and two data channels. The two data channels, transmitting at 64K bps each, are combined to provide the 128K bps. In most areas, ISDN is still charged at the higher business rate tariffs, but in some cases, the lower residential rate tariffs are being offered.

**Cost Affordability.** Personal, desktop video equipment costs between $2000 and $5000, depending on the manufacturer, phone company rebates or subsidies, and needed equipment. Several manufacturers such as PictureTel, Apple, and Intel offer equipment for the desktop computer.

The Internet is playing a role in videoconferencing software. Cornell University's CU-SeeMe is free simply by downloading it from the FTP site (master distribution site is http://baby.indstate.edu/msattler/sci-tech/comp/CU-SeeMe/index.html). The only expense is the cost of a digital camera and sound card.

Videoconferencing is an exciting new technology. But, because of the expense, many employers may balk at implementing it in every telecommuter's home. For those applications or situations where visual contact is necessary, videoconferencing could be cost justifiable. In the other cases, the benefits of telecommuting are just as real without it.

## Online Connections

Commercial online services such as CompuServe, Prodigy, and America Online have been around for some time. These services, outgrowths of simpler bulletin board systems, provide a wealth of information, services, and value. Simply by signing onto a service, a subscriber gains an electronic mail account. For companies that cannot afford to run an internal system and support remote users, online services provide excellent means to support remote (alias, telecommuting) workers.

Most online services provide either 800-number access or a local telephone number. Regardless of whether the telecommuter is stationary at home or travels, processing E-mail and gathering information is inexpensive.

As time passes, these online services will continue to enhance their service offerings, providing a one-stop shopping experience for all communications. A subscriber may be able to receive not just E-mail, but faxes and voice messages. Instead of having to access several locations and services to retrieve all communication types, the subscriber can access a universal inbox where all types of messages are stored.

## Peripheral Equipment

The computer in the home office should be as equally equipped as the office computer. If the telecommuter is expected to produce at least as much at home as in the office, the computer must be up to the task. In addition, the peripheral equipment must be present. CD-ROMs, large memory banks, large hard disks, and sound cards all come standard on today's computers. The same should hold true for the home office.

## Words of Advice

The telecommuter should avoid the hand-me-down situation of aging equipment. By using less adequate machines, the telecommuter will not realize the productivity gains desired. Subtle but costly changes to the equipment, such as smaller monitors or less memory, may impact the work of a page layout artist or a CAD/CAM specialist. Simply requiring the telecommuter to switch from one screen format to another may cause frustration and lower productivity.

Many telecommuters expect the company to supply the equipment for the home office. This is not always the case. For those telecommuters who are required to purchase their own equipment, they should not be tempted into buying equipment less capable than the office equipment. As software evolves, it requires more computing power to run effectively.

To get the most from the investment, the telecommuter should buy the most powerful equipment affordable. The employer may be willing to finance the cost with a low interest rate and easy payment plans. The telecommuter's equipment is the tool of his or her trade. It is cheaper to buy the better quality and more powerful equipment than to have to replace it in a year or two.

**HOME OFFICE WORK ENVIRONMENT**

The home office is an important piece of the telecommuting puzzle. After all, the telecommuter will be spending a considerable amount of time in the home office. It is important that the office be comfortable and conducive to productive work. The office should be situated so that interruptions and distractions are minimized, the surroundings and equipment are comfortably accessible, and the telecommuter enjoys the home office experience.

The office space design is dictated by three variables: the telecommuter's preferences and tastes, the house or apartment layout, and the employer's requirements. A successful office will consider all three viewpoints.

When determining the office location, the telecommuter should ask the following questions to determine a suitable location:

- Will distractions be kept to a minimum?
- Is it close to a bathroom and far enough away from the kitchen?
- Is the space comfortable year-round?
- Are the ambient temperature and humidity computer friendly?
- Is there sufficient lighting?
- Is the electrical wiring adequate to support the office needs?
- How difficult would it be to run a phone line(s) into the space?
- Is there room for everything needed (e.g., desk, file cabinet, computer, printer, and chair)?
- Is there enough storage space?

## Office Space

Home office space will follow one of three patterns: separate office space, shared room space, and hideaway. Many books exist to help telecommuters design the "ideal" office for themselves.

**Separate Office Space.** Separate office space is ideal for the home office. This space can be dedicated solely for the use of work. At the end of the day, the telecommuter can close the door to the office and leave work behind. Separate office space can be a spare bedroom, a renovated basement or attic, a loft above the garage, or a garage converted to an office. Some telecommuters go the extra mile and build an addition to their house so that there is a physical barrier between home and work.

Regardless of where the separate space is, it should be away from the normal family traffic flow to minimize the interruptions and other distractions. Noises coming from the remainder of the house should be shielded from the office space through use of heavier doors and sound-absorbing carpet and ceiling tiles. The office decor can be easily modified to be an office environment without clashing with the home furnishings of the rest of the house. Additionally, the clutter of an office can easily be hidden from unexpected guests and secured from curious youngsters.

**Shared Office Space.** The shared office space is a combination of normal living quarters and the office. For many telecommuters, the shared space becomes the best method.

Following the guidelines described previously, the telecommuter must determine the best shared location. The location could be a desk and computer in the corner of the living room. For others, it may be a portion of the bedroom or dining room.

The shared office space can be designed in two formats. Some telecommuters will use bookshelves to separate the office from the remaining room space. In this manner, the work area can be contained within that section. If there are other family members around, the separation helps them understand that the telecommuter is working while in that space. Interruptions should be kept to a minimum, if possible.

In other cases, the telecommuter coordinates the office furniture with the other furniture in the room. In this style of office, the separation between office and living quarters is indistinguishable. It may be difficult for the telecommuter to avoid interruptions and other distractions.

**Hideaway Office Space.** For some telecommuters, having a "permanent" office location is impractical. Some telecommuters will use a cabinet on wheels that contains all the office paraphernalia and that can be rolled into place at the start of the workday and rolled out of sight at the end of the day. Other telecommuters will use a filing cabinet next to the kitchen or dining room table to store all work. At the end of the day, the "office" is stored in the cabinet.

**Furniture**

The office location may determine the type of furniture that can be used. If the telecommuter has a hideaway office, the furniture used will be the same furniture used for everyday life. If the telecommuter has a shared office space, the desk and chair style and size may be determined by the other furniture in the room. The office furniture style many need to blend with the regular home furnishings as well.

The telecommuter using separate office space has the widest choice of furnishings. Deciding the type of furnishing for the office may require some forethought. If the telecommuter will be spending a lot of time in the office, a good-quality desk and chair are essential.

The desk should be solid and able to endure everyday office use. Many desks on the market today appear to be solid, but they can quickly wear out from heavy use. A good-quality desk will have sturdy drawers with metal glides and heavy rollers that can support a lot of paper files. The desktop should be solid and not wobble from use. If a computer is being used, a computer keyboard tray should be attached. Most desk heights average 29" to 30" from the floor — the proper height for writing, but the wrong height for typing. Keyboard trays should be between 23" and 26" from the floor.

The chair is another critical piece of equipment. Since most telecommuters are informational professionals, they sit in chairs most of the time. A good chair will have proper lumbar support, tilt controls, five leg bases, solid padding, and an armrest to support the shoulders and arms during extended typing sessions. The telecommuter should expect to spend over $500 for a quality chair that will last. So-called economy chairs wear out quickly, which results in more costs over time.

Other furniture includes filing cabinets, printer stands, computer hutches, and other tables. Although these items are not as critical as the desk and chair, the telecommuter should be concerned about the quality. Better-quality furniture lasts longer.

**Multifunctional Equipment**

Several companies have produced products specifically designed for the limited space of home offices. The most-often-required equipment are fax machines, copiers, scanners, and printers. Buying those devices separately and trying to find room for them all in a confined home office can make for a challenging afternoon. The manufacturers have combined these products into one unit that takes up about the same space as just one of the devices mentioned.

The cost of the multifunctional device may be close to the price of one device. Devices range in price from $700 to over $1500. The concept of the single device that does all is compelling to the home office.

While choosing a multifunctional device, the telecommuter must consider several issues. For example:

- If the fax portion of the machine is receiving a fax, can the printer print?
- If the printer is printing, can the fax receive a fax?
- If the printer is printing and the device cannot receive a fax, does the sender receive a busy signal so that they can try again later?
- Is the quality of the individual functions equal to the quality of the stand-alone device (e.g., is the laser printer of the same quality and speed as a stand-alone laser printer)?
- What functions are blocked while other functions are working?

Understanding these types of issues will help the telecommuter determine any impact on productivity, and increased waiting times and frustration levels.

**Electrical Power Concerns**

Telecommuters rarely consider the electrical power needs when choosing their home office location or the equipment they will place in it. While no single device will overload circuits placed in homes today, the combination of the office equipment may put a strain on house circuits.

Having enough electrical service to the office location will lower risks of fire hazards, circuit breaker trips, and loss of data. The telecommuter is wise to contact a professional electrician to help determine the power needs for the office space. If necessary, the electrician will recommend running separate circuits to handle the load of the equipment.

**Test Your Environment.** If a telecommuter has already supplied the chosen office location with a bevy of equipment, he or she can perform a quick check to determine the load on the circuitry. Start all the equipment and have it perform the functions it was designed to do.

For example, the computer is computing, the monitor is being updated, the printer is printing, the fax is faxing, the lights are lit, and the answering machine is answering. Run to the circuit box and listen very carefully for any humming noises. Feel the circuit breaker(s) that services the office area. Are they humming? Are they overly warm? If so, the circuits are at or over capacity and additional circuits are needed. A new circuit may cost a shiny penny, but it will be much cheaper than replacing the lost data from a breaker trip or a lost house from an overheated wire.

Exhibit 29-5 lists common office equipment and their power requirements. Equipment from different manufacturers vary, so the telecommuter's values may vary:

**Exhibit 29-5. Common Office Equipment and Power Requirements.**

| Equipment | Power requirement |
|---|---|
| Answering machine | 0.20 amps |
| Computer processor | 2.00 amps |
| Copier | 12.00 amps |
| Cordless phone | 0.06 amps |
| External modem | 0.20 amps |
| Fax machine | 1.50 amps |
| Laser printer | 7.50 amps |
| Lights (500 watts) | 4.50 amps |
| Office radio | 0.15 amps |
| Postage meter | 1.60 amps |
| Small color TV | 0.70 amps |
| Video display | 0.50 amps |
| *Total* | *30.91 amps* |

Most circuits in today's homes support between 15 and 20 amps. From the information in the table, power consumption is a major concern.

## Lighting

Improper lighting causes eye strain, headaches, and other maladies that are counterproductive and lower the enjoyment of working from home. Glare on computer monitors, shadows while writing, and insufficient light while reading may be subtle, but all can cause great physical and mental stresses.

The computer monitor should be positioned so that there is no glare from lights, windows, or other bright objects. The telecommuter should be particularly careful about light coming from windows, overhead skylights, and room lights. Fluorescent lights flicker at similar rates to the computer monitor. Although not detectable directly by the eyes, the flickering causes severe eye strain and headaches very quickly. Other forms of lighting should be considered.

Several books have been written describing proper lighting techniques. These books can be found at the library, bookstores, or home improvement centers. They describe the types of lighting and the benefits and drawbacks of each.

### Security

Security should be considered when developing a telecommuting program. The employee will have valuable equipment and, more important, valuable company information stored at home. What is the likelihood and what are the ramifications if the house was burglarized and the equipment and information stolen? Should the telecommuter install a security system? Should they carry extra homeowner insurance to cover the equipment loss? Is the proposed office space in a high-risk area, such as a glassed-in porch where everything is in open sight, or is it in an obscure part of the house? A security plan should be devised to consider various problem scenarios.

### Zoning Laws

Some communities consider any type of work-for-hire to be a business. The local zoning laws may not permit such activities. The telecommuter is responsible for abiding by the local ordinances and obtaining proper variances if necessary. In many communities, information-based businesses are not considered to fall within the zoning ordinance's jurisdiction. In those cases, no additional action must be taken.

### Support Services

The telecommuter may need two types of support services: administrative and psychological. Unlike the commuting employee who might have access to support services at the office, the telecommuter more than likely will not. For example, who will provide the labor for making 100 copies of the 60-page document, double-sided and stapled? Who will make travel arrangements or meeting arrangements?

Telecommuters work in an environment very different to what they might be accustomed. As a result, they need a support network of people to maintain the socialization, information flow (grapevine), and other social aspects of the office life. This support can be accomplished by having regular social events, the manager feeding information (both important and grapevine) to the telecommuters, and others making the telecommuter a part of the team.

### Family Issues

The family can be the biggest reason for and against telecommuting. Desiring to have more quality time with the family drives many telecommuters to working from home. At the same time, if the family constantly interrupts or the telecommuter is not self-disciplined to manage family matters, the family becomes a hindrance to the desired results. Thus, the telecommuter and family must be trained to be considerate of the work hours and to keep the interruptions to a minimum.

Telecommuting is not a substitute for childcare. The telecommuter should not try to provide childcare and work at the same time. One will gain at the expense of the other. Likewise, a spouse should not consider the telecommuter to be home and available for chores or other activities. Just as the commuter is absent while at work, the telecommuter is really not home during work hours. It may be difficult having the family understand this principle. In that case, it would be better for the telecommuter to commute to work.

## CASE STUDY: MAJOR U.S. FINANCIAL SERVICES FIRM, TECHNOLOGY DIVISION

This case study was first reported in *The Telecommuting Handbook* from the Electronic Messaging Association (May 1995). (For more information, contact the EMA at [703] 524-5550.)

### The First Telecommuting Pilot

The financial services firm started looking at telecommuting informally in late 1993. Some of the drivers were the typical ones: Clean Air Act requirements and the desire to better balance the professional and personal lives of employees. Another driver was the desire for better equipment in employee homes to use for ad-hoc support requests during nonbusiness hours (e.g., late-night batch runs, early-morning checkouts).

**Scope.** The firm officially launched the first pilot for telecommuting in March 1994. The pilot ran for 6 months, primarily to test out the technical feasibility in terms of response time, bandwidth available to each residence, and productivity.

This first pilot involved six people working at home one day a week. Most of these six users were communicating over ISDN lines at 128K bandwidth (before compression) into a TCP/IP client/server distributed computing environment at the office. Most of the telecommuters were using X-Windows terminals from Network Compute Devices (NCD) at home, X-hosting all of their applications from the office. For the ISDN equipment, Gandalf 5240 IP bridges were tested.

The firm also created a Telecommuting Working Group to manage the growth of telecommuting and to serve as its champion. The Telecommuting Working Group met with several other companies that had active or nascent telecommuting programs already under way.

**Findings.** By the end of the first pilot, the initial findings were summarized as follows:

- ISDN works technically, but takes a long time to get installed and debugged.
- NCD X-terminals worked reasonably well, but Tektronix X-terms were considered to be better.
- Support issues for telecommuters immediately surfaced.
- Initial costs for wiring and recurring connect charges were collected.

Overall, the first pilot was considered to be a success. This led to a proposal, placed before the senior managers of the technology division, to expand the pilot into a second stage, increasing the participants to include another 30 employees (as a policy statement, it was determined that only full-time or part-time professional employees would participate; no contractors or outside consultants).

In addition to expanding the number of participants, the second pilot also expanded the mix of home equipment and communication methods. This second pilot would go beyond mere technical feasibility to test the more general issues involved in institutionalizing telecommuting as a recognized method of working at the firm.

## The Second Pilot

The second pilot began in October 1994 and continued through March 1995. After meeting with senior managers and establishing the goals, 30 people were selected from a wide range of positions. Some were entry-level programmers, some were senior programming specialists. Also included in the expanded pilot were several managers and a few nontechnical administrative personnel. Like the first pilot, most telecommuters worked one day per week from home, although some extended to two days per week.

In terms of communication methods, ISDN, frame relay, and high-speed analog modems were all employed by various participants. By this time, Gandalf 5242 IP bridges were used, which were superior for home telecommuters. In terms of hardware, Suns, PCs, Macs, and X-terminals were all used.

An initial meeting of all 30+ participants was held in October 1994 to explain the goals of the second pilot, the rules under which the pilot would be conducted, and the methods for telecommuters to use to get help while at home. Participants were also advised that a contract of understanding between them and the firm would be drafted for their signatures.

**Key Issues Surface.** During the second pilot, several issues became more sharply focused. These issues were categorized as managerial, social/cultural, career-related, legal, and technical scalability/support-related.

Managerial issues included:

- Methods and mechanics for managing remote workers
- Criteria for deciding who could or could not telecommute
- Costs per telecommuter (both startup and recurring)
- "Trust" issues (e.g., is the employee working and productive?)
- Internal customer perceptions of response time to problems
- Questions that remain open (e.g., should aspiring managers telecommute?)

Social/cultural issues surfaced related to:

- Isolation
- Reluctance of co-workers to call telecommuters at their homes
- Reduced "face-time" with boss and peers

One major career-related issue emerged. That was a concern that telecommuters were "out-of-sight, out-of-mind" in terms of promotions. The firm has a meritocracy culture with a bonus component to pay: some telecommuters fear financial penalty despite stated policy to the contrary. To some degree, interpersonal networking leverages the value of each employee's technical contribution.

The chief legal issue that emerged concerned coverage for workers' compensation and insurance.

Finally, bandwidth management and provision of remote support was the leading technical scalability/support issue.

**Findings.** At the end of the second pilot, separate meetings were held with the telecommuters and then with their managers to discuss issues and findings. The majority of telecommuters and their managers felt they were more productive while telecommuting. Technical support, scheduling conflicts, and social and career concerns also surfaced.

*Hardware/Software.* On the technical side, the firm concluded that they would not use low-end (less than 386 class) PCs for productivity and response reasons. The Tektronix X-terms were considered to be very good. Few problems occurred with the Mac users. Sun users were generally pleased, although the security software used during the pilot (SecureID cards) did not give them the full range of access they wished to receive.

*Communications Networks.* On the communications side, ISDN orders still took one to two months to be processed. Once installed, however, response time was considered good.

For the frame relay users, only 56K of bandwidth was provided, which was not sufficient for the majority of telecommuters to achieve full productivity. Currently, the firm is using both Cisco and Access frame relay access devices (FRADs). To get sufficient throughput, additional compression is required on the 56K-bps line, or alternatively more bandwidth, perhaps closer to a fractional T1 line. Most of the high-speed analog (14.4 to 28.8K bps) users found the service to be sufficient for their needs.

*Telecommuting Policy.* On the policy side, a new question arose: should participation be restricted by performance or salary level? Although not yet firmly resolved, the sentiment is to restrict participation to exempt (off-the-clock) employees above some pay level. Participation would also be restricted to employees who received at least a "good" annual review.

At the end of the second pilot, a presentation to the firm's senior management, including the head of the technology division and his manager, was given to further articulate the findings, concerns, and next steps for the telecommuting effort as a whole. Based on this session, the firm is now looking to expand into a third phase, extending the number of participants, extending the program beyond the technology division, and introducing desk-sharing as a means of offsetting the costs of the program. The initial target is currently proposed to be another 30 telecommuters each quarter.

**Overall Lessons Learned**

Following is a list of the lessons learned to date from this firm's telecommuting experience:

1. The main impediments to implementing a telecommuting program in this environment were nontechnical.
2. Setting up a telecommuter almost always takes longer than planned.
3. Remote technical support for telecommuters is more difficult than office-based support.
4. Productivity can be increased by working at home.
5. It is shortsighted to use low-end equipment or communication links for telecommuters.
6. Management and employees must have a trust-based relationship from the beginning.
7. Using a third-party VAR to deal with the actual home installations and telephone company coordination worked well.
8. For the firm, limiting telecommuting to one to two days per week maximum made sense. Less-formalized but nonetheless important, communication sources available only to those workers present in the office were considered critical in terms of productivity, team-building, and career advancement.

**FORMALIZING THE TELECOMMUTING ARRANGEMENT**

A Telecommuting Acknowledgment form provides a framework of understanding between the telecommuter and the company. The form establishes the rules for insurance liabilities, company property and intellectual property protection, and reimbursable expenses. The use of the form ensures that each party is adequately represented and understood. A sample form is shown in Exhibit 29-6.

Some companies may use a Telecommuting Rules document (in addition to the Telecommuting Agreement). A sample Telecommuting Rules document is given in Exhibit 29-7. This contract in effect outlines the employee's and the firm's responsibilities for telecommuting. Another sample Telecommuting Agreement form is presented in Exhibit 29-8.

*Caveat:* In all cases, these are only sample agreements. A particular company's policies may dictate some changes. The agreement should be reviewed by corporate legal council for applicability and local law compliance.

### The Best of Both Worlds

The following letter was written by a person who would have preferred to telecommute, but due to work duties and managerial tasks, was required in the office. One day, this individual happened to work from home in the morning, arrive at work later, and go home at a normal time. Reflecting on the experience, this manager decided to institute the practice in a formal program that meets the organization's needs:

> A funny thing happened on the way to the office today. I decided to cut my morning commute in half, work longer hours, and was happier because of it. The credit belongs to a variation on what I usually think of as telecommuting.
>
> I typically get up at 6 a.m., leave my house around 7 a.m., and wave to the company's parking lot guard at 8 a.m., tense from an hour of competitive lane changing. Today I started working from home around 7 a.m. by dialing into my E-mail and working some spreadsheets at home. At 9, I was still answering E-mails, doing spreadsheets, and talking on the phone to vendors and fellow employees. I left my front door at 9 a.m. and waived at the guard at 9:30. Both traffic and stress were light.
>
> The company and I both gained a half-hour that day. At 9:30 when I parked my car, I was relaxed, prepared for the day, and already feeling productive, having eliminated five tasks from the to-do list. The 30-minute drive was refreshing rather than draining. I had avoided an hour of lane changing.
>
> Unfortunately, my schedule does not always permit my combining telecommuting with commuting, and other's jobs do not permit this flexibility either. However, for those that do, moving the morning commute "inside" the workday can yield a generous return.

## Exhibit 29-6. Sample Telecommuting Acknowledgment Form.

The undersigned employee of the Firm, including its subsidiaries and affiliates, (the "Company") agrees to participate in a telecommuting work arrangement on the following terms:

1. I understand and agree that the telecommuting arrangement will begin on a trial basis for a period of 60 to 90 days, which period may be extended at the discretion of my manager. During the trial period I will telecommute in accordance with a schedule to be determined by my manager. I understand that my manager or I may terminate the trial at any time.
2. I understand that regardless of my telecommuting arrangement, business requirements may necessitate my being in the office during times when I would normally work at home. In such circumstances, I will adjust my work site accordingly.
3. I understand that this work arrangement will have no effect on my base compensation, except that if I am working part-time, my compensation will be adjusted accordingly and my other benefits will be affected in accordance with Company policy (e.g., medical, dental disability benefits). In any event, discretionary bonus compensation will continue to be based on a variety of factors, including performance and productivity, which may be affected by the telecommuting arrangement.
4. I understand that this work arrangement will have no effect on my at-will employment status, and that my employment will continue to be subject to termination at will by the Company or myself at any time.
5. I agree to execute and abide by the terms of the attached agreement concerning any equipment provided to me by the Company for use in my home.
6. I understand that I will be reimbursed on a monthly basis for business-related phone charges incurred in connection with my telecommuting work arrangement. However, the Company will not reimburse me for any other home-related expenses such as heat, light, electricity, or insurance.
7. I agree that I am solely responsible for (i) carrying appropriate homeowner's or renter's insurance to cover third-party injuries or property damage arising out of or relating to home office use; and (ii) maintaining my at-home work area in a safe and secure condition.
8. I understand and agree that (i) Workers Compensation liability and any disability benefits for work-related injuries or illness will continue during the approved at-home work schedule in my at-home work area; (ii) the Company will not be liable for any injuries to family members, visitors, and others in my home; and (iii) the Company will not be liable for any property damage in my home.

Employee Name: _____
(please print neatly)

Employee Signature: _____ Date: _____

Manager Signature: _____ Date: _____

Human Resources Representative: _____

Signature: _____ Date: _____

**Exhibit 29-7. Sample Telecommuting Rules (In Addition to the Telecommuting Agreement).**

The undersigned employees of the Firm agrees to these telecommuting rules outlining the employee's and the Firm's responsibilities:

1. Telecommuters are the Firm's employees, not consultants.
2. Telecommuters should work from home 1 (one) day per week (preferably the same day every week).
3. Telecommuters are NOT to be simultaneously working and providing primary dependent care.
4. Telecommuters are not to hold Firm's-related meetings at their home while telecommuting.
5. Telecommuters should strive to set themselves up in a separate room, away from distractions.
6. Pagers (*beepers*) are to be worn while working at home. Additionally, voice mail should be forwarded to the home.
7. The Firm will pay for hardware and communications costs.
8. Telecommuters will pay for occupancy expenses (rent, electricity, insurance) associated with working from home.
9. Weekend use of telecommuting equipment does not qualify for $50 reimbursement.
10. Telecommuters are advised to keep track of the number of days they telecommute in a given year for tax purposes.
11. Telecommuters should expect to someday share desks in the office. We hope to offset telecommuting costs by reducing occupany costs.

I have read the Telecommuting Rules document and agree to abide by them while participating in the Telecommuting pilot.

Employee Name: _____

Employee Signature: _____ Date: _____

Manager's Name: _____

Manger's Signature: _____ Date: _____

---

Maybe there is some interesting middle ground (between *commuting* and *telecommuting* to the office) that can bring some of us the best of both worlds. I will continue my practice of mixing telecommuting and commuting.

Signed,

A Commuting Telecommuter

## GUIDELINES FOR THE PROSPECTIVE TELECOMMUTER

Telecommuting in general has proven to provide positive results in productivity, employee morale, and quality of life, as well as helping corpora-

# LEVERAGING THE ELECTRONIC MESSAGING INFRASTRUCTURE

### Exhibit 29-8. Sample Telecommuting Agreement Form.

This agreement, effective _____, is between the Company and _____, an employee of the Company. Except for those additional conditions expressly imposed on Employee under this Agreement, the conditions of Employee's employment with the Company remains unchanged.

This document does not constitute a contract of employment, either expressed or implied. At the Company, there is no fixed duration to the employment relationship. Employees can terminate their employment whenever they wish and for whatever reason they might have, just as the Company may terminate an employee at any time for any lawful reason. This is known as "employment-at-will."

I have read the following documents and agree to follow the policies and procedures outlined in them:
- Company telecommuting policy and related documents.
- Company code of conduct, employee handbook.
- Company security instructions.

The location from which I will telecommute is (give full address):
_____
_____
_____

My work area at the above location will be as follows (describe room):
_____
_____
_____

In establishing the home work area, I have determined that all common safety practices have been followed, and that this area provides a safe work environment for myself and others who may enter it.

My telecommuting schedule on a weekly basis will be as follows:

Day _____ Hours _____ Home office _____

Day _____ Hours _____ Company _____

Day _____ Hours _____ Home office _____

Day _____ Hours _____ Company _____

If not scheduled on a weekly basis, describe the telecommuting schedule:
_____
_____

## Exhibit 29-8. Sample Telecommuting Agreement Form. *(Continued)*

During scheduled telecommuting times, I can be reached at (phone number) _____ and, if applicable at (E-mail address) _____

I agree to obtain my telephone messages at least \_\_\_\_ times on each scheduled workday while telecommuting.

Work assignments that I will work on and output, I will produce while telecommuting are:

_____
_____
_____

The Company will provide the following equipment for my telecommuting arrangement:

_____
_____
_____
_____

In addition to those listed in the Company telecommuting policy, reimbursable expenses include:

_____
_____
_____

If this Agreement is part of a telecommuting trial or if the supervisor has agreed to this telecommuting agreement for a predetermined period of time, the termination date of this Agreement is _____

This Telecommuting Agreement may be terminated at any time by the supervisor provided the employee receives at least 30 days notice before the termination date.

Employee's Signature: _____ Date: _____

Supervisor's Signature: _____ Date: _____

Supervisor's Name (Printed): _____

Supervisor's Title: _____

Company (Department, Division, Organization): _____

---

tions contribute to cleaning our air. This chapter has so far discussed the characteristics for successful telecommuting arrangements. The following questions (adapted from articles that appeared in *Electronic Messaging Update,* January/February 1995, Vol. 3 No. 1) will help prepare a prospective telecommuter for success:

## Are You a Good Candidate for Telecommuting?

- Do you have the temperament and skill to work without guidance or assistance?
- What functions will you perform at home?
- Can these functions be scheduled for certain days of the week?
- Will customers be favorably affected by your telecommuting?
- Are you strongly motivated to telecommute?
- Is your work area at home conducive to working productively?
- Do you consider yourself extroverted or introverted?

## Organizational Approval

- Does your company have a telecommuting policy? If so, get a copy of it and familiarize yourself with its provisions.
- Prepare your proposal for your supervisor's approval. Anticipate his or her questions. When will you telecommute? What will you do? How will this affect the company's business, its customers, and your co-workers?
- Are you set up for work at home? What additional equipment will you need? Who will pay for it?
- Be prepared to answer questions about applicable laws and regulations (e.g., workmen's compensation, zoning laws, and tax considerations).
- Be prepared to discuss reservations your supervisor may harbor about telecommuting (e.g., will you be watching television rather than working? In actuality, the real problem is that telecommuters in general do not know when to stop working).
- Complete the telecommuting agreement ahead of time. Use it to make sure that you cover all of the relevant topics and reinforce your seriousness about telecommuting.
- If your supervisor says no, ask why. The reasons may be temporary and you can reapply at a later date.

## Preparations for Telecommuting

- Check local zoning laws if you have not already done so.
- Order the installation of a second phone line.
- Set up your work area. If it is not in a separate room, make sure that you have enough room to accommodate the equipment and papers that you will be using.
- Talk to your insurance agent for your home owners' policy about obtaining liability insurance for your home office. Your policy may already provide coverage.
- Get the home office supplies you will need. (Keep it to a minimum until you see what you need.)

- Make sure you understand any new security arrangements your company may require.
- Discuss your telecommuting plans with your family, your co-workers, and your customers.
- Install and test your equipment, software, and skills.

**Getting Started**

- Establish your routine. Plan your workday and discipline yourself to stick to it. Do not forget to schedule your breaks. Schedule your quitting time. If you find yourself working past your planned quitting time (a common occurrence), take a long break for family and dinner and return to work later in the evening.
- Organize the work that has to be done when you return to the office (e.g., faxing, copying, and printing).
- Keep your supervisor informed of your work progress. Remember, one of the concerns of a supervisor is that he or she is relinquishing control. (Never use telecommuting as an excuse for not meeting deadlines.)
- Make a to-do list, prioritize the activities, and use it to measure your productivity.
- Address misconceptions with your family promptly. Telecommuting provides more flexibility in your personal life, but you cannot be expected to be a babysitter, handyman, or errand runner any more than you would be if you were sitting in your office.

**Assessment**

- After you have telecommuted for a few weeks, ask yourself: Am I at least as effective as I was before I began telecommuting? If the answer is yes, be specific. What are your accomplishments? You will need to discuss that with your supervisor.
- How can the telecommuting arrangement be improved upon? You may now be able to justify more equipment (e.g., a fax machine for your home office).
- What effect has telecommuting had on customers and co-workers and colleagues? Ask them. They will tell you.
- What bad habits, if any, have you picked up? What are you going to do to eliminate them?
- Does your supervisor know of the positive results of telecommuting? Has there been any negative impact on the supervisor's work? If so, modify the arrangement and change the telecommuting agreement to reflect it. As a general rule, every 3 or 6 months schedule with your supervisor a discussion on your telecommuting arrangement.

# LEVERAGING THE ELECTRONIC MESSAGING INFRASTRUCTURE

**SUMMARY**

Telecommuting to some is an old practice that started simply as working from home. Others view it as a new method of work that gives workers the flexibility needed to accomplish a holistic approach to family and profession. Some businesses use a formal method of instituting the work-from-home methodology, while others simply have a handshake agreement with their supervisor.

Telecommuting provides many benefits to the telecommuter. Aside from shortened commute times, telecommuters are free to structure their day according to their best work times. The lowered stress levels and fewer interruptions increase the worker's productivity. The less formal surroundings puts the worker at ease.

Managers benefit because work objectives are stated as milestones and not time estimates. The communications between manager and employee are more formal because face-to-face meetings happen less frequently. The clearer communication channels help all aspects of projects.

Companies are realizing that the bulk of today's work effort is information development and transmission. Employees are no longer required to "go" to work in order to accomplish work. In fact, because of a more satisfying work environments to employees, companies benefit from higher productivity. Lower absenteeism and the employee's willingness to work the "commute" hours actually increase the average work hours per employee, without added expense to the company.

Telecommuting depends on communication with others. Technological advancements support telecommuting arrangements as the office of the future. Technologies available today permit people to work in remote locations without losing human contact. Voice mail, fax, E-mail, and pagers keep people constantly in tune with events happening elsewhere. Electronic online services makes information gathering easier. Electronic mail provides the highways to transport the information.

Because communications is the lifeblood of the telecommuter, the enterprise infrastructure must be engineered to meet the needs of the telecommuter. Voice mail should be flexible enough to forward calls and messages to the telecommuter's home. E-mail should have a remote or mobile mode so that continuous connection to the office LAN is not required. Connection mediums may range from analog dialup lines to ISDN or direct connections, depending on work requirements.

# Section III
# Managing Electronic Messaging Systems

The implementation and value-added use of the messaging system are two fundamental steps in creating an infrastructure for information delivery. But, it all goes for naught if the entire environment is not managed effectively and efficiently. This section of the handbook focuses on how to manage and protect this complex messaging milieu.

Messaging management is the measurement, monitoring, and control of all the services, resources, and components of the entire messaging system environment. Management can be as basic as tracking an errant message from a sender through multiple gateways and networks to a recipient on another E-mail system. Or it can be as complex as establishing service-level agreements with users and senior management, along with the measurement capabilities to ensure that the required levels are met.

Messaging managers always want to know how well their messaging system is performing. Are all of the messages being delivered? Do they get to their intended recipients within the timeframe promised?

The results of the monitoring and measurement of the messaging system can be analyzed using statistical process control. The analytical approach to performance measurement discussed in this section allows messaging managers to better tune their systems for optimum performance and provides a higher level of reliability and availability to the end-users.

Growth ensures that the messaging systems installed today will need to be changed at some point. The methodology used to select a new messaging system is explored, as well as how to implement a messaging infrastructure across a diverse organization.

The messaging system, like any computer resource, is highly vulnerable to security threats. The issue of information security is addressed in this section as we examine techniques and methods to protect messaging systems against spamming, hackers, worms, and other attacks.

## MANAGING ELECTRONIC MESSAGING SYSTEMS

The notion of privacy as applied to electronic messaging has been litigated in the courts for several years. In this section, we review the current legal ramifications of privacy in the workplace and provide a mechanism for the development of privacy policies as applied to the messaging environment.

# Chapter 30
# Introduction to Messaging Management

*Roger Mizumori*
*Sue K. Lebeck*
*Ed Owens*

Electronic messaging in its most simple form has become ubiquitous in corporations of all sizes around the world. Its growth, however, has been determined more often from the bottom up than from a statement of corporate goals or objectives. Now that these messaging systems are recognized as mission critical and that crucial company information is traveling on the messaging system, it is apparent to senior management that some order must be obtained and some mechanism for ensuring reliable delivery capability must be available. To this end, the IS departments of most companies are becoming more involved in the day-to-day administration of messaging systems. These departments are used to — and often require — extensive information on the operation of their systems. In the past, this information has not been available for LAN and distributed messaging systems.

In addition, the costs of administering the many disparate messaging systems have far outweighed the initial costs of the system. Although most would feel that those costs do not outweigh the benefits of messaging, some quantification (and eventual reduction) of those costs is required. To this end, many large E-mail software providers have turned their attention to the messaging management problem. Many E-mail service providers are also wrestling with the issues of providing consistent services across heterogeneous environments.

Although many similarities exist between physical network management and messaging management, there are more areas where messaging management must be treated differently (e.g., the corporation requiring messaging management may not own or control the underlying network on which it runs, at least not through every connection). Two of the greatest challenges to managing a messaging service are to manage user expectations of the service and to manage the scope of responsibility for the service. Accordingly, this chapter outlines a mechanism for stating goals in relation to a messaging service and a process for attaining those goals.

## MESSAGING MODELS AND PARADIGMS

In understanding the users expectations of a messaging system, it is important to review the evolution of a mail service. This starts with the paper model and transitions to the current realm of multiple messaging forms and variations. It is the consistency and reliability of the paper mail system that sets the basis for electronic mail expectations.

### Paper Model (Managing the Paper Mail Environment)

Ever since Ben Franklin invented the post office, people have been learning how to manage the smooth transport of information. The paper mail model has evolved considerably since then. Normal service in the paper model started with a collection of mailboxes in a common post office. Mail was sorted and made available to each individual mailbox. Delivery services were added to help speed end-to-end receipt of the letter. Size and shape of the letter were initially constrained only to what the mail handlers could wrestle with.

As delivery services became more institutionalized, mail gained rain-sleet-snow reliability. Next-day delivery of local mail was expected. Improving transport mechanisms allowed for highly reliable classes of service. The Pony Express enabled transcontinental delivery. Shortly after that, railroads made that obsolete and added better package handling. Telegraphs added another immediate delivery class of service. At the start of this century, air mail opened the possibilities for global delivery.

The postal system has also been a fairly forgiving system. If a name was misspelled or numbers transposed in an address, the local postal carrier often could determine the real intended destination and would make the adjustments with no other intervention. Exception handling was done with human agents.

Unfortunately, all of these friendly aspects of the service have diminished as the exploding volume of mail exceeded the capacity of the human mechanisms to process them. Zip codes were introduced to smooth sorting. Address correction is done much less frequently. Delivery expectations now are two to three days for local delivery. International postcards are never received before the vacation is over.

### Messaging Today

The total volume of mail for the world has been growing at a geometrically increasing rate. Economics are driving business processes to become faster, which means more transaction cycles occur within any given period of time. Much of this traffic is moving from the paper letter paradigm to other mechanisms.

Fax traffic has been exploding with no linkage to postal systems. With express delivery, private couriers are now carrying 80% of the traffic. In the face of all of this, regular paper mail traffic is still seeing significant growth. The evidence is clear that the volume of information flow is growing at orders-of-magnitude increases.

Electronic forms of messaging have been developing and changing over the last 10 to 15 years. Telex was the evolution from telegraphs. Facsimile, voice mail, and electronic mail have arrived, each along its technological path. Electronic mail started on mainframes with data center protections. Security and integrity were ensured by the operational procedures of a closed environment. Then, local variations emerged on the desktop and local area networks. These initial implementations were rudimentary and constrained by the limits of the early operating systems. Now, the capabilities of the operating systems on the desktop have grown and have the rigor to support greater security and integrity in a client/server fashion.

Accessibility to messaging capabilities is at an all-time high and increasing daily. A plethora of messaging-related products are available to meet the needs and tastes of a varied audience. An assortment of messaging technologies has been deployed, and linked together, to form the (sometimes patchwork) infrastructures underlying these products.

The goal of supporting distributed commercial applications across this internetwork messaging framework is within reach, and is already being experienced first-hand. With the opportunities of this variety of technologies, however, comes the confusion of the Tower of Babel.

The reliability and manageability aspects of the many and varied messaging technologies differ widely, both in their approaches and in their comprehensiveness. Messaging technologies are mapped to each other in a variety of ways, often differing from vendor to vendor, and these transformations through the network further exacerbate the ability to manage this heterogeneous morass.

Messaging users must be able to send messages to other messaging users. Unfortunately, those address formats are defined in a manner and universe independent of their own, and no common enforced method exists to map these formats. Add to this an evolving and many-faceted network management environment and what is left is what bad dreams are made of!

## SERVICE MANAGEMENT MODEL

This chapter describes a three-part model for provisioning a messaging service. It outlines a process and structure for maintaining sufficient order in the chaos to achieve the business objectives. The three principal aspects are the service architecture, technical architecture, and service plan.

**Service Architecture**

The service architecture describes what must be done to ensure a future service environment that meets strategic business needs. It starts with the business requirements and strategies that determine what kind of service is needed to achieve the business objectives. These requirements then drive the service definition. When the characteristics of the required service are known, it becomes much more straightforward to establish the management approach, the support approach, and the policies for providing the service. These latter aspects must be defined in a joint fashion because interdependencies among them exist, including the following:

- *Requirements and strategies.* The messaging service must be supportive of a business need. Whether considered from the perspective of an end-user or a service provider, the service must have a business rationale for its existence. This drives many business decisions relative to the ability to provide the service. In most corporations, the cultures are shifting to empower employees. This usually means providing information to the employees so that they can make more of the operational decisions. To improve the quality of the decisions, the quality of the information must be high. It cannot be corrupted in transit and it should not be obsolete.
- *Service definition.* Once the business drivers for a messaging service are determined, it becomes clearer how the service should be defined. This ranges from general interaction with the end-user to the detailed technical issues that specify the service configuration (e.g., elements of service or definition of management domain). To ensure the reliability of the information transport, service-level agreements should be established to ensure that the service providers understand the parameters of user expectations. Is first-class postal quality sufficient or is courier reliability with status tracking needed? Once the service is defined, it is possible to establish how it is to be managed and supported and what policies need supporting.
- *Management approach.* When the messaging service is defined in business terms, it establishes many of the criteria for making decisions relative to managing and administrating the service. These include business aspects such as jurisdictional topology (e.g., centralized, decentralized, distributed, and out-sourced), settlement (e.g., pricing and billing), level of service (e.g., quality, security, integrity, and reliability), as well as such technical issues as performance, backup, audit trails, and encryption.
- *Support approach.* The business definition also provides guidance as to the level of support that the user should expect as part of the serv-

ice. This includes trouble management (e.g., helpdesk and escalation), training (e.g., on-site, executives, and service personnel), and consulting (e.g., local or call-in support).
- *Policies.* Understanding the business purpose of the messaging service also facilitates policy definition around its use. This includes such aspects as retention, privacy, and appropriate use. All of these are also impacted by legislation and government policies and procedures.

## Technical Architecture

The technical architecture describes what can be done to transition current technologies to selected future technologies. These technical opportunities must first be organized through a technical framework to understand the relative and complementary effects of the various technologies. The long-term viability of a technology is not decided by any single company, but by a collective marketplace and by industry initiatives. Therefore, it is critical to understand industry directions, messaging management standards, and the current status and future expectations about the relevant technologies.

**Technical Framework.** The technical framework is a description of how all the relevant technologies of the service are expected to interplay as they emerge. It must be described both in terms of what has to be managed as well as the tools or structure used to manage. The technical framework should illustrate the role each technology plays as well as its interfacing requirements. To maximize the flexibility of such a framework, these interfacing requirements should be expressed in terms of standards, whether they be *de jure* or *de facto*. Standard interfaces enable smoother transitions when business conditions change (e.g., acquisitions and mergers) or technology changes.

Electronic messaging systems, especially interpersonal E-mail systems, have now been in existence since the 1960s. Over that period of time, the technical architectures of competing systems have varied greatly, but have settled into three general categories of systems: those based on a single mainframe computer providing all messaging services to a large group of users; those based on a local area network (LAN) configuration where messages in transit are stored on a shared file server; and, more recently, those based on a client/server architecture that splits the responsibility for message handling between an intelligent workstation (the client) and an equal or more powerful computer (the server); see Exhibit 30-1.

## CATEGORIES OF MESSAGING AND MANAGEMENT

When considering the different resources that comprise the messaging system, and therefore the resources that require management, the follow-

# MANAGING ELECTRONIC MESSAGING SYSTEMS

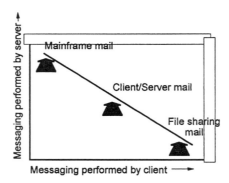

Exhibit 30-1. The Technical Framework.

ing four categories can be used: messaging-specific resources, network resources, systems resources, and messaging application resources.

## Messaging-Specific Resources

The most obvious of components of a messaging system, these specific resources include the actual client workstation programs (e.g., user agents), the programs that run the messaging servers (e.g., message stores, message transfer agents), gateways between messaging systems, and all the ancillary tools for keeping the messaging systems running. This most obvious resource is always considered when calculating the expense of a messaging system.

The type, quantity, and characteristics of this component vary between the three architectures outlined earlier but are fairly similar within the architectures. There may be requirements for these components on several different hardware platforms (e.g., user agents for Microsoft Windows and Apple's Macintosh, server software for UNIX, and OS/2 and Windows NT servers). Understanding the capabilities and maintenance requirements of each component is essential to providing adequate service to the enterprise. Messaging-specific resources, however, are only a small piece of the overall messaging component list and often represent only a small portion of the overall expense of the messaging system.

## Network Resources

All three architectures described earlier require some mechanism for communication between devices. This requirement on the corporate network can often have an impact both on the performance of the messaging system and the performance of other applications that are using the network bandwidth. Each architecture presents differing loads to the network.

## Introduction to Messaging Management

Mainframe-based messaging systems usually only require relatively small amounts of information to be passed to the terminal for viewing with relatively long intervals between transmissions.

Shared-file messaging systems require a great deal of interaction between the messaging client (doing all of the messaging work) and the file server that is only storing and retrieving raw "ones" and "zeros" of information. These transmissions on the network may not contain much information, but many such transmissions need to be made.

Client/server messaging systems attempt to optimize the network traffic by having the client make a single request for a large amount of information, having the server perform the processing and data retrieval to satisfy that request, and then passing back to the client, in one or several larger messages, the required results.

In all architectures, however, as the number of users increases and as the size of messages increases, additional load is placed on the underlying network. The net result of this increase in use is usually a slow-down of all activity on the network, especially at peak times. Performance, and therefore success, in meeting the messaging goals of the corporation, can usually be improved by increasing the network bandwidth. When this is not possible (e.g., where existing infrastructure cannot be replaced, as is the case where the network passes through a satellite link), messaging alternatives are limited.

Changing the messaging architecture to one that demands less of the network is one solution (e.g., going from file-sharing to client-server). Another option may be to isolate segments of the messaging system to manageable network bandwidth sections. The disadvantage of this approach is an increase to overhead. Communication between these segments may have to be achieved by lower bandwidth mechanisms (e.g., telephone lines and modems).

The mechanism by which multiple computers attached to the same physical network manage to communicate with each other and actually ensure that data is received as it was transmitted are beyond the scope of this chapter. These mechanisms, however, do have an effect on the messaging system and the variety of choices users have in selecting a messaging architecture.

If, for example, users choose to replace a mainframe-based messaging architecture with a file sharing-based one, they will need to consider the effect on the actual network structure. The traditional mainframe-to-terminal network does not readily support the file-sharing network protocols required. For another example, if users are migrating an existing file-sharing messaging system to a client/server-based system, the actual physical network will probably work fine but they may have to install software to support the required client/server communications on each client workstation as well as on the server. If a network and its attached computers already support the ITU-TSS OSI protocols or the Internet TCP/IP protocols, a client/server messaging architecture is an obvious choice and will fit well with the existing network.

### Systems Resources for Messaging

It is often difficult to determine exactly what physical hardware is actually being used by the messaging system. Desktop computers used for messaging are also used for a multitude of other tasks. If users are in the habit of saving messages for a long period of time, the resource requirements for those saved messages must be understood and allocated appropriately.

In a filesharing messaging architecture that has grown substantially, it is often the case that a complete file server is dedicated to the task of managing the message storage. In the case of a client/server messaging architecture, some computer on the network must become the server machine. It is often appropriate to dedicate this machine to the messaging task. When moving from one to the other, however, it may not be possible to use the same computer, or it might require changing the operating system running on that computer.

One aspect of messaging system resources that is often overlooked is the staffing requirements for running and maintaining a messaging system. Although it may not be a full-time task, some operations staff must be used for monitoring the messaging system, performing appropriate backup and maintenance tasks, and responding when problems occur in the system. This staff may not be dedicated solely to the messaging system; however, some understanding of their workload and time commitments must be included when considering the overall cost and impact of a messaging system. In addition, training in the system-level operations of the messaging system for this staff must be supplied.

### Messaging Application Resources

If messaging has become mission-critical for a company, it is likely that several applications have been written that rely on the underlying messaging infrastructure to pass information between individuals and programs, or between programs within separate areas of the company. These message-reliant applications may not have been written or controlled by the management team of the messaging system. They often rely on application program interfaces (APIs) that are supplied by the specific messaging vendor for access to the messaging system. These APIs are often very specific to the actual vendor's implementation. When considering the overall messaging system, some attention must be given to the mechanism by which programs interact with the messaging infrastructure.

More recently, standards have evolved in the messaging API arena. The X.400 API Association (XAPIA), in collaboration with X/Open, created a standard API for interfacing with the ITU-TSS X.400 and X.500 messaging and directory protocols. These APIs have now achieved wide implementation in this particular market segment. The XAPIA has also responded to a request from messaging users that a more generic API be developed that could be implemented for any messaging system. The result was Common Messaging Calls version 1.0 (now called simple CMC) and Common Messaging Calls version 2.0 (which includes simple CMC). Over time it is expected that messaging vendors will include these APIs to allow for a standard interface to messaging systems.

When analyzing an existing messaging system, users must consider the effect of message-reliant applications currently in use or that will be created. If a change in messaging infrastructure is envisioned, the effect on these programs should be carefully considered. If the programs are written to a standard API, users must be sure that their new messaging system supports those standards. If these programs are written to a proprietary API, the effect of rewriting them to either a new proprietary API or to a standard API must be considered.

## MANAGEMENT COMPONENT ARCHITECTURE

The management component architecture provides the technical blueprint for constructing a manageable system. It includes the managed resources themselves, the management information and functions supported by the managed resources, management agents, management consoles, and management communication mechanisms.

### Managed Resources

The manageability of a messaging system begins with the managed resources themselves. Within the messaging-specific resources, for example, are user agents, message transfer agents, message stores, and sometimes gateways. The resources also include subfunctions within a resource (e.g., a routing module). The fundamental manageability of a managed resource is driven by the reliability, availability, and sound behavior of the functioning resource itself — no amount of problem management can substitute for a robust resource.

### Management Information and Functions

Beyond the fundamental manageability deriving from reliable and robust resources, additional management for messaging systems includes such things as: system and network configuration, process control, dynamic monitoring of each resource's operation and performance, management of individual connections, management and tracking of individual messages, access control, error logging, accounting, and the detection and correction of problems.

To offer this management, each managed resource must provide instrumentation to accept, collect, and generate management information, and must provide some sort of access to that management information. For example, relatively static configuration information must be accepted and used; this includes topology information, addressing information, and operational parameters (e.g., time-out values). In addition, dynamic information must be generated and made available; this includes information reflecting the operational status of the running system (e.g., statistics, connection and message queue status, and error logging).

The managed resource must also accept commands and perform appropriate control functions in response to those commands. For example, it must accept a shutdown command and cleanly close down operation. Static information, especially information to be shared by multiple resources, is often stored in a directory of some sort, whether proprietary or standardized. This information is then retrieved and utilized by the running system, and by the management components themselves.

Dynamic information is emitted by the managed resource, usually in some proprietary way. This emitted information can, if appropriate, be converted into one or more standardized forms by a management agent, and communicated, perhaps by a standardized mechanism, to a management console. Control functions are generally initiated by commands issued through a management console, and communicated to the managed resource by the management agent; a standardized mechanism and format may be used to communicate this control function.

### Management Agent

A management agent interfaces to the proprietary interfaces of a managed resource, and exchanges information with a management console, first converting the information into a standardized form if appropriate.

### Management Console

A management console provides the management application interface to the operations staff in charge of managing the system. This console provides a visual, usually graphical, interface for interactive use by the operator. It often also provides a programmatic or scripting interface for automated use.

### Management Communication Mechanisms

A communication mechanism of some sort is used to exchange information between a management console and a management agent. Standard-

*Introduction to Messaging Management*

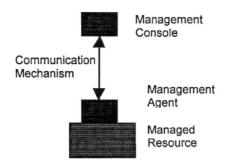

**Exhibit 30-2. Management Communication Mechanisms.**

---

ized mechanisms are often employed, allowing management agents to integrate into a standardized management application (or console) environment (see Exhibit 30-2). Examples of standardized communication mechanisms include the Simple Network Management Protocol (SNMP) from the IETF, the Common Management Information Protocol (CMIP) from OSI, and standardized electronic mail formats from the Electronic Messaging Association.

## INDUSTRY DIRECTIONS

Trends in messaging reflect a much greater attention to management needs. The whole subject of messaging management has become a significant priority. Greater management function is being demanded, management approaches are becoming consistent across messaging technologies, and more efforts are being invested in establishing effective cross-vendor standards.

### Management. A Priority Again

As discussed, the technical architectures of messaging systems vary widely. The good news is that enterprisewide manageability is again coming into vogue. Mainframe mail systems generally have enjoyed all along the inherent manageability and consistent behavior that comes with centralized and traditional formalized IS operational practices. In contrast, the newer LAN-based file-sharing mail systems have tended to suffer the inherent unmanageability and inconsistent behavior that comes with a distributed collection of localized, often ad-hoc operational practices.

Conspicuous by its absence within today's popular file-sharing mail systems, manageability has recently come back into the limelight. The movement today toward client/server mail and messaging systems brings with it an enterprisewide perspective. Vendors of these new-generation mail systems, often also the creators of the previous generation's file-sharing mail systems, have been persuaded by experience and by customer complaint, to give a greater emphasis to manageability.

### Richer Functional Ability

With a renewed emphasis on manageability, messaging vendors are beginning to compete in the manageability arena. This results in greater attention to operational concerns and more richly featured systems. Competition attracts hype, however, so it is important to evaluate asserted capability carefully.

### Increased Consistency

The importance of consistent management throughout an enterprise is also being recognized. Consistent manageability allows for a single-system view of a sometimes mixed-technology messaging environment. Increasingly, messaging vendors are developing integrated messaging environments, often including their proprietary messaging system with gateways or connectors to standardized messaging technologies. These integrated messaging environments are often managed by a single management scheme. Sometimes this management scheme is in turn integrated into the overall management environment of the platform.

## EMERGING STANDARDS

The need for consistent management is also leading to the development of messaging management standards. Although some standards activities are driven by a particular messaging technology, more often the emerging standards address cross-technology, cross-vendor environments. These messaging management standards provide an important level of consistency that cannot be achieved within a single-vendor integrated messaging environment.

Standards are particularly critical when addressing problems that extend outside of single-vendor environments. The problem of tracking a single message from submission to delivery is a classic example of this, because a given message will often span multiple vendor environments. The problem of monitoring a mail environment to assess its general health is another example, because the mail environment managed by a single operations staff often includes more than one vendor's software. As discussed in the following sections, these two problems have received the most attention from standards-defining communities.

*Introduction to Messaging Management*

**Messaging Management Standards**

To address the cross-technology that cross-vendor messaging management needs, industry standards are required. Vendor groups as well as more formal standards groups are and have been working to define standards that will address the challenges of management across diverse products and jurisdictions. The simple ability to monitor current statistical information across diverse messaging components can go a long way toward enabling broad-brush health checks of the inter-networked messaging infrastructure.

The first standards group to recognize and act swiftly on this is the Internet Engineering Task Force (IETF). The IETF created the messaging and directory management (MADMAN) initiative, to develop draft Internet standards to monitor mail and other network applications. The management information base (MIB) resulting from this work defines a table of information objects, which can be navigated using IETF's Simple Network Management Protocol (SNMP) standardized communication mechanism.

The recognized need for cross-product management also led to a multi-vendor initiative called the Messaging Management Council (MMC). This council was formed with the goal of developing a near-term industry standard addressing both mail monitoring and message tracking across multiple messaging systems. Sponsorship of this council was transferred in July 1994 to the Electronic Messaging Association (EMA), which formed the EMA's first Technical Subcommittee (TSC), called the Messaging Management TSC (MM-TSC). The work of the MM-TSC was aimed at establishing a near-term standard, for roll-out in the industry over the next year and a half and leveraged the work of IETF, and of ISO/ITU, IFIP, and others as appropriate.

The work focused on mail monitoring and message tracking across any type of MTA or gateway, seen as the most pressing management requirements for heterogeneous environments. The mail monitoring work extended the earlier IETF MADMAN work, and defined an additional standardized communication mechanism employing E-mail and a text-based format. The message tracking work extended some earlier EMA work to define a way to query the current status of one or more messages, or the path traveled by an individual message. Messages tracked in this way may have traversed many different messaging systems, coming from different vendors and adhering to different technologies and messaging standards.

Maintaining unique identification of a message as it crosses inter-system boundaries, however, represents a significant challenge that no previous standardization activity has addressed. This task is perhaps the MM-TSC's greatest challenge and, if successful, may have major significance in the industry. The message tracking work includes definitions for two communication mechanisms: an electronic-mail-based mechanism and an SNMP-based mechanism.

Currently, the MM-TSC is working to define common objects and attributes that may be stored in a common directory structure so that there can be a consistency of terminology, language, and use across all environments. Other standards groups are addressing more long-term and comprehensive aspects of messaging management. The international standards bodies ISO/IEC, and ITU-T (formerly CCITT) are jointly developing a series of standards to handle comprehensive messaging management, including logging, fault management, configuration management, performance management, and security management. This work uses an object-oriented model, and follows a philosophy well suited to use the OSI Common Management Information Protocol (CMIP) mechanism.

The IETF is currently going through an update of the MADMAN mail monitoring and related MIBs; as part of this work, the additional attributes identified through the MM-TSC work are being evaluated for inclusion within the IETF proposed standard. The International Federation of Information Processing (IFIP) E-Mail Management group, under the auspices of WG 6.5 (Messaging) and WG 6.6 (Network Management), has done important work to be a liaison across the various groups developing messaging management standards, and to generally raise awareness of the need for messaging management that spans messaging technologies.

The Open Systems Environment (OSE) Implementers Workshop (OIW) are post-standards groups dealing with messaging, directory, network management, and other open systems areas. This activity is sponsored by the National Institute of Standards and Technology (NIST). The OIW, together with other regional Implementers Workshops, are working with members of the academic research community, primarily at the University of Missouri-Kansas City, to develop "Ensembles" based on the ISO/IEC and ITU-T standards work. These Ensembles specify how to apply those draft standards to the problems of mail monitoring and message tracking.

**CURRENT STATUS AND FUTURE EXPECTATIONS**

The new generation client/server-based systems are starting to include stronger management features than their LAN-based, file-sharing predecessors. In addition, products are starting to implement standard features. Slowly but surely, the manageability of messaging systems is being improved. The functional ability available with today's systems is spotty, however, and sometimes not all that one would hope. Integration between vendor-specific environments continues to be difficult. Looking forward, the increasingly mission-critical demands on message systems will continue to apply pressure, and observable progress will continue to be made.

## SERVICE PLAN

The service plan describes what will be done, given tactical constraints and requirements, to provide a service consistent with the service architecture. This includes the operational plans for the service, its funding, its staffing, and near-term service upgrades.

The service plan defines the distinction between what is desired from the service and what can be afforded. This is where user expectations are merged with the realities of what is doable. Although it is important that users be involved in each of these three aspects, it is critical that they are part of defining the service plan. It is here that decisions of what is done and what is deferred are made. By having the user help make the decisions and understand why they resulted as they did goes a long way in developing the customer rapport to achieve the longer-term goals.

Derivative from the service architecture, the service plan should follow a similar approach with a focus on the operating period. Each of the following components should be stated in a measurable fashion:

- *Requirements and strategies*. This section should focus on the business requirements of the pending operating year and how the messaging system will function to support them. Measurement can be in the form of customer survey or specific milestones.
- *Service definition*. This section should specify service features expected of the service in the pending operating year. Examples of these might be multimedia support, fax integration, and external messaging connectivity. It should also define the specific service-level parameters (e.g., delivery times, availability, or concurrent use support).
- *Management*. This section should describe the management and administrative structure for the messaging service as it will be deployed in the pending operating year. This includes cost management and billing, working agreements among the various messaging domains and environments connected, and service monitoring. It should also include the specific plans for providing the service as specified in the service definition.
- *Support*. This section should state the type and levels of support that will be provided with the messaging service. This includes helpdesk response time, maintenance upgrades, training, and consulting.
- *Policies*. This section should be a compilation of all relevant policies around use and provisioning of the messaging service. This includes retention, archival, administrative monitoring, privacy, and security.

## SUMMARY

By applying these approaches, a consistent service can be achieved. User expectations can be uncovered and managed while critical features are identified and implemented. It establishes a partnership between the service user and the service provider.

# Chapter 31
# Implementing Electronic Messaging Systems and Infrastructures

*Dale Cohen*

Implementing a messaging system infrastructure requires taking small steps while keeping the big picture in mind. The complexity of the endeavor is directly affected by the scope of the project. If implementing messaging for a single department or a small single enterprise, a vendor solution can probably be used. All users will have the same desktop application with one message server or post office from that same application vendor.

By contrast, integrating multiple departments may require proprietary software routers for connecting similar systems. When building an infrastructure for a single enterprise, the IT department may incorporate the multiple-department approach for similar systems. Dissimilar systems can be connected using software and hardware gateways.

If the goal is to implement an integrated system for a larger enterprise, multiple departments may need to communicate with their external customers and suppliers. The solution could implement a messaging backbone or central messaging switch. This approach allows the implementers to deploy common points to sort, disperse, and measure the flow of messages.

If an organization already has an infrastructure but needs to distribute it across multiple systems connected by common protocols, the goal may be to make the aggregate system more manageable and gain economies of scale. Implementations can vary widely, from getting something up and running to reducing the effort and expense of running the current system.

## HOW TO ACCOMPLISH ROLLOUT AND MANAGE CONSTRAINTS

Messaging is a unique application because it crosses all the networks, hardware platforms, network operating systems, and application environments in the organization. Plenty of cooperation will be necessary to accomplish a successful rollout. The traditional constraints are time, functionality, and resources, though implementers must also manage user perceptions.

**Resource Constraints: Financial**

In an international organization of 5000 or more users, it is not unreasonable to spend $200,000 to $500,000 on the backbone services necessary to achieve a solution. The total cost — including network components, new desktop devices, ongoing administration, maintenance, and end-user support — can easily exceed $2500 per user, with incremental costs for the E-mail add-on at $300 to $500 per year.

The initial appeal of offerings from Lotus Development Corp., Novell Inc., and Microsoft Corp. is that a component can be added at a low incremental cost. In reality, the aggregate incremental costs are huge, although most of the purchaser's costs are hidden. For a corporate PC to handle E-mail, the corporatewide and local area networks and support organizations must be industrial strength.

Although this investment may at first glance seem prohibitively high, it allows for add-ons such as Web browsers or client/server applications at a much lower startup cost. Vendors argue that they make it possible for the buyer to start small and grow. It is more likely that an organization will start small, grow significantly, and grow its application base incrementally. In the long run, the investment pays for itself repeatedly, not only for the benefits E-mail provides but for the opportunities the foray offers.

**Resource Constraints: Expertise**

It is easy to underestimate the expertise required to operate an efficient messaging infrastructure. Most IT departments are easily able to handle a single application in a single operating environment. Multiple applications in multiple operating environments are a different story.

Messaging systems must be able to deal with multiple network protocols, various operating systems, and different software applications — all from different vendors. Given these facts, it is difficult to understand why already overburdened LAN administrators would take on the significant systems integration responsibilities of a messaging system rollout.

When confronted with problems during a messaging system integration, the staff must be able to answer the following questions:

- Is it a network problem or an application issue?
- Is it an operating system-configured value or an application bug?
- Can the problem be handled by someone with general expertise, such as a front-line technician or a support desk staff member?

**Skill Sets.** Individuals performing the rollout must be technically adept, have strong diagnostic skills, and understand how to work in a team environment. They must be adept with multiple operating systems and understand the basics of multiple networks. Ideally, they understand the difference between a technical answer and one that solves the business issue at large.

Many organizations make the mistake of assigning first-tier support staff to an E-mail project when systems integrators are called for. The leanest integration team consists of individuals with an understanding of networks and their underlying protocols, operating systems, and two or more E-mail applications. Database knowledge is very useful when dealing with directories and directory synchronization. A knowledge of tool development helps automate manual processes. Application monitoring should occur alongside network monitoring because nothing signals a network error as well as an E-mail service interruption.

**Cross-Functional Integration Teams.** The most efficient way to coordinate a rollout is through cross-functional teams. It is important to incorporate E-mail implementation and support into the goals of the individuals and the teams from which they come. Many organizations do this informally, but this method is not always effective. A written goal or service-level agreement is extremely helpful when conflicting priorities arise and management support is needed.

When creating the core messaging integration team, it is very helpful to include individuals from WAN and LAN networking, systems, operations, and support desk staff, in addition to the individual application experts from each E-mail environment.

## Functionality and Scope

At any point in the project, network administrators may find themselves trying to implement an enterprisewide solution, a new departmental system, a corporatewide directory service, or a solution for mobile E-mail users. When building a house, it is commonly understood that the plumbing and waste systems must be installed before hooking up the bath fixtures. This is not the case with messaging.

A messaging system rollout should start with a basic infrastructure "plumbed" for future expansion, and be followed directly with reliable user functionality. Results should be monitored and measured, and original infrastructure issues should be revisited as appropriate. Project success comes with regular reports on what has been delivered and discussions of incremental improvements in reliability and services.

# MANAGING ELECTRONIC MESSAGING SYSTEMS

## Supporting Internal and External Customers

No matter how good the features of any product or set of products, if the system is not reliable, people cannot depend on it. If the system is perceived as unreliable, people will use alternative forms of communication.

To satisfy user needs, the IT department should separate internal customers from external customers. Internal customers are those that help provide a service. They may be IT management, support personnel, or networking staff — they could be considered an internal supplier. Because of the nature of most organizations, internal customers are both customer and supplier. They need to be provided with the means to supply a service. For example, IT management may need to create step-by-step procedures for the operations staff to carry them out. If the information technology group cannot satisfy the requirements of internal customers, it probably will not be able to satisfy the needs of external customers.

External customers are the end-users. If they are in sales, for example, external customers may include the enterprise's customers from other companies. It is the job of the IT staff to provide external customers with messaging features, functionality, and reliability so they can do their job.

## IMPLEMENTATION MODELS AND ARCHITECTURES

It is helpful for network managers to know how other enterprises have implemented messaging systems. The next few sections describe the various components of the infrastructure, common deployment architectures, and how to plan future deployments.

## Infrastructure vs. Interface

Often, messaging systems are sold with the emphasis on what the end-user sees. Experienced network managers know that this is the least of their problems. The behind-the-scenes components, which make the individual systems in an organization work as a near-seamless whole, include:

- Network services
- Message transfer services
- Directory services
- Management and administration services

**Network Services.** The network services required for a messaging rollout involve connectivity between:

- Desktop and server
- Server to server
- Server to gateway
- Gateway to foreign environment

It is not unusual to have one network protocol between a desktop device and its server, and a second protocol within the backbone server/gateway/router environment. Servers may communicate via WAN protocols such as TCP/IP, OSI, DECnet, or SNA, and the desktops may communicate over a LAN protocol such as IPX or NetBIOS. WAN connections may occur over continuous connections or over asynchronous dialup methods.

The network administrator's greatest concern is loss of network connectivity. It is important to understand how it happens, why it happens, how it is discovered, and what needs to be done on an application level once connectivity is restored.

If the network goes down, E-mail will be faulted. Weekly incident reports should be issued that cite direct incidents (i.e., an E-mail component failure) and indirect incidents (i.e., a network failure), as well as remote-site issues (i.e., a remote site lost power). Such information can help to clarify the real problem.

**Message Transfer Services.** The message transfer service (also termed the message transport system) is the most visible part of the messaging infrastructure. The message transfer service is responsible for moving a message from point A to point B. This service consists of one or more message transport agents and may be extended to include gateways and routers. The most popular services are X.400 and SMTP international standards, and IBM's SNA Distributed Services (SNADS) and Novell's Message Handling Service (MHS) proprietary industry standards.

*X.400.* More widely used in Europe than in North America, X.400 is popular because it:

- Provides universal connectivity
- Has a standard way of mapping features
- Is usually run over commercial WANs so it does not have the security problems associated with the Internet

*SMTP.* Simple Mail Transfer Protocol's allure is its simplicity. Addressing is easier and access to the Internet is relatively simple compared with establishing an X.400 connection. Because it is simple, there is not much that can go wrong. However, when something does go wrong, it is usually monumental.

**Directory Services.** The directory service is critical to a company's E-mail systems, but it is also problematic. The problems are a result of the difficulty in keeping directories up-to-date, resolving redundant or obsolete auto-registered entries, and failures of directory synchronization.

The directory serves both users and applications. End-users choose potential recipients from a directory. The directory should list enough information for a user to distinguish between the George Smith in accounting and the George Smith in engineering. Some companies include in their directory individuals who are customers and suppliers. The ability to distinguish between internal users and external users is even more important in these cases.

**Management and Administration Services.** Management refers to scheduled maintenance and automated housekeeping procedures that involve system-related tasks such as reconfiguration and file maintenance. The constant I/O on messaging components leads to disk and sometimes memory fragmentation. Regular defragmentation procedures, including repro/reorg, tidy procedures, and checkstat and reclaim, are required. Whatever the environment, such procedures should be done more often than is recommended to prevent problems from occurring.

*Alerts and Alarms.* Alerts and alarms are extremely helpful because the system can tell the user if there is a potential problem. Alerts generally refer to warnings such as "too many messages in queue awaiting delivery." Alarms are a sign of a more serious problem, such as a disk full condition.

*Mail Monitoring.* Mail monitoring is typically an administrative function. One way of monitoring a system is to send a probe addressed to an invalid user on a target system. On many systems, the target system will reject the message with a "no such addressee" non-delivery message. When the initiating system receives this message, it indicates that mail flow is active.

Timing the round-trip provides a window to overall system performance. A message that does not return in a pre-established timeframe is considered overdue and is cause for further investigation.

*Reporting.* Reporting is used for capacity planning, measuring throughput and performance, chargeback, and statistical gathering. At initial implementation, network administrators will generally want to report breadth of coverage to demonstrate the reach of the infrastructure. Breadth can be measured by counting users and the number of messaging systems within each messaging environment.

Performance can be measured by reporting the volume — the average number of messages delivered per hour, or messages in each hour over a 24-hour period. This measure can be divided further by indicating the type of message (i.e., text only, single/double attachments, read receipts). This information gives network managers a measurable indication of the kind of features the user community requires.

For network planning purposes, it may be useful to measure volume or "system pressure," ignoring the number of messages sent and focusing on the number of total gigabytes sent per day.

## IMPLEMENTATION SCENARIOS: A TIERED APPROACH

Manufacturing environments have long used a tiered approach to messaging for distributing the workload of factory floor applications. As environments become more complex, the tiered approach offers additional flexibility.

An entire enterprise can be considered a single department, indicating the need for a one-tier system where clients are tied into a single server or post office. Multiple departments in a single enterprise or a single department communicating with multiple enterprises require routers and gateways to communicate with the outside world. When multiple departments need to communicate with each other and with multiple enterprises, a messaging backbone or messaging switch is called for.

The following table summarizes the implementation scenarios discussed in this article:

|  | ENTERPRISE | |
|---|---|---|
|  | Single | Multiple |
| Single department | One-tier single system | Two-tier similar systems |
| Multiple departments | Two-tier dissimilar systems | Three-tier cross-enterprise systems |

### One-Tier Messaging Model

A single department in a single enterprise will most likely deploy a one-tier messaging model. This model consists of a single messaging server or post office that provides all services. It may be as large as an OfficeVision system on a mainframe or a Higgins PostOffice on a Compaq file server running NetWare. The department need only concern itself with following corporate guidelines for networking and any naming standards.

Caution should be observed when using corporate guidelines. It is often simple to apply mainframe conventions when standardizing PC LAN-based applications. Many large organizations tend to forget that the whole reason for deploying desktop computers is to move away from mainframe conventions (e.g., 8-character user IDs) that are nonintuitive for users. Exhibit 31-1 shows a typical one-tier model within a single department of an enterprise.

### Two-Tier Model: Multiple Servers

As the number of E-mail users grows, or multiple departments need to be connected, an organization will probably deploy multiple servers. This two-tier model can consist of integrating similar messaging systems from

## MANAGING ELECTRONIC MESSAGING SYSTEMS

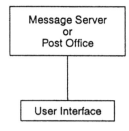

**Exhibit 31-1. One-Tier Model.**

the same vendor or from different vendors. Exhibit 31-2 illustrates a connection between two departments using the same vendor software connected via application routers.

In a typical PC LAN environment using a shared-file system such as cc:Mail or Microsoft Mail, the router acts the same way as the PC. The post office is completely passive. When users send messages, their workstations simply copy the message to the file server as an individual file or as an insertion into a file server database. In either case the PC workstation actually does the work — the post office simply serves as a shared disk drive. The router is also an active component, but has no user moving mes-

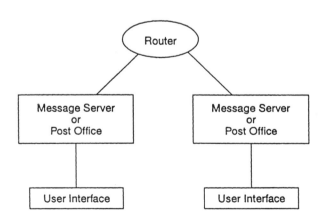

**Exhibit 31-2. Two-Tier Model.**

## Implementing Electronic Messaging Systems and Infrastructures

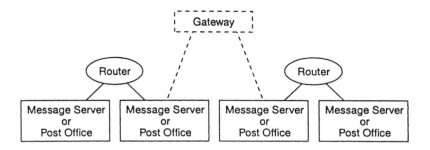

Exhibit 31-3. Using Application Gateways.

sages. It periodically moves messages from one post office to another without user interaction.

### Application Gateways for Integrating Dissimilar Systems

Many enterprises have different departments that have chosen their own E-mail systems without a common corporate standard. To integrate dissimilar systems, application gateways can bridge the technical incompatibilities between the various messaging servers (see Exhibit 31-3).

A simple gateway can translate cc:Mail messages to GroupWise. A more complex gateway can bridge networks (e.g., Ethernet to Token Ring), network protocols (i.e., NetWare to TCP/IP), and the E-mail applications.

Converting one E-mail message to the format of another requires a lot of translation. Document formats (i.e., DCA RFT to ASCII), addressing formats (i.e., user@workgroup@domain to system::user), and message options (i.e., acknowledgments to read or deliver receipts) must all be translated. Gateways can emulate routers native to each environment. They perform message translations internally. The alternative to this approach is to place the gateway between the routers as opposed to between the post offices — this is not an end-user design, it is merely a function of the vendor software (see Exhibit 31-4).

If an enterprise is large, network administrators may want to make use of economies of scale to handle common administration, common gateways to X.400, and Internet networks. The network administration staff may simply need points in its network where it can measure progress. Gateways from each environment to every other environment can be provided, but this solution becomes costly and difficult to maintain. A better approach would be to use a central switching hub or a distributed backbone, as shown in Exhibit 31-5.

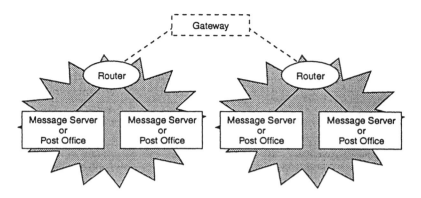

Exhibit 31-4. Placing a Gateway Between Routers.

**Distributed Hubs.** The central switch or hub allows for a single path for each messaging environment to communicate with all other messaging environments. The central hub, if it is relatively inexpensive, can be expanded into the distributed model. This is often done as the aggregate system grows and requires additional performance and capacity.

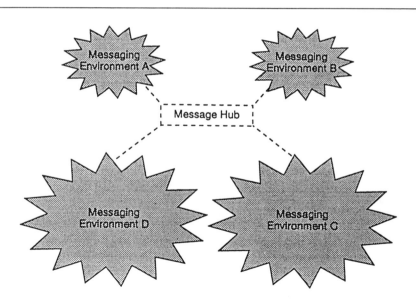

Exhibit 31-5. A Central Switching Hub.

*Implementing Electronic Messaging Systems and Infrastructures*

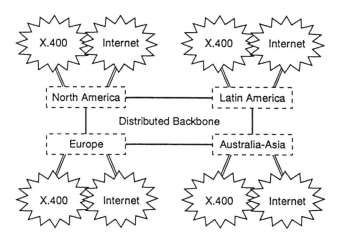

**Exhibit 31-6. Worldwide Distributed Hubs.**

However, this implementation can be taken to an extreme, as seen by the number of companies that have grown PC LAN/shared file systems beyond their original design. It is inexpensive to grow these systems incrementally, but difficult to provide end-to-end reliability. Most organizations plug the technical gaps in these products with the additional permanent and contract personnel to keep the multitude of routers and shared-file system post offices up and running.

Some organizations have taken this distributed hub approach to the point where they have multiple connections to the Internet and the X.400 world (see Exhibit 31-6). Some organizations offer the single message switch for their global environment, and their messages are more well-traveled than their administrators. A message sent from Brussels to Paris may stop in Los Angeles on the way because of the central switching mechanism. In addition to local switching, the distributed hub allows for redundancy.

## THREE DEPLOYMENT ARCHITECTURES AND OPTIONS

Most companies deploy E-mail systems using variations of three architectures: a common platform, where all E-mail systems are identical; a multiple backbone where each E-mail environment has its own gateways; or a common backbone where all systems share common resources. The following sections describe these architectures along with the advantages and disadvantages of each.

## Common Platform Architecture

For years, a major automotive manufacturer delayed PC LAN E-mail deployment in deference to the purported needs of the traveling executive. Senior managers wanted to be able to walk up to any company computer terminal, workstation, or personal computer anywhere in the world and know that they would be able to access their E-mail in the same manner. This implies a common look and feel to the application across platforms, as well as common network access to the E-mail server. In this company's case, PROFS (OfficeVision/VM) was accessible through 3270 terminal emulators on various platforms. As long as SNA network access remained available, E-mail appeared the same worldwide. This IBM mainframe shop had few problems implementing this model.

The common platform model is not unique to IBM mainframe environments. Another manufacturer used the same technique with its DEC All-In-One environment distributed across multiple VAX hosts. As long as a DECnet network or dialup access was available, users could reach their home systems. The upside of this approach is that an individual's E-mail files are stored centrally, allowing for a single retrieval point. The downside was that the user had to be connected to process E-mail and was unable to work offline.

This strategy is not limited to mainframe and minicomputer models. A number of companies have standardized on Lotus Notes, Microsoft Mail, or Novell's GroupWise. None of these products are truly ready for large-scale deployment without IT and network staffs having to plug the technical gaps.

## Multiple Backbone Model

The multiple backbone model assumes that an organization integrates its E-mail systems as though it were multiple smaller companies. The OfficeVision/VM system may connect via Advantis to reach the Internet and X.400 world. The cc:Mail WAN may have an SMTP gateway for access to the Internet and an ISOCOR MTA for access to the Message Router/X.400 gateway. All the various E-mail environments may have a proprietary Soft*Switch gateway for access to the IBM/MVS host so that everyone who needs to can access their OfficeVision/400 systems (see Exhibit 31-7).

On the surface, this hodgepodge of point-to-point connections may seem a bit unwieldy, but it does have advantages. Users of cc:Mail can address Internet E-mail users by filling out an SMTP template rather than waiting until the cc:Mail administrator adds recipients to the cc:Mail directory. OfficeVision/VM users can fill out a simple address block within the text of their message to reach an Internet user. AS/400 users can send mail to an application that forwards the message on their behalf. The trouble occurs when the recipients of the AS/400 users try to reply — they end up

*Implementing Electronic Messaging Systems and Infrastructures*

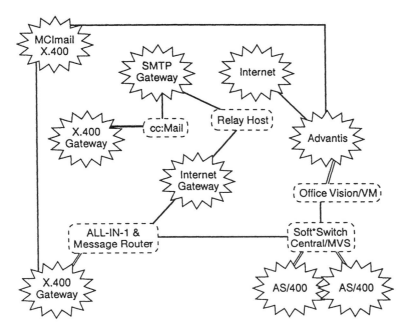

Exhibit 31-7. The Multiple Backbone Model.

---

replying to the application that forwarded the message rather than the original sender, or originator, of the message.

This architecture may still work. If each E-mail environment had its own gateway, network administration could offer multiple connections to the Internet.

**Common Backbone**

The common backbone takes two forms:

- A central E-mail hub or message switch on a single system that serves as the common denominator among all E-mail environments
- A distributed model where all backbone components run a common software protocol

The common hub involves a single switch that serves the users' applications, thus serving their needs indirectly. Each E-mail environment has an application gateway that converts its environmental format to that of the common hub. Other systems are attached to this hub in a similar manner. Messages destined for dissimilar environments all pass through this central point to be sorted and delivered to their final destinations.

## MANAGING ELECTRONIC MESSAGING SYSTEMS

The distributed backbone takes the central hub and replaces it with two or more systems sharing a common application protocol. This solution offers the ability to deploy two or more less expensive systems rather than a single, more expensive system. Any system connected to any point in the backbone can use any other service (e.g., gateway) connected to that same backbone.

Network managers may decide to purchase a single hub and gradually add systems to form a distributed backbone. Should you decide to use a common backbone protocol like X.400 or SMTP, there is an advantage. Because these protocols are available from a number of vendors, the cc:Mail/X.400 gateway could connect to an X.400 system running in an HP9000, DEC/Alpha, or Intel/Pentium system — all running the same protocols. It is possible to change distributed servers without having to change the gateways to these servers. Exhibit 31-8 illustrates three-tier flexibility.

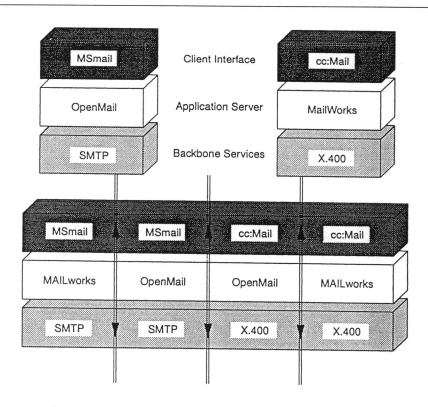

**Exhibit 31-8. Three-Tier Flexibility.**

A third approach is to use one central server or a distributed backbone of similar systems. In the central server/central hub approach, all E-mail environments use application gateways to connect to the central switch. There they are routed to their target environment.

Two-tier models may seem most convenient because they can use the offerings of a single vendor. One problem is that the system must use that vendor's protocols for a long time. Three tiers allow the layers in the model to be changed, which allows for ease of transition.

Under most application scenarios, changing one component of the messaging environment entails changing all the pieces and parts with which it is associated. It may be necessary to provide adequate support staff and end-user training, or hire consultants to handle the need for temporary staff during the transition — a significant business disruption.

For example, in one environment, users have Microsoft Mail on their desktops and a traditional MSmail post office is used, as well as message transfer agents (MTAs), to route mail between post offices. The engineering department uses OpenMail. The IT group would like to begin consolidating systems. With minor changes to the desktop, IT can retain the Microsoft Mail user interface, remove the back-end infrastructure, and use the same OpenMail system as the OpenMail desktop users by consolidating the second tier and simplifying the support environment. The client changes somewhat because it is using a different directory server and message store, but it appears as a minor upgrade to the users — no significant training is necessary.

Likewise, IT can change the back-end and still allow the OpenMail systems to communicate with the MAILworks and ALL-IN-1 systems without locking into a single vendor solution. This is a feasible option. Today, users can plug an MSmail client into a MAILworks or OpenMail server. Novell recently announced the ability to plug a cc:Mail or MSmail client into its GroupWise 5.0 server. A Microsoft Exchange client plugs into various servers, and Lotus's cc:Mail can plug into anything.

## ESTABLISHING MESSAGING POLICIES AND PROCEDURES

An organization can prevent misunderstandings, conflicts, and even litigation if it publishes its policies and procedures for messaging applications at the outset. Most important are privacy and confidentiality.

### Privacy

A privacy policy serves two purposes: to properly inform employees that their messages may not be private and to protect the organization from legal liability. Most organizations create a policy that cautions users as follows:

## MANAGING ELECTRONIC MESSAGING SYSTEMS

All electronic data is company property and may be viewed by designated personnel to diagnose problems, monitor performance, or for other purposes as the company deems necessary. While you normally type a password to access your E-mail and you may feel that your messages are private, this is not the case. The E-mail you create, read, or send is not your property nor is it protected from being seen by those other than you and your recipients.

Organizations can contact the Electronic Messaging Association (EMA) in Arlington, Virginia for a kit to aid in developing a privacy policy.

### Proprietary and Confidential Information

E-mail appears to ease the process of intentional or inadvertent disclosure of company secrets. If this is a concern, an organization could try the following:

- Let users know that the IT department logs the messages that leave the company.
- Perform periodic audits.
- Apply rules or scripts that capture E-mail to or from fields, making it possible to search on competitor address strings.

Some systems insert a header on incoming E-mail that says: "WARNING: This message arrived from outside the company's E-mail system. Take care when replying so as not to divulge proprietary or confidential information."

A company may also specify that proprietary information should not be sent to Internet addresses if security measures on the Internet are inadequate for the company's needs. Users may be asked to confirm that only X.400 addresses are used. It is helpful to incorporate any such E-mail ground rules — for example, that the tranmission of proprietary information without a proper disclosure agreement is grounds for dismissal — as part of the new employee orientation process.

### RECOMMENDED COURSE OF ACTION

One of the most important elements of a successful messaging system rollout is a staff that is well versed in the workings of the network, operating system, backup procedures, and applications.

### Network Connections

An implementation needs individuals that can set up network connections efficiently. A messaging system needs procedures in place to notify users when a network link is unavailable. If the network goes down, often one of the first applications blamed is E-mail. It is the job of the network staff to diagnose the problem quickly and have the right people to remedy the problem.

## Operating Systems

Many E-mail groups have their own systems and servers and operate them as their own. Consequently, many successful organizations pair systems programmers or senior software specialists with systems engineers who can provide installation services and upgrade support.

## Backup

Most messaging support organizations are not set up to provide 24-hour support. It is important to borrow methodologies from the mainframe support environment and staff an operations center that can answer phone calls, fix problems, and backup and archive applications regularly.

## Applications Support

This function demands staff members with:

- Excellent diagnostic skills
- Excellent communication skills
- Database and business graphics experience
- Cross-platform network experience
- Basic understanding of the operating environment of each of the platforms

E-mail integration by its nature involves cross-platform expertise. Most applications are fairly straightforward. In the case of an integrated infrastructure, an organization may need people familiar with NetWare, SNA, TCP/IP, and LAN Manager. They may also need to understand Mac/OS, UNIX, OS/2, and VMS.

When staffing an implementation, the key is to match expertise across the various groups within the company. The team should be application-centric with contributors from across the enterprise. If an implementation is properly staffed, and the implementers keep in mind the big picture as well as the daily objectives, the messaging system rollout is far more likely to be a success.

# Chapter 32
# Selecting Electronic Messaging Products
*Rhonda Delmater*

The process for selecting an electronic messaging product is similar to the process used to select almost any other kind of product. The effort that is appropriate for the product evaluation project is relative to the implementation cost and organizational effects. Factors to consider include the authority needed to implement the project, along with the implementation timeline, complexity, and end-user effect.

In other words, it would be silly to spend several months to evaluate a $200 software package for a single implementation that could easily be replaced with a different one if it did not perform as expected. At the opposite end of the spectrum, it would be unwise to select a user agent (i.e., interpersonal messaging client software) for 5,000 users without a thorough evaluation, no matter how low the product cost.

This chapter defines the steps in the product selection process in general, and provides detailed examples that pertain to electronic messaging. Depending on the size of an organization, a single product selection may not be adequate to satisfy all of its electronic messaging requirements. If this is the case, selecting products that lend themselves to integration with other products is particularly important. The product evaluation methodology is a waterfall approach, with each sequential step building on results of the previous phase. Occasionally, it is appropriate to return to a previous step.

## PRODUCT SELECTION PROCESS

Assuming the make or buy decision has already resulted in a buy decision, or that a make decision would only be made if no acceptable product could be found in the marketplace, the product selection process is further explained in Exhibit 32-1.

### Exhibit 32-1. The Product Selection Process.

| No. | Process steps | Activities |
|---|---|---|
| 0 | Project definition and planning | Form team |
|   |   | Estimate resources |
|   |   | Develop project timeline |
| 1 | Define requirements | Review literature for common features |
|   |   | Brainstorm additional desired features |
|   |   | Organize requirements |
|   |   | Generate requirements document |
| 2 | Weight requirements | Determine scheme for assigning importance "weights" |
|   |   | Assign importance factors or "weights" to requirements |
| 3 | Determine product configuration | Allocate requirements to multiple products (if applicable) |
|   |   | Develop product "specifications" |
| 4 | Identify applicable products | Survey the market |
| 5 | Conduct paper evaluation | Determine scheme for rating capabilities |
|   |   | Select products for paper evaluation |
|   |   | Review meeting |
| 6 | Select live evaluation product(s) | Select live evaluation product(s) |
| 7 | Conduct live evaluation | Test product and validate results |
|   |   | Validating COTS product form/fit/function/performance claims |
|   |   | Review meeting |
| 8 | Determine product selection | Determine product selection |
| 9 | Publicize results | Publish results |

## Project Definition and Planning

The product selection recommendation is likely to meet with greatest acceptance if the selection is based on requirements that have been generated by a cross-section of the anticipated user community. The role of the team members is to ensure that the requirements of their constituencies are considered in the product selection, and given appropriate priority (i.e., weights).

For example, electronic data interchange (EDI) capability may be very important to the purchasing organization, but somewhat less important for human resources. The team should also include appropriate technical representation (e.g., the telecommunications and operations departments, as well as the information systems department).

The team leader selection is particularly important. The team leader must encourage participation of the team members, build consensus, and

*Selecting Electronic Messaging Products*

lead the team toward timely completion of the deliverables. In addition, the team leader is likely to be the focal point for vendor contact and should be capable of soliciting vendor cooperation in terms of providing the needed information and evaluation products.

The product evaluation effort should be proportional to the scope of the evaluation. Other factors that indicate the scope of the projects include the following:

- What existing electronic messaging systems are already in place?
- Will it be necessary to interface to the existing system during the migration phase? On a long- term basis?
- What is the anticipated messaging server platform (e.g., LAN, mini, mainframe)?
- What services will the electronic messaging system support (e.g., interpersonal messaging, E-forms, workflow, electronic data interchange, and multimedia)?
- What external connectivity requirements are there?

In addition to providing information to help estimate resources required to conduct the product evaluation project, the answers to some of the questions will help provide the scope of requirements for the define requirements phase. A project timeline, such as the one in Exhibit 32-2 (which illustrates the phases and activities of Exhibit 32-1) is recommended, along with milestones representing the completion of deliverables (e.g., the documents including requirements list, weighted requirements, and vendor list).

**Define Requirements**

Because of the wealth of IS and networking trade publications, a great deal of time and effort can be saved by reviewing trade publications to identify desirable product features. This review can include vendor advertisements, published product reviews, case studies, etc. The information can be used as a straw man or for validation of completeness.

A useful icebreaker for the team is for each member to describe the current and anticipated uses of electronic messaging by their constituencies. This will also help capture requirements. Brainstorming is also recommended as a technique to generate additional requirements.

Once all possible requirements are gathered, they should be organized into meaningful groupings that will facilitate later evaluation. Some of the topic areas that are applicable to electronic messaging include:

- Training
- Field support

# MANAGING ELECTRONIC MESSAGING SYSTEMS

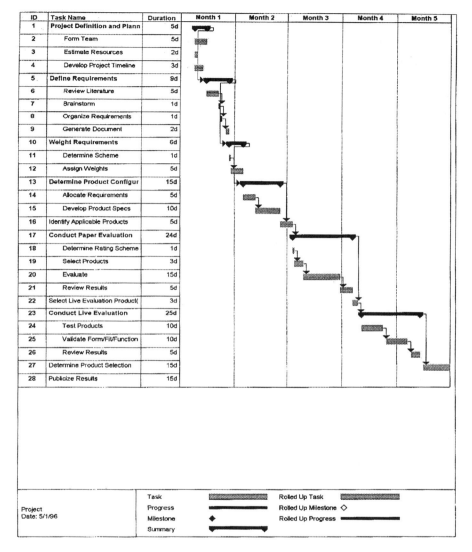

Exhibit 32-2. Sample Project Timeline.

- Messaging architecture
- Standards
- Functional ability
- Product and service trends
- Platforms
- Organizational impact

## Weight Requirements

As many schemes exist for weighting and rating as there are product evaluators. It is best to use the smallest number of meaningful choices (e.g., high = 3, medium = 2, and low = 1) to avoid wasting time in the process of assigning weights. If a requirement is weighted as zero, or no importance, it does not belong on the requirements list. Numerical weights can be factored with evaluation ratings to generate a score. This technique helps give higher credit to the products that perform the most important functions well.

Another technique that can be used along with the weighted requirements is to identify requirements that are mandatory. These may not have a numerical rating, but a product that does not meet the mandatory requirements is automatically removed from consideration.

Once the weighting scheme is determined, the team assigns the weights to the requirements (see Exhibit 32-3). This can be a very time-consuming process to conduct as a team. Each member should individually make a determination of the importance of each requirement. These recommendations can then be discussed as a team, or if meeting time is at a premium, it is sometimes useful to calculate the mean, median, and mode for the individual weights. Then, if there are some requirements that everyone agrees to, they do not need to be discussed in a group session.

## Determine Product Configuration

Depending on the scope of the electronic messaging system, and in light of the emerging electronic messaging architectures, it may be appropriate to allocate requirements to multiple products. This allocation could be as simple as client and server, or far more complex to include multiple types of gateways and message transfer agents (MTAs). The configuration may include an outside service (e.g., America Online, Prodigy, or CompuServe). If so, the same process can be used to select the service provider that most closely satisfies your organization's requirements. A sample product configuration is shown in Exhibit 32-4.

Once the configuration is known, the requirements list should be separated into those requirements that are applicable to each component piece. The requirements may not be specific to a single component. Particularly, ease of use, standard architecture, and training desired would apply to all components. A requirements document can now be generated for each product type to be evaluated. Steps 4 through 8 in Exhibit 32-4 are executed for each component product. Publicizing the results (step 9) is best done after all evaluations have been completed and the results synthesized.

| Numbers Category<br>Requirement statements | Req't weight |
|---|---|
| 1. User Interface | |
|   1.1. Consistent interface across platforms | 2 |
|   1.2. Advanced word processing features | |
|     1.1.1. Global search and replace | 3 |
|     1.1.2. Spell checking | 3 |
|     ⋮ | |
| 2. Addressing | |
|   2.1. Public mail lists are supported | 3 |
|   2.2. Private mail lists are supported | 2 |
|   2.3. User names supported up to 50 characters | 3 |
|   2.4. User names supported up to 75 characters | 2 |
|   ⋮ | |
| 3. Architecture | |
|   3.1. Client platforms supported | |
|     3.1.1. MS DOS | 2 |
|     3.1.2. MS Windows 3.1 | 3 |
|     3.1.3. MS Windows 95 | 2 |
|     3.1.4. Macintosh | 2 |
|     3.1.5. X-windows | 1 |
|     3.1.6. Telnet | 1 |
|     ⋮ | |
|   3.2. Server platforms supported | |
|     ⋮ | |

**Exhibit 32-3. Sample Requirements Document with Weights Excerpt.**

## Identify Applicable Products

Preliminary activity to identify applicable products should be ongoing from the onset of the evaluation project. Starting with the literature survey in phase one, all of the team members should be taking opportunities to identify potential solutions. At the same time, they must avoid letting any one solution drive the requirements definition or configuration design.

Excellent resources are available to identify electronic messaging products, such as the annual *Global Guide to Messaging Products and Services*, published jointly by the Electronic Messaging Association, the European Electronic Messaging Association, the Japan Electronic Messaging Association, and the Electronic Messaging Association of Australia. This document is a compendium of information provided by the electronic messaging vendors, and also features a product index matrix that cross-references the companies with a list of product and service categories. The *Guide* features over 250 companies cross-referenced to 31 product categories.

Many conferences and exhibits feature electronic messaging products and services. Because electronic messaging is such a dynamic market, it is important to gather as much information about new releases as possible,

*Selecting Electronic Messaging Products*

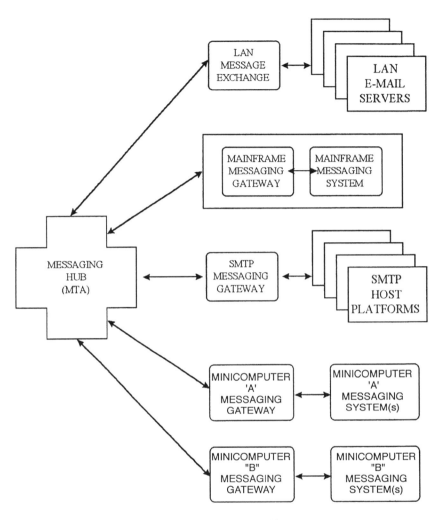

**Exhibit 32-4. Logical Product Configuration Diagram.**

---

through trade shows and current periodicals. The character of organizations varies regarding the product maturity sought, which may also be reflected in the requirements list (e.g., number of existing sites and years first installed).

## Conduct Paper Evaluation

As with the requirements weighting, a scheme is also needed for rating product performance. Similar to requirements weighting, using the smallest number of meaningful choices facilitates the process. The highest possible rating represents a product fully satisfying the requirement. A score of zero is also useful in this case, as a product may not meet a requirement at all. Intermediate ratings can be used to identify various degrees of satisfying the requirement.

As an example, *Consumer Reports* uses symbols that represent good, poor, and in-between. Of course, most evaluations will go beyond the *Consumer Reports* approach, because companies will be scoring the product performance according to their own customized weighted requirements list.

Once the population of applicable vendors is identified and a rating strategy is determined, an approach for completing the paper evaluation must be selected. If an organization has sufficient purchasing clout, it may be able to get the vendors to provide answers to its requirements list through a formal approach (e.g., request for information, or RFI) or some less formal mechanism.

If for some reason it is not feasible to get the vendors to provide information regarding how well they satisfy the requirements, it will most likely be necessary to select a subset of the total product population for evaluation. This can be based on market leadership, new technology, or some other criteria. In any case, each team member should be assigned one or two products for which to complete the paper evaluation.

Several review sessions will be needed for team members to discuss how their assigned products meet the requirements and to compare results. Once these are completed, the results are tabulated in a matrix as illustrated in Exhibit 32-5.

**Select Live Evaluation Products**

The number of products selected for live evaluation will depend on the results of the paper evaluation. If a clear leader exists, it may be logical to evaluate that product alone. If the product scores are clustered, it would be appropriate to evaluate those in the top group, if resources are available.

**Conduct Live Evaluation**

The live evaluation is used to validate the results of the paper evaluation and determine whether documentation used to develop the paper evaluation was accurate. Sometimes it is useful to get a few additional guinea pigs involved in trying out the product to see how intuitive it is for various classes of users.

If multiple products are being evaluated, it is useful to assign responsibility for evaluating each product to a group of team members. Each of these groups presents the results of their live evaluation to the team along with a

### Exhibit 32-5. Sample Paper Evaluation Matrix.

| Numbers Category Requirement statements | Req't weight | Product A Raw rating | Product A Weighted rating | Product B Raw rating | Product B Weighted rating | Product C Raw rating | Product C Weighted rating |
|---|---|---|---|---|---|---|---|
| 1. User Interface | | | | | | | |
| 1.1. Consistent interface across platforms | 2 | 3 | 6 | 4 | 8 | 5 | 10 |
| 1.2. Advanced word processing features | | | | | | | |
| 1.1.1. Global search and replace | 3 | 5 | 15 | 4 | 12 | 4 | 12 |
| 1.1.2. Spell checking | 3 | 4 | 12 | 4 | 12 | 4 | 12 |
| ⋮ | | | | | | | |
| 2. Addressing | | | | | | | |
| 2.1. Public mail lists are supported | 3 | 5 | 15 | 0 | 0 | 4 | 12 |
| 2.2. Private mail lists are supported | 2 | 3 | 6 | 0 | 0 | 4 | 8 |
| 2.3. User names supported up to 50 characters | 3 | 5 | 15 | 5 | 15 | 5 | 15 |
| 2.4. User names supported up to 75 characters | 2 | 0 | 0 | 0 | 0 | 5 | 10 |
| ⋮ | | | | | | | |
| 3. Architecture | | | | | | | |
| 3.1. Client platforms | | | | | | | |
| 3.1.1. MS DOS | 2 | 5 | 10 | 5 | 10 | 5 | 10 |
| 3.1.2. MS Windows 3.1 | 3 | 5 | 15 | 5 | 15 | 5 | 15 |
| 3.1.3. MS Windows 95 | 2 | 5 | 10 | 0 | 0 | 5 | 10 |
| 3.1.4. Macintosh | 2 | 0 | 0 | 0 | 0 | 5 | 10 |
| 3.1.5. X-windows | 1 | 0 | 0 | 0 | 0 | 0 | 0 |
| 3.1.6. Telnet | 1 | 0 | 0 | 0 | 0 | 0 | 0 |
| ⋮ | | | | | | | |
| 3.2. Server platforms supported | | | | | | | |
| 3.2.1. MS NT | 3 | 5 | 15 | 5 | 15 | 5 | 15 |
| 3.2.2. Novell Netware | 3 | 5 | 15 | 5 | 15 | 5 | 15 |
| ⋮ | | | | | | | |
| 3.3. APIs supported | | | | | | | |
| 3.3.1. MAPI | 3 | 5 | 15 | 5 | 15 | 5 | 15 |
| 3.3.2. VIM | 2 | 0 | 0 | 0 | 0 | 5 | 10 |
| Total: | | 55 | 149 | 42 | 117 | 71 | 179 |

demonstration. The matrix generated should be revised to reflect any changes in scoring that result from hands-on experience with the product.

## Determine Product Selection

If multiple products can be used, a company is in a great position to negotiate with the vendor.

## Publicize Results

In a large organization there are often some specialized users that have their own pet products they would like to see adopted organization-wide. Providing a thorough, objective head-to-head analysis of the competing products sometimes helps them understand and accept a product other than their first choice, if it is the best overall choice and supporting multiple products is not practical.

MANAGING ELECTRONIC MESSAGING SYSTEMS

## SUMMARY

The process described in this chapter has been used very successfully to evaluate a broad range of information systems and office automation software, including electronic mail systems. It does require a significant effort, but it is well worth the investment when considering the user impact of implementing the best, most reliable electronic messaging system.

# Chapter 33
# Performance Measurement
*Clyde F. Haggard*

It is easy to confuse performance factors with workload factors that are readily measured. Workload factors are important influences on performance, as increases in workload generally reduce performance of the messaging system. Some workload factors can be measured and reported so that as the work increases, the measurement drops below the goal or objective. To the unsophisticated, such measurements appear to be a drop in performance. In these situations, the messaging function can be trapped into justifying or explaining measurements over which it has no control.

Performance factors and workload factors should be reported together, clearly separated into their respective classifications, because increasing workload can be used to justify additional resources. This chapter helps distinguish performance factors from workload factors.

## MESSAGING PERFORMANCE

Every messaging system has certain performance characteristics, which can be distinguished from the software features and workload factors. Performance characteristics are simply five factors that will always generate complaints if they are outside the bounds of the users' expectations:

- Availability
- Addressability
- Reliability
- Delivery time
- Consistency

*Availability* is whether the messaging system is available for use. Is the message store hardware and software functioning and available for use? Network availability is closely related but usually measured separately.

*Addressability* is whether the user can address a message correctly to the desired recipients. Is the directory correct and complete?

*Reliability* is whether a message will be delivered to the correct recipients. Lost messages are very difficult to measure.

*Delivery time* is the time it takes a message released from one message store to be available to the recipients in the same or other message stores. It may seem instantaneous in some cases, or take many hours or days in others. What delivery times are acceptable depend on user expectations.

*Consistency* is the range of delivery times. The difference between the slowest and fastest messages affects user expectations of delivery time. The range of delivery times should be a low multiple of the average delivery time.

## MESSAGING WORKLOAD

Workload is the quantifiable measurement of the messaging systems functions, such as the number of accounts that can be serviced, the number of messages sent, the size of messages, etc. Messaging workload is consumption of the resources used to create, send, receive, and store messages and to manage and administer the messaging systems.

Resources such as hardware, software, and network components of the system have capacity limits. When capacity limits are reached, performance begins to suffer, often very suddenly. Capacity planning hinges on careful workload measurement and analysis.

### Link to Capacity Planning

Because workload factors are directly related to quantifiable capacity, they are the easiest messaging factors to measure. Often, hardware and software have explicit built-in limitations. Everyone measures the number of accounts on the messaging system, and usually by individual message store.

Some messaging systems count the number of messages, recipients, or the size of the messages they process. This can be very helpful in determining if the number and/or size of messages being processed by the system are increasing.

Most installations can count the number of simultaneous network connections, for comparison with capacity limits. Sometimes the messaging software limits the number of active licenses, which relates to the number of connections that can actually be supported.

Administrative workload is important in managing a messaging system. When people are migrating between multiple messaging systems, moves can consume much of the administrative resources. Administrative functions need to be tracked to assist in planning for new software releases, mi-

## Performance Measurement

gration to different software platforms, and integration with different mail systems due to mergers and acquisitions.

Satisfactory messaging system performance depends on all components of the system having excess capacity. When a capacity limit is reached, it often introduces a bottleneck that backs messaging traffic up through other parts of the system. Comprehensive workload measurements make it much easier to find the true cause of the performance degradation.

Some capacity limits are obvious, or specified limitations of the hardware or software. Others, however, must be learned by experience. In either case, workload measurements provide a basis for capacity planning. For example, if the number of accounts per message store is approaching capacity, administrators need to increase capacity or limit the number of accounts. Effective capacity planning requires constant workload measurement.

### PERFORMANCE REPORTING

Messaging system functions need performance reporting to help them manage the system and justify resource requirements and budgets. As long as performance is satisfactory to those using the system, very little external reporting is required. However, once performance fails to satisfy the users, there will be constant pressure to produce performance measurements along with goals. If messaging system management has already selected and measured performance characteristics, then it will have no difficulty in reporting them externally.

Performance is often measured only for certain hours. In the examples that follow, prime time is defined as the 10 hours from 7:00 a.m. to 5:00 p.m., Monday through Friday. Measurements may be reported daily, weekly, monthly, or for other periods. The examples use a weekly reporting period.

### Availability

Availability is defined as whether the messaging system is available for use (e.g., is the message store hardware and software functioning and available for use?).

Message stores are hosted on all sizes of computers. Only on larger systems, however, is performance measurement software available to record availability. Most often, someone must record all downtime for each message store. Availability is usually reported as a percentage and computed as:

$$\frac{\text{Time available}}{\text{Time available} + \text{Time not available}}$$

# MANAGING ELECTRONIC MESSAGING SYSTEMS

*Example:* If the message store was not available for 1 hour in the week, the formula would be: 49 ÷ (49 + 1) = 98% and it would be reported as: Prime time availability = 98%.

Depending on the performance goal, this might or might not be acceptable. Availability can be aggregated for multiple message stores, but each individual is only concerned about availability as seen by one account. When consolidating availability, it should be weighted for the number of accounts on each message store.

Network availability is closely related but usually measured separately. It may be reported separately for individual LAN segments or protocols.

There is virtually no way to measure availability where it counts the most — at the client level. Reasons for inability to access the message store can be at the client computer or terminal, anywhere along the network or phone connection to the message store, or at the message store itself. The difficulty can be hardware or software related and can depend on the relationship of any combination of hardware or software components. Objections to reported availability should be expected.

## Addressibility

Addressibility is whether the user can address a message correctly to the desired recipients. It is a difficult characteristic to measure.

In homogeneous mail systems, addressibility may never come into question because all valid addresses are available. With more complex systems, there are problems in synchronizing directories so that valid addresses are presented in each message store's directory. One formula for measuring addressibility is:

$$\frac{\text{Time} \times \text{Accounts with correct directories}}{\text{Time} \times \text{Accounts with correct directories} + \text{Time} \times \text{Accounts with incorrect directories}}$$

*Example:* If there are two message stores with 250 accounts on each, and one message store had incorrect directories for 4 hours, the formula would be:

$$\frac{(50 \text{ hours} \times 250 \text{ accounts}) + (46 \text{ hours} \times 250 \text{ accounts})}{[(50 \text{ hours} \times 250 \text{ accounts}) + (46 \text{ hours} \times 250 \text{ accounts})] + (4 \text{ hours} \times 250 \text{ accounts})}$$

This formula works out as 24,000 ÷ 25,000, or 96%. It would be reported as: Prime time addressibility = 96%.

## Performance Measurement

Whether or not 96% is acceptable or poor performance depends on the messaging system and the users' expectations. Performance goals should be attainable.

**Factors Affecting Addressibility.** Some of the factors affecting addressibility include the size of the directories, the number of directories to be synchronized and the synchronization process, and various limitations on address lists. Large directories can exceed capacity limitations, so that the message store is unable to rebuild them correctly for lack of space.

As the number of directories and mail systems increases, the synchronization process becomes more complex and prone to failure. When some portion of the directory synchronization fails, one or more message stores will have incorrect directories, and some mail will not be addressed correctly. Some messaging systems reassign internal addresses whenever an account is added or deleted from the message store. Directory synchronization on such systems is critical after any changes.

A related problem is encountered with messaging systems that do not automatically update personal address books or distribution lists. In large messaging systems, where there are many directory changes, personal address books or distribution lists rapidly become out of date. Thus, users should validate the addressees each time they use such lists to address a message. (This problem is a fault with the features of the messaging system and not a performance characteristic.)

## Reliability

Reliability refers to whether a message will be delivered to the correct recipients. Many messaging systems only send undeliverable messages back to the sender, without tallying the number of messages or invalid recipients. In those cases, reliability cannot be measured. When the necessary information is available to the system administrator, reliability is computed as:

$$\frac{\text{Total number of messages} - \text{Undeliverable messages}}{\text{Total number of messages}}$$

*Example:* If there are 12,300 messages sent during the week (prime time is ignored), and 246 are returned undeliverable, the formula is: $(12,300 - 246) \div 12,300$ equals a Reliability rate of 98%.

There is a trap in reporting reliability if there are multiple messaging systems and one reports undeliverable messages and others do not. Undeliverable message counts may include messages that were not deliverable on either system, but which are returned to senders on the system that keeps the statistics. If the total message count is only for one system, but the rejected message count reflects both systems, the reliability will appear to be much lower than it should be.

For example, suppose system A with 1000 accounts exchanges messages with system B of 9000 accounts, and 2% of all messages are undeliverable. If each account sends out messages to 10 recipients, then system A would report 10,000 messages but would count about 400 undeliverable messages: 200 or 2% of messages sent by system A, and 200 or 2% of messages sent to system A from system B. Thus, reliability will appear to be 96% rather than the actual 98%.

**Delivery Time**

Delivery time is the time it takes a message released from one message store to be available to the recipients in the same or other message stores. The ideal statistic is the average delivery time, but most systems do not measure and accumulate such data.

A true average can only be constructed if the actual time each and every message is sent and delivered is known. This can be distorted by inaccurate clocks or systems that record the time the message was created, not the time it was sent.

An acceptable substitute is to take sample messages and average their delivery times. However, it is very difficult to determine a representative sample of the entire messaging system when there are several different message stores.

Average delivery times for sample messages often are representative only of the specific path between two message stores, and not the messaging system as a whole. Subsets of sample data should be weighted by the number of messages in the population from which they are taken. For example, the sample data has the following results:

| From message store | To message store | Sample delivery time (in minutes) | Number of sample messages | Total messages |
| --- | --- | --- | --- | --- |
| A | A | 1 | 10 | 30000 |
| A | B | 5 | 10 | 2000 |
| B | B | 1 | 10 | 7000 |
| B | A | 3 | 10 | 1000 |

Here, a simple average is 2.5 minutes (100 sample minutes ÷ 40 sample messages), but a weighted average is 1.25 minutes (50,000 message minutes ÷ 40,000 messages).

Messages are usually delivered to recipients on the same message store in a matter of moments. Fast delivery times do not raise performance questions. Moving messages from one message store to another, particularly between dissimilar messaging systems or through gateways, introduces delays and brings requests for performance measurements. The only way to get reliable data in these situations is to send sample messages. When using the times from sample messages, clocks at both ends must be reasonably accurate. Otherwise, the recorded delivery times can be extremely inaccurate, even negative.

Subjective measurements of delivery times will be made by the users of the messaging systems. They will remember the messages that took 6 hours and forget the ones that were delivered quickly. It seems that the slowest messages are the most urgent. Moreover, users will always remember old experiences, even if the delivery times have improved. Well-constructed sample data can counter subjective measurements.

Delivery times are usually reported for a particular time period, whether weekly or monthly, and in minutes. If actual average delivery times are not available, then average delivery times between message stores may be reported as: Average delivery time (from A to B) = 5 minutes.

The changes in daily average delivery times over a reporting period are important. This is usually shown best in graphical form, preferably with a similar chart showing the range of delivery times for each day.

**Consistency**

Consistency is the range of delivery times. In any messaging system, the users of the system develop expectations of delivery time and base their use of the system on those expectations.

A wide range of delivery times is not consistent with user expectations and will be regarded as poor performance. The difference between the fastest and slowest messages should be a low multiple of the average delivery time.

Whether average delivery time is 5 minutes or 30 minutes is less important than the consistency. If the average delivery time is 5 minutes, but some messages regularly take 60 minutes, then the system is not working properly. On the other hand, a system that usually delivers messages in 30 minutes can regularly take 60 minutes and still be considered to be working just fine.

The range of delivery times is a byproduct of the process that measures delivery times. The messaging software may provide the information, or sample messages can be used to compute it. Consistency can be reported as a ratio of the range to average delivery time. For example: Consistency (from A to B) = 1.75 times average delivery time.

# MANAGING ELECTRONIC MESSAGING SYSTEMS

The daily range can also be reported graphically. This works best when there is an accompanying chart showing the daily average delivery times for the same messages.

## WORKLOAD REPORTING

Workload factors tell how much work the system is processing, while performance tells how well the work is being processed. Each part of the messaging system is capable of processing a certain quantity of work. By measuring and reporting workload factors, bottlenecks and capacity problems can be identified and fixed.

### Accounts

Accounts supported by the messaging system is the basic workload statistic. Even if no other workload statistics are reported, the number of accounts on each message store provides a basis for comparing changes in the system's load. The account workload is usually reported as a number:

$$\text{Accounts (message store A)} = 2857$$

### Simultaneous Connections

Simultaneous connections refers to the number of network or dialup connections that the hardware and software can process at one time. Generally, the number of accounts is greater than the number of simultaneous connections that can be established. When the limit is reached, new connections will be rejected. On some mail systems, this may cause programs that connect in order to pick up or deliver messages to be unable to function. Thus, significant delays in delivery times will result.

Simultaneous connections can be limited in many ways. Limitations can be by hardware constraints, either in the network or at the message store. There may be software license restrictions that cause connections to be rejected when a limit is reached. Network software or message store operating systems can impose limitations on connections.

The average number of connections is available on some systems, for the message store, the host or server, or for the network. This is reported as:

$$\text{Average simultaneous connections} = 234$$

### Accessibility

Other systems report the number of failed connections and the reasons the connections failed. This allows the workload to be reported as accessibility, or the percentage of successful connections. Successful connections are sometimes called *logins*. This is computed as:

## Performance Measurement

$$\frac{\text{Successful connections}}{\text{Successful connections} + \text{Unsuccessful connection attempts}}$$

Accessibility is easily confused with performance factors because a typical objective is accessibility = 98%. Accessibility depends on the capacity of the system, not on how well the system performs its work. It should be confined to the workload section of any reports as: Accessibility = 97.5%.

### Messages

Messages have several characteristics that define the messaging workload that can be processed. First, the number of messages is important. In a system where each message is stored once and recipients are given a pointer to the actual message, the number of messages is an adequate statistic.

There is also the concept of *half messages,* in which the number of messages sent is counted separately from the number of messages received. Thus, a message with 10 addressees would count as 11 half messages. This is an important statistic when there are multiple messaging systems, gateways, and switches.

In addition to the number of messages or half messages, the size of the messages is also an important measure of workload. More and more messages are including voice, images, or large spreadsheets or presentations. Both average message size and maximum message size are usually reported. There may be capacity limitations for the maximum message size, so it may not be a significant statistic. Messages and average message size are typically counted for a fixed period (e.g., weekly or monthly) and reported as:

| | |
|---|---|
| Messages | 5432 |
| Half messages | 20,789 |
| Average message size | 19,532 bytes |
| Maximum message size | 1,866,777 bytes |

### Storage

Storage is often a critical workload factor in messaging systems. If storage is permitted on the message store, it must be managed so that adequate space is always available.

Even if messages are to be stored on client computers, it is not unusual for some accounts to retain messages in the mailbox or inbox after they have been read. Since these files are on the message store, space requirements must still be managed.

Storage can be reported both as an average space required per account, or as a percentage of space available. For example:

>   Average storage per account        5.2 megabytes
>   Storage used (message store A)     87.3%

Various messaging systems may provide other workload measurements. Sometimes it is important to manage *CPU utilization* or *memory utilization*. These types of measurements are often unique to the specific computer system or software. Their usefulness depends on whether or not system tuning to improve messaging performance is possible.

The administrative workload is often overlooked in workload analysis and reporting. There are many activities performed in support of a messaging system that consume labor resources. What is simple or automated on one system may be time-consuming or manual on another. Many tasks are relatively simple, such as password resets or account renaming. At a minimum, count and report adds, moves, and deletes.

Moves can be counted as adding an account to one message store and deleting an account on another. However, if there are additional tasks involved to move stored messages, then moves should be counted separately. Each message store's administrative tasks should be reported separately, as follows:

>   New accounts (message store A)       18
>   Deleted accounts (message store A)    6
>   Mail moves                            4

Directory systems can require significant administrative resources, especially in complex systems. Repetitive tasks should be counted and reported.

## PRESENTATION

Performance and workload factors can be reported as simple lists. It is normal to attach goals or objectives to all of the reported performance factors. Workload factors have assumed or capacity limits. A sample weekly report might look like this:

| Performance Factor | This Week | Objective |
|---|---|---|
| Prime time availability | 98% | 97% |
| Prime time addressibility | 96% | 97% |

## Performance Measurement

| Performance Factor | This Week | Objective |
|---|---|---|
| Reliability | 98% | 98% |
| Average delivery time (from A to B) | 5 minutes | 10 minutes |
| Consistency (from A to B) | 1.75 | 2.0 |

| Workload Factor | This week | Capacity |
|---|---|---|
| Accounts (message store A) | 2857 | 3000 |
| Average simultaneous connections | 234 | 300 |
| Accessibility | 97.5% | 98% |
| Messages | 5432 | 8000 |
| Half messages | 20,789 | 30,000 |
| Average message size | 19.5 kilobytes | 40 kilobytes |
| Maximum message size | 1. megabytes | 2 megabytes |
| Average storage per account | 5.2 megabytes | 10 megabytes |
| Storage used (message store A) | 87.3% | 90% |
| New accounts (message store A) | 18 | 50 |
| Deleted accounts (message store A) | 6 | 25 |
| Mail moves | 4 | 15 |

The static picture this table presents can be clarified by using charts to report the same information over time. By showing how workload and performance have changed over the period, the effects of workload on performance may become clearer.

Sometimes, it is helpful to add trend lines to smooth out rough charts. In any event, changes over time are much easier to see and understand. Exhibit 33-1 is a sample chart showing improvement in availability as the messaging system was stabilized over a 15-month period.

Messaging can be viewed as a process, with the delivery times of messages as the result of the process. Statistical process control (SPC) techniques, common in manufacturing industries, can be used to manage the message delivery process.

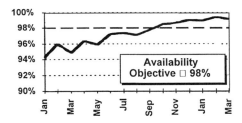

Exhibit 33-1. Performance Measurement Charts.

# MANAGING ELECTRONIC MESSAGING SYSTEMS

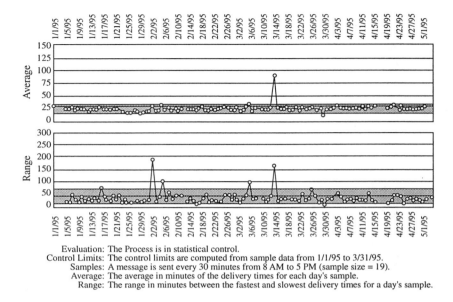

Evaluation: The Process is in statistical control.
Control Limits: The control limits are computed from sample data from 1/1/95 to 3/31/95.
Samples: A message is sent every 30 minutes from 8 AM to 5 PM (sample size = 19).
Average: The average in minutes of the delivery times for each day's sample.
Range: The range in minutes between the fastest and slowest delivery times for a day's sample.

**Exhibit 33-2. Statistical Process Control Charts.**

The two most common SPC charts are the Average and Range charts (as shown in Exhibit 33-2). On these charts, the average and range of daily delivery times are used to compute the process's capability limits. The charts clearly show changes and trends in the process, often giving early warning of problems affecting message delivery.

## SUMMARY

Messaging system managers must measure and report both performance and workload factors. Performance factors tell how well the system is processing messages. Workload factors tell how much work is being performed and how much capacity is being used. Both types of information are useful in managing the messaging system. When performance declines, workload changes can be used to explain the decline and justify additional resources.

# Chapter 34
# Using Statistical Process Control to Manage Message Delivery

*Clyde F. Haggard*

Statistical process control (SPC) is a practical tool that can be used to manage message delivery. It is comprised of many mathematical techniques used to control processes. SPC commonly refers to methods based on statistical sampling and the concept of standard deviations.

A strong mathematical background is not required to apply SPC in a messaging application. Most often used in manufacturing, SPC can be used in any area where a repetitive process with a measurable result occurs. It is well known in the quality control and quality assurance areas.

Message delivery is a process. Cross-platform messaging is delivery from one message store to another. If the message stores are on different messaging systems, the process is much more complex and difficult to manage. The delivery times of individual messages are the process result. Because delivery times for every message cannot be measured, sample messages must be used. This chapter details how samples of message delivery times can be analyzed and graphed using SPC.

The benefits of SPC for messaging managers include establishing process capability. For example, is the process capable of delivering messages within 30 minutes between message store A and message store B? If improvements are made in the process over time, they will show up in the SPC charts, communicating accomplishments. The SPC charts also provide early warning of process changes and indicate changes to the process.

# MANAGING ELECTRONIC MESSAGING SYSTEMS

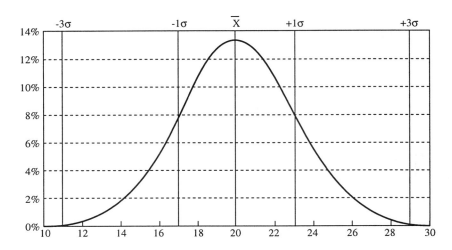

Exhibit 34-1. The Normal Distribution.

## MATHEMATICAL FOUNDATION

SPC is actually several mathematical techniques used to analyze processes. This chapter concentrates on the most useful techniques based on the normal curve (i.e., the bell-shaped curve and normal distribution). It is a distribution occurring throughout nature (e.g., crop yields per acre, rainfall amounts, weights of cattle, or men's heights). It is assumed that message delivery times also fit the normal distribution. Although that assumption might not be valid, it is close enough to be useful.

To analyze messaging processes, certain statistical values are computed from the delivery times of sample messages. Because the normal distribution is well understood, mathematicians have provided tables that make it easy to compute the useful values (see Exhibit 34-1). The mean is the average of all the data values. It is the midpoint of the normal distribution and is represented by an X with a bar over it. Standard deviation ($\sigma$) is a measure of the spread of normal distribution. Because normal distribution is symmetric around the mean, 68.2% of the values will fall within one standard deviation on either side of the mean. Similarly, 95.4% will be within two standard deviations of the mean, and 99.7% will be within $\pm 3\sigma$.

A process has natural variations, and every result is not the same. Every component in the process has something that influences the result. In delivering messages, there are different traffic levels on the network, or one message may spend more time in a queue than the one before it. If no variations existed, every message would be delivered in the same length of time.

The capability of a process is the area within three standard deviations ($3\sigma$) of the average of the sample sizes. This is the area within which 99.7%

## Using Statistical Process Control to Manage Message Delivery

of the sample values will fall. If the midpoint (Xbar) of the normal curve is 20 minutes, and a standard deviation is 3 minutes, the process is capable of producing delivery times from 11 to 29 minutes. Sample average delivery times are expected to fall in that range. For message delivery, the lower capability limit is typically ignored, because faster delivery is better.

The capability is not the tolerance. If it is decided that delivery times must be 15 minutes or less, that is the upper limit of the tolerance. If a process has an upper capability limit greater than 15 minutes, the process must be changed to bring its capability within the tolerance. In manufacturing, products whose measurements are outside the tolerance are rejects and are either scrapped or reworked. With electronic messages, the rejects are used, but the customers will let management know that tolerances are being exceeded.

### Range and Average Charts

The two most useful and common SPC charts are the range and average charts (see Exhibit 34-2). Both charts are graphs of data computed from samples.

To be useful, a sample should consist of at least five values. For a messaging system, sample messages every half hour from 8 a.m. to 5 p.m. will produce 19 values per day in a daily sample. If a few values are missing, the remaining values can still be used as the daily sample.

For a given sample, the range is the difference between the largest and smallest values. The average is also computed. For ease of use, the values are in minutes from the time the message was sent until it was delivered to the recipient message store. By plotting computed average and range values instead of the individual sample values themselves, some of the variability of the data is reduced, and the number of points to be plotted is minimized. The average of a sample from a population is fairly stable from one sample to another. Although the average gives the center of the sample, the range tells how wide it is.

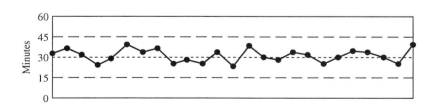

Exhibit 34-2. Range and Average Chart Format.

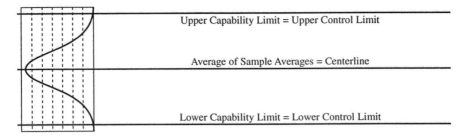

**Exhibit 34-3. Capability Limits.**

The capability limits of a process are just the 3σ limits of the normal distribution. When the normal curve is turned on its side and extended over time, the capability limits define the range in which the values will fall (see Exhibit 34-3). Statistical formulas help compute the capability limits from the samples themselves.

The first step is to choose a sample size (e.g., 19 values as outlined above). Then, 20 or more samples are taken to give a base to calculate the capability limits. Some statistical process control books have tables with some factors needed. The appropriate values depend on the size of the samples used. The books also have a simple formula for each of the following:

- Average chart: upper control limit
- Average chart: lower control limit
- Range chart: upper control limit
- Range chart: lower control limit

The formulas are not provided in this chapter because they contain factors that are dependent on sample size. Once the sample size is determined and the factors obtained from the appropriate statistical tables, it is easy to put them in a spreadsheet and never worry about them again.

Many satisfactory SPC books are available in company, public, and university libraries. The section on average and range charts provides the information needed, and usually includes a table that lists these values for various sample sizes: A2, D3, D4. The books may be listed under statistical process control, quality, quality control, quality assurance, statistics, or process control.

## CONTROL CHART ANALYSIS

The primary purpose of control charts is to analyze the variation that occurs within a process over time (see Exhibit 34-4). Control charts pro-

## Using Statistical Process Control to Manage Message Delivery

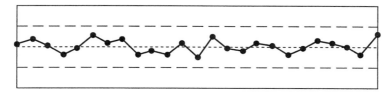

Exhibit 34-4. Control Chart for a Process in Control.

vide information as to whether a process is being influenced by common or special causes of variation. Common causes refer to the many sources of random variation within a process. Common causes are evidenced on the control charts by the absence of points beyond the control limits, and by the absence of nonrandom patterns or trends within the control limits.

A process is said to be in statistical control when influenced strictly by common causes. A special cause is a source of variation that affects the process in unpredictable ways. When special causes are present and making the process unstable and unreliable, the process is said to be out of statistical control.

The characteristics of a process in statistical control should be noted. Just as expected from a normal distribution, most of the points are near the centerline (mean). A few points are out near the control limits. No points, or only a rare point, are beyond the control limits. No patterns or trends are evident in the points plotted — they occur randomly.

About three points out of a thousand will fall outside the control limits if the process is under control (see Exhibit 34-5). Most of the time, it is possible to identify something that affected the process. For message delivery, this might be a network problem or a hardware or software failure. The cause of any point outside the control limits should be identified to be sure problems are addressed promptly.

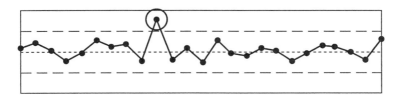

Exhibit 34-5. Points Outside the Control Limits.

563

# MANAGING ELECTRONIC MESSAGING SYSTEMS

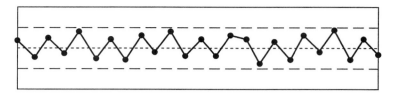

**Exhibit 34-6. Systemic Variability.**

A predictable pattern, such as where a low point is always followed by a high point, or vice versa, is an example of systematic variability (see Exhibit 34-6). It indicates that something that is not random is affecting the process. If the cause is found, fixing it will often reduce the spread between the control variables and lower average delivery time or range.

**Patterns to Watch For.** A run of seven successive points above or below the centerline (mean) is an out-of-control condition (see Exhibit 34-7). Such patterns should be watched for to determine if the process has changed or if something new is affecting it.

A run of seven intervals up or down is a sign of an out-of-control trend (see Exhibit 34-8). This is a classic early warning of a problem with the process. The process does not have to result in points outside the control limits before the problem can be addressed.

Unnaturally large fluctuations, characterized by erratic up and down movements, are an indication of instability in the process (see Exhibit

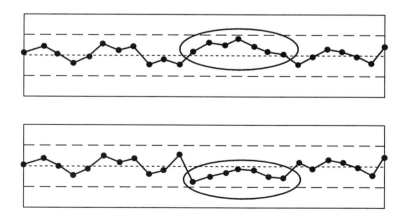

**Exhibit 34-7. Run of Seven.**

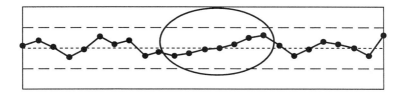
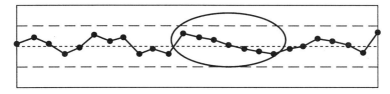

**Exhibit 34-8. Trends.**

34-9). An unstable process is out of statistical control. The sources of variability should be located to eliminate them.

If an unusual number of points are near the edge of the chart, the chart may show a mixture (see Exhibit 34-10). In this case, it appears that the result is coming from two different processes. Such a process is out of statistical control.

Stratification, where too many of the points are close to the centerline (mean), is too good to be true (see Exhibit 34-11). This indicates that the points are coming from a different distribution than the one used to compute the standard deviation and control limits. A sudden shift in level indicates a change in the system or process (see Exhibit 34-12). The mean, standard deviation, and control limits should be recomputed to reflect the new process.

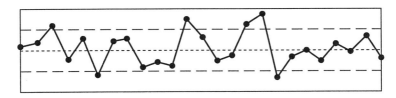

**Exhibit 34-9. Instability.**

## MANAGING ELECTRONIC MESSAGING SYSTEMS

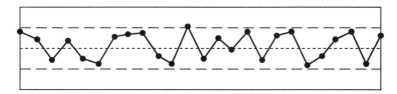

**Exhibit 34-10. Mixtures.**

Short trends in data that occur in repeated patterns are cycles (see Exhibit 34-13). This is the result of some special cause that must be corrected for maximum efficiency. Freaks are sudden abnormal conditions (see Exhibit 34-14). These are most often seen as the result of some external cause, such as a power failure delaying messages.

### CROSS-PLATFORM MESSAGES

SPC is best applied to certain individual pathways between message stores. A message sent from message store A to message store B will be processed differently than a message from A to C.

A pathway is the unique combination of hardware and software that processes a message between one message store and another. It may involve storing the message at various points, translating it from one message protocol to another, and reasserting the message. If each message store can send messages to every other message store, the number of unique pathways is equal to the number of message stores squared.

SPC can be used to identify message delivery problems. Because messages delivered to the same message store is rarely a problem, it is generally not necessary to be concerned with the time to deliver messages from message store X to itself. In fact, message delivery times between similar message stores is a not a common problem.

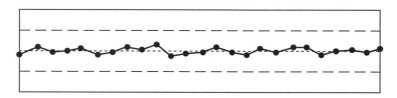

**Exhibit 34-11. Stratification.**

## Using Statistical Process Control to Manage Message Delivery

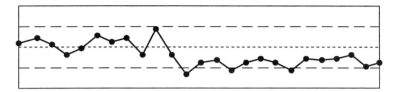

**Exhibit 34-12. Sudden Shift in Level.**

For example, mail from one Microsoft Mail PC Post Office to another on the same system is usually quick and simple. SPC is unlikely to offer many benefits on homogeneous systems. Messages from a message store that go through a gateway to another messaging system may encounter a variety of delays. These types of pathways are the ones that generate the most complaints and that benefit the most from SPC.

### SPC AS AN EARLY WARNING SYSTEM

SPC can provide early warning of message delivery problems. Delivery times may begin to increase, or the ranges begin to widen. Are such problems caused by a temporary fluctuation or is the process not capable of delivering mail consistently?

SPC will not substitute for more in-depth analysis, but the charts offer clues to the nature of the problem. Temporary fluctuations may show up as freaks, or a short period of higher values. Messages should be monitored to see if a run of seven or a trend is developing.

Customers sometimes want messages delivered faster than the process is capable of delivering. SPC can show this in graphical form. If the centerline (mean) is at 35, and the upper control limit is at 55, less than half of the messages will be delivered in under 30 minutes, but almost all of them will be delivered in less than an hour. The message delivery process can

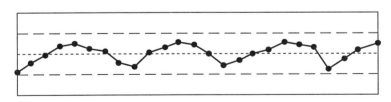

**Exhibit 34-13. Cycles.**

567

# MANAGING ELECTRONIC MESSAGING SYSTEMS

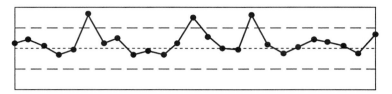

**Exhibit 34-14. Freaks.**

change without any warning. Network modifications can easily speed up or slow down the process. Capacity limitations may suddenly cause message delivery to slow down. SPC will show these quickly.

Variability is the enemy of SPC. A primary goal is to eliminate as much variability as possible from processes. Daily problems must be fixed, not just corrected when they occur. Message delivery should occur like clockwork. The goal is to have so few problems that constant firefighting is eliminated. A process that is subject to a new difficulty every day is not under control.

Change control is essential. Modifications to the message delivery processes should be planned and well thought out. This minimizes unexpected changes to the processes. Dedicated messaging hardware and software are required for a consistent message delivery process. Using the messaging systems for other processes introduces too many variables for consistent message delivery.

## APPLYING SPC TO PROCESS IMPROVEMENT

To apply SPC, administrators must choose the pathways to be measured, send sample messages along the pathways, collect the data, construct the charts, and analyze the charts. Many more pathways between message stores exist than can be measured.

Because data collection can be tedious, the pathways to be measured should be carefully selected. First, administrators should choose the pathways that generate the most complaints. Customers are quick to recognize pathways that have slow message delivery. For customer relations, addressing complaints using SPC offers big dividends.

Representative pathways can be used by choosing one pathway for each group of similar source and destination message stores. For example, if five Quarterdeck Mail Post Offices are on Macintoshes, and eight Microsoft Mail Post Offices are on PCs, and a single pathway in each direction. One pathway might be from Quarterdeck Mail Post Office A to Microsoft Mail Post Office 7. The other pathway might be from Microsoft Mail Post Office

## Using Statistical Process Control to Manage Message Delivery

2 to Quarterdeck Mail Post Office E. It is important not to measure too many pathways — the amount of data collected can be overwhelming and difficult to automate.

When pathways share components (e.g., a software program, network segment, or gateway), problems may show up on those pathways simultaneously. This can be helpful in identifying the causes of the problems. For example, if all the pathways passing through an Alisa switch suddenly show long message delivery times, but other pathways do not, the first place to look for a cause is evident.

Programs or scripts can typically be written to generate sample messages automatically at the desired intervals. Each message should identify the time it was sent in its subject or body. The tedious part of statistical sampling is recording the actual delivery times. Sometimes a program can be written to collect the data. Other times each message must be read to determine when it was sent and when it was delivered. The number of minutes required for delivery of each message should be computed and entered on a worksheet. These worksheets are then entered onto a daily spreadsheet similar to Exhibit 34-15.

It is helpful if outages, incidents, and changes are recorded on the daily spreadsheets for use in later analysis of the SPC charts. The daily spreadsheets also can be published for the messaging operations staff. A summary spreadsheet can be used to collect the range and average data for each month.

### Real-World Examples

The following examples are from an actual multiplatform messaging system. Microsoft Excel was used to create the SPC charts from the range and average data. The example SPC charts in Exhibits 34-16 through 34-20 were

| From Message Store: | A | A | B | B | C | C |
|---|---|---|---|---|---|---|
| To Message Store: | B | C | A | C | A | B |
| Time | | | | | | |
| 8:00 | 1 | 3 | 48 | 45 | 10 | 12 |
| 8:30 | 2 | 6 | 45 | 44 | 13 | 18 |
| 9:00 | 2 | 5 | 40 | 42 | 15 | 17 |
| 4:30 | 7 | 8 | 29 | 27 | 23 | 18 |
| 5:00 | 9 | 10 | 25 | 22 | 32 | 19 |
| Low | 1 | 3 | 25 | 22 | 10 | 12 |
| High | 9 | 10 | 48 | 45 | 32 | 19 |
| Range | 9 | 7 | 23 | 23 | 22 | 7 |
| Average | 5 | 6 | 34 | 32 | 15 | 16 |

**Exhibit 34-15. Daily Spreadsheet.**

# MANAGING ELECTRONIC MESSAGING SYSTEMS

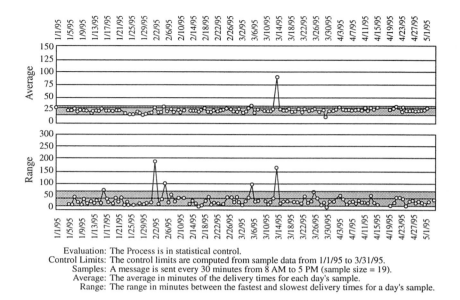

Evaluation: The Process is in statistical control.
Control Limits: The control limits are computed from sample data from 1/1/95 to 3/31/95.
Samples: A message is sent every 30 minutes from 8 AM to 5 PM (sample size = 19).
Average: The average in minutes of the delivery times for each day's sample.
Range: The range in minutes between the fastest and slowest delivery times for a day's sample.

**Exhibit 34-16. Process in Good Control.**

---

created in an Excel workbook. A new Microsoft Excel workbook was prepared each month, containing the worksheets and charts for the monitored system. Each pathway has a worksheet containing three months of historical data, one month of current data, formulas to compute the SPC data, and the average and range charts.

The three months of historical data are used to compute the process average and control ranges. Processes change over time, and the goal is to make the centerline lower (i.e., fewer average minutes to deliver a message) and the control limits closer together (i.e., smaller standard deviation). When current month data are entered, the charts are monitored on screen for out of control conditions.

Four months of data are collected on each set of SPC worksheets. When a new set of worksheets is started, three months' data are rolled up to leave room for the new month. New capability ranges and a new centerline are computed for the coming month based on the three months of historical data. The fourth month is then free to enter current data as it is collected. Exhibits 34-16 through 34-20 are actual SPC charts illustrating some interesting process conditions.

**Processes In and Out of Control.** In Exhibit 34-16, the process is in good statistical control. The freaks stand out clearly. The freak that is on both

*Using Statistical Process Control to Manage Message Delivery*

the average chart and on the range chart was a disk failure. The others were software problems in the Alisa switch that were fixed by the vendor. The average delivery time (centerline) is just over 25 minutes. There is a built-in delay in this process because the two All-In-One mail systems are on different networks. Mail is transferred through a third VAX that connects to one network to send and receive mail, and then disconnects. After an interval, it connects to the other network to send and receive mail, and then disconnects. This prevents the networks from being connected to one another, but introduces an average 15-minute delay in the mail delivery process. Overall, however, this process is very consistent with a narrow control range.

The process in Exhibit 34-17 had a great deal of variability in the first month; however, great improvements were made during the first 40 days or so on this chart. The centerlines and control limits are high and wide because of the wide fluctuations in delivery times in February and March. It is evident from the data that the average centerline should be about 25 minutes during the last half of this period. As the earlier months roll off, the centerline and control limits automatically adjust themselves. The fluctuations at the right are the result of testing some accounting software on the Alisa Mail Information Switch. The system clearly did not have enough capacity to process the mail and keep the accounting statistics as well.

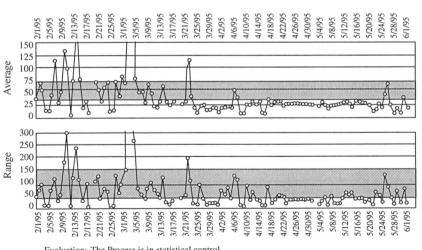

Evaluation: The Process is in statistical control.
Control Limits: The control limits are computed from sample data from 2/1/95 to 4/30/95.
Samples: A message is sent every 30 minutes from 8 AM to 5 PM (sample size = 19).
Average: The average in minutes of the delivery times for each day's sample.
Range: The range in minutes between the fastest and slowest delivery times for a day's sample.

**Exhibit 34-17. Process Coming into Control.**

## MANAGING ELECTRONIC MESSAGING SYSTEMS

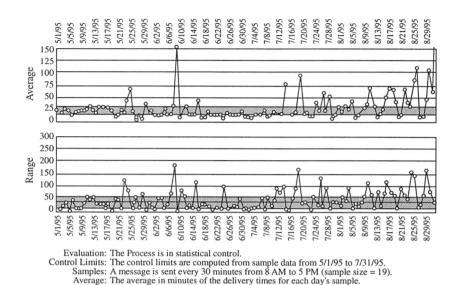

Evaluation: The Process is in statistical control.
Control Limits: The control limits are computed from sample data from 5/1/95 to 7/31/95.
Samples: A message is sent every 30 minutes from 8 AM to 5 PM (sample size = 19).
Average: The average in minutes of the delivery times for each day's sample.
Range: The range in minutes between the fastest and slowest delivery times for a day's sample.

**Exhibit 34-18. Process Going Out of Control.**

In July, this process began to go out of control (see Exhibit 34-18). Analysis indicated that all the available network connections on the destination post office were used up during the day, and the program that picked up and delivered mail was unable to log on. Consequently, mail delivery to that post office often waited for hours until a network connection was open. Two Microsoft Mail PC post offices, each with about 450 users, are on the POP1 server. Upgraded PC post office servers were ordered to permit conversion to Windows NT and an increase in simultaneous network connections from 300 to 1000.

This process had progressively worse performance from July until November 12th. What was done on the 12th to get such dramatic improvement (see Exhibit 34-19)? A fire in a transformer cut off all power for most of the day. Every post office server, every gateway, every network component, and most of the client PCs and terminals were shut down. It was learned that some network changes were made during the power outage, but there is no real explanation for the improvement. As Exhibit 34-20 shows, the problem began to recur in December.

In early December, the process in Exhibit 34-20 starts to go out of control again. Then, on December 16th, a sudden dramatic improvement to the process occurred. This was the date of replacement of the hardware sup-

## Using Statistical Process Control to Manage Message Delivery

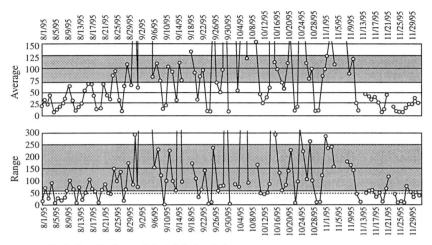

Evaluation: The Process is in statistical control.
Control Limits: The control limits are computed from sample data from 8/1/95 to 10/31/95.
Samples: A message is sent every 30 minutes from 8 AM to 5 PM (sample size = 19).
Average: The average in minutes of the delivery times for each day's sample.
Range: The range in minutes between the fastest and slowest delivery times for a day's sample.

**Exhibit 34-19. Process With Unexpected Improvement.**

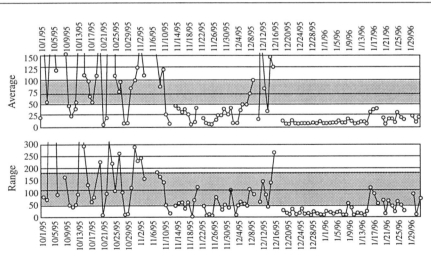

Evaluation: The Process is in statistical control.
Control Limits: The control limits are computed from sample data from 10/1/95 to 12/31/95.
Samples: A message is sent every 30 minutes from 8 AM to 5 PM (sample size = 19).
Average: The average in minutes of the delivery times for each day's sample.
Range: The range in minutes between the fastest and slowest delivery times for a day's sample.

**Exhibit 34-20. Process Changes.**

573

porting Alisa Systems Information Switch software, and conversion to a new version of the software. A new DEC Alpha 5/250 and Release 5 of Alisa removed a bottleneck in processing and provided improvements to all the pathways that use Alisa. In January, the plots still show some problems with the network connections. Some of the HP servers have difficulty in supporting two 3Com network cards.

**SUMMARY**

It is more difficult to address the problems that SPC points out than it seems. Many times, no quick fix is available, or the real cause of the problem proves elusive. The preferred procedure is to review all points that fall above the capability band (upper control limit). Those exceptions are then matched up to known problems and corrective actions. Exceptions that do not have a match are investigated as far as possible. Recurring problems get special attention to be sure that appropriate efforts are being made to eliminate them. Exceptions between pathways must be correlated to see if there are common problems. It is normal for a problem with one component to affect several different pathways.

Emphasis on process improvement must continue. Customers always expect better service at lower cost. SPC helps manage delivery expectations and measure improvements in the process. The average and range charts provide visual evidence of what processes are capable of achieving, and show that room for improvement exists.

Comparison of SPC charts from different pathways offers clues as to where capacity restraints might be eliminated. Above all, administrators must strive to reduce the variability in mail delivery. Customers should be able to count on delivery within a specific period of time, whether it is 1 hour, 15 minutes, or 30 minutes.

The effectiveness of messaging management must be communicated to customers. There is no other way to get the funding and resources required to provide quality messaging management. The first thing SPC provides is a way to picture process capability. The range and average charts are used to show clients the capability of messaging processes. Trends, good or bad, show up and can be used to justify additional resources if required. Most important, the charts show that messaging management is not simple.

SPC can show that capability can be improved. More resources can be applied to reduce average delivery times and variability. Process changes, or the lack of unplanned changes, can add to the capability of the messaging system. Software and hardware changes may be immediately reflected in the SPC charts. SPC can also show capability that is being reduced. If the system is growing with more clients, more mail messages, or larger mail messages, systems capability is being reduced. SPC charts show the effect of these increases in demand.

Many factors can affect process capability, and SPC can help identify them. External changes also can reduce the capability of the messaging system. Network modifications or increases in network load from other systems may show up in the SPC charts. Software or hardware changes may have an adverse effect on mail delivery times. SPC charts should be used to keep customers and messaging management up to date with changes to the messaging processes.

Process control is an important part of effective messaging management. Range and average charts should be used as management aids. Administrators should identify problems affecting processes, identify trends, allocate problem correction resources, and monitor the effects of process changes. Without adequate information, how can process improvements required to maintain the desired process capability be justified? Satisfactory message delivery times may require additional resources, monitoring, problem-solving capability, or processing capacity.

# Chapter 35
# Information Security

*Phillip Q. Maier*
*Lynda L. McGhie*

The electronic messaging, communications, and electronic commerce industries are making well-planned and orchestrated moves toward the expanded use of open network technologies and the convergence with private network systems. With that migration, traditional information security threats increase in scope, as more opportunities grow exponentially for data compromise, destruction, corruption, misdirection, and user impersonations.

Information security has continued to gain its well-deserved place in the information technology environment. This position, however, has not evolved without considerable frustration and compromise to traditional security goals and objectives. Modern technology and the media have facilitated a heightened awareness within management and user communities alike. The principle issue has not changed — protecting the integrity of information and information systems — but the methodologies, control mechanisms, and the approach have changed appropriately. Integrity, confidentiality, and privacy are still the primary concerns and objectives of today's information security architectures and supporting policies and procedures.

## INTEGRITY AND PRIVACY POLICIES

Information integrity is achieved when an electronic file or document is transmitted across an open public or unprotected network and is received by the intended recipient without disclosure or modification. Information confidentiality can be defined as the act of keeping things hidden or secret. This includes data on some type of storage medium, automated data resident on computer systems, and transmitted data over communications and networking systems. Typical control mechanisms include physical controls, software controls, and transmission controls.

Information privacy is the guaranteed or implied protection of information using technical control mechanisms or electronic security, primarily authentication and encryption. Most companies today are covering their legal risks by publishing and enforcing corporate or company management policies informing employees of their implied or explicit rights to security versus the company rights.

This means employees should be notified and then acknowledge the receipt of the company's privacy policies. Based on the company's risk profile, business environment, and other variables specific to its requirements, the privacy policy may contain or mandate strong prohibitions (e.g., the phrase "all information created for business purposes, created on company-owned computing assets and during the employee's paid workweek, are considered the property of the company and all rights to the information is exclusively owned by the company"). In this example, employees are not guaranteed any degree of privacy over their personal information created on company computing resources, or business-related information created during the course of their job.

The extreme opposite of this strict policy would be to do nothing, treat each case or problem as an individual incidence, and not publish a company-wide privacy policy. There are many degrees of prohibitions and approaches in between; however, policies in this middle-ground area are much more difficult to enforce and require more interpretation and case-by-case approvals. Examples of categories and types of information under privacy rights, individual or company, are also required to support any policy in between. Appropriate use policies (AUPs) are another popular mechanism for communicating the company position relative to the use of electronic messaging and communication systems, including public and private networks (e.g., the Internet).

**Risk Trade-offs**

Efforts are underway at the national level to legislate the protection of electronic information and assign areas of responsibilities to individuals, carriers, and institutions. The Telecommunications Bill will no doubt be debated and interpreted for some time to come. It attempts to define indecent material, using laws that are ancient and have not been used for years, much less applied to the today's complex communications environment. A House-sponsored amendment adds criminal penalties for transmitting objectionable material, subjecting the provider of the transport mechanism to severe criminal penalties, but who makes the decision as to what objectionable material really is, and how it is defined, is open to much debate.

Who should determine what constitutes objectionable material? And presuming that we can assign that responsibility to the correct body or individuals, how does it get implemented or applied globally to today's communications infrastructure or electronic communications environment? Most people abhor child pornography and would want to restrict it, but the trade-off is that censorship won't stop there. Industry and commercial companies must track the progress of this legislation and ensure that their opinions, directions, and requirements are represented.

Current telephone ordering systems transmit credit card numbers in clear text. Online transactions using credit card information are not any more or less secure. The hype surrounding the inherent risks of privacy and integrity over public networks has assisted in management and customer awareness, but budget allocations toward secure distributed systems implementations, including secure messaging systems, are still slow to evolve. The trade-offs here are similar to general risk management methodologies where we must continue to maintain an equitable balance between level of risk, protection of information, and access to that information.

All risks cannot be eliminated, and a more holistic approach to security is advocated and tied to the overall corporate information security architecture for electronic messaging.

Security for security's sake can be cumbersome to implement and very costly. Living in fear of using the benefits of electronic commerce as a result of not understanding the risks is often the path of choice for many senior managers and financial officers. Sitting back to see how technology evolves or waiting for effective security mechanisms to be created to protect and ensure privacy of electronic messaging is also not an effective approach. As with all protection plans or privacy policies, the fundamental guiding component or tool is a risk assessment, resulting in identification of corporate resources requiring protection. Legal aspects and liabilities must also be factored in.

Using an overall automated risk management approach to electronic information protection and privacy can be an effective approach by enabling the end-user to make the final decisions as to the protection requirements of each message or transaction. All resultant information and outcomes should be aggregated as input to the overall corporate risk and vulnerability model as well as the information security architecture. Too much security at one company could very well be too little at another.

## SECURITY REQUIREMENTS FOR ELECTRONIC MESSAGING SYSTEMS

Information security requirements for electronic messaging systems are consistent with security requirements in any of today's highly interconnected, global, mobile, and distributed computing and networking environments. The challenge is to be proactive rather than reactive to changes to risks and vulnerabilities based on changing technology and systems, communications, and application changes or enhancements.

In the traditional mainframe environment, or the security mentality and environment of yesteryear, information security involved building fences around sensitive or valuable resources. In today's environment, we cannot continue to take this approach, but must be more open and receptive toward looking for connectivity solutions that exploit today's tools and technologies while still maintaining secure computing environments. Consideration should be given to conformance to the overall information security architecture and overall corporate-level information security policies and procedures when designing and implementing new security systems (e.g., those bundled solutions required to secure and protect electronic messaging systems).

Information security requirements can be categorized as identification, authentication, authorization, audit, and administration. Secure and reliable authentication and encryption are the fundamental requirements for public or open network access, electronic commerce, and electronic messaging systems. The following section addresses the risks and vulnerabilities inherent in electronic messaging systems, as well as the overall enterprise supporting communications and E-mail systems. Following the discussion of risks and vulnerabilities, the previously mentioned security requirements and components of sound and secure electronic messaging systems are discussed in detail, with recommendations for an integrated overall information security architecture.

## THREATS AND VULNERABILITIES WITHIN OPEN NETWORKS

As with any type of communication using an electronic medium, the potential for security vulnerability is great, especially in the growing arena of electronic messaging. The standard data threats exist: data integrity, data compromise, and destruction, as well as threats specific to the messaging environment including misdirection of electronic messages and no authentication of the sender or receiver.

Some forms of electronic messaging have addressed these threats with closed environment models, but as will be discussed later, this type of environment may not be suitable for the broad open messaging networks in place today.

Understanding the emergence of the open network model of today's business environment is key to performing a risk assessment of a messaging environment and developing a corresponding security architecture for a messaging system (see Exhibit 35-1). An open network model is usually defined as a network that is beyond the direct control of the sender and receiver. Certain parties on the network may not always be intended recipients of any or all of the messages.

### Multilevel Security

An open network could be a network with as few as three users (i.e., Users A, B, and C in Exhibit 35-1), where user A needs to send a private mes-

# Information Security

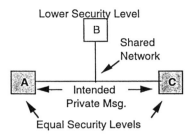

**Exhibit 35-1. Messaging in an Open Network.**

---

sage to User C, and only User C, and where user B has no right to be able to read the message. In government security terms, this would be referred to as a *multilevel security environment*.

Multilevel security exists where users on the same network have multiple levels of security requirements. In the example given, Users A and C would share the same security level, with User B at a lower or different security level (i.e., no authority to read A and C's messages). In open networks, this implies that the network is open to multiple levels of users, not all of whom should have access to all the data or messages traversing the network.

This concept is of paramount interest and, of course, the motivation for developing some secure methodology to conduct electronic messaging. Historically, closed networks were put into place where like parties exchanged messages in a closed loop, where all shared a common interest in the messages, and potentially this multilevel security requirement was not prevalent. For the purpose of this discussion of messaging threats, it is important to understand the open network model as an environment where multiple users share the same network medium, but where not all users on the network need to know the message content of each and every message.

The electronic messaging industry is definitely making the move to the expanded use of open networks, as the convergence of private networks continues to occur in the messaging environment. But with the move to open networks, the common data threats increase in scope. More opportunities now exist in the transport medium for data compromise, destruction, or misdirection, as well as user impersonations.

## Data Compromise, Destruction, Misdirection, and Impersonation Threats

Recognition of these threats is important in developing the appropriate protection model in an open network environment. The data compromise threat is probably the largest area of vulnerability in an open network environment where no safeguards are implemented (e.g., message encryption). The prime example of this type of environment is the Internet mail system. There are millions of parties on the open Internet, all sharing the same medium, but each with their own levels of security requirements based on the message content being exchanged. With clear text (unencrypted) messages traversing the Internet, the capability of unintended parties intercepting and reading the message content is great.

Using the Internet open network as an example, the other threats become quite apparent. Data integrity is a critical threat in this type of environment; it is difficult to maintain some form of assurance that the message sent is not corrupted, intentionally or unintentionally, before reaching its intended receiver. When sending clear-text messages, there is the possibility of interception (data compromise), alteration of the content, and the resending of the data to the intended receiver without the receiver knowing that any of this corruption took place.

Following these threats are the additional threats of data destruction, misdirection of messaging, and impersonation of an authorized sender or receiver for the purpose of distorting the truth or casting unfavorable representations of the person or organization being impersonated.

Impersonation of a messaging party is becoming a greater threat than data compromise. In today's electronic medium, whatever appears in print carries great weight for believability. As an example, an impersonator may generate a message or purchase order under someone else's identity or organization's identity, which is believed to be true by the receiver. Without a sophisticated authentication scheme, the receiver acts on the order only to have it canceled or unpaid by the impersonated party. This type of impersonation is also referred to as spoofing, which originated with spoofing of devices on a network, but now has become interchangeable with spoofing the identity of a network user.

### Traffic Analysis Threats

Without going to a more detailed level, additional threats exist in the open network model. They include traffic analysis where outside parties monitor an organization's or person's traffic flow, identifying who they are communicating with, and developing a profile on their traffic patterns. This data profile can reveal an organization's business partners or potential business partners during the relationship-forming stages. These types of threats can target various levels or sources, from competition-sensitive materials, to government levels (local or international, depending on the scope of the open network), information warfare motivations, and outright malicious mischief by interested third parties.

The concept of an open network is one of the single largest threats to a secure messaging environment. As the overwhelming advantage of external communication increases, however, the drive to use open networks for messaging outside the enterprise outweighs the threats.

## INTERNAL AND EXTERNAL SECURITY REQUIREMENTS

Electronic messaging security requirements span both the private internal network, including carrier-provided leased lines, and public networks (e.g., the Internet). Security concerns are similar in both environments, although most companies secure their internal network with firewalls. In both environments, users must be authenticated by some trusted network security server/services, and sensitive information must be encrypted using either symmetric or asymmetric key encryption services.

Authentication and encryption requirements are not unique to electronic messaging. Security solutions should be consistent with other internal and external security architectures (e.g., mainframe, client/server, local area network, and electronic data interchange). Security services (e.g., integrated key management, authentication, encryption, and administration services) should be planned, implemented, and administered centrally to ensure consistent and cost-effective management.

There are other considerations providing input to the overall information security architecture and electronic messaging security. These include an understanding of the value of the corporation's information, an information classification scheme, an up-to-date threat and vulnerability assessment, an overall understanding of user access and information and electronic resource sharing requirements, and an understanding of security technologies, products, and viable integrated solutions.

## ELECTRONIC MESSAGING SECURITY MECHANISMS

Information security is typically divided into several major categories relating to the overall goals and objectives of the security policy and architecture. These higher-level guidelines are further implemented at the functional and operational support level through control mechanisms comprised of hardware, software, and administrative processes. The following sections give examples of typical categories and technical control mechanisms that implement the company's overall privacy objectives.

## IDENTIFICATION (LOGONIDS)

Most corporations employ standard userid configurations to enable secure and efficient logonid maintenance, including the ability to ensure unique identification and ease of database management. Options for unique logonid configurations include employee number or Social Security number. A less-structured userid configuration empowers users and recognizes employee individuality and creativity. An organization may adopt a userid configuration standard much less defined or stringent (e.g., enabling system administrators to select the userid naming conventions or employees/users to select their own logonid).

E-mail logonids frequently do not follow conventional userid naming standards or corporate standards because many E-mail systems are LAN-based or outside the IS-controlled backbone network and mainframe computing environments. Typical naming standards in less-structured environments may include some combination of user's name, (i.e., last name, first name; last name, first initial; or first name, last name).

As noted, consideration and planning to interface to the overall information security architecture should acknowledge the need for strict conformance to standards to enable and support single signon, and database reconciliation to ensure the prompt removal of userids when employees terminate or access is no longer required. Other problems include the need to maintain userid reference tables and directory synchronization across disparate mail systems and userid databases.

## AUTHENTICATION SERVICES

Authentication is the process used to verify the identity of a user, device, or other computer- or network-based entity. Origin authentication depends on the validity of E-mail header information indicating the originator of the message.

Most E-mail users take for granted the integrity of the originator or sender's address. It is possible to forge the header so that it indicates that the message was sent or originated from another address (referred to as E-mail address spoofing). Sound or secure electronic authentication systems prohibit or deter this type of forgery. Origin authentication ensures that the originator of the message is indeed who he or she claims to be. The authentication system is similar to a public notary system for physical documents ensuring the authenticity and legality of the signature.

### Static Passwords

Implementing standard effective password management practices, including allowing users to select their own passwords, ensures that the user can easily remember it and therefore will not compromise by writing it down. The organization should provide education and training on effective password management practices (for system managers), including the selection of secure passwords (e.g., passwords with embedded numerics or special characters, recognition of upper and lower case, and other syntax enhancements). The operating system (OS), network operating system (NOS), E-mail system, or other application may not enforce site-specific password configurations (e.g., enforcing password changes, password expiration, minimum and maximum password length, and violation count incremented toward automatic userid suspension). If so, scripts, utilities, or manual processes can be used to enhance password management deficiencies and coordinate password management practices across the overall systems, network, and application architecture.

## Kerberos

Kerberos is an open systems-based distributed authentication system developed to support Project Athena at MIT. Kerberos provides a trusted third-party authentication service using Data Encryption Standard (DES) private key encryption. Its fundamental assumption is that unprotected workstations wish to access servers distributed across the enterprise.

In an unprotected network, client workstations cannot be trusted to identify users correctly or to pass on that authenticated process to other servers on the network. The Kerberos authentication server knows the passwords of all authorized registered clients, and stores the password in a centralized secure database.

The authentication server also shares a unique private encryption key with each client that has previously been distributed physically or in another secure manner. A ticket granting service (TGS) issues tickets to users who have been authenticated by the central authentication server. The ticket includes the user's logonid, the network address of the client, and the logonid of the TGS. The password is never transmitted across the network.

The service-granting ticket has a time stamp and a predefined lifetime, thereby enabling the user to access previously authenticated servers and services without reinitiating the logon process. Kerberos also enables a single sign-on process and provides a mechanism to encrypt sessions using the secure private key, which authenticates clients to the application server and key distribution center (KDC).

## RSA

RSA is a public-key cryptography system developed by Rivest, Shamir, and Aldeman. The Internet standard for privacy-enhanced mail uses the RSA system. Public-key systems have a publicly known encryption key, so that virtually anyone can send an encrypted message. A person or organization publishes its public key in a directory that is accessible to anyone wishing to transmit encrypted data to recipients in the directory. Messages and other data may by encrypted by this public key. The owner of the public key also has a private key (known only to the owner), which is the mechanism used to decrypt the encrypted data back to clear text. Both keys are generated together, but cannot be derived from one another. Public keys can freely be distributed without compromise to the secrecy of the private key.

## Privacy Enhanced Mail (PEM)

PEM provides an automated means to encrypt a message before transmitting on insecure or unprotected network systems (i.e., open). No separate procedure is needed to encrypt the message and once the mail message is encrypted, it cannot be compromised or obtained by an unauthorized user. E-mail is usually sent in by way of the Internet (public network) or the internal network using simple mail transfer protocol (SMTP). SMTP transmits data in clear text format. SMTP can be used to transmit ASCII text data only.

To encrypt SMTP messages, an indirect encryption mechanism must be used. First, the electronic message must be encrypted to convert it to a binary file. SMTP cannot be used to transmit binary data; it transmits text-only data. The sender must then encode the binary data as text to transmit encrypted SMTP messages. A popular way to do this on the Internet or insecure networks is to use a utility known as UUENCODE. UUDECODE utilities can be used to convert the encrypted files back to readable text. There are issues to consider regarding the distribution and management of encryption keys.

## Smartcard Technologies

Smartcard technologies employ the use of challenge response systems or mutual authentication protocols. These protocols authenticate communicating parties to one another by exchanging private session keys.

Smartcards or hand-held portable (HHP) devices have microprocessor input-output ports, and a few kilobytes of nonvolatile memory. The user must have one of these devices in his possession to be able to logon to the target system. Authentication is based on something a user knows and something a user has — a two-factor identification/authentication process. The host computer prompts the user for a value, obtained from a Smartcard, then the user is prompted for a password by the computer. In some cases, the computer provides the user some piece of information that the user has to enter into the Smartcard. The Smartcard then displays a response that must be entered into the computer. If the response is accepted, the session will be enabled.

## Certified Digital Signatures

Message authentication protects two parties who exchange messages from any third party; but it does not protect the two parties against each other. The digital signature function includes this authentication function.

The government-established standard for this mechanism is outlined in FIPS PUB 186, the Digital Signature Standard (DSS). This particular standard is not extensively used in commercial industry, but is a mandatory standard for the federal government. Until it becomes a more recognized and accepted standard, most commercial entities will continue be slow to incorporate its benefits.

### X.509 Directory Authentication Services

X.509 is based on the use of public-key cryptography and digital signatures. The standard recommends the use of RSA and is part of the X.500 distributed directory services model.

The directory is a user database and authentication server. The database includes a mapping of userid, network address, and other attributes and user-defined information fields. The directory serves as a repository for public-key certificates. Each certificate contains the public key of a user and is signed with the private key of a trusted certification authority. In addition, X.509 defines alternate authentication protocols based on the use of public-key certificates. The user certificate is assumed to be created by some trusted certification authority (CA) and placed in the directory by the CA or by the user. The directory server itself is not responsible for the creation of public keys or for the certification function; it merely provides an easily accessible location for users to obtain certificates.

## AUTHORIZATION SERVICES

Although the major components of electronic messaging security are authentication and encryption services, authorization to information, electronic resources, and services are traditional security components. Such techniques and technical control mechanisms should be integrated into the overall information security architecture.

Secure authorization services require a preauthenticated logonid to ensure the integrity and security of the access control management process through authorization to computer and network information and services. The authentication process integrates and coordinates the operating system authorization services, the network operating system, security software systems (e.g., CA-ACF2, RACF, and TopSecret), and application systems security.

## ADMINISTRATION SERVICES

Account administration processes should be automated and standardized to the extent feasible depending on the size of the administrative support organization and budget allocated to the function. Procedures for registering users and requesting privileges should be established with the goal of making the process as cost-effective, timely, and secure as possible. Administrative processes, policies, and procedures and automated utilities should be created, maintained centrally, and ported to distributed systems managers and administrators on a regular basis as updates and enhancements are applied.

Policies for shared and guest accounts should be developed, keeping in mind the need for standard processes and the criticality, sensitivity, and integrity required for independent and collective systems. Adherence to the overall information and information security architectures should also be considered.

An approval hierarchy should be developed for levels of access, including classes for privileged users, security administrators, systems administrators, data owners, and application developers. Consideration should be given to added security awareness training or possibly background investigations for privileged users or security and administrative support personnel. Where possible, network administrators should limit special privileges, regularly audit privileged activities, and reconfirm access requirements minimally at six-month intervals.

## AUDIT AND ALARMING SERVICES

Enterprise systems typically require the logging of all transactions, including access to information, transactions, utilities and commands, and logon. A compromise to this policy is the logging of violations, access to sensitive data/transactions/commands, or privileged access. Traditionally, logging files are not regularly reviewed because the volume of data is impossible to keep up with, and the ability to associate anomalies is also difficult given the great numbers of records for review.

Automated audit processes, including rule-based and knowledge-based audit systems, while expensive to implement and time intensive to develop, provide return on the initial investment over the long term in the operational implementation life cycle. Violation reconciliation procedures should be developed, supported by strong company or corporate-level management policies with adequate levels of clout, enforcement, and support. Strict enforcement and penalties should be clearly defined and enforced. Users should be made aware of these cases, violations, and corrective and punitive actions taken.

## PRIVATE MESSAGING NETWORKS VS. PUBLIC NETWORKS

No discussion of messaging security is complete without introducing the much publicized issue of the benefits of closed/private messaging networks over more open and accessible public networks. Using the Internet, a public network, as an example, and such private message handling systems (MHSs) as a proprietary value-added network (VAN), the comparison and contrast of the two point out vast differences from a security perspective.

The traditional messaging VAN, can be considered a closed network, at least in comparison with the Internet as it exists today. A VAN offers a more centrally run network, with very specific boundaries. Many are operated by a centralized controlling party, with very defined operating procedures, protocols and member lists. The users share a common interest in communicating in this medium. Historically, VANs emerged for vertical markets further highlighting the commonality of the participants.

In direct contrast to this is the Internet — no centralized controlling mechanism, no clear boundaries, and an unknown and diverse member list with varying interests and motivations to communicate or participate in this medium. The issue facing many organizations today is whether to use the open Internet as their messaging or transport medium, instead of their more closed and perceived secure private networks.

The Internet offers a globally connected and accepted standard, with practically no boundaries, and a growing member list. Additionally, the Internet functions as a diverse communications medium, offering more then just plain old E-mail or static messaging, as the explosion of the World Wide Web has proven in recent years.

The argument for staying within the confines of a private VAN environment are based on perceived security as a result of the limited messaging types, limited membership list, common goals in communicating, and somewhat centralized control. The fact that outsiders are minimized through targeted membership may change, especially as VANs build gateways to the Internet. These gateways are generally implemented in a controlled fashion, where only the Internet-bound traffic and replies are passed through the gateway, maintaining the protection over the wholly internal VAN traffic.

Many organizations are recognizing the need to use the Internet or are being driven by market forces to use the Internet as their messaging environment because of its global and market acceptance. The Internet's very benefits, however, present the biggest risks or threats to conducting business communications over it on a broad scale. Its openness makes it conducive to message interception, corruption, destruction, messaging impersonation, and possibly even misdirection if the corruption is deep enough into the Internet infrastructure. In addition, the mere action of connecting an organization's enterprise to the Internet puts the enterprise itself at risk through unauthorized access from Internet members.

The risks can be categorized on two levels, one being the need to protect the message as it traverses the open network (in this case the Internet) and two, the need to protect the enterprise when making the connection to the open network.

## MANAGING ELECTRONIC MESSAGING SYSTEMS

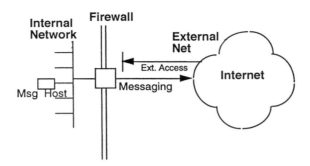

**Exhibit 35-2. Firewalled External Connection.**

---

The following section addresses the security mechanisms that can protect messaging in an open network, but the second issue of protecting the enterprise is a major issue that also must be recognized.

Connecting any private information enterprise to the Internet requires a strict security architecture implementation. The common standard of transmission control protocol/Internet protocol (TCP/IP) of the Internet makes connectivity almost too easy. Unless very specific gates or network security firewalls are implemented between a private information enterprise and the Internet, vast exposures can be created into the private network.

A full discussion of network security firewalls is beyond the scope and intention of this chapter, but suffice it to say that the firewall component must be an integral part of any plan to connect a private enterprise network to the Internet or other open network. Exhibit 35-2 illustrates a simple firewalled Internet connection from a private information enterprise network. The specific data flows between these two enterprises would of course be defined in the security policy statement for the organization.

### DIGITAL MESSAGE CERTIFICATION AND KEY MANAGEMENT

Once the decision has been made to subject an organization's messaging system to the capabilities and threats of an open network, specific security precautions must be invoked to mitigate the risks identified in this type of environment. The fundamental mechanism for providing security in open networks is encryption.

With encryption, however, comes the complexities of managing an encryption system (e.g., choosing an encryption methodology that is right for an environment both from a data protection standpoint and environment compatibility perspective). As an example of environment compatibility, a full end-to-end encryption methodology is not acceptable or functional in a public network because the routing data of the message would not be discernible by the rest of the wide area network infrastructure.

From a functional standpoint, a weak data encryption methodology is not worth the effort. Simple encrypted messages often become targets. They become a challenge to many users to decrypt and publicize for peer recognition, financial gain, or good neighborliness, to point out the shortcomings of an encryption method or product.

In light of these decision factors, the generally accepted security model is one of the various methods of public key/private key secure encryption architectures. As previously stated, there are various implementations, with the leading products for E-mail and messaging being PEM and PGP. Both are based on similar public key/private key architecture. Internal administration, with respect to certifying authorities, differs between the two implementations. A critical component of both these architectures is the implementation of the digital certificate. The digital certificate closes the loop in protecting messaging in an open network.

Review of the various components of the public/private key architecture indicates how each method or component provides protection against the threats identified:

- A message encrypted under the public key/private key architecture protects it from compromise/disclosure to unauthorized parties.
- A message encrypted with a sender's private key and inclusion of the sender's digital signature authenticates the sender.
- Incorporating a secure hash function protects against data corruption, providing a degree of data integrity in an open network environment.
- The use of a certified digital certificate protects against impersonators or spoofing of another's identity.

To illustrate how the digital certificate closes the loop in this security architecture, a rudimentary knowledge of the public key/private key architecture must be assumed. Given that a direct relationship exists between the user's public key and his/her private key, a message encrypted with one of these keys can be decrypted with the other. The public key is made available to all users, hence the name public key.

At issue with this basic architecture is the ability of others to purport to be someone else, register themselves as someone else, and post or publicize a public key as such. Any messages sent to or from this impostor under the posted public key could be interpreted to be true, without some form of trusted, certifying authority attesting to the validity of this key as belonging to the named user. By incorporating a trusted authority, a digital certification can be generated that contains not only the user's information in support of the public/private key architecture, but also the electronic

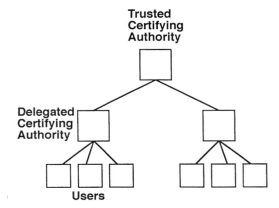

Exhibit 35-3. PEM Hierarchical Certification.

signature of the certifying authority indicating that the key truly belongs to the named user.

Under this architecture, the issue is who or what entity should be named as the certifying authority. In the RSA realm, a spin-off company has been formed to fill this void. In the Internet community, the Internet Activities Board (IAB) fills this function. The point is that visible, trusted parties fill the role as the certifier of these digital certificates. Anyone bearing their authorized certification can be trusted to represent the named owner.

In the PEM public key/private key architecture, this is established in a hierarchical method with one single certifying authority at the top. The certification process is then delegated hierarchically down the chain. In this manner, an organization can become a certifying authority, if it is empowered with this ability in the hierarchy. This gives them the ability to certify users within their organization. Exhibit 35-3 illustrates the hierarchical certifying authority structure.

The PGP certifying authority structure has a different philosophy on the certification process. It operates in more of a distributed mode. Instead of building a hierarchy conducive to large organizations, PGP allows any one (or more) member with a PGP certificate to certify additional users in the PGP certification web. The theory being that: if User A certifies User B's certificate, and User C receives User B's certificate, and he or she knows A, but not B, by determining that B has been certified by someone he knows (User A), he or she has some assurance that User B's certificate is valid. Exhibit 35-4 provides a graphic illustration of the PGP certifying relationship as discussed here.

*Information Security*

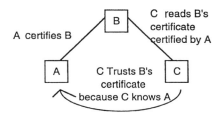

**Exhibit 35-4. PGP Distributed Certification.**

Once the certificate process has been established and followed with recognized certifying authorities, the digital certificates can be freely distributed. They are no longer prone to counterfeiting, as they now bear the mark or certification of another party in the process (or in the PGP architecture, multiple marks, or certifications).

From an organizational viewpoint, the hierarchical process is generally more fitting and controllable as a method for certification of their members. For example, organization AA establishes itself as a certifying authority in the certifying hierarchy, and now has the ability to certify its members under its own certification authority. An organization's AA-certified members can now exchange certifications and secure messages within the hierarchy, not only in the organization but outside of it, with other certified members, as illustrated in Exhibit 35-5.

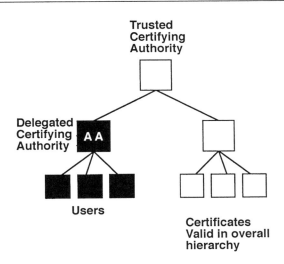

**Exhibit 35-5. Delegated Certification.**

593

The certification process provides the needed protection of the public-key features in an open network messaging architecture. Still to be addressed, however, is the task of establishing the basic components needed to make a user certifiable, namely the initial generation of the key pair (the public key and private key).

The assignment and verification of the assigned or named user's identity to this key pair and the distribution of the user's private key in a secure manner are also concerns. In a large organizational environment, this can be performed in a manner similar to establishing user accounts, through a centralized medium, with distributed points of contact for end-user verification. This process would entail generation of a key pair, assignment and verification of the users identity (and association of user name, E-mail address, and other user attributes), generation and certification of the user's certificate with this information, distribution of the private key, and posting or registration of the certification in the certificate directory for the messaging system the key pair is to be used.

A secure infrastructure must be built to support this process, to ensure the secure distribution of the new user's private key. This is where the system is used to build on itself. Designated remote or distributed points of contact are established in the system with a public/private key pair, so that they can now exchange secure messages with the central site where new key pairs are securely generated. When new User X wishes to enter the system, and their nearest point of contact is User C, a key pair can be generated centrally and distributed securely through the remote point of contact (User C), who already has secure messaging capability. These remote-distributed points of contact can also be used to verify the identities of the remote users, so they do not all require physical verification at some single central site.

As evidenced by this discussion, a secure messaging infrastructure can be built with the components of the public key/private key architecture. Attention must be paid to the various risks identified, but the tool sets exist to mitigate the risks and make them acceptable in today's business environment if appropriately applied. All of this relies on the industry acceptance of industry directory standards and public-key architectures as stated.

## COMPUTER CRIME REPORTS

A recent Datapro Information Services Group survey revealed that many companies today still do not have adequate and up-to-date corporate security policies. Of the 150 companies that responded to the Michigan State survey of 600 large companies, 148 said that they had suffered from crimes that included the theft of credit-card numbers, trade secrets and software, and snooping by employees into confidential computer files.

Statistics still consistently reveal that over 90% of computer crime comes from a company's own internal employees. Most companies today spend their predominate allocation of security revenues in the protection of external networks. In the same survey, more than 43% said they had been victimized 25 times or more, the survey being conducted by the university's School of Criminal Justice in East Lansing. Most conclude that companies should pay more attention to hiring, training, and management practices rather than focusing so much on technological safeguards (e.g., network firewalls and encryption).

**SUMMARY**

Safeguarding a messaging environment requires the same attention as nearly any other data processing environment, and possibly more because of the explosive growth and interconnectivity of today's messaging architectures. The development and implementation of protection mechanisms must be conducted with the overall security architecture of the enterprise in mind, so as not to overprotect the environment and make it unusable or highly inefficient.

As this chapter points out, many risks and conditions must be considered, but there are toolsets available that when used properly can achieve this balance of the appropriate amount of security while still maintaining the ability to openly communicate in externally connected enterprise environments.

# Chapter 36
# Securing Electronic Messages
*Gilbert Held*

Because of differences among businesses, the need for message security varies with the organization. In addition, different types of data transmitted by different groups or departments in an organization can require varying levels of message security.

For example, some organizations (e.g., banks and brokerage firms) require a very high level of security for wire transfer operations, which can represent the movement of a considerable amount of funds. For the transmission of research reports, customer orders, and administrative data, a lower level of security may be sufficient to meet both the potential threat to the information to be transferred as well as its value to the organization. For other organizations (e.g., insurance companies), the transmission of customer policy information may not warrant the implementation of any security measures. Therefore, the degree of security needed is based on the type of data transmitted and the potential threat to that data.

Three major areas of message security must be considered:

- Message privacy
- Message integrity
- Authentication

Each area may require a different set of implementation measures, depending on such factors as the type of transmission facility used, the availability of a computer for use by message originators and message addressees, and funds available to acquire security-related hardware and software.

## MESSAGE PRIVACY

Message privacy involves measures taken to ensure the contents of a message are not understandable to unauthorized persons. The most common method used to implement message privacy — addressing — is also the least secure method of effecting message privacy. This is true because the address is used for delivery routing, and although it may ensure that the message reaches its correct electronic destination, it does not ensure that the message is not read by unauthorized persons either through an illicit intercept or from the inadvertent or purposeful observation of the contents of a message when it reaches its destination.

**Codes vs. Ciphers**

More sophisticated methods used to ensure message privacy supplement addressing and include the use of codes and ciphers. Codes involve the substitution of one word for another word or phrase. Though popularly employed with the transmission of information over telegraph, telex, and teletypewriter exchange service systems, codes are used only by old-timers on modern electronic transmission systems because of the development of hardware and software encryption-performing products.

In an encryption system, each character in a message other than addressing information is replaced by another character through the use of a pseudorandom number generator. The generator uses a key to initiate a sequence of random numbers. Those numbers are used in a modulo arithmetic operation with the plaintext characters in a message to form ciphertext characters. The transmitted message then appears to represent a sequence of random characters without meaning. At the electronic destination of the message, the recipient uses the same key to generate a similar sequence of random numbers through the use of specialized hardware or software. The hardware or software performs a modulo arithmetic operation on each of the received ciphertext characters in the plaintext message.

**Encryption Methods**

Two methods can be used to encrypt and decrypt electronic messages — online and offline. Online operations can be performed by hardware or software, though offline operations are primarily restricted to the use of software.

**Online Operations.** Online encrypting and decrypting of electronic messages requires hardware or software to automatically encrypt the body of a message while recognizing the message header and permitting the envelope to remain in plaintext. Doing so ensures that the message can be correctly routed through an electronic mail system. Otherwise, the encryption of the message header would impair the ability of the electronic mail system to route the message to its correct destination.

Ideally, encryption and decryption operations are either transparent or near-transparent to the user, only requiring the encryption key to be changed on a periodic basis.

One of the major problems associated with online encryption systems (e.g., numerous commercially available devices based on the federal Data

Encryption Standard) is that such systems generate encryption that is unsuitable for transmission over many electronic mail systems. In addition, the file representing the encrypted message may not be retrievable if stored on certain types of personal computers. Understanding the cause of these problems requires an explanation of computer codes.

All products that comply with the Data Encryption Standard (DES) use a key to generate a sequence of pseudorandom numbers. Those numbers are used in a modulo arithmetic operation performed on each character in a plaintext message, resulting in an equal probability of generating any of the 256 characters in the 8-bit character set.

Unfortunately, many electronic mail systems are limited to supporting 7-bit ASCII character sets. This means that a message whose characters were encrypted into 8-bit characters cannot be transmitted over a 7-bit electronic mail system. Even when a message encrypted using the DES algorithm is transmitted over an 8-bit electronic mail system, problems may be encountered when the retrieved message is stored on certain types of personal computers. This occurs because one of the encrypted characters that a DES-enciphering process can generate is the CTRL-Z character. Unfortunately, DOS uses that character as the end-of-file mark. Therefore, unless the encryption system automatically decrypts messages before their storage on a personal computer using DOS, the message or the portion of the message following the CTRL-Z character may be irretrievable.

**Offline Operation.** The second method of encrypting messages — offline encryption — is accomplished through the use of a software program that encrypts a message before transmission. The program encrypts the contents of a file as an entity. Then, the file is transmitted using an electronic mail system in which the message originator manually enters the destination address that is used by the mail system as an envelope to transport the file to its destination.

In a manner similar to using an online encryption system, the recipient of a message encrypted offline must have the key used for the encryption operation. In addition, the recipient must have a copy of the program used to perform the encryption operation because most software programs use propriety methods, and messages encrypted by one vendor's program cannot be correctly decrypted by another vendor's program.

## Public vs. Private Keys

Two types of keys are used with encryption systems — public and private. Before the use of each type of key is discussed, their general use in enciphering and deciphering operations is reviewed.

Exhibit 36-1 illustrates the use of a key to generate a sequence of random binary numbers of encryption and decryption operations using modulo 2

```
Encryption
  key = 10110101
Plaintext                        10110100...
Random sequence                  10101010...
                                 --------
Modulo 2 addition produces
Encrypted text                   00011110...
                                 --------
Decryption
  key = 10110101
Ciphertext                       00011111...
Random sequence                  10101010...
                                 --------
Modulo 2 subtraction produces
Decrypted text                   10110100...
```

**Exhibit 36-1. Encrypting and Decrypting Using Modulo 2 Arithmetic.**

---

arithmetic. The key, usually in the form of several hexadecimal characters, is entered into hardware or software and converted into a binary number that generates a pseudorandom number sequence.

As shown at the top of Exhibit 36-1, the resulting pseudorandom number sequence is added to the binary representation of the plaintext through modulo 2 addition. At the receiving end, the user enters the same key to generate the same pseudorandom number sequence. The number sequence generated by the key is then subtracted from the binary value of the encrypted text through modulo 2 subtraction, resulting in the decrypted text having the same value as the plaintext before its encryption. This process is illustrated in the lower portion of Exhibit 36-1.

The use of a single key to encrypt and decrypt messages requires the key's value to be known by both the message originator and receiver. This type of key is known as a *private key* because its disclosure enables any person gaining access to the encrypted message to obtain the ability to read the message.

The most common problem associated with a private key concerns its exchange between the message originator and message recipient. Unless key values are predefined for initial use, which can be a complex process if many people require message security, the alternative is to initiate the first key exchange over a nonsecure communications link that lets users exchange additional keys at predefined times in a secure mode.

The private-key concept is the basis for DES-compatible hardware and software products that represent the vast majority of security products available for use. However, it is not the only type of key.

**RSA Systems.** A second key-based system was developed by Ronald Rivest, Adi Shamir, and Len Adleman in 1977. Then professors at MIT, they formed the company RSA Data Security, which markets the RSA Public Key Cryptosystem for commercial and governmental use.

The RSA system uses a matched pair of encryption and decryption keys. Each key generates a sequence of pseudorandom numbers that, through a mathematical process, perform a one-way transformation on the data. Each key generates a mathematical process that is the inverse of the process generated by the other key. One key is known as the RSA public key and can be made available by its owner to other persons. The second key, known as the RSA private key, is kept secret.

Under the RSA system, a message originator uses the recipient's public key to encrypt a message. That message is then decoded by the recipient using his or her private key. Because the process is reversible, a person using the RSA system can encrypt a message using the private key that can be decrypted by another person using the public key. Therefore, the RSA system permits secure communications without requiring any previous relationship between persons (e.g., entering predefined keys into a hardware product or software program).

## MESSAGE INTEGRITY

When a message is transmitted, its contents should be received correctly and without error. The contents can be either plaintext or ciphertext; the integrity of the message is separate from its privacy.

Most electronic mail systems provide end-to-end data integrity through the subdivision of messages into blocks of data on which an error-checking algorithm is performed. The algorithm generates a cyclic redundancy check (CRC) pair of characters that are added to the end of each data block.

As the data block flows through the electronic mail network, each node performs a similar error-checking algorithm, generating a CRC that is compared with the CRC contained at the end of the data block. If they match, the block is considered to have been received without a bit error occurring. If they do not match, one or more bit errors is presumed to have occurred, and the receiving node requests the transmitting node to retransmit the data block. Therefore, errors are corrected by retransmission to ensure the integrity of data as it flows through the network.

# MANAGING ELECTRONIC MESSAGING SYSTEMS

Although almost all electronic mail networks provide data integrity, a major weakness in using those networks involves transmission into the network and extraction of messages delivered by the network. Most transmission into and extraction of messages from electronic mail systems occurs by asynchronous transmission. Unfortunately, asynchronous transmission does not include an error detection and correction mechanism. Therefore, it is extremely important to use error detection and correction modems to transmit and receive data from electronic mail systems. Doing so eliminates the weak link in message integrity and provides error-free transmission from the message originator through the electronic mail system to the message recipient.

## AUTHENTICATION

The authentication process verifies that a message was transmitted by the person who claims to have originated it. Similar to message integrity, message authentication is a separate security consideration. Depending on the method used to obtain message privacy, however, a built-in message authentication capability may be obtained.

Authentication can also be defined as the process that separates bogus from legitimate messages. Several common authentication systems are used with financial messages. The most common authentication systems are used with financial messages. The most common system is used by the Society for Worldwide Interbank Financial Telecommunications (SWIFT). This proprietary system is available only to members of SWIFT; its method is not disclosed to the general public.

If a mechanism is required to authenticate messages, a system that is both easy to use and difficult for an unauthorized user to duplicate should be developed. For example, the Julian date is easy to use but would also be fairly obvious if several messages were intercepted by a person who desired to send a bogus message. If instead of the Julian date, an equation was used (e.g., $J*5 - D*7$, where J is the Julian day and D is the numeric value of the day of the week), it would be considerably more difficult for an unauthorized user to determine the method used to develop an authentication number appended to the end of a message.

If the message is encrypted, it contains built-in authentication in the form of the key used to encrypt the message. In other words, if the message recipient cannot successfully decrypt the message, either the distribution of keys between the message originator and recipient was incorrect or the message originator was someone other than who he or she claimed to be. In the second situation, a person may be using available hardware or software to send a bogus message using an incorrect key because that person is not privy to the composition of the correct key.

Therefore, encryption usually ensures the authentication of a message, though an interloper gaining access to hardware or software previously correctly set using a valid key could transmit bogus messages that would appear to be valid. To prevent this situation from occurring, the physical security methods and procedures that govern the access of employees to equipment and software used for message privacy should be reviewed.

**SUMMARY**

Electronic message security is built on a three-tier foundation that must consider the privacy, integrity, and authentication of messages. Although each tier can be independent of the others, it is important to ensure the use of appropriate equipment and procedures to obtain all three elements. This will help to guarantee that the contents of messages are known only to appropriate persons, that the messages arrive intact, and that the recipients know that the messages were not bogus.

# Chapter 37
# E-Mail Security and Privacy
*Stewart S. Miller*

The majority of electronic mail, or E-mail, is not a private form of communication. E-mail is often less secure than sending personal or business messages on a postcard. Many businesses monitor employee computer files, E-mail, or voice mail. Some corporations monitor E-mail to ensure that trade secrets are not being communicated to the outside world. Because E-mail often travels through many computers, it is easy to intercept messages.

Bulletin board systems, college campus networks, commercial information services, and the Internet are mainly open information systems where hackers can easily tamper with E-mail. Whenever mail is sent over the Internet, the message first arrives at the Internet service provider's (ISP) outgoing mail server. Once there, anyone using that provider can read the mail as it goes out to its destination.

Passwords do not protect E-mail. Most major E-mail and groupware products that combine messaging, file management, and scheduling allow the network administrator to change passwords at any time and read, delete, or alter any messages on the server. Network monitoring programs, including AG Group's LocalPeek, Farallon Computing's Traffic Watch II, and Neon Software's NetMinder, allow network managers to read files sent over the Internet. In fact, these products mimic tools specifically designed for surveillance used primarily on mainframe systems. Encryption is a key element in secure communications over the Internet.

Pretty Good Privacy (PGP) software encrypts E-mail-attached computer files, making them unreadable to most hackers. PGP is a worldwide standard for E-mail security. Anonymous remailers allow users to send E-mail to network newsgroups or directly to recipients so that they cannot tell the sender's real name or E-mail address. For business communications, one of the motivations behind the use of PGP is to prevent the sale of company business plans or customer list information to competitors.

## ESTABLISHING SECURE E-MAIL STANDARDS

Secure Multipurpose Internet Mail Extension (SMIME) is a standard for secure E-mail communications that will soon be built into most E-mail products. A secure E-mail standard allows users to communicate safely between separate or unknown mail platforms. SMIME guarantees security end-to-end using digital signature technology. SMIME can be used for applications such as processing business transactions between trading partners over the Internet.

SMIME is one of the only ways to prove a user sent what he or she claims to have sent. The SMIME specification was developed to allow interoperability between various E-mail platforms using Rivest-Shamir-Adleman (RSA) encryption. This standard permits various encryption schemes, various key lengths, and digital signatures. SMIME also supports VeriSign's digital certificates, which is a form of identification used in electronic commerce.

Verification of E-mail services is a key component in preventing fraudulent messages. Netscape Navigator supports VeriSign's Digital IDs and SMIME. Qualcomm, the manufacturer of Eudora, plans to include an application programming interface (API) layer to SMIME in Version 3.0 of its E-mail program. In addition to encryption, SMIME modules link into the Eudora translator and offer PGP. WorldTalk, a manufacturer of gateway mail software, is building SMIME support into its network application router to permit cross-communications throughout disparate E-mail packages in addition to a centralized mail management and Internet access.

When a gateway supports SMIME, businesses can audit files as they enter and leave the company. SMIME can replace software based on PGP code. The difference between the two security technologies is that SMIME uses a structured certificate hierarchy; PGP is more limited because it relies on precertification of clients and servers for authentication.

Businesses are working closely with agencies such as the Internet Engineering Task Force (IETF) to achieve effective security for E-mail on the Internet. The lack of a common E-mail security standard is a hurdle to electronic commerce efforts. The Internet has failed to achieve its full potential because of the lack of secure transmission standards. However, standards are continually being proposed at meetings such as the E-Mail World conference, at which businesses work toward ensuring interoperability between E-mail vendors' implementations.

### Secure Directories

Many companies are developing directories of businesses on the Internet. Banyan Systems Inc. has released Switchboard, a highly scalable directory that allows Internet users to locate electronic addresses and other information for businesses worldwide. Switchboard appears to be the biggest Internet address directory in existence. This system also offers safeguards to protect privacy and permit secure communication.

When Internet E-mail users express concerns over privacy, Switchboard implements a feature much like Caller-ID that alerts a listed person whenever anyone asks for the person's address. The recipient, who will be given information about whomever is seeking the address, can then decide whether to allow access.

### Privacy Enhanced Mail (PEM)

The IETF is working to establish a standard for encrypting data in E-mail, designed to be a stable specification on which vendors can build products that can work together. Once the specifications have been clarified, the proposed standard is adopted as a final standard. The IETF standard for encryption includes Privacy Enhanced Mail (PEM) technology.

PEM encrypts E-mail into the Multipurpose Internet Mail Extension (MIME), which is the standard for attaching files to an E-mail message. PEM provides a utility called nonrepudiation, in which an E-mail message is automatically signed by the sender. Therefore, privacy is assured so that the author is unable to deny that he or she sent the message at a later point in time. PEM uses the Digital Encryption Standard (DES) and public key encryption technology to ensure that messages are easy for legitimate users to decrypt, yet difficult for hackers to decode.

PGP uses the RSA algorithm along with an enhanced idea encryption algorithm. Although the draft standards for PEM are not yet widely supported, they will probably gain acceptance as the language of the draft is clarified to remove ambiguity regarding the manner in which users are named and certified.

## LEADING E-MAIL COALITIONS

### Internet Mail Consortium (IMC)

The IMC is a new union of users and vendors interested in developing E-mail standards for the Internet. The group formed because its members feel that present organizations have not acted quickly enough to adopt standards for the Internet. The IMC acts as a link between the E-mail users, vendors, and the IETF.

The IMC plans to build consensus on conflicting Internet mail security protocols by holding informative workshops. This group's goal is to establish one unifying system that will ensure privacy in E-mail communications. The IMC's four founding members are Clorox Co., First Virtual Holdings Inc., Innosoft International Inc., and Qualcomm Inc.

For the IMC to attain its goal, it needs users and vendors to come together and discuss Internet mail issues. The IMC will most likely have even more influence than the Electronic Messaging Association on technical and business issues involving E-mail and the Internet.

**Electronic Messaging Association (EMA)**

The EMA is one source for users to consult if they have problems with their E-mail. The EMA's primary purpose is assigning standards to E-mail message attachments. In the EMA's efforts to regulate E-mail, security issues have been most prevalent regarding Internet communication. The EMA is making strides toward secure file transfers, thanks to the advent of PGP's success with encryption of E-mail file attachments.

The EMA is composed of corporate users and vendors of E-mail and messaging products. The EMA focuses heavily on the X.400 standard and has recently established a workgroup to research interoperability between Simple Mail Transfer Protocol (SMTP) and X.400 systems.

The EMA formed a Message Attachment Working Group in 1993 whose purpose was to develop a standard for identifying file attachments transmitted from one vendor E-mail system to another. The group was set up to use the IETF's MIME and the X.400 File Transfer Body Part (FTBP) as the method for identifying different attachment types. The FTBP defines an attachment by the application that created it. The EMA formed tests that were considered successful if the attachment was received without experiencing data loss.

Attachment transfers are simple when done from within one vendor's mail system, but difficult when performed across systems. File attachments from Microsoft, Lotus, and WordPerfect were used to make certain that the specification developed was capable of transferring attachments.

The next step is integrating MIME and SMIME support into future E-mail packages. It is typical for X.400 to be used as a backbone system for E-mail connectivity. There is a high degree of interest in developing secure methods of transmitting attachments in SMIME. MIME is the current standard for Internet file attachments, and SMIME is well on its way to becoming the secure standard for E-mail communications.

## CORPORATE SECURITY: PROBLEMS AND SOLUTIONS

One of the biggest security problems an organization faces involves how to implement a secure server yet allow access to the applications and resources from the corporate intranet. If an organization's Web or mail server is not protected by a firewall or kept on a secured part of the network, then data is open to hacker attacks. In terms of commercial transactions on the Internet, such a security breach can have lasting repercussions that could make customers lose faith in a company. A hacker could easily alter shipping data, create bogus orders, or simply steal money or products directly from a company's online site.

Security is a full-time job — E-mail is vulnerable to eavesdropping, address spoofing, and wiretapping. A security breach can be anything from unauthorized access by an employee to a hacker break-in. Attacks are not always conducted in a piecemeal fashion. Sometimes they occur on the entire system and focus on stealing or destroying the total assets of a company. Security breaches have resulted in corporate losses ranging from several hundred to several million dollars. Many organizations are not even aware that security breaches occur many times. A hacker can enter and exit a system undetected. Only when data or E-mail becomes lost, stolen, or tampered with do companies start to realize how much money an organization can actually lose in the process.

**Internal Precautions**

The first step toward preventing data loss is to take internal precautions. Defunct user accounts should be deleted right away, users should not log on at unauthorized or nonbusiness hours, and of course, users should be warned against posting their passwords in easily accessible places. Employees should be up to date concerning corporate security measures. Internal attacks can sometimes be thwarted simply by alerting all users that there are stringent security measures in place. Also, many workers do not realize the value of the data they have access to. Users can be encouraged to be more vigilant if they are made aware of potential losses due to breaches in security.

**Firewalls**

Firewalls provide an excellent means of keeping data integrity safe. They have the power to block entry points into the system — if an intruder does not have an account name or password, he or she is denied access. When configured correctly, firewalls reduce the number of accounts that are accessible from outside the network, and as a result, make the system much less vulnerable.

Firewalls are an excellent method of keeping attacks from spreading from one point in the network to another. Firewalls restrict users to one controlled location in the network — access is granted (or denied) at one highly guarded point. Firewalls stop hackers from getting close to security defenses and offer the best protection when placed near a point where the internal network or intranet connects to the larger Internet. Any network traffic that comes from or goes out to the Internet must go through the firewall, which then approves each transmission and deems it acceptable or unacceptable.

## Preventing E-mail Flooding and Denial of Service

Corporations sometimes fall prey to E-mail "bombs," which essentially flood an E-mail account with several hundred or thousand messages. This overwhelms the entire system and disrupts network services so that other messages cannot get through.

One solution some vendors provide is E-mail filters. If the E-mail bombs are originating from one domain or a few domains, the recipient can simply input those domains into the filter to be screened and deleted before cluttering up an E-mail account.

While an E-mail flood is interrupting service, a hacker can entirely disable or reroute services. These attacks can be combated by programming the system to shut out connections or questionable domains that repeatedly try to log into the system unsuccessfully. Attackers are therefore prevented from inputting multiple passwords in an attempt to gain access and shut down service. However, if repeated attempts on each user account result in shutting the account down, a hacker can effectively deny service to multiple people in an organization by simply trying to access all of the user accounts unsuccessfully. This would effectively deny service to most of the users.

When a security breach is successful, the hacker gains complete access to a user account, and assets are in jeopardy. One effective method of preventing an attack is to run a secure gateway such as Netscape's Commerce Server, which makes it very difficult for hackers to breach Internet security.

## Encryption

Most Internet E-mail security measures are accomplished using *encryption* and *authentication* methods. The Internet Privacy Enhanced Mail standard is the method of encrypting E-mail recommended by the IETF.

Another way to secure E-mail contents is the *digital signature* method, which identifies, stores, and verifies hand-written signatures electronically. This process is accomplished when users sign their names using a digitized pen on a computer. The service can record the specific signature metrics, the speed at which the signature is written, and the order of the unique hand-written strokes. The information is used as a basis to match with any computerized document signed by the same individual again. The comparison can determine the identity of the sender for submitting online payments or securing confidential data.

Administrators often encrypt data across wide area networks in addition to using digital signatures in E-mail packages to determine a user's true identity. The combination of encryption and digital signatures helps slow hacker attempts at gaining network access. However, companies still need to guard against the many methods of hacking, including phone tampering and remote access authentication.

The good news is that encryption is becoming universally accepted on the Internet. In some cases, users do not even realize that the latest Netscape Navigator Web browser employs encryption to secure both documents and E-mail messages. It uses the secure socket layer (SSL) protocol, supported by all of the major Web browsers and servers, to accomplish this goal and provide a safer means of communication.

### Unlisted Sites

One method of protecting an E-mail site is to take the *unlisted* approach. This is an effective security model that works only if no one knows that a particular site exists. If no one knows the site is there, no one will try to hack it. Unfortunately, this model is only good as long as the site remains a secret; and because the Internet is an open system with multiple search engines, the site will probably not remain secret for long.

### An Integrated Strategy

An organization concerned with network security may want to control access at each host and for all network services, which is more effective than the piecemeal approach of securing each service individually. This solution can involve the creation of firewalls for internal systems and the network, as well as incorporating detailed authentication approaches such as a password that expires after each use. Encryption can also be implemented on various levels to protect important data as it travels throughout the network.

### SUMMARY

The Internet evolved as an open system; however, it is this very openness that makes this venue of communicating so risky. The expansion of the Internet has promoted the growth of E-mail as a relatively quick, low-cost, easy-to-use method of communication. However, the irrefutable fact is that information is power, which makes this particular form of communicating an attractive target for thieves. This chapter has dealt with the problems surrounding E-mail privacy and proposed a few practical, easy-to-implement solutions.

# Chapter 38
# Creating Policy Relative to E-Mail
*Carroll M. Pearson* *

Many companies search for an effective policy governing the use of company-owned electronic mail (E-mail) facilities. The issue of privacy relates to employee messages that reside within and are transmitted internally by the E-mail system. Policy must be established and communicated to employees for two reasons: it provides legal protection for the company and informs employees as to proper business conduct.

A discussion of E-mail privacy quickly leads to other issues: privacy versus confidentiality; employee workplace privacy versus prudent business practices; ethics, morality, and legality; business-sensitive information; security; and enabling technology. This chapter attempts to clarify these issues relative to the central theme of E-mail privacy.

Guidelines are provided for companies wishing to formalize their E-mail privacy policies. In addition, companies are advised to monitor proposed legislation concerning workplace privacy, in particular E-mail privacy, and make their legislators aware of the need to preserve a company's right to regulate company-owned resources. Concerns should be addressed to members of the House and Senate Labor committees.

## OLD E-MAIL NEVER DIES

E-mail came into being in the late 1970s when companies with large mainframe computers introduced E-mail to the business. The computer-based facility allowed people to communicate with one another through electronic messages. One important business advantage was the ability to leave written messages without a hardcopy delivery system or actually reaching someone on the telephone. As technology advanced, the E-mail systems moved to the local area networks (LANs).

---

*Carroll Pearson has been instrumental in advising Lockheed Martin about policies associated with electronic mail or messaging. However, the content of this chapter and the advice given is that of Pearson's. None of the material contained herein is a statement by Lockheed Martin.

# MANAGING ELECTRONIC MESSAGING SYSTEMS

Both the mainframe and LAN-based E-mail systems collect the messages destined for or saved by individuals in data files that are now called *message stores,* which are located on a central server. These systems provide individuals with a means to store their messages in logical files similar to storing written correspondence in a file cabinet. Entire files or specific messages within a file may be marked for deletion by the owner of the E-mail account.

Systems administrators for the computer systems supporting E-mail make backup copies of files at intervals — some each night, others weekly — to provide recovery of the information in case of a system malfunction. Because of this system backup provision, information that an individual deleted could, and still can, reside on the system backup devices (usually magnetic tapes) until the backup tapes are reused or erased. The most noted example of the presence of E-mail on backup devices was the discovery of the Ollie North E-mail data on the White House computer backup tapes during the Iran-Contra affair.

In addition, systems administrators often delete messages after a specified period of time to manage storage space on the system. In general, the individual users are responsible for the content of the E-mail data and may retain data beyond a specified length of time. Individual data retention from LAN-based E-mail systems is usually done by copying the data to files at the individual's local desktop device.

## PUBLIC VS. PRIVATE E-MAIL

In 1986, the Electronic Communications Privacy Act became part of the law. This Act updated the federal wiretap law to protect the user against unauthorized access to the information being transmitted and stored by public service providers. The law does not include information being transmitted or stored by private companies or entities. Private companies or business entities own the resources associated with their E-mail systems and, as such, can regulate their use.

There is a significant difference between public and private E-mail. Public is that which is available to the public sector. Individuals or companies may buy the E-mail services from public providers. The companies that sell public E-mail services come under many government regulations associated with the right to privacy.

Private or company E-mail is provided by the company. The components of the service — the computers and the network — are owned, leased, or acquired by the company. Company E-mail systems do not come under the same regulations as public E-mail providers.

Exhibit 38-1 shows an individual using a public E-mail provider. The access to the data transmitted over the public communications lines and to

*Creating Policy Relative to E-Mail*

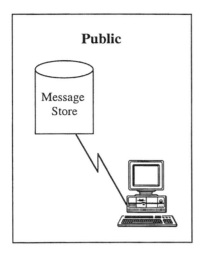
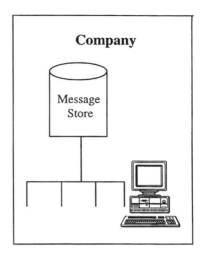

**Exhibit 38-1. Public vs. Company Messaging.**

the data stored in the public message store is regulated according to the Electronic Communications Privacy Act. Using a public system, an individual has data privacy rights. On the right of the same exhibit, the message store and the communications lines are provided by and under the control of the company. Employees using these resources come under the policy established by the company.

**Internet Mail**

The information superhighway, or the Internet, provides people with widespread E-mail opportunities. Many companies have gateways to the external Internet and use Internet facilities internal to their company. The internal Internet runs on computing facilities that are owned by the company, so the use and policies are set by the company. For external Internet E-mail, the messages flow through a loose federation of Internet nodes, very much a volunteer environment with no direct fee for use. This is a "use at your own risk" environment. There is no guarantee of E-mail delivery or privacy protection of the messages.

E-mail coming from a source external to the company is treated as internal company E-mail once it enters the bounds of the company-owned E-mail system. For example, if someone gives an organization both their company E-mail address and their private E-mail address — a system for which they, as individuals, have a personal subscription — the messages arriving at their company E-mail message store fall under the policy established by the company; the messages in the message store provided by the public service

# MANAGING ELECTRONIC MESSAGING SYSTEMS

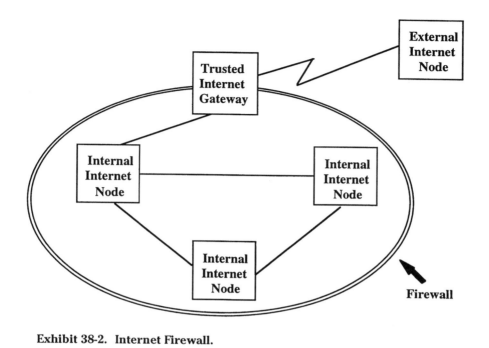

Exhibit 38-2. Internet Firewall.

provider of E-mail for the personal subscriber are subject to governmental laws and regulations under the Electronic Communications Privacy Act.

Exhibit 38-2 illustrates the internal (i.e., within the firewall) and external E-mail routing using Internet facilities. The connections between internal and external Internet communications are called firewall gateways. They permit Internet technology to be used in a company-secure mode behind the protection of the firewall.

Because systems administrators are necessary for the maintenance of E-mail systems, it is virtually impossible for a company to ensure that an employee's E-mail files are completely private, should a company desire to provide employees with privacy on the company E-mail system. In general, system-level access to the E-mail files assigned to an individual can be limited to the people with systems administration responsibilities. Each individual usually has a sign-on password that provides protection from indiscriminate access to their files. This protection, when implemented with proper password procedures, prevents unauthorized access to business data. The technology of encryption and reduced cost of implementation may make information privacy a reality in the near future, but it is not a prevalent feature on E-mail systems today.

## EMPLOYER VS. EMPLOYEE: CONFLICTING VIEWS ON PRIVACY

Privacy is often confused with confidentiality. Both employees and companies expect that the information contained within their E-mail messages will be treated with confidentiality. An employee expects a memo about a personnel matter to remain confidential. The company expects to be able to use the E-mail communication tool to discuss product information of a competitive nature without concern of disclosure to competitors.

Privacy is the constitutional right of privacy. When you send a letter through the U.S. Postal Service, your rights of privacy are protected. There are stringent laws and regulations associated with access to the contents of U.S. Mail. Because internal company E-mail is somewhat analogous to the U.S. Mail, the expectations of privacy exist.

The 1986 Privacy Act clearly gave requirements for the public sector; however, it remained silent on the private sector. If you remain silent on something, what happens? Companies assume that they have complete control (i.e., "We own it. We pay for the resources. We can specify who can and who cannot access electronic mail files.") Employees, however, assume that their E-mail files have the same degree of privacy as perhaps their desks, their briefcases, their purses, or the U.S. Mail. When an issue arises over management accessing an employee's E-mail files, the employees say, "Is this right? Is this moral? What kind of a company am I working for anyway? Is this big brother watching me?"

In short, employees will assume a degree of privacy. They will assume that the company's treatment of privacy matches the practiced behavior of management in accessing other forms of your correspondence. For example, does your company routinely open your mail that arrives through the company? If so, then employees would likely assume that someone might monitor their electronic mail as well. If hardcopy mail is not monitored routinely, even if it is company policy to do so, then employees would not expect their E-mail to be monitored.

### Practice vs. Policy

This brings up the issue of practice vs. policy. Even if your company has a policy concerning E-mail privacy, and if you do not practice that policy, then the policy is subject to challenge in the event of an issue. Can policy be enforced? Will it be enforced?

Employers want to be sure that no illegal or unethical activity is being conducted using the company E-mail system. They want to use any measurement tools at hand to evaluate the quality and quantity of employee performance. Employers use the tools provided with the E-mail systems to monitor employee's E-mail to look for employee misconduct. Companies

## MANAGING ELECTRONIC MESSAGING SYSTEMS

need to guard against the leaking of information to competitors via E-mail. They want to ensure that the E-mail system is being used for legitimate business purposes unless the company policy allows for personal E-mail use.

Employees need to understand the necessity for management's control and monitoring; however, they do not want their privacy invaded in the process. Employees may incorrectly perceive that their internal company E-mail is private. Thus, the conflict.

The key word is *perceive*. There is no right of privacy associated with company-provided E-mail services. As E-mail has grown over the years, the perception that an individual's E-mail files are private has also grown.

An employee needs to feel confident that an E-mail report to a supervisor of a personal nature will be treated with the same confidentiality as a written report. This is a most reasonable expectation, and management needs this degree of confidentiality as much as the employees. Take, for example, the manager who is working with the supervisors on performance reviews for the group. The interchange via E-mail is of a very confidential nature. This is an issue of confidentiality, not one of privacy.

Most employees use the E-mail systems for both business and non-business purposes. Examples of non-business use might be exchanging hobby information, making lunch or dinner dates, and scheduling after-work sports activities. Most companies do not object to this incidental use of the company E-mail system, but they want to draw the line when it comes to running an outside business, offering products for sale, soliciting membership in organizations, or conducting illegal or unethical activity, to name a few.

### POLICY RECOMMENDATIONS

The simple recommendation is: put a policy in place! A policy can be anything that a company judges is necessary to support the business interests of the company. Employees do not necessarily have rights of privacy associated with E-mail messages sent within or stored on a company-owned E-mail system. However, if the company policy concerning E-mail privacy does not exist or is not clearly spelled out and conveyed to the employee, a court may decide a privacy issue in favor of the employee, especially if the perception of privacy can be established.

Management has the right to regulate, monitor, and supervise the use of the company-owned E-mail resources, much as it regulates the use of company-owned or -controlled vehicles. Sound business practices indicate that companies should take reasonable precautions to preclude unauthorized access to employee E-mail files and should, through policy, protect individuals from abuse or harassment by others using the E-mail system in the same way that hardcopy material or telephone conversations are treated. Systems administrators should be treated as employees with confidential jobs similar to employees who deal with personnel files. Companies should train administrators to handle the confidentiality of information to which they may inadvertently become exposed.

The Electronic Messaging Association (EMA) provides excellent resources to assist companies in formulating their E-mail privacy policy. *Privacy Tool Kit: Access to and Use and Disclosure of Electronic Mail on Company Computer Systems*, prepared by David R. Johnson of Wilmer, Cutler & Pickering and John Podesta of Podesta Associates, Inc., which may be ordered from EMA, contains examples addressing issues within corporate E-mail privacy policy. These examples cover companies desiring to state the most stringent control of internal E-mail systems to companies desiring to provide as close to actual privacy for the employee as the E-mail systems permit. The *Privacy Tool Kit* may be ordered from the EMA at 1655 North Ft. Myer Drive, Suite 850, Arlington, Virginia 22209. The EMA member price is $20 and the non-member price is $45.

In 1991, the Aerospace Industries Association published a guideline for electronic mail privacy to its membership. With the permission of the AIA, that document is included at the end of the chapter in the Appendix for additional reference. The guidelines and recommendations presented in this chapter demonstrate advanced consideration regarding the subject of E-mail privacy on company-owned E-mail systems.

**Raising Employee Awareness and Responsibility Level**

The company E-mail privacy policy should be consistent with the other policies of the company dealing with employee and workplace privacy. The E-mail policy should not be the vehicle for documenting policy on employee behavior. Behavior or conduct policy should be independent of the media, whether it be telephone, E-mail, use of fax machines, copiers, or any other company-provided resource. How a company permits employees to use internal E-mail is a separate policy issue from the privacy of E-mail messages. Use separate policies for clarity.

One technique for ensuring employee awareness of the company E-mail policy is to have the employees indicate through positive action, like a signature, that they have been made aware of the policy. This simple method does add administrative cost to a business.

Another technique frequently used is to display a privacy message to the user at the time of messaging or E-mail system logon. Including the E-mail privacy awareness along with other company processes like annual security or ethics training would minimize the cost.

A resource on policy for employee behavior may be found in the 1994 copyrighted publication by the Regents of the University of California, called *A Model Employee Handbook for California Businesses*, prepared by

Margaret Hart Edwards and Barbara D. Stikker. This model specifies a strict policy of business-only E-mail use. While companies have every right to take this approach to the use of company resources, many companies are realizing that employees are more productive and learn the technology faster if permitted personal use of company resources on a nondisruptive basis. It is impossible to control incoming E-mail when E-mail addresses are on business cards or to separate a business message when a personal message is included. Enforcing a policy of "absolutely no personal use" in today's environment of the Internet may be very difficult to do and counterproductive. Policies that promote the efforts of employees to be responsible users are more useful and productive in the long run.

**Examples.** A policy that cannot be enforced is like having no policy at all. For example, a company may create a policy that states that individual employee files on company E-mail will be protected from unauthorized access. Then the company does not provide an access control mechanism such as password protection. If an employee were adversely impacted by exposure to E-mail files, then the company may be liable for damages.

For another example, a company may create a policy that states that company E-mail is to be used for business purposes only. Subsequently, the company does not check to see if personal, non-business messages are present within the message store. Furthermore, the company finds some innocuous E-mail and ignores it. If an employee is dismissed for using the E-mail system for non-business purposes, conditions illustrated by this example can cause the policy to be considered invalid because it was not enforced consistently. As a result, the company may lose a wrongful termination suit.

It merits repeating: E-mail privacy should be treated the same as other workplace privacy issues, including telephone use, fax use, paper files, and employee work areas.

The overriding message is to get policy in place and to tell people what that policy is. It does not matter if your policy states that management has the right to monitor any and all messages, or that employees' E-mail files are completely private and will not be accessed by management under any circumstances. Get a policy in place that your company can and will consistently enforce.

## POLICY TEMPLATES

### Example 1

The following template provides a guideline for establishing Company Employee Conduct Policy associated with use of company resources, including E-mail:

*Creating Policy Relative to E-Mail*

Employees are prohibited from using the Company's resources, for example, electronic mail, voice mail, fax, telephone, internal hardcopy mail, and other information technology systems, in any way that may be disruptive or offensive to others, including, but not limited to, the transmission of sexually explicit or harassing material or cartoons; sexist, ethnic, or racial slurs; or anything that may be construed as harassment or disparagement of others.

Employees are prohibited from use, for other than authorized business purposes, of access codes and passwords of other employees to gain access to their assigned company resources, such as, E-mail accounts, voice-mail accounts, and other information technology systems.

Employees are prohibited from unauthorized use of company resources that result in personal profit for the individual or benefit to another profit business.

Employees are prohibited from using company resources in any way that is out of compliance with the company's Conflict of Interest or Business Ethics Policy.

Employees are prohibited from using company resources in any way that violates existing federal, state, and local laws and regulations or company and corporate policies.

Employees are permitted to use company resources if the following conditions are met:

1. The above prohibitions are not violated.
2. The personal use of the resources is not done in a way that incurs additional cost to the company or the company's customers for labor or resources.
3. The personal use of the resources does not interfere with company business requirements or operating efficiency.
4. The personal use of the resources does not reduce the useful business life of the resource.
5. Any incidental supplies used in significant quantities must be provided by the employee at the employee's expense.

## Example 2

The following template provides a guideline for establishing company policy for Access to Information with emphasis on electronic storage systems such as E-mail:

The resources used for the storage of information, hardcopy files, electronic mail, voice mail, or other automated information systems storage facilities, are the property of the company and, as such, the employee has no rights of privacy from authorized company examination.

## MANAGING ELECTRONIC MESSAGING SYSTEMS

Employees have been given individual access codes and passwords to E-mail, voice mail, and other computer systems for the purpose of establishing sound business practices in protecting the company's products and processes. The company has the right to access this material at any time and the material may be subject to unannounced inspections and be given to third parties for audit and other business purposes. Additionally, backup copies of the data are maintained for system recovery purposes and those data copies may be accessed by the company.

The company shall take reasonable steps to protect the data in individual employee files on electronic systems where access codes and passwords are used to prevent unauthorized intrusion by other employees, including unauthorized management. (Note: Policy should define conditions for authorized access.)

Access to information stored under the control of an individual employee shall be authorized under, but not limited to, the following conditions:

1. In response to a subpoena, court order, or order of an administrative agency.
2. To a governmental agency as part of an investigation for company compliance with applicable law or regulation.
3. As part of a company investigation for compliance with existing regulation, company policy, and corporate policy.
4. In a lawsuit, administrative proceeding, grievance, or arbitration in which the employee(s) and/or the company are parties.
5. As part of routine monitoring procedures.

Access to the files in the possession and under the control of individual employee(s) may be made by authorized company representatives with or without the knowledge of the employee(s).

Messages received by the company E-mail or messaging system from an external source shall be treated as internal company information subject to the company policy associated with access to information.

## Example 3

The following template is for procedures relating to E-mail privacy and use:

Management shall advise employees of the Employee Conduct Policy and the Access to Information Policy.

Management shall obtain positive acknowledgment that the employee has been advised of the policy.

Management shall brief systems administrators on business confidentiality.

## SUMMARY

There is no right of privacy for employees associated with messages stored on or transmitted internally with a company-provided E-mail system. However, in the absence of clear policy, companies will take unnecessary risk of interpretation and, thus, conflict with employees. The policy concerning personal privacy of company E-mail use is at the discretion of the company. That policy should be consistent with other employee and workplace privacy policies. Policies must be validated by practice.

It is worth stressing once again that organizations should create policy, tell the employees, and document their awareness. Then follow policy with appropriate action, but be consistent.

# Appendix to Chapter 38
# Guidelines for Electronic Mail Privacy

## From the Aerospace Industries Association (AIA) E-Mail Panel

The widespread use of electronic mail within and between companies in the aerospace industry inevitably raises privacy and ownership issues. The underlying assumptions and expectations of both the employer and employee need to be recognized, and a clear policy should be established and communicated to avoid misunderstandings. The employer usually views electronic mail as a resource owned and provided by the company to enhance performance and productivity. Employees, however, might assume that electronic mail and their saved files are as private as personal mail deliver to their home.

In an effort to assist AIA member companies in dealing with these potentially conflicting viewpoints, the AIA E-mail Panel has developed the following guidelines for consideration by member companies in formulating their own policy statements.

For the purpose of this document, voice mail is considered the same as electronic mail.

### DEFINITIONS

Definitions are an essential part of any policy statement. The following definitions are suggested:

*Privacy:* The protection of electronic mail from unauthorized access.

*Electronic mail:* Correspondence transmitted and stored electronically using software facilities called "mail" or "messaging" systems.

# MANAGING ELECTRONIC MESSAGING SYSTEMS

*Public electronic mail:* Electronic mail services provided to the public for a fee; similar to public telephone services. For purposes of these guidelines, the Internet is considered public.

*Company electronic mail:* Electronic mail services provided by a company where the company controls the delivery of the service, electronic storage, and the computer systems.

*Voice mail:* Voice messages transmitted and stored for later retrieval from computer systems.

## ASSUMPTIONS

The following assumptions underlie the guidelines suggested for consideration:

- Statements concerning electronic mail privacy do not imply compliance with DoD security requirements.
- Public electronic mail privacy is covered by the 1986 Federal Electronic Communications Privacy Act; however, the Act does not cover internal electronic mail systems as the company or corporation is considered one entity and thus private (not offered to the public).
- In the absence of a specific electronic mail privacy policy, companies will probably assume that the policies or procedures established for employee interpersonal communication will apply.
- If no specific policies for electronic mail privacy exist, employees will probably assume that such communications are private and not subject to company scrutiny.

## GUIDELINES

### Basic Recommendations

A. The policy concerning the personal privacy of company electronic mail and voice mail systems should be at the discretion of the company providing the service, and should be consistent with other company personal privacy policies

B. Employees shall be effectively advised of their company electronic mail personal privacy policies.

### Specific Recommendations

1. Electronic mail privacy for employees shall be controlled, but not assured, by policies and facilities to protect against unauthorized access.
2. Companies shall treat electronic mail transmission and files as private information. Access authorization techniques shall be in place.
3. Companies may monitor the use of electronic mail systems for system performance/utilization analysis.

4. Personnel who support or maintain computer systems for electronic mail files shall be considered to have limited operational access to employee electronic mail files. Such personnel shall be briefed on their responsibilities for the protection of employee files.
5. Companies shall "seize" electronic mail data for investigation only with the active participation of the company investigative services and the office responsible for computing security.
6. Company electronic mail is for the primary purpose of correspondence relating to business.
7. Misconduct on the part of employees or management associated with the use of electronic mail shall be treated in accordance with company conduct policies.
8. The company or corporation will not routinely monitor electronic mail for the purpose of discovering misconduct.

The following noninclusive list of restrictions and guidelines should be considered by member companies for incorporation with their conduct policy information.

Examples of misuse of electronic mail or voice mail include, but are not limited to:

1. Use of electronic mail or voice mail for any unlawful endeavor.
2. Requesting or providing any copyrighted material in a way which would infringe on the rights of the copyright holder.
3. Threatening, insulting, obscene, or abusive language; for example:
   a. Derogatory remarks based on race, religion, color, sex, handicap, or national origin.
   b. Remarks that are defamatory toward any person.
   c. Remarks that constitute sexual harassment.
4. Use of electronic mail for advocacy of religious or political causes without company approval.
5. Use of electronic mail as any part of an employee's off-the-job pursuits without management approval.
6. Use of electronic mail for commercial purposes or to advertise items, services, activities, or discounts offered by a commercial enterprise, or by an individual conducting a commercial enterprise without company approval.

# Chapter 39
# Ethical Management of Employee E-mail Privacy

*Janice C. Sipior*
*Burke T. Ward*
*Sebastian M. Rainone*

E-mail enhances communication and information access. However, the benefits organizations stand to gain cannot be realized without a full consideration of the associated costs and responsibilities.

Organizations have an obligation to themselves, their employees, business partners, customers, and society at large to act in an ethically responsible manner regarding their policies on E-mail privacy. This chapter seeks to raise awareness of the ethical issues of E-mail privacy by explaining the vulnerabilities to which E-mail is subject and the privacy expectations of both employers and employees. Finally, a strategy for ethics management, which integrates ethical behavior and legal compliance, is applied to E-mail privacy policy.

## SAFEGUARDING THE COMPANY OR SPYING?

Ethics deal with processes for determining reasonable standards of moral conduct. A primary concern in ethical behavior is actual or potential harm experienced by an individual or group. Ethics try to resolve actual or potential conflicts of interest.

The determination of what is ethical or unethical is not simple or straightforward. Employers and employees may see the ethical and legal issues associated with E-mail privacy differently. For example:

- The ethical position/*employer's perspective* says: Employees have no expectation of privacy, because E-mail is a company resource.

# MANAGING ELECTRONIC MESSAGING SYSTEMS

- The ethical position/*employee's perspective* says: E-mail monitoring is an invasion of privacy.
- The legal position/*employer's perspective* says: E-mail monitoring is a property right.
- The legal position/*employee's perspective* says: A "right to privacy" is assumed when there is, in fact, no E-mail privacy protection.

An understanding of these differing perspectives may provide insight necessary to formulate an ethical approach to managing E-mail communications in today's networked organizations.

**Employer's Perspective on E-mail Privacy**

Corporations may mistakenly assume that external E-mail networks, used to communicate with employees, consultants, information services, suppliers, or customers, have a set of rules, penalties, and controls. Work that spans geographically dispersed employees, business partners, or customers subjects sensitive company information to the risk of interception by competitors, corporate spies, or hackers when transmitted. This concern is especially evident with wireless transmissions, which are inherently vulnerable to interception. When Princess Diana's private cellular telephone conversations were intercepted, recorded, and subsequently broadcast by the news media, all the world seemed to unanimously agree on the unethical nature of the interceptors' actions. Such actions may be just as unethical when they affect commoners and business organizations.

Amoco Corp., for example, limits the content of E-mail communications with branch offices to matters that are not mission-critical. Does this self-imposed limitation provide practical and adequate protection from the unethical conduct and consequences of message interception? Are there instances where the interception of E-mail communications would be ethically acceptable to organizations? The answer may depend on whether the organization is intruded upon or is the intruder.

A view from within the organization reveals a different perspective on intrusions into E- mail communications. Given that an organization expects the privacy of its external E-mail correspondences to be secure, is it ethical for an employer to monitor internal E-mail communications? E-mail monitoring may be viewed not only as a right, but a necessity.

As owners of information resources, organizations commonly grant themselves the right to search employee computer files, voice mail, E-mail, and other networking communications. Such monitoring may or may not be ethically acceptable or legally permissible, although the reasons organizations give for engaging in this practice are considered by them to be both ethical and legal. Reasons include:

## Ethical Management of Employee E-mail Privacy

- Prevention of personal use or abuse of company resources
- Prevention or investigation of corporate espionage or theft
- Cooperation with law enforcement officials in investigations
- Resolution of technical problems, or other special circumstances

**Validity of Tracking Performance through E-mail.** A grayer area is encountered when E-mail monitoring is undertaken for the purpose of keeping track of worker productivity, performance, and conduct. For example, employees may use E-mail to assist in making a sale or to meet deadlines in completing work. If not for the record of correspondence and time and date stamps associated with E-mail messages, recognition for productivity accomplishments may otherwise go unnoticed.

Consider another example of publicly traded companies, wherein it is the company's responsibility to ensure that employees abide by Securities and Exchange Commission rules. In this instance, a company failing in its responsibility to monitor performance by examining E-mail messages sent to external destinations may be negligent in its duty to protect trade secrets and proprietary information.

To achieve the positive results presented in these two examples, all messages sent and received by employees would be subject to scrutiny. Information that an employee intended to keep private and confidential may be examined.

An employer could argue that this is not an unethical disregard for an employee's privacy. Because the E-mail system is owned by the employer and is to be used for the employer's purposes, the employee should not expect communications to be private. The employer would conclude that E-mail monitoring is not unethical since no privacy expectation was invaded. Conversely, employees may argue such monitoring is indeed an invasion of an individual's privacy. Because the potential for negative consequences resulting from the organization's action is present, the practice of E-mail monitoring may be viewed as unethical.

Clearly, monitoring can be a double-edged sword to those subject to it. Productive and high-performing employees (as well as those who are not productive) can be readily identified throughout the organization. Whether the practice is unethical in the workplace may depend more on the employer's reason for monitoring its employees E-mail, rather than the intrusive nature of the monitoring itself.

### Employees' Perspectives on E-mail Privacy

Employees may not even be aware that the content of their E-mail messages may not actually be a confidential correspondence between the sender and receiver(s). This perception seems reasonable, given that E-mail is accessed as a facility within a user's computer account, to which access is password-controlled, providing an illusion of confidentiality.

The fact that users do mistakenly make a reasonable assumption that E-mail messages are private is underscored by the increasing number of legal cases filed by employees. These legal cases serve as a means of understanding the employees' perspective and are highlighted in the following discussion.

**Do Privacy Cases Hold up in Court?** Electronic surveillance, purportedly for the purpose of improving job performance, quality, and productivity, exists in many forms. Employee computer screens may be viewed, or messages may be printed from the E-mail system, a user's hard drive, or backup storage, without either the employee's knowledge or consent.

For example, at Epson America, Inc., an E-mail system administrator claimed to have been terminated for protesting the routine practice of intercepting and printing employees' MCI E-mail. In the 1990 case, *Shoars v. Epson America, Inc.*, the administrator claimed the invasion of privacy and wrongful termination violated California law. In a 1993 case, *Bourke v. Nissan Motor Corp.*, two software specialists contend they were forced to quit after a supervisor read their personal E-mail correspondences, which contained sexual statements. The claims were again invasion of privacy and wrongful termination in violation of California law. In both cases, the companies' rights to manage their E-mail systems were legally recognized.

Repeated and explicit assurances by an employer that E-mail will not be monitored would certainly provide employees with a heightened expectation of privacy. In a Pennsylvania case, the Pillsbury Co. had repeatedly assured employees through a formal E-mail privacy policy that all E-mail was confidential. Relying on this policy, a regional operations manager responded to E-mail he received at home from his supervisor. Contrary to the stated policy, the E-mail correspondence was intercepted and the employee was terminated for inappropriate statements in his messages. In the resulting wrongful discharge suit, *Smyth v. The Pillsbury Co.* (1996), the Pennsylvania court found the company's interest in preventing inappropriate or even illegal message content outweighed any privacy the employee may have had.

Many users have the expectation that the delete command, when issued, results in the actual deletion of a message. However, user deletions are often archived and stored for years. A record of all messages ever transmitted — including the user names of both the sender and recipient(s), the subject of the correspondence, and the date — may exist somewhere on a file within the system network for retrieval. In one case, an employee of Borland was suspected of using a company-supplied MCI mail account to divulge trade secrets to his future employer, and Borland rival, Symantec Corp. Borland requested MCI Communications Corp. to retrieve the former employee's deleted messages. This intrusion was considered a property right. Because Borland paid for the E-mail service, the employee's account became Borland's property on his departure. As a result, criminal investigations of both the sender and recipient, as well as a civil suit, are pending.

Although the employer E-mail monitoring was found to be legal in the example cases, was it ethical? Disagreements over such monitoring will undoubtedly increase as the number of companies employing such practices was estimated to be as high as 40% in 1990 and growing. Given the differing perceptions of employers and employees regarding E-mail privacy and the potential problems that may result from privacy intrusions, there is a need for organizations to address ethically responsible E-mail management.

## ETHICALLY RESPONSIBLE MANAGEMENT OF E-MAIL

The issue of E-mail privacy remains ethically unresolved. The legal issue depends, in part, on whether the intrusion was into a place or type of information for which the individual had a reasonable expectation of privacy. There is no agreement on what constitutes a reasonable expectation of privacy.

Since the existing legal system has not kept pace with the ethical issues accompanying the use of E-mail, the responsibility to address issues of privacy in employees' E-mail communications falls on organizations to create their own internal policies.

Addressing privacy issues in the management of E-mail systems are increasingly being regarded as a necessity as organizations rely more and more on electronic communications. As the number of users and different systems grow, the complexity of managing connections, message archival and retrieval, access security, directory integration, and privacy increase proportionately. Organizations cannot afford to rely solely on the legal system to provide guidance. Rather, the implementation of a management strategy that integrates ethical behavior and legal compliance is preferred.

A comprehensive approach, reported in *Harvard Business Review* ("Managing for Organizational Integrity," by L.S. Paine, March-April 1994), goes beyond legal compliance alone to guide an organization's strategy for ethics management. The application of this strategy, developed by a management ethics specialist, assures an integrated combination of legal compliance with an emphasis on ethically responsible management. The implementation of the strategy for ethics management, adapted in Exhibit 39-1, first entails the development of a *compliance strategy*. The ideal for the

**Exhibit 39-1. Strategy for Ethics Management.**

| | Compliance strategy | | Integrated strategy |
|---|---|---|---|
| Ideal | Conformity with externally imposed standards | Ideal | Self-governance according to chosen standards |
| Objective | Prevent illegal conduct | Objective | Enable responsible conduct |
| Leadership | Lawyer driven | Leadership | Management driven with aid of lawyers and others |
| Methods | Education, reduced discretion, auditing and controls, penalties | Methods | Education, leadership, accountability, controls, organizational systems and decision processes, auditing and controls, penalties |
| Behavioral | Autonomous beings, guided by material | Behavioral | Social beings guided by material self-interest, values, ideals, peers |
| | **Implementation** | | **Implementation** |
| Standards | Criminal and regulatory law | Standards | Company values and aspirations<br>Social oblications under the law |
| Staffing | Lawyers | Staffing | Executives and managers with lawyers and others |
| Steps | 1. Develop compliance standards<br>2. Train and communicate<br>3. Handle reports of misconduct<br>4. Conduct investigations<br>5. Oversee compliance audits<br>6. Enforce standards | Steps | 1. Develop company values and standards<br>2. Train and communicate<br>3. Integrate into company systems<br>4. Provide guidance and consultation<br>5. Assess values performance<br>6. Identify and resolve problems<br>7. Oversee compliance |
| Education | Compliance standards and system | Education | Decision-making and values<br>Compliance standards and system |

compliance strategy — conformity with externally imposed standards — must be lawyer driven. This is followed by the implementation of an *integrated strategy,* a self-governance according to chosen standards, which extends the legal basis, forming the corporation's policy for ethical behavior.

## The E-Mail Privacy Policy

By following this framework, an organization is able to develop an E-mail privacy policy. Discussed next are the seven activities identified as necessary to implement the ethically and legally integrated strategy.

**Step 1. Develop Company Values and Standards.** Organizations should determine and formulate an E-mail policy in accordance with specific organizational needs, employee expectations, considerations for outside entities, and a balancing of other interests. Input from various organizational perspectives, to promote agreement on policy formulation, is achieved by securing the involvement of managers, users, and IS personnel, with the continued assistance of lawyers. Incorporating legal counsel in privacy policy formation should reduce the risk of future litigation and employer liability. Corporate culture may dictate the orientation of the policy applied. Among the companies that *protect* the privacy of employee E-mail are Citibank, General Motors, Hallmark Cards, McDonnell Douglas, and Warner Brothers.

The opposite position, in which the organization regards employee E-mail as corporate property, is taken by American Airlines, DuPont, Federal Express Corp., Eastman Kodak, Epson America, Inc., Pacific Bell, and United Parcel Service, Inc.

Formal E-mail privacy policies have been established by a number of companies, including Hewlett-Packard Co. and a coordinated effort by the 54 members of the Aerospace Industry Association of America, Inc. Among the major tenets expressed in these policies are as follows:

- E-mail is considered private, direct communication between sender and recipient(s).
- Company E-mail is for the primary purpose of correspondence relating to business.
- E-mail will not be monitored, observed, viewed, displayed, or reproduced in any form by anyone other than the sender or intended recipient(s).
- In the case of employment termination or when an employee is absent for an extended period of time, work-related mail is forwarded to the most appropriate employee. Personal messages are forwarded to the intended recipient. If that is not possible, the messages are destroyed. Messages are not examined further than is necessary to determine in which category they fall.

Specific organizational considerations may differ, underscoring the importance of continued involvement by legal council. Another aid for organizations is the corporate E-mail Privacy Policy Kit, available from the Electronic Mail Association (EMA) in Arlington, Virginia for a nominal fee.

**Step 2. Train and Communicate.** Training and communication of the E-mail privacy policy is an extremely important implementation activity. It is critical that the policy and specific technological forms of protection available be communicated to employees in an explicitly clear manner.

The policy should be disseminated through traditional outlets, such as employee training sessions, handbooks, or computer manuals, as well as through more overt methods, such as:

- Adding a policy statement to screen menus every time the E-mail facility is accessed or at intermittent intervals
- Placing stickers on equipment
- Requiring employees to sign an agreement indicating they understand their rights and responsibilities in using E-mail systems

**Step 3. Integrate the Policy into Company Systems.** To integrate the policy into company systems, technical forms of privacy protection can be implemented within the company's E-mail system. Guidelines have been developed for the provision of Privacy Enhanced Mail (PEM), an initiative of the Privacy and Security Research Group of the Internet Activities Board. PEM is an effort to protect the privacy dimensions of integrity, origin authentication, nonrepudiation, and confidentiality. Specifically:

- The integrity feature assures the recipient that the message received contains the identical content of the message sent, protecting against interception of the message while in transit.
- The authenticity facility provides reliable determination of the identity of the sender of a message. However, in the case of a forwarded message, it is the forwarder, not the originator of the message, who is identified.
- Nonrepudiation combines the integrity and authenticity facilities by verifying both the originator and contents of a forwarded message.
- Confidentiality safeguards protect against loss or unauthorized access, destruction, use, or disclosure of message content, either while in transit or stored in the user's mailbox. The most common technical form of affording protection for all four dimensions is *encryption*.

Additional privacy features, beyond those available through PEM, can also be implemented within the system as necessary. For example, anonymity is especially important to organizations promoting open communications that cross the traditional channels defined by organizational hierarchies. Collaborative work, entailing activities such as decision-making, negotiation, problem resolution, brainstorming, rank ordering, or voting, can benefit from the opportunity to focus on the task at hand, rather than on personalities, professional position, or status.

Users should be warned not to presume facilities for various dimensions of privacy exist, because not all systems necessarily achieve or surpass the PEM standard. A recognition of the dynamic nature of the PEM standard is also necessary. For example, the Multipurpose Internet Mail Extensions (MIME) Internet standard for the transmission of multimedia allows graphics, images, video, and sound to be sent to users worldwide.

**Step 4. Provide Guidance and Consultation.** For any policy to be effective, it is critical that a mechanism be present to provide continuing guidance and consultation to employees in interpreting and applying the policy to their particular job responsibilities. An awareness of the consequences to employees resulting either from actions taken, or not taken, in regard to employee use of E-mail systems is necessary. The absence of clarification for an E-mail policy may cause ill-will, employee discontent, or confusion, all of which could have a negative effect throughout the organization.

Conversely, elaborate rules and restrictions may discourage individuals from taking responsibility for their own behavior. It is important to offer legally and ethically sound advice, in accordance with the E-mail policy. Be careful not to demoralize employees or devalue E-mail when addressing their privacy concerns. A communication service is more highly valued when privacy can be assured. The perceived disregard of E-mail privacy may diminish the value of this increasingly relied-upon resource, resulting in users seeking alternative channels of communication. Such a disruption in organizational communication could have reverberating effects throughout the organization, reducing employee job satisfaction, performance, and productivity. Providing guidance and consultation for the policy may not only improve employee relations by building employee trust, but may also forestall legal action on the part of employees.

**Step 5. Assess Values Performance.** Once an E-mail privacy policy has been formulated and placed in effect, it is important to perform an evaluation of that policy. A technique that may be used is a periodic E-mail privacy audit to review the enforcement and assess the efficacy of the policy in effect. The policy formulated must be evaluated both in terms of the performance of the system in supporting organizational communications requirements as well as user satisfaction.

The evaluation of the impact the E-mail privacy policy has should be directed toward both external and internal environments. Organizations do not function in isolation. In their dealings with other organizations, individuals, and society as a whole, it is necessary that they act in a responsible manner by contributing to the definition of, and abiding by, both ethical and legal standards for the information age. With regard to internal environments, the policy adopted should serve to make the organization less vulnerable to privacy breaches, employee discontent, legal liability, and misuse of company resources.

This evaluation should be extended to consider the global consequences as well. With the increasing intensity of global competition, organizations can no longer afford to be shortsighted by considering only the national implications of an E-mail privacy policy. The necessity for an international outlook is underscored by the European Community's serious consideration of adopting a directive that would restrict the transmission of personal information from a member country to another country. The restrictions would apply unless the target country has "adequate" protection for personal data.. Other areas of the world, including the former Soviet Union and Eastern Europe, are studying the value of privacy and are considering the right of privacy for inclusion in their new constitutions.

**Step 6. Identify and Resolve Problems.** The E-mail privacy audit is also an important mechanism for the identification and resolution of problems. The acceptability of a policy that previously worked (or was expected to work) well for an organization may change. Advances in technology, new legislation, judicial decisions, changing social mores, or even changes in personnel may have an impact on what is regarded as appropriate. It is therefore imperative that a policy be updated to respond to change.

**Step 7. Oversee Compliance Activities.** Finally, to ensure compliance with the E-mail privacy policy, it may be beneficial to identify one individual — a messaging systems manager — or a number of individuals — an E-mail system steering committee — whose responsibility it is to oversee all facets of E-mail system management. Included among the responsibilities would be the allocation of resources, monitoring program functionality, and development of security and control measures in line with the organization's E-mail privacy policy.

The organizational communications objectives set forth must be considered in conjunction with what can be achieved through the application of current technology. The functions, capabilities, and components of the E-mail system that serve user needs throughout the organization must be identified and carefully considered. Furthermore, the manager or steering committee would be responsible for training and communication to the user community. This should promote an awareness and understanding of the organization's perspective concerning E- mail monitoring. In this capacity, the manager or steering committee serves as a liaison for coordination across the organization.

## SUMMARY

The evolution of ethical principles for employee E-mail use requires a balance of the employee's expectations of personal privacy with an organization's proprietary and access interests. From a managerial perspective, it is important that E-mail, an increasingly important organizational resource touted as having such an important impact on worker productivity, be managed and used in an ethically acceptable manner. Because clear guidance is not yet provided by the U.S. legal system, organizations must formulate their own internal E-mail privacy policies. Without a formally established policy, employees' expectation of privacy may differ from the employer's perspective.

Even with a formal policy, this balance is not easy to achieve, as was demonstrated by *Smyth v. The Pillsbury Co.* As previously discussed, in this case, the company's stated policy was that all E-mail was confidential; the company further assured that it could not be intercepted and used against employees as the basis for either reprimand or termination. Using corporate E-mail privacy policy as a guide for his actions, the employee in this case responded to E-mail from his supervisor — only to be terminated for transmitting what were deemed inappropriate and unprofessional comments. Pillsbury had violated its own E-mail privacy policy by intercepting the employee's E-mail correspondence!

In assessing the employee's claim under the Pennsylvania common law regarding invasion of privacy, specifically the tort of "intrusion upon seclusion," the court concluded the employee had no reasonable expectation of privacy in voluntary E-mail communications, despite assurances by the employer to the contrary. Furthermore, the company's interest in preventing inappropriate or unprofessional comments, or even illegal activity, outweighed any privacy interest the employee may have had.

Although the employer's actions in the Smyth case were found to be legal in the commonwealth of Pennsylvania, were they ethical? An employee who relies on stated organizational policy should not be subject to negative consequences. This case underscores the need for a comprehensive approach to manage E-mail privacy that goes beyond legal compliance to embrace ethically responsible management.

# About the Editor

NANCY A. COX is the senior technology editor for messaging and groupware for *Network Computing* magazine. Previously, she worked for Lockheed Martin as a messaging systems architect, integrating large-scale E-mail systems. She co-authored the *LAN Times Guide to Multimedia Networking* (Osborne McGraw Hill, 1995) and the *LAN Times E-Mail Resource Guide* (Osborne McGraw Hill, 1994). She authored *Building and Managing a Web Services Team* (Van Nostrand Reinhold, 1997). She was the technical editor of the X.400 API Association's Common Messaging Call 2.0 specification, published in June 1995.

# Glossary

## A

**Access Control List** — The means by which permission and denial of service for users and hosts is managed.

**Access Unit (AU)** — A component of an X.400 network that enables users of one service to intercommunicate with message handling services, such as the Interpersonal Personal Messaging service.

**Actual recipient** — In X.400 the potential recipient for which delivery or notification of receipt takes place.

**ADDMD** — X.500 Administrative Directory Management Domains are part of the X.500 directory managed by the commercial entity.

**Address** — Computer notation representing the location of a device, host, user, or information on a network. Also, the unique identifier of a mailbox.

**Address mapping** — The process by which an alphabetic Internet address is converted into a numeric IP address and vice versa.

**ADMD** — In X.400, the Administrative Domain that offers messaging transport services to individuals or organizations. MCI and AT&T are examples in the U.S. of ADMDs.

**Administration** — In X.400, a commercial administration or a recognized private operating agency (RPOA).

**Administration domain name** — In X.400, the standard attribute that identifies an ADMD relative to the country that is denoted by a country name.

**Advanced Research Projects Agency Network** — A leased line, packet network funded by the federal government, which eventually became the Internet.

**Advertisement** — Content providing advertising as well as entertainment for the viewer.

**.aif** — PC file extension for a sound file.

**Aliases** — Used to reroute browser requests from one URL to another.

**Algorithm** — A computing procedure designed to perform a task such as encryption, compressing, or hashing.

**Alternate recipient** — A user or distribution list to which the originator can request that a message be conveyed if and only if it cannot be conveyed to the preferred recipient.

**American Standard Code for Information Interchange** — A standard character-to-number encoding scheme in wide use in the computer industry.

**Anchor tags** — Hyperlinks in HTML documents that enable a user to jump from one screen or page to another.

**Anonymous FTP** — A server that permits users to download files without having to supply a userid and password to gain access to the remote computer.

**ANSI** — American National Standards Institute, a standards-making body.

**AOCE** — Apple Open Collaborative Environment.

**API** — Application Program Interface.

**Applet** — A small computer program, a mini-program.

**AppleTalk** — A network protocol developed by Apple Computer, Inc., enabling Apple devices to communicate with each other.

**Application Program Interface** — Computer code specifications that detail how one application may be invoked and/or accessed by another application. Three APIs most prevalent in Electronic Mail are Common Message Calls (CMC), Message Application Programming Interface (MAPI) and Vendor Independent Messaging (VIM).

**Archie** — A search mechanism for use on anonymous FTP servers.

**Architecture** — A framework for the design of a workable computer service or system.

**ARPANET** — Advanced Research Projects Agency Network.

**ASCII** — American Standard Code for Information Interchange.

**ASN.1** — Abstract Syntax Notation One.

**Asynchronous Transfer Mode** — A high-speed networking standard using fixed cells and providing dynamic bandwidth allocation.

**ATM** — Asynchronous Transfer Mode.

**Attribute** — A component of an attribute list that describes a user or distribution list.

**Attribute list** — A data structure that constitutes an O/R address.

**Attribute type** — An identifier, that is a part of an attribute, that denotes a class of information such as personal names.

**Attribute value** — The value of an attribute type.

**.au** — PC file extension for a sound file.

**Authentication** — The process of establishing the true identity of a person or process.

**Authorization** — The process of granting or denying access to a user or process.

**Auto-answer** — A mechanism on the user's message store that automatically responds to received messages with usually a short predetermined message.

**Autoforward** — A mechanism on the user's message store that automatically resends a received message to a predetermined address

**Automated Directory Update (ADE)** — The directory update protocol used in cc:Mail.

# B

**Bandwidth** — The difference between the highest and the lowest frequencies of a transmission circuit. More commonly, the total amount of data that can be carried on a transmission circuit.

**Basic service** — In X.400, the sum of standard features offered by the messaging service.

**BBS** — Bulletin Board System

**Bit** — The smallest unit of computer data represented by either a 0 or a 1. An acronym for "binary digit."

**Body** — One of four possible components of a message. Other components are the heading, attachment, and the envelope.

**Body part** — One of the parts of a message body. The body is a standard part of an electronic mail message.

**Bookmarks** — Notes a favorite Web page for a user for quick access.

**Browser** — The user interface or client for the World Wide Web, enabling search and retrieval of information.

**Bulletin Board System** — A computer software application that provides messaging, file archival, and other services of interest to the users of the system.

**Byte** — A sequence of 8 bits.

## C

**CERN** — The European Laboratory for Particle Physics. CERN developed HTTP and HTML, which form the foundation of the World Wide Web.

**CGI** — Common Gateway Interface.

**Chameleon** — Client software that enables connectivity to TCP/IP networks.

**Character sets** — Collections of symbols that may be used for the various parts of a message. The address may allow different character set use than that used in the message body. Examples are ASCII or IA5.

**CIX** — Organization of commercial Internet providers (http://www.cix.org)

**Client** — Software enabling users to interface with servers, making requests and receiving responses. Mosaic and Netscape are examples of World Wide Web clients commonly known as browsers.

**Client/server architecture** — A computer processing arrangement in which both the workstation and the server share the workload.

**CMC** — Common Messaging Calls application programming interface specification produced by the XAPIA (X.400 API Association), an industry consortium.

**Common Gateway Interface (CGI)** — A specification for how a server should communicate with server gateway programs.

**Common Name** — In X.400, the standard part of an O/R address form that identifies a user or distribution list such as "Ben Smith" or "All Managers." Components include surname, given name, initials, generation qualifier, etc.

**Compiler** — A program used to translate source code into executable program code.

**Content** — A word with multiple meanings in messaging. In SMTP, content is equivalent to the X.400 body. An X.400 body may have several parts; each part may be called a content. In X.400, the content is neither examined nor modified, except for conversion, during its conveyance of the message.

**Content type** — In X.400, an identifier, on a message envelope, that identifies the type of the information in the content. Information such as Binary or Group 4 Fax.

**Conversion** — An event in which an MTA transforms parts of a message's content or body part from one encoded information type to another. Conversions such as spreadsheet type A to spreadsheet type B.

*Glossary*

**Copyright** — The exclusive right to publish, sell, reproduce, and distribute the contents of a literary or artistic work for a specified number of years.

**Country name** — A standard attribute of an X.400 and X.500 name form that identifies a country. A country name is a unique designation of a country for the purpose of sending and receiving messages.

**CU SeeMe** — A videoconferencing application developed by Cornell University that runs on the Internet.

**Cyberspace** — A term describing the virtual world of computers.

**D**

**Daemon** — In UNIX, an independent, automated background program that performs specific functions.

**DAP X.500 Data Access Protocol** — DAP X.500 Directory Access Protocol is used by the Directory user agent to gain entry or access the Directory DSA.

**DECnet** — A proprietary networking protocol developed by Digital Equipment Corporation enabling DEC devices to communicate with each other.

**Delivery** — A message transfer step in which an MTA conveys a message to the MS, UA or AU of a potential recipient of the message.

**Delivery report** — Notification that acknowledges delivery, non-delivery, export, or affirmation of the subject message or probe, or distribution list expansion.

**DGN/DEN** — Distribution Group Name/ Distribution Element Name. The SNADS addressing string.

**DIA** — Document Interchange Architecture from IBM.

**Dial-up** — A temporary network connection established over a telephone line.

**DIB** — X.500 Directory Information Base containing all the data in the X.500 directory.

**Digital signature** — The act of electronically affixing a seal or token to a computer file or message in which the originator is then authenticated to the recipient.

**Directory** — A collection of systems cooperating to provide directory services.

**Directory System Sgent (DSA)** — A part of the directory, whose role is to provide access to the directory information base to DUAs and/or other DSAs.

**Directory User Agent (DUA)** — An application process that represents a user in accessing the directoy. Each DUA serves a single user so that the directory can control access to directory information on the basis of the DUA names.

**DISP** — X.500 Directory Information Shadowing Protocol.

**Distribution List (DL)** — An object that represents a prespecified group of users and other distribution lists that may be addressed by name.

**Distribution List Expansion** — An event in which an MTA expands a distribution list to its component addresses.

**Distribution List Name** — The name allocated to a distribution list.

**DIT** — X.500 Directory Information Tree defines the layout of the directory information.

**DIU** — Distribution Interchange Unit is part of the IBM document interchange architecture.

**DL** — Distribution List.

**DLL** — Dynamic Link Libraries.

**DNS** — Domain Name System.

**Domain** — A managed community of users and host systems in the Internet hierarchy, such as .com or .org.

**Domain-defined attributes** — In X.400, optional special parts of the X.400 O/R name that are used by the destination delivery system.

**Domain name** — A unique, alphabetic name following the @ symbol in an Internet address, identifying an Internet site.

**Domain name system** — A system used to look up and resolve host IP addresses.

**DOS** — Disk Operating System.

**Download** — A process used to transfer or copy files from a host computer to your own computer.

**DSA** — X.500 Directory System Agent is a database containing the X.500 data. Many of these make up the X.500 directory.

**DSP** — X.500 Directory System Protocol is used by a DSA to exchange information and requests with other DSAs.

**DUA** — Directory User Agent is the software used by the user to search directories.

*Glossary*

**Dynamic indexing** — The capability of automatically listing and sorting a group of URLs, records, etc.

**Dynamic Link Libraries** — DLL

## E

**EDI** — Electronic Data Interchange.

**Edutainment** — Content that is both educational and entertaining for the viewer.

**Electronic mail** — Application program that enables the exchange of computer generated messages between users over a network.

**Element of service** — A describable part of all the services offered by the electronic mail system.

**EMA** — Electronic Messaging Association

**Encoded Information Type (EIT)** — In X.400, an identifier, on a message envelope, that identifies one type of information contained in the message body, such as IA5 text or group 3 facsimile.

**Encryption** — Computer software used to scramble data located in files so as to make it unintelligible.

**Envelope** — Part of a message that is used by the MTAs to convey and handle the message. In many but not all systems, it is the only part of the message that is modified by the MTAs.

**Ethernet** — A network technology used in local area networks providing coaxial cable connections for devices and a network speed of 10 Mbps.

**Explicit conversion** — A conversion in which the originator selects both the initial and final encoded information types.

## F

**FAQ** — Frequently Asked Questions.

**FDDI** — Fiber Distributed Data Interface.

**Fiber distributed data interface** — A standard for local area networks using fiber optic media and offering a network speed of 100 Mbps.

**File transfer** — The process of copying a file from one computer to another over a network.

**File Transfer Protocol** — Format and methodology, based on TCP/IP, used to exchange files with remote computers.

**Finger** — An Internet resource that displays logged on users and information about them.

**Firewall** — Computer hardware and software that serve to permit or deny access to network resources.

**Forms** — Fill-in text fields on home pages that provide interactive query and response for users seeking information or wishing to provide feedback.

**Front porch** — The access point to a secure network environment also known as a firewall.

**FTP** — File Transfer Protocol

## G

**Gateway** — Computer hardware and software that translate between two disparate application programs or networking protocols.

**GIF** — Graphics Interchange Format

**Gopher** — A text-based distributed information system developed at the University of Minnesota.

**Graphical User Interface** — Computer program code that resides between the user and the application program enabling quick and easy access to information and program features using such techniques as graphics, icons, buttons, and pull-down menus.

**Graphics interchange format** — A commonly used image file format

**GUI** — Graphical User Interface

## H

**Heading** — Component of an electronic message containing addressing information. Other components are the envelope, attachment, and the body.

**Helpdesk** — Single point of contact for internal users for problem resolution.

**Heterogeneous network** — A communications network using more than one protocol such as IP, DECnet, and AppleTalk.

**Hierarchy** — A stratification of objects all having relationships that descend from the top most object, called the root, to lower levels of more and more specialized objects.

**History list** — A list of URLs and titles of documents accessed during a user's Mosaic session.

*Glossary*

**Home page** — The initial screen of information displayed to the user when initiating the client or browser software or when connecting to a remote computer. The home page resides at the top of the directory tree.

**Host** — A computer that enables users to communicate with other user computers on a network.

**Host address** — The IP address of the host computer.

**Host name** — The name of the user computer on the network.

**Hotlist** — A list created by the user of URLs within a particular Web document.

**HTML** — Hypertext Markup Language.

**HTTP** — Hypertext Transport Protocol.

**Hyperlink** — A location in a document that enables a user to jump to a location within the same document or another document. Links are identified by highlighted, underlined, or colored text or images.

**Hypermedia** — Documents containing several information types such as text image, audio, and video.

**Hypertext Markup Language** — A specific programming language, based on SGML, that enables the development of documents that can be accessed and displayed by WWW client software. HTML documents use the .html or htm PC file extension.

**Hypertext Transport Protocol** — The rules, based on TCP/IP, that govern the exchange of information between WWW servers.

# I

**IA5** — International Alphabet Number 5.

**Icons** — Small images, generally uniform in size and appearance, that link users to other locations.

**Imagemap** — An image, such as a still photograph with embedded hypertext links, enabling users to click on the image and be transported to another Web page.

**Implicit conversion** — In X.400, a conversion in which the MTA, not the user, selects both the initial and final encoded information types.

**Indirect submission** — In X.400, a transmittal step in which an originator's UA gives the message to the MS for submission to the MTA.

**Infobots** — Software agents that perform specified tasks for a user or application.

**In-line image** — A graphic image such as a still photograph or logo that is displayed within an HTML document.

**Integrated Services Digital Network** — A digital network for voice and data transmission offering high speed and increased bandwidth.

**International Standards Organization** — An organization composed of interested parties in various countries who collaborate to create standards in such areas as communications and computers.

**International Telecommunications Union (ITU)** — International standards-making body.

**Internet** — An international network of networks connecting computers in government, educational institutions, businesses, and other entities.

**Internet Protocol** — The basic set of rules that govern the exchange of data on the Internet.

**Internet Relay Chat** — An Internet resource that enables users to communicate with each other in real-time over the network.

**Internet voice** — A product enabling users to place a telephone call using Internet facilities.

**Interpersonal Messaging Service (IPM)** — Messaging that extends between humans, rather than between machines.

**Intranet** — The internal network within an enterprise.

**IP** — Internet Protocol.

**IP address** — The 32-bit representation of the location of a device on the Internet. Takes the form of 111.111.111.111 and is also known as a dot address, host address, Internet address, or network address.

**IRC** — Internet Relay Chat.

**ISDN** — Integrated Services Digital Network.

**ISO** — International Standards Organization.

**ITU** — International Telecommunications Union.

**J**

**Joint Photograph Experts Group** — A methodology used to store images such as still photographs in a compressed digital format.

**JPEG** — Joint Photograph Experts Group.

*Glossary*

# L

**LAN** — Local Area Network.

**LDAP** — See Lightweight Directory Access Protocol.

**Lightweight Directory Access Protocol** — A directory protocol, defined in RFC 1777, designed to provide access to the X.500 Directory while not incurring the resource requirements of the Directory Access Protocol (DAP).

**Links** — See Hyperlinks.

**Listserv** — A program that automatically send messages to a predefined distribution list.

**Local Area Network** — A network connecting multiple computers within a single location.

**Lycos** — An Internet search engine.

# M

**MacTCP** — Apple's network software that enables Macintoshes to connect to TCP/IP networks.

**Mailbots** — Software agents that perform specified electronic messaging functions on behalf of a user or application.

**Mailbox storage** — An area in an electronic mail network used for inbound user messages.

**Management Domain (MD)** — In the context of message handling, a set of messaging systems — at least one of which contains, or realizes, an MTA — that is managed by a single organization. It is a primary building block used in the organizational construction of MHS. It refers to an organizational area for the provision of services.

**Management Domain Name** — Unique designation of a management domain for the purpose of sending and receiving messages.

**Management Information Base** — In an OSI network, a repository designed to facilitate communications between network devices in open systems.

**MAPI** — Message Application Programming Interface from Microsoft.

**Masquerade** — A type of security threat that occurs when an entity successfully pretends to be a different entity.

**MBONE** — Internet's multicast backbone

**Message** — An electronic communication transmitted from one end user to one or more recipients composed of an envelope, attachment, and body.

**Message Handling System (MHS)** — All the components, MS, UA, and MTAs, that form the electronic messaging environment.

**Message Store (MS)** — A component of the MHS that provides a single direct user with capabilities for message storage. It is always available to the MTAs for delivery of messages.

**Message Switch** — An integrated multiple message gateway.

**Message Transfer Agent (MTA)** — A component of the MTS that actually conveys information objects to users and distribution lists.

**Message Transfer System (MTS)** — The MTS consists of one or more message transfer agents that provide store-and-forward message transfer between user agents, message stores, and access units.

**Message User Agent** — Software that handles the flow of communication between the messaging system and the user.

**MHS** — Message Handling System. A Novell product.

**MHS** — Message Handling System. The electronic messaging system defined in the X.400 standards.

**MIB** — Management Information Base

**MIME** — Multipurpose Internet Mail Extensions

**Modification** — A type of security threat that occurs when is content modified in an unanticipated manner by a nonauthorized entity.

**Mosaic** — A computer program developed at the National Center for Supercomputer Applications that provides a simple graphical interface to the Internet.

**Motion Picture Experts Group** — A methodology used to store movie files in a compressed digital format.

**MPEG** — Motion Picture Experts Group

**MUA** — Message User Agent. Often simply referred to as UA.

**Multimedia** — An integrated collection or presentation of various information types such as text, audio, video, animation, graphics, 3-D images, etc.

**Multimedia messaging** — Electronic messaging that contains the multimedia information types such as text, audio, video, animation, graphics, 3-D images, etc.

**Multimedia networking** — The capability of transmitting multimedia information types over traditional and emerging data communication paths.

**Multipurpose Internet Mail Extensions** — A protocol that allows a mail system to attach binary files such as graphics and spreadsheets.

*Glossary*

# N

**Name resolution** — Mapping a name to its address.

**Naming Authority** — A recognized body responsible for the allocation of names.

**NAPs** — Network Access Points.

**National Center for Supercomputing Applications** — Originator of the Mosaic WWW browser client software located at the University of Illinois in Urbana — Champaign, Illinois.

**NCSA** — National Center for Supercomputer Applications.

**NetSearch** — An Internet search engine.

**Network** — A data communications system connecting computing devices over a physical medium.

**Network Access Points** — Nodes providing entry to the high-speed Internet backbone system.

**Network Address** — In X.400, a standard attribute of an O/R address form that gives the network address of a terminal.

**Network File System** — A set of rules that permits access to files contained on remote hosts as if they were on local disks.

**Network News Transfer Protocol** — A set of rules for the distribution of news articles over the Internet.

**NFS** — Network File System.

**NIST** — National Institute of Standards and Technologies.

**NNTP** — Network News Transfer Protocol.

**Non-delivery** — An event in which an MTA determines that it cannot deliver a message to one or more of its immediate recipients.

# O

**Open Systems Interconnection (OSI)** — An international standard for data communications networks.

**O/R Address** — The name format, called Variant 1.1, used most often in X.400. Also, an attribute list that distinguishes one user or DL from another.

**O/R Name** — An O/R name distinguishes one user or distribution list from another.

**Organization Name** — Standard attribute of an O/R address as a unique designation of an organization for the purpose of sending and receiving of messages.

**Organizational Unit Name** — Standard attribute of an O/R address as a unique designation of an organizational unit of an organization for the purpose of sending and receiving of messages.

**Originator** — The user that is the ultimate source of a message.

**OS/2** — IBM's operating system.

**OSI** — Open Systems Interconnection.

**OSI Directory** — The X.500 directory.

**OSI Reference Model** — The seven-layer architecture designed by ISO for open data communications networks.

# P

**P1** — Protocol One defined in the X.400 standards. It is used between the MTAs.

**P2** — Protocol Two defined in the X.400 standards. It is used between the end-user mail user agents.

**P3** — Protocol Three defined in the X.400 standards. It is used between the message store and the user agent and the MTA.

**P7** — Protocol Seven defined in the X.400 standards. It is used between the user agent and the message store.

**Packet** — A unit of data sent across a data communications network.

**Patent** — Exclusive right granted to an inventor to produce, sell, and distribute the invention for a specified number of years.

**PEM** — Privacy Enhanced Mail.

**PERL** — Programming language based on C.

**Personal Name** — In X.400, a standard attribute of an O/R address form that identifies a person.

**Point of Presence** — Location where access to a telecommunications network may be obtained.

**Point-to-Point Protocol** — A set of rules defining the manner in which packets are transmitted over a serial point-to-point link.

**POP** — Point of Presence. Post Office Protocol.

*Glossary*

**Post Office** — Post Office is usually part of the server and temporarily holds the inbound and outbound mail.

**Post Office Protocol** — A set of rules defining the manner in which messages from a server can be read by a single user host.

**Postscript** — A page description language developed by Adobe Systems.

**PPP** — Point-to-Point Protocol.

**PRDMD** — Private Directory Management Domain. An entity that manages an X.500 directory.

**Private Domain Name** — In X.400, a standard attribute of an O/R address form that identifies a PRMD relative to the ADMD denoted by an administration domain name.

**Private Management Domain (PRMD)** — In X.400, a management domain that comprises messaging system(s) managed (operated) by an organization other than an Administration or RPOA.

**Protocol** — A set of rules and formats governing the manner in which a transfer of data is conducted over a network.

**Proxy** — Software residing on a Web server or firewall that enhances security by creating a filter for authorized use.

**Public Key Encryption** — An encryption scheme where two pairs of algorithmic keys, one private and the other public, are used to encrypt and decrypt messages, files, etc.

# Q

**QuickTime** — Apple's movie and audio file digital storage method.

# R

**RDN** — Relative Distinguished Name.

**Receipt** — A step in which a UA conveys a message or report to its direct user, or the communication system that serves an indirect user conveys such an information object to that user.

**Replay** — A type of security threat that occurs when an exchange is captured and resent at a later time to confuse the original recipients.

**Report** — A notification generated by the MTS that reports the outcome or progress of a message's transmittal to one or more potential recipients.

**Request for Comments** — The set of documents describing the Internet's protocols, standards, and other relevant items.

# MANAGING ELECTRONIC MESSAGING SYSTEMS

**RFC** — Request for Comments.

**Root** — In a hierarchy of objects, the top-most object, the object having no others above it.

**Route** — The path that network traffic takes from its source host to its destination host.

**Router** — A computing device that relays data communications traffic between other devices on a network.

## S

**Scripts** — Executable programs used to perform specified tasks for servers and clients.

**Search engine** — Software that permits the lookup of information on the Internet, such as Excite or Yahoo.

**Security capabilities** — In the context of message handling, the mechanisms that protect against various security threats.

**Serial Line Internet Protocol** — A set of rules defining the manner in which packets are transmitted over a serial network link such as a telephone line.

**Server** — A computer that responds to requests from client software, such as Internet browsers.

**SGML** — Standard Generalized Markup Language.

**Simple Mail Transfer Protocol** — The set of rules and formats defining the manner in which electronic mail messages are transmitted between computers over the Internet.

**Simple Network Management Protocol** — The set of rules and formats defining the manner in which management information about computing devices on a network is collected.

**SLIP** — Serial Line Internet Protocol.

**SMDS** — Switched Multimegabit Data Service.

**SMTP** — Simple Mail Transfer Protocol.

**SNMP** — Simple Network Management Protocol.

**SONET** — Synchronous Optical Network.

**Standard Generalized Markup Language** — An international standard for encoding textual information that specifies particular ways to annotate text documents separating the structure of the document from the information content. HTML is a specialized form of SGML.

**Subject** — The information part of the header that summarizes the content of the message as the originator has specified it.

**Switched Multimegabit Data Service** — A high-speed network service offered by the telephone companies.

**Synchronous Optical Network** — A high-capacity network for fiber optic media providdng high-speed data transmission.

# T

**T1** — Network transmission speed of 1.544 megabits per second.

**T3** — Network transmission speed of 45 megabits per second.

**Tagged Image File Format** — A method of storing image files.

**TCL** — Programming language.

**TCP/IP** — Transport Communications Protocol/Internet Protocol. Transmission Control Protocol/Internet Protocol. A set of rules, based on IP, governing the reliable transfer of data between computers on the Internet.

**TIFF** — Tagged Image File Format.

**Telnet** — An internet standard protocol used to access remote hosts.

**Threats** — Risks of security intrusions into the network.

**Token ring** — A type of local area network in which the devices are arranged in a virtual ring and in which the devices use a particular type of message, called a token, in order to communicate with one another.

**Topology** — The arrangement of computing devices on the physical media of the network.

**Trademark** — A registered word, letter, or device granting the owner exclusive rights to sell or distribute the goods to which it is applied.

**Traffic analysis** — A type of security threat that occurs when an outside entity is able to monitor analysis traffic patterns on a network.

**Transfer** — In the context of message handling, a transmittal step in which one MTA conveys a message, probe, or report to another.

**Transmission Communications Protocol/Internet Protocol** — A set of rules, base on IP, governing the reliable transfer of data between computers on the Internet.

**Trojan horse** — A type of computer virus in which the malicious code hides inside an innocuous-looking file or executable.

**Twisted pair** — A type of network physical media made of copper wires twisted around each other. Ordinary telephone cable.

## U

**UA** — User Agent. Component of MHS with which the user interacts.

**Uniform Resource Locator** — The address of a source of information located on the WWW. URLs take the form of "http://host/directory or file name" as in http://www.scifi.com/listings.

**UNIX** — A multitasking computer operating system developed by AT&T.

**URL** — Uniform Resource Locator.

**USENET** — A collection of newsgroups active on the Internet.

**User** — A component of the message handling environment that engages in (rather than provides) message handling and that is a potential source or destination for the information objects an MHS conveys.

**User Agent** — In the context of message handling, the functional object, a component of MHS, by means of which a single direct user engages in message handling.

## V

**VAN** — Value-Added Network. A commercial electronic messaging network that carries electronic mail or EDI messages.

**Veronica** — Search software for filenames on Gopher servers.

**Viewer** — A software program invoked when the file received is not supported on the client's workstation. For example, a user receiving a postscript file must invoke a viewer in order to display the contents of the file.

**VIM** — Vendor Independent Messaging. API from a consortium led by Lotus Development

**Virtual Reality Markup Language** — A computer programming modeling language for use in rendering 3-D simulated "worlds" on Web sites.

**Virus** — A malicious software program that replicates itself, causing damage to files and utilities in a computer system.

**Voice body part** — A body part sent or forwarded from an originator to a recipient that conveys voice-encoded data and related information.

**VRML** — Virtual Reality Markup Language.

# W

**W3** — World Wide Web.

**WAIS** — Wide Area Information Server.

**WAN** — Wide Area Network.

**Web crawler** — A software program that searches the Web for specified purposes such as to find a list of all URLs within a particular site.

**Whois** — An Internet resource that permits users to initiate queries to a database containing information on users, hosts, networks, and domains.

**Wide Area Information Server** — Software that permits the searching of huge Internet indices by means of keywords or phrases.

**Wide Area Network** — A network connecting multiple computers at multiple locations, often geographically dispersed.

**Windows** — A computer operating system developed by Microsoft.

**World Wide Web** — An information system created at CERN enabling easy access to distributed hypertext information.

**WYSIWYG** — "What you see is what you get."

**WWW** — World Wide Web.

# X

**X.25** — A data communications protocol within the OSI suite.

**X.400** — An international standards specification for the exchange of electronic mail messages.

**X.435** — A set of standards defined by the International Telecommunications Union that describe Message Handling System support of EDI.

**X.500** — An international standard specifying a model for a distributed directory system.

**XAPIA** — X.400 Application Programming Interface Association.

**XBM** — X bit map, a black and white image format.

# Y

**Yahoo** — An Internet search utility.

# Index

## A

AARP, *see* American Association of Retired Persons
Abstract service, 190
Acceptance test plan (ATP), 239
Access Control Decision Function (ACDF), 194
Access Control Information (ACI), 195
Access to Information, company policy for, 621
Access units (AUs), 246
Accounting database, 377
Account-tracking system, 55
ACDF, *see* Access Control Decision Function
ACI, *see* Access Control Information
ACP, *see* Allied Communications Publication
ACSE, *see* Association Control Service Element
Activity room, 436
ADDMDs, *see* Administrative Directory Management Domains
Address
   book, 30, 141, 432
   conversion, 98
   list (AL), 114
   mapping, 217
ADMD, *see* Administrative Management Domain
Administration
   domain name, 167
   facilities, 233
   management, 250
   model, 73
   services, 587
Administrative Authority, 175, 191
Administrative Directory Management Domains (ADDMDs), 175
Administrative DUA (ADUA), 167
Administrative Entry (AE), 174
Administrative Management Domain (ADMD), 106
Administrative Points (AP), 174
Administrative staff, 21
Administrative users, remote, 30
ADUA, *see* Administrative DUA

AE, *see* Administrative Entry
Aerospace Industries Association (AIA), 625
AIA, *see* Aerospace Industries Association
AL, *see* Address list
Alias names, 189
Alisa Mail Information Switch, 571
Allied Communications Publication (ACP), 101, 112
Allied Message Handling (AMH), 123
Alternate recipient assignment, 108
Alternate routing, 234
American Association of Retired Persons (AARP), 127
American National Standards Institute (ANSI), 190, 374
America Online (AOL), 270, 277, 296, 321, 541
AMH, *see* Allied Message Handling
ANSI, *see* American National Standards Institute
Answering machines, 423, 476
AOL, *see* America Online
AP, *see* Administrative Points
API, *see* Applications programming interface
Application(s)
   client/server, 414
   crashing of, 19
   design, 46
   development, fast, 45
   directory services, 135
   line-of-business, 161
   mail-enabled, 325
   management tools, xiii
   messaging, 83, 99
   -to-person message, 431
   programming interfaces (APIs), 5, 18, 64, 217, 333, 510
   software, 220
   support, 457
   vendor, 519
Appropriate use policies (AUPs), 578
Association Control Service Element (ACSE), 118
Asynchronous transfer mode, 256
ATMs, *see* Automated teller machines
ATP, *see* Acceptance test plan
AT&T Interchange, 276

663

*Index*

ATTMail, 6
Attribute(s), 172
 types, 173, 189
 Value Assertion (AVA), 173
Audit services, 588
AUPs, *see* Appropriate use policies
AUs, *see* Access units
Authentication, 137, 198, 388, 577, 597, 610
 buyer, 388
 client, 364
 data, 14
 definition of, 602
 protected simple, 200
 services, 197, 584
 simple, 199
 strong, 200
 unprotected simple, 199
Authorization services, 197, 587
Autoforwarding, 108
Automated teller machines (ATMs), 349
AVA, *see* Attribute Value Assertion
Average chart format, 561

**B**

Back Office: Exchange Server, 9
Banking, 10, 319
Basic rate interface (BRI), 481
BBN, *see* Bolt Beranek and Newman Inc.
BBS, *see* Bulletin board system
BCSs, *see* Business communication services
BeyondMail, 34
Biometric test, 362
Bolt Beranek and Newman Inc. (BBN), 322
BPR, *see* Business process reengineering
Brainstorming, 539
BRI, *see* Basic rate interface
Broadcast media, 277
Brokerage firms, 597
Browser
 -based mail, 97
 launching, 328
 platform, 94
 telephone-based, 382
Browsing, 178, 311
Bulletin boards, 8, 44, 415
Bulletin board system (BBS), 293
Business
 -to-business transactions, 391
 communication services (BCSs), 257
  comparison chart, 262
  connectivity, 260
  message conversion of, 260
  network access by, 261
  reliable delivery of, 260
  selection criteria, 261
 -to-consumer transactions, 391
 correspondence, 433
 online service, 276

process reengineering (BPR), 405
 tools, 181
Buzzwords, 148

**C**

CA, *see* Certification Authority
Calendar(ing), 89, 93, 342
 information, 73, 74
 integration, 99
 programs, 447
 software, 457
Call back folder, 336
Caller identification, 78
Call forwarding, 436, 476
Capability
 links, 562
 ranges, 570
Car mileage, reduced, 471
CASE, *see* Computer-aided software engineering
CAW, *see* Certification authority workstation
CBT, *see* Computer-based training
cc:Mail, xii
 client, 83
 integration, 341
Cellular phones, 456
Central application management, 46
Centralized directory, 167
Central switching mechanism, 529
Central synchronization service, 141
Certificate
 repository, 364, 365
 revocation list (CRL), 365
Certification
 Authority (CA), 201, 202, 362, 364, 587
 authority workstation (CAW), 103, 105
 delegated, 593
 path, 202
 PGP distributed, 593
 process, 594
Certified product list (CPL), 111
CGI, *see* Common gateway interface
Chaining, 193
Chat rooms, 272
Childcare, telecommuting and, 489
CICS, *see* Command Information Control System
Circadian rhythm, 467
CIX, *see* Commercial Internet Exchange
Clear service indicator, 114
Client
 information display, 87
 /network model, 453
 platform support, 333
 replication, 335
 software, 394, 455
 /server
  architectures, xiv

# Index

computing 39
deployment, 162
download encryption, 87
model, 451
support, 53
Client/server messaging, mobile worker and, 445–460
  client/server flexibility, 458–460
    mobile user support, 458–459
    security, 459–460
  effect on messaging components, 453–458
    applications, 457–458
    command and control, 456–457
    media modules, 454
    transport, 454–456
  evolution of communication, 446–447
    1800s, 446
    1900s, 446–447
    approaching 2000 and beyond, 447
  file sharing vs. client/server messaging, 448–453
  number of mobile workers, 448
Clinical graphics, 132
CLNP, *see* Connectionless Network Protocol
Clustering technology, 92
CMC, *see* Common Messaging Calls
CMIP, *see* Common Management Information Protocol
Collaboration, 7, 8, 57
Collaborative computing, xiii, 76, 81
College students, 267
Combination topology, 18
Command Information Control System (CICS), 129
Commercial Internet Exchange (CIX), 283
Commercial networks, 23
Commercial online services comparison chart, 272
Commercial services, 24
Common gateway interface (CGI), 285
Common Mail Call, 65
Common Management Information Protocol (CMIP), 513, 516
Common Messaging Calls (CMC), 47, 233, 511
Communication(s)
  authority, local, 215
  circle of influence, 446
  control of incoming, 429
  costs, 286, 287
  cross-platform, 332
  everyday, 381
  evolution of, 446
  facilitation of, 160
  infrastructures, 446
  link, failure of, 19
  long-distance voice, 314
  management services, 77, 82
  protocols, lower-level, 224
  redefined, 275

  strategies, 408
  ultimate in, 317
  wireless, 444
Company
  culture, 465
  electronic mail, definition of, 626
  Employee Conduct Policy, 620
  policies, 473, 474, 493
  values, 635
CompuServe, 276, 541
Computer
  -aided software engineering (CASE), 405
  -based patient record (CPR), 128
  -based training (CBT), 26
  monitor, 487
  notebook, 451, 455, 460
Computing, architectures of, 452
Conferencing, 93
Configuration management, 250
Conformance testing, 162
Connectionless Network Protocol (CLNP), 106
Connection scenarios, 448
Connectivity
  availability, 350
  options, unlimited, 129
  support, 269
Connect rate, 308
Consumer
  -to-consumer transactions, 391
  market, 282
  shopping, 359
  transaction, 356
Content security, 361
Control
  chart analysis, 562
  limits, points outside, 563
Convergence, 384
Copy precedence, 113
Core engine technology, 71
Corporate directory, 183
Corporate gateways, 23
Corporate network, 399
Corporate security policies, 594
Counterfeiting, 593
CPL, *see* Certified product list
CPR, *see* Computer-based patient record
CPU utilization, 556
CRC, *see* Cyclic redundancy check
Credit card
  information, 382
  transactions, 281, 357
Crisis management, 470
CRL, *see* Certificate revocation list
Cross File (XFL), 23
Cross-platform messages, 566
Cross-references, 208
Cryptography, 200

665

## Index

Cryptosystems, symmetric and asymmetric, 201
Customer
  account tracking, 55
  service, 290, 379
  support, 55, 80, 177, 258
  technical issues, common, 41
  tracking, 40, 45
Cyberhome, 435
Cyberspace, investing in commercial, 318
Cyclic redundancy check (CRC), 601

### D

DACD, see Directory Access Control Domain
DAP, see Directory Access Protocol
Data
  analysis capabilities, 413
  backups, 459
  center environment, 92
  communications network engineers, xiv
  compromise, 581
  confidentiality, 203
  delivery, 390
  Encryption Standard (DES), 356, 599, 607
  integrity, 204, 602
  management, 177, 186
  overload, 76
  stream encryption, 89
  transfer, 270
  transmission integrity, 390
  warehouse, 129
Database(s), 15
  accounting, 377
  backups of, 20
  businesses that maintain, 318
  corruption, 20
  design, 86
  development, 398
  human resources, 337
  low-level, 149
  for message storage, 69
  monitor, 339
  Notes, 338, 398, 419
  PBX, 154
  problem in, 70
  publicly accessible, 364
  servers, Lotus Notes, 98
  software companies, 397
  tools, 394
da vinci eMail, 11
DCE, see Distributed computing environment
DDN, see Defense Data Network
DDS, see Distributed directory services
DEC All-In-One, 33, 72
Decision-making, 76, 81
Dedicated server, 287
Defense Data Network (DDN), 102

Defense Information Systems Agency (DISA), 101, 106
Defense Information Systems Network (DISN), 102
Defense Message System (DMS), 101–124, 180
  characteristics, 107
  components, 111
  messaging systems, 116
  object classes, 122
  program overview, 102–111
    availability of DMS-compliant products, 111
    DMS background, 102–103
    DMS components, 103–106
    DMS deployment schedule, 111
    DMS operational environment, 106–107
    rationale for X.400-based DMS, 109–111
    unique DMS characteristics, 107–109
  standards and protocols, 112–123
    ACP 123, military message content types, 112–115
    ACP 123 U.S. Supplement 1, 115–116
    DMS directory schema, 120–123
    DMS security, 118–119
    international standardization of ACP 123 message extensions, 117
    Message Security Protocol, 119–120
    users, 118
Defragmentation procedures, 524
Delivery
  notifications, 116, 229, 248, 253
  services, institutionalization of, 504
  time, 552
    of individual messages, 559
    range of, 553
    subjective measurement of, 553
Department of Defense, 411
Deployment architectures, 529
Dereferencing, 195
DES, see Data Encryption Standard
Design problems, 236
Desktop devices, intelligent, 10
Diagnostic images, 132
Dial-in subscribers, 119
Dial-up modem, 268, 450
DIB, see Directory Information Base
Digital encryption, standards-based, 48
Digital signatures, 42, 202, 330, 363
Digital Signature Standard (DSS), 118, 586
Directory
  Access Control Domain (DACD), 195
  Access Protocol (DAP), 14, 143, 165, 166, 168, 204
  applications, 179
  authentication framework, 200
  centralized, 167
  distributed, 167
  elements of, 150
  entry, 171, 186

*Index*

facility, 229
group, 404
information, 15
  model, 163
  structure, 171
Information Base (DIB), 121, 165, 171, 186, 198
Information Shadowing Protocol (DISP), 151, 170
Information Tree (DIT), 144, 151, 165, 169, 170, 187, 188
infrastructure, 179
interrogation operations, 190
levels of, 387
management, 211
Management Domains (DMDs), 174
of messages, 7
name, X.500, 173
objects, protection of, 198
OpenMail, 88
Operational Binding Management Protocol (DOP), 170
OSI, 191
products, types of, 388
replication, 42
security, X.500, 196
service(s), 7, 100, 206, 522
  agent, 149, 191, 205
  building of, 146
  controls, X.500, 195
  enterprise-scale approach to, 85
  expectations of, 148
  low-risk White Pages, 145
  rise of, xii
storage within, 150
synchronization, 1, 72, 91, 96, 149–157, 551, *see also* Directory synchronization, elements of; Directory synchronization, meta-directory
  agent (DXA), 50
  automatic, 68
  automation of, 157
  continuous, 70
  cross-platform, 142
  cycle, 153
  messages, 63
  problem, 151–152
  protocol, transport mechanism of, 153
  solutions, 152, 157
  system, 155
  two-way, 98
system
  agent (DSA), 14, 104, 142, 166, 167, 247
  global, 212
System Protocol (DSP), 143, 165, 168
universal, 164
user agent (DUA), 14, 103, 142, 247
uses, generic, 178

Directory synchronization, elements of, 149–151
  case study, 156
  directory service agent, 149–150
  directory user agent, 150
  protocol, 151
  storage, 150–151
Directory synchronization, meta-directory, 152–156
  elements of directory synchronization, 152
  error alerts and recovery, 155–156
  filtering, 155
  name and object mapping, 154–155
  protocol, 153
  transport mechanism, 153–154
DISA, *see* Defense Information Systems Agency
Discordance, 384
Discussion-list server, 295
Disintermediation, 384
Disk mirroring, 59
DISN, *see* Defense Information Systems Network
DISP, *see* Directory Information Shadowing Protocol
Distinguished names (DNs), 144, 189, 362
Distributed computing environment (DCE), 143
Distributed directory, 167
Distributed directory services (DDS), 33
Distributed operations, 205
Distribution code, 114, 117
DIT, *see* Directory Information Tree
DMDs, *see* Directory Management Domains
DMS, *see* Defense Messaging System
DNs, *see* Distinguished names
DNS, *see* Domain name service
Document management, 79, 93, 416
DOD, *see* U.S. Department of Defense
Domain name service (DNS), 139
DOP, *see* Directory Operational Binding Management Protocol
DOS platform, 35
Downloading
  performance, 311
  selective, 343
Downsizing, 54
DSA, *see* Directory system agent
DSP, *see* Directory System Protocol
DSS, *see* Digital Signature Standard
DUA, *see* Directory user agent
DXA, *see* Directory synchronization agent
Dynamic Web pages, 264

**E**

EC, *see* Electronic commerce
Echo program, 21

667

## Index

Economy chairs, 485
ECP, *see* Emergency command precedence
ECS, *see* Electronic commerce services, developing trusted infrastructure for
EDI, *see* Electronic data interchange
Editor box, 426
EFT, *see* Electronic funds transfer
EIS, *see* Executive information system
Electrical power concerns, 486
Electronic calendar, 437
Electronic catalogs, 323
Electronic commerce (EC), 7, 93, 125, 142, 256, 325, 345–360, 373
  basics, 345–346
  business-to-business, 378
  consumer-to-consumer, 383
  expansion, 360
  interchange participants, 346–347
    consumer, 347
    distributor, 346
    manufacturer, 346
    merchant, 347
    supplier, 346
  merchant-consumer electronic purchase payment process, 354–359
    certificates, 358–359
    electronic signatures, 357–358
    encryption's use in electronic commerce, 356–357
    payment processes, 355–356
    public/private key encryption, 358
  participants, 347
  process, 348
  product chain, 346
  simplified electronic commerce participants, 347–350
    negotiating product and relationships, 348
    order fulfillment, shipping, and delivery of product, 348–349
    ordering process, 349–350
    paying, 349
  status, 353
  technology issues, 350–354
    bandwidth and connectivity requirements for EC, 350–352
    connecting consumer's device, 350
    human interface requirements and needs, 352–353
  world of, 373–391
    business transformed, 375–376
    definition of electronic commerce, 376–384
    directories, 385–389
    evolving electronic commerce landscape, 384–385
    history, 373–375
    value-added networks vs. Internet and Web, 389–390
Electronic commerce services (ECS), developing trusted infrastructure for, 361–372
  available technologies for electronic commerce, 369–371
    secure E-mail, 370
    secure open EDI, 370–371
    secure World Wide Web, 370
  public key certificate infrastructure, 371
  service attribute authority, 361–368
    certificate authority, 362–364
    certificate repository, 364–366
    electronic postmark, 366–367
    return receipts, 367–368
    storage and retrieval services, 368
    use of commercial exchange services, 368–369
Electronic Communications Privacy Act, 616
Electronic data interchange (EDI), 9, 31, 91, 127, 159, 203, 258, 325, 538
  movement of into global market, 375
  network provider, 370
  security services, 370
  topology, 379
  transactions, 116, 376
  translation, 380
Electronic forms, 7, 8, 38, 79
Electronic funds transfer (EFT), 203, 350, 377, 380
Electronic mail privacy, guidelines for, 625–627, *see also* E-mail
  definitions, 625–626
  guidelines, 626–627
    basic recommendations, 626
    specific recommendations, 626–627
Electronic marketplace, portal to global, xii–xiii
Electronic meeting system (EMS), 402
Electronic messages, securing, 597–603
  authentication, 602–603
  message integrity, 601–602
  message privacy, 597–601
    codes vs. ciphers, 598
    encryption methods, 598–599
    public vs. private keys, 599–601
Electronic messaging, strategic growth of, 3–10
  messaging architecture, 3–6
    not so simple SMTP, 4
    PC E-mail packages, 4–6
    proprietary E-mail applications, 3–4
  messaging functional ability, 7–8
  messaging standards, 6–7
  organizational impact, 9
  product and service trends, 8–9
Electronic Messaging Association (EMA), 10, 210, 250, 515, 542
Electronic Messaging Association of Australia, 542

# Index

Electronic messaging infrastructure, leveraging, 325–326
Electronic messaging products, selecting, 537–546
 conducting of live evaluation, 544–545
 definition of requirements, 539–540
 determination of product configuration, 541
 determination of product selection, 545
 identification of applicable products, 542–543
 paper evaluation, 543–544
 project definition and planning, 538–539
 publicizing of results, 545
 selection of live evaluation products, 544
 weight requirements, 541
Electronic messaging systems, implementing infrastructures and, 519–535
 deployment architectures and options, 529–533
  common backbone, 531–533
  common platform architecture, 530
  multiple backbone model, 530–531
 establishing messaging policies and procedures, 533–534
  privacy, 533–534
  proprietary and confidential information, 534
 how to accomplish rollout and manage constraints, 519–522
  functionality and scope, 521
  resource constraints, 520–521
  supporting internal and external customers, 522
 implementation models and architectures, 522–524
 implementation scenarios, 525–529
  application gateways for integrating dissimilar systems, 527–529
  one-tier messaging model, 525
  two-tier model, 525–527
 recommended course of action, 534–535
  application support, 535
  backup, 535
  network connections, 534
  operating systems, 535
Electronic postmark, 366
Electronic product exchange, 359
Electronic publishing, 322
Electronic purchase payment process, 354
Electronic signatures, 357
Electronic surveillance, 632
Electronic village, 324
Element of service (EOS), 112, 115, 218
EMA, *see* Electronic Messaging Association
E-mail
 address(es), 296
  collection of workstation, 138
  public, 231
  searching for, 297
 administration, 177
 attaching files, 27
 bombs, 610
 communications, intrusions into, 630
 CompuServe, 419
 connectivity
  of BCS, 261
  external, 130
 decisions regarding, 327
 directories, accuracy of, 21
 editing capability, 26, 27
 facilities, 613
 -to-fax translations, 272
 files, access to, 616
 forms, 29
 groups, 535
 hate, 293
 hub, central, 531
 IDs, 432
 implementation, 327
 as interactive resources, 295
 management, 249
 messaging, 279
 networks, 630
  enterprise, 385
  homogeneous, 130
 Notes, 419
 opportunities, 615
 packages, 135
 policy, employee awareness of company, 619
 post office, 449
 privacy, 611, 622, 629
  audit, 638
  employees' perspectives on, 631
  policy, 635
 private, 24
 programs, 295
 protocol, 243
 providers, 614
 resources, company-owned, 618
 responsible management of, 633
 serious problem in, 25
 service providers, 109
 software providers, 503
 spam, dealing with, 280
 spell checker, 26
 state of, 244
 system(s)
  administrator, 632
  company, 617
  directories, 30
  LAN-based, 41, 614
  -level goals for, 131
  limitations of, 102
  policy recommendations, 618
 technology, 419
 tracking performance through, 631
 use, ethical principles for employee, 638

669

## Index

E-mail, creating policy relative to, 613–623
  employer vs. employee, 617–618
  old E-mail, 613–614
  policy recommendations, 618–620
  policy templates, 620–622
  public vs. private E-mail, 614–616
E-mail, exploiting and extending, 327–343
  enhancements to infrastructure, 339–343
    calendaring and scheduling, 342
    complete remote connectivity, 343
    global service and support, 343
    integration with workflow, authoring tools, and document libraries, 341–342
    powerful and consistent user interface, 341
    seamless Internet integration, 339–340
    standards support, 340–341
  goals of enterprisewide messaging, 327–328
  lower cost of ownership, 339
    extensible, replicated directory, 333–334
    integrated management and administration, 336–339
    replication and synchronization capabilities, 334–335
    user agents, 335–336
  reliability and architecture, 329–333
    broad cross-platform communication, 332–333
    fault tolerance with store-and-forward and least-cost routing, 331
    multilevel security, 330–331
    object store flexibility, 331–332
    superior scalability, 329–330
  where messaging, groupware, and Internet meet, 328–329
E-mail privacy, ethical management of employee, 629–637
  ethically responsible management of E-mail, 633–638
  safeguarding company, 629–633
    employees' perspectives on E-mail privacy, 631–633
    employer's perspective on E-mail privacy, 630–631
E-mail security and privacy, 605–611
  corporate security, 608–611
    encryption, 610–611
    firewalls, 609
    integrated strategy, 611
    internal precautions, 609
    preventing E-mail flooding and denial of service, 610
    unlisted sites, 611
  establishing secure E-mail standards, 606–607
    privacy enhanced mail, 607
    secure directories, 606–607
  leading E-mail coalitions, 607–608
    Electronic Messaging Association, 608
    Internet Mail Consortium, 607–608
E-mail systems, popular, 23–35
  E-mail services, 29–32
    directories, 29–30
    fax gateways, 30
    gateways, 31–32
    message notification, 30–31
    security, 31
  features and functions, 24–29
    features and services checklists, 26–27
    features for creating messages, 27–29
    X.400 and SMTP, 25
    X.500 directory service, 25–26
  historical perspective of electronic messaging, 23–24
  popular messaging systems, 32–35
    BeyondMail, 34–35
    DEC All-in-One, 33
    Fisher TAO, 34
    GroupWise, 35
    HP Open Desk Manager, 33
    IBM Office Vision 400, 32–33
    Lotus cc:Mail, 32
    Lotus Notes, 34
    Memo, 34
    Microsoft Exchange, 34
    QuickMail, 35
  primary electronic messaging system categories, 24
Emergency command precedence (ECP), 116
EMO, *see* Enterprise Messaging Operation
Emoticons, 303
Employee(s)
  recruiting, 469
  retention, 461, 469
  right of privacy for, 623
  workplace privacy, 613
Employer responsibilities, 462
EMS, *see* Electronic meeting system
Encryption, 330, 577, 595, 610, 636
  methods, 598
  schemes, types of, 356
  services, asymmetric key, 583
  technology, 354
End-user
  client interfaces, 85
  forms designed by, 46
Engineering benchmarks, 252
Enterprise communications, 84
Enterprise directory service, 1, 135–148
  application vs., 135–136
  flexibility of, 148
  roles and responsibilities, 145–148
    operations management, 146
    product management, 146–147
    project management, 147
    service management, 147–148

*Index*

service provider's perspective on X.500, 143–145
  database issues, 144
  directory information tree structure, 144–145
  information management issues, 144
what goes on in, 136–143
  authentication/confidentiality services, 140–141
  directory synchronization services, 141–142
  information solicitation services, 137
  naming/identification services, 139–140
  registration services, 137–139
  X.500, 142–143
  X.500 infrastructure coordination services, 143
Enterprise Messaging Operation (EMO), 83
EOS, *see* Element of service
Error
  conditions, 219
  handling, 232
  logging, 91, 512
Ethics management, strategy for, 634
European Electronic Messaging Association, 542
Exchange Server, 83
Executive information system (EIS), 413
  capabilities, 418
  future of Notes for, 421
  manager, 417
  role of Notes in, 416
  software, 417
Expense reporting, 77, 79, 80
External customers, 522

**F**

Facsimile solutions, 93
Family issues, telecommuting and, 488
FAQs, *see* Frequently asked questions
Fault
  management, 250
  tolerance, 331
Fax
  applications, network, 62
  gateways, 29, 30, 32
  input, 26
  machines, 423, 473
  modem, 445
  output, 26
  receipt, 445
  technology, 477
Federal Information Processing Standard (FIPS), 124
Federal Telecommunications System (FTS), 109
Federal wiretap law, 614
Feedback access, direct, 383

FIFO, *see* First-in, first-out
File
  attachment support, 232
  compression, 28
  security, 317
  server, 13, 15
  sharing, 448
    architecture, 456, 510
    systems, 454
  Transfer Body Part (FTBP), 26, 52, 608
  transfer protocol (FTP), 4, 137, 259, 285
    mail, 301
    mechanics of, 302
    passwords, 311
    use of to download files, 300
Financial data, 441
Finger, 314
Fingerprinting, 362
FIPS, *see* Federal Information Processing Standard
Firewall, 609
  corporate, 95
  protection of, 616
  second-generation, 93
First-in, first-out (FIFO), 228
Fisher TAO, 34
Flaming, 300
Flextime, 466
Font differences, 397
Foreign form, 374
Forwarding, 230
FQDN, *see* Fully qualified domain name
FRADs, *see* Frame relay access devices
Frame relay access devices (FRADs), 491–492
Freeware software, 145
Frequently asked questions (FAQs), 285, 305
FTBP, *see* File Transfer Body Part
FTP, *see* File transfer protocol
FTS, *see* Federal Telecommunications System
Fully qualified domain name (FQDN), 297

**G**

Gatekeeper, 19
Gateway
  application, 527
  components, 219
  configuration models, 220
  corporate, 23
  deploying, 238
  fault tolerance of, 237
  fax, 29, 30, 32
  to foreign environment, 522
  functions, basic, 216
  GroupWise, 67, 72, 75
  hardware, 519
  initial testing of, 239
  Internet mail, 225

671

## Index

large-scale, 222
maintenance of, 240
MDI, 420
MFI, 108
model
   colocated, 221, 222
   distributed architecture, 223
   stand-alone, 221
multiple, 227
news service, 305
placing of between routers, 528
product
   flexibility of, 236
   hardware model independent, 220
   qualities of, 235
scalability of, 237
SMTP, 25
solutions, 93
technology, 213, 215, 241
trial service of, 239
two-way, 227
GCCS, *see* Global Command and Control System
GE Information Services (GEIS), 24
GEIS, *see* GE Information Services
Global Command and Control System (GCCS), 119
Global competition, xi, 638
Global computing environment, 163
Global electronic directory, 183
Global management, single interface for, 51
Global networking vision, 60
Global Network Navigator (GNN), 321
Glossary, 643–661
GNN, *see* Global Network Navigator
Gopher, 309
GOSIP, *see* Government Open System Interconnection Profile
Government
   mandates, 461
   Open System Interconnection Profile (GOSIP), 6
   services, 183
Graphical user interface (GUI), 4, 39, 288, 310, 317
Graphics design, 281
Group
   calendaring programs, 18
   communication strategies, 408
   directory, 404
   scheduling, 8, 38, 40, 43, 56
Groupware, 215, 401, 447
   applications, 42, 56, 396, 405
      custom, 77, 78
      easy-to-create, 45
   directories, 98
   meetings with, 403
   package, 408
   redefining, 37
   session, 407, 409

solution, next-generation, 76
tools, 403, 410
GroupWise 5, 35, 75
GUI, *see* Graphical user interface

## H

Hackers
   firewalls and, 609
   messaging systems protected against, 502
   networks safe from, 376
Half messages, 555
Hand-held portable (HHP) devices, 586
Hard drive, 13
Healthcare
   administrator, 125
   organizations, 133
   providers, 125
   suppliers, 319
   vision for, 133
Healthcare industry, electronic messaging in, 125–133
   case study, 128–132
      anatomy of E-mail network, 129–130
      electronic forms and goals, 131–132
   emergence of telemedicine, 126–127
Helpdesk activity, 458
HHP devices, *see* Hand-held portable devices
High-speed dedicated line, 480
Home-grown applications, 144
Home office environment, 475, 483
Home pages, 124
Hot-links, 381
HP Open Desk Manager, 33
HP OpenMail, 83–100
   architecture and function, 85–93
      complementary technologies, 93
      EDI/electronic commerce, 91
      forms and workflow, 90
      management, 91–92
      messaging, 85–90
      performance and reliability, 92
      platforms, 93
      public folders, 90–91
      server architecture, 85
   market positioning and advantages, 83–85
      product strategy, 85
      value to customers, 84
   solutions, 93–100
      application messaging, 99–100
      cc:Mail environments, 98
      host-based E-mail interoperability, 98–99
      intranet, 93–96
      Lotus Notes, 98
      Microsoft environments, 96–98
HP OpenView Operations Center, 92
HTML, *see* Hypertext markup language
HTTP, *see* Hypertext transport protocol

*Index*

Hub(s)
 central switching, 528
 routing, 331
 -spoke
  routing, 17
  topology, 16
 worldwide distributed, 529
Human resources organization, 182
Hybrid approach, disadvantages to, 398
Hybrid environment, 397
Hyperlinks, 353
Hypertext links, 310, 381
Hypertext markup language (HTML), 43, 264, 277, 394
Hypertext transport protocol (HTTP), 264, 311, 396

# I

IAB, *see* Internet Architecture Board
IBM
 mainframe environments, 530
 OfficeVision, 72
 Office Vision 400, 32
Idea
 analysis, 403
 generation, 403, 406
IEA, *see* International Electrotechnical Association
IETF, *see* Internet Engineering Task Force
IFIP, *see* International Federation Information Processing
IIS, *see* Internet Information Server
IMAP, *see* Internet Mail Access Protocol
IMC, *see* Internet Mail Consortium
Immediacy, 384
Implicit conversion, 113
Importance indication, 113
Importing text, 28
Industrial Revolution, 461
Information
 access, 329, 629
 batch-level replication of, 151
 cached, 195
 choices for transporting, 255
 clearinghouse for, 439
 confidential, 534
 credit card, 382
 directory, 156
 dissemination, 400
 employee, 140
 encrypted, 108
 exchange tools, 402
 executive demand for, 414
 explosion, 23
 flow of, 447
 HTML, 398
 interoperability of multivendor, 164
 low-cost transport of, 267
 message, 78
 movement of, 50
 about non-employees, 138
 about nonpeople, 138
 object, 172
 overload, 18
 privacy, 577
 publication, 144
 routing, 438
 search and retrieval, 258
 server disk usage, 337
 Services Business Division (ISBD), 23
 solicitation, 136
 starvation, 76
 status, 413
 streaming, 436
 systems
  evaluation of, 546
  managers, 161
 transmission, computer desktop, 481
Information security, 577–595
 administration services, 587–588
 audit and alarming services, 588
 authentication services, 584–587
  certified digital signatures, 586–587
  Kerberos, 585
  privacy enhanced mail, 586
  RSA, 585
  Smartcard technologies, 586
  static passwords, 584–585
  X.509 directory authentication services, 587
 authorization services, 587
 computer crime reports, 594–595
 digital message certification and key management, 590–594
 electronic messaging security mechanisms, 583
 identification, 583–584
 integrity and privacy policies, 577–579
 internal and external security requirements, 583
 private messaging networks vs. public networks, 588–590
 security requirements for electronic messaging systems, 579–580
 threats and vulnerabilities within open networks, 580–583
  data compromise, destruction, misdirection, and impersonation threats, 581–582
  multilevel security, 580–581
  traffic analysis threats, 582–583
Initial Operational Capability (IOC), 111
In-line addressing, 230
Insurance companies, 597
Integration-internetworking, 384
Intelligent forms, 131
Intelligent messaging, 259
Intelligent networks, 440

673

# Index

Intelligent post offices, 436
Intelligent services, 430, 441
Interactive voice response (IVR), 382, 478
Interface technology, 181
Internal reference, 207
International Electrotechnical Association (IEA), 101
International Federation Information Processing (IFIP), 249
International Standards Organization (ISO), 101, 142, 160
International Subject Matter Experts (ISME), 123
International Telecommunications Union Telecommunications Standards Sector (ITU-TSS), 25
Internet
  Activities Board, 592
  Architecture Board (IAB), 284
  built-in support for, 48
  business enthusiasm for, 258
  collaboration services, 94
  connectivity, 37, 259
  customer service solution, 97
  definition of, 294
  direct connections over, 49
  Engineering Task Force (IETF), 204, 244, 245, 284, 371, 606
  explorer, novice, 291
  finding information on, 315
  globalization of, 266
  growth of, 39
  Information Server (IIS), 49
  international providers of, 265
  Mail Access Protocol (IMAP), 47
  Mail Consortium (IMC), 607
  newsgroup(s)
    data, 48
    support for, 49
  protocol (IP), 105
  realignment of socialization by, 266
  Registry (IR), 284
  Relay Chat (IRC), 294, 314
  server, personal, 287
  service provider (ISP), 49, 96, 225, 280, 283,
    see also Internet service providers, selecting
    connectivity support, 269
    future direction of, 286
    indexing of, 318
    list of, 290
    outgoing mail server, 605
    topics carried by, 304
    use of by trading partner, 390
  shorthand, 303
  Society (ISOC), 283, 284
  standards, 57
  success on, 320
  support, integrated, 48

Internet resources, using, 293–315
  anatomy of electronic mail address, 296–304
    domains, 297–298
    file transfer protocol, 300–303
    Internet shorthand, 303–304
    mailing lists, 298–300
    searching for specific E-mail address, 297
  basics, 293–294
  Gopher, 309–310
  helpful tools, 313–314
    Finger, 314
    IRC, 314
    videoconferencing, 313–314
  Telnet, 308
  tools of trade, 294–296
  Usenet news, 304–307
    FAQs, 305–306
    reaching communities through Usenet, 304–305
    Usenet filters, 307
    Usenet netiquette, 306–307
  wide area information system, 312–313
  World Wide Web, 310–312
    download and search performance, 311
    links, 310–311
    uniform resource locators, 311–312
Internet service providers, selecting, 283–291
  role of commercial online services, 290–291
  starting point, 284–290
    big list of ISPs, 290
    cost of connection, 286–287
    defining purpose, 285
    doing homework, 285–286
    finding best connection, 287–288
    Internet utilities, 284–285
    managing Internet, 284
    qualities of good ISP, 288–290
Interoperability testing, 162
Interprocess communication (IPC), 71
Intranet(s), 93, 264, 393–400
  concept of, 394
  corporate, 399
  existing investments, 400
  hybrid environment, 397–398
    administrative benefits, 398
    drawbacks, 398
  implementation issues, 398–399
    fit with existing technology, 398–399
    how connected, 399
    learning and training, 399
    management, 399
  long-term thinking, 400
  Lotus Notes in enterprise, 394–395
  Notes vs., 395–397
    browser-based intranets, 396–397

*Index*

Notes-based intranets, 397
solutions, NT-based, 98
Intrusion upon seclusion, 639
Inventory management, 182
IOC, *see* Initial Operational Capability
IP, *see* Internet protocol
IPC, *see* Interprocess communication
IR, *see* Internet Registry
IRC, *see* Internet Relay Chat
ISBD, *see* Information Services Business Division
ISDN line, 480
ISME, *see* International Subject Matter Experts
ISO, *see* International Standards Organization
ISOC, *see* Internet Society
ISP, *see* Internet service provider
ITU-TSS, *see* International Telecommunications Union Telecommunications Standards Sector
IVR, *see* Interactive voice response

## J

JANAP, *see* Joint Army, Navy, Air Force Publication
Japan Electronic Messaging Association, 542
Job descriptions, 406
Joint Army, Navy, Air Force Publication (JANAP), 102
Journaling, 41

## K

KDC, *see* Key distribution center
Kerberos, 585
Key
  distribution center (KDC), 585
  pairs, 358
Knowledge economy, 384

## L

LAN, *see* Local area network
Laptop computers, 67, 431
LDAP, *see* Lightweight Directory Access Protocol
Leadership style, 408
Lead tracking, 77
Leased-line connections, 378
Legacy environment, 226
Legal council, corporate, 493
Legal services, 319
Library services, 183
Lighting, 487
Lightweight Directory Access Protocol (LDAP), 2, 88, 143, 185

significance of, 204
uses and limitations, 204
Links, types of, 310
LISTSERV, 295, 299, 304
LLC, *see* Logical Link Control
Local address books, system directories vs., 29
Local area network (LAN), 1, 213, 507
  administrator, 12, 520
  E-mail post office consolidation, 97
  environments, small, 386
  file servers, 16
  -to-LAN connections, 270, 271
  manager, 535
  networking, 521
  protocols, 249
  /WAN infrastructure, 20
Local area network (LAN) messaging, 11–21
  components of LAN messaging system, 13–15
    directories, directory system agent, directory user agent, 14–15
    message store, 13
    message transfer agent, 13
    user agent/user interface, 13
  features of LAN messaging system, 15–19
    application programming interfaces, 18
    directory updates and synchronization, 18
    file sharing vs. client/server architecture, 15–16
    filters, rules, and agents, 18–19
    gateways, 17–18
    message routing topology, 16–17
  myths and realities of LAN messaging system, 19–21
    management, 21
    planning, 20–21
    reliability, 19–20
    scalability, 20
  popularity of LAN messaging systems, 11–12
    ease of setup, 12
    ease of use, 11–12
    low initial investment, 12
Local law compliance, 493
Local networks, 25
Logical Link Control (LLC), 106
Logonid maintenance, 583
Lotuc cc:Mail, 32, 11
Lotus Notes, use of in executive information systems, 413–421
  collaborative work support, 420–421
  EIS capabilities provided by Notes, 418–420
    extracting, filtering, compressing, and tracking critical data, 419–420
    presentation of graphical, tabular, and textual information, 419

675

# Index

presentation of status and trend information, 419
support for electronic communications, 419
support for soft information, 418
EIS evolution, 413–414
future of Notes for EIS, 421
limitations of Notes, 421
Lotus Notes, 414–416
  Notes applications, 415–416
  what Notes can do, 414–415
ten-company study, 416–418
  reason for using Notes, 416–417
  use of Notes, 417–418
Lurking, 298

## M

MAC, see Media Access Control
Macintosh, 78
Mail(ing)
  list, 7, 298, 299, 304, 315
  agent (MLA), 103, 104
  expansion, 120
  monitoring, 24, 339
  systems, homogeneous, 550
  total volume of world, 504
  transport agent, 244
  user agents (MUAs), 244
Mailbox, 425
  subsets, downloading of, 343
  searching, 7
Mainframe
  centralized, 213
  computer, 449
  environment, 452
  hardware reliability, 19
  hosts, older, 225
  mail systems, 513
  -to-mainframe transactions, 391
  -server model, 448, 449
Maintenance upgrades, 517
Managed care, 125
Management information base (MIB), 249, 515
Management workstation (MWS), 103, 105
Manual reentry errors, 379
Manufacturer-merchant exchange, 351
Manufacturing schedules, 380
MAPI, see Messaging applications programming interface
Market information, access to, 40
Materials management transaction set, 128
Matrix analysis, 404
MCI, 6
MDI gateway, see Medium-dependent interface gateway
Media Access Control (MAC), 106
Media Circus, 322

Medium-dependent interface (MDI) gateway, 420
Memo, 34
Memory utilization, 556
Merchant-consumer purchase and payment process, 356
Merchant-consumer relationship, 351
Mesh routing, 331
Message(s)
  archiving, 234
  authentication of, 603
  box, 425, 429
  delivery
    process, 567, 568
    times of, 559
  digest, 366, 368
  exchange, 217
  filing, 7
  handling
    international standard for, 243
    system (MHS), 34, 61, 246, 523, 588
  instructions, 114, 117
  integrity, 597
  maintenance upgrades, 517
  management, 69, 248, 251
  media, integration of, 428
  notification, 26, 29, 30
  originator, 601
  printing, 7
  priority, 218, 427
  privacy, 597
  reception processing, 216
  rendering, 435
  repair, 232
  replies, 427
  retrieval, 234, 434
  rooms, 433
  routing, 61, 217
  Security Protocol (MSP), 102, 108, 110, 116, 118
  server, features of, 70
  services, 424
  status tracking, 72
  storage, 459
  store (MS), 13, 87, 103, 105, 246, 549
  tracking, 78
  Transfer Service (MTS), 116
  transfer agent (MTA), 13, 50, 51, 104, 246
  transmission processing, 216, 219
  transport agent, 68
  waiting indicator (MWI), 477
Messaging
  applications programming interface (MAPI), 47, 48, 233
  architecture, 3, 448, 450, 509
  backbone, 519
  built-in, 395
  business-quality, 223
  capabilities, accessibility to, 505
  client/server, 458, 460

## Index

electronic forms of, 505
enterprisewide, 327
environment
  diversity of, 84
  safeguarding, 595
hardware, dedicated, 568
impersonation, 589
intelligent, 259, 444
management, 92, 501, *see also* Messaging management, introduction to
  effectiveness of, 574
  on intranet, 94
  service plan, 517
  standards, 514
Management Council (MMC), 515
mobile, 7, 259
model
  one-tier, 525, 526
  two-tier, 525, 526
networks, monitoring of, 21
party, impersonation of, 582
personal, 83, 94, 433
processes, analysis of, 560
products, LAN-based, 73
requirements, 537
server, 16
service
  architecture, 506
  plan, 506, 517
software, 553
-specific resources, 508
support organizations, 535
switch, central, 519
system(s), 508
  addressibility of, 547
  availability of, 547
  client/server, 509
  consistency of, 547, 548
  day-to-day administration of, 503
  delivery time, 547, 548
  first LAN-based, 11
  forward-based, 68
  global, 84
  LAN-based, 139
  mainframe based, 19
  manageability of, 511
  managers, 558
  multiplatform, 569
  problems with, 551
  reliability of, 547, 548
  remote access to, 73
  resources for, 510
  shared-file, 509
  store-and-forward, 163
  user expectations of, 504
tasks, automation of, 341
technical architecture, 506, 507
technologies, 505
third-generation, 453
ultimate abuse of, 280

vendors, 514
workload, 548
Messaging gateways, 17, 213–241
  assessing messaging environment's complexity and costs, 213–215
  basic functions, 215–219
    element of service handling, 218
    message reception processing, 216–217
    routing and address mapping, 217–218
    screening, 219
    transmission processing, 219
  components, 219–220
    hardware components, 219–220
    software components, 220
  configuration models, 220–222
    colocated model, 221–222
    distributed processing model, 222
    stand-alone or dedicated model, 221
  facilities and features, 228–235
    management and administration facilities and features, 233–235
    message handling facilities and features, 228–233
  qualities to look for in gateway product, 235–238
    documentation, 236
    ease of installation and configuration, 236
    fault tolerance, 237
    flexibility, 236
    reliability, 236
    scalability and ease of expansion, 237
    throughput, 237
    value, 237–238
  scenarios, 222–228
    electronic mail hub, 227–228
    linking LAN messaging to outside world through network service provider, 223–225
    tying together legacy and modern messaging environments, 225–226
  steps to deploying gateway, 238–240
    do-it-yourself vs. outsourcing, 239
    initial testing, 239
    live service and maintenance, 240
    picking vendor or product, 238–239
    trial service and user feedback, 239–240
    up-front analysis and planning, 238
Messaging management, introduction to, 503–518
  categories of messaging and management, 507–511
    messaging application resources, 510–511
    messaging-specific resources, 508
    network resources, 508–510
    systems resources for messaging, 510
  current status and future expectations, 516–517
  emerging standards, 514–516

677

## Index

industry directions, 513–514
  increased consistency, 514
  management, 513–514
  richer functional ability, 514
management component architecture, 511–513
  managed resources, 511
  management agent, 512
  management communication mechanisms, 512–513
  management console, 512
  management information and functions, 511–512
messaging models and paradigms, 504–505
  messaging today, 504–505
  paper model, 504
service management model, 505–507
  service architecture, 506–507
  technical architecture, 507
Meta-Directory, 149, 152, 153, 154
MFI, *see* Multifunction interpreter
MHS, *see* Message handling system
MIB, *see* Management information base
Microsoft
  environments, multiple, 96
  Excel, 341
  Exchange, 34, *see also* Microsoft Exchange Server, introduction to
  Mail, xii, 572
    Connector, 52
    first version of, 38
    for PCs, 11
  Network (MSN), 278
  Office 97, 41
  Outlook, 40
Microsoft Exchange Server, introduction to, 37–57
  building a business strategy around Microsoft Exchange Server, 54–56
    connecting multisystem environments, 54–55
    downsizing, 54
    upgrading current Microsoft mail systems, 55–56
  easy and powerful centralized administration, 50–54
    easy migration, 54
    easy-to-use graphical administration program, 50
    management of all components from single seat, 53–54
    Microsoft Exchange Server components, 50–51
    reliable movement of information, 50
    single interface for global management, 51–53
  infrastructure for messaging and collaboration, 41–42
    fast, secure, and reliable, 41–42
    remote client access, 42
    scalable, 42
    tight integration with desktop applications, 41
    Universal Inbox, 41
  Internet connectivity, 48–49
    direct connections over Internet for mobile users, 49
    integrated Internet support, 48
    outlook Web access, 49
    support for Internet newsgroups and discussion groups, 49
  redefining groupware, 42–48
    bulletin boards, 44
    easy-to-create groupware applications, 45–47
    group scheduling, 43
    messaging application programming interface, 47–48
    offline folder synchronization, 45
    outlook forms, 44
    public folder replication, 44–45
    public folders, 43
  trends in messaging and collaboration, 38–41
    Microsoft Exchange product family, 40–41
    unifying LAN- and host-based E-mail, 38–40
Military message (MM), 112, 113, 119
Military Message Handling System (MMHS), 113, 247
MIME, *see* Multipurpose Internet Mail Extensions
MISSI program, *see* Multilevel Information Systems Security Initiative program
MLA, *see* Mail list agent
MM, *see* Military message
MMC, *see* Messaging Management Council
MMHS, *see* Military Message Handling System
MMTA, *see* Multitasking message transfer agent
Mobile employees, 267
Mobile messaging, 7, 259
Mobile users, 81
Mobile worker, 447
Molecularization, 384
Monitoring tools, 50, 53
MPR, *see* MultiProtocol Router
MS, *see* Message store
MSN, *see* Microsoft Network
MSP, *see* Message Security Protocol
MTA, *see* Message transfer agent
MTS, *see* Message Transfer Service
MUAs, *see* Mail user agents
Multicasting, 193
Multifunctional device
  choosing, 486
  cost of, 486

## Index

Multifunctional equipment, 485
Multifunction interpreter (MFI), 103, 104
Multilevel Information Systems Security Initiative (MISSI) program, 118
Multilevel security environment, 581
Multimedia, 56, 539
Multinational corporation, 214
Multiple backbone model, 530, 531
Multiple network protocols, 520
Multiple servers, 525
MultiProtocol Router (MPR), 65
Multipurpose Internet Mail Extensions (MIME), 7, 37, 52, 109, 244, 637
Multisystem environments, connecting, 54
Multitasking
    message transfer agent (MMTA), 9
    operating systems, 19
MWI, see Message waiting indicator
MWS, see Management workstation

## N

NAB, see Name and Address Book
NADF, see North American Directory Forum
Name
    and Address Book (NAB), 333, 334
    -to-address mapping, 162
    registration, 190
Naming
    authority, 175
    contexts, 176
    convention, 211
    service, 139
    tree, 63
National Electronic Directory, 183
National Institute of Standards and Technology (NIST), 117, 253, 516
National Science Foundation, 317
National Security Agency, 372
Native addressing, 230
Navigation commands, cryptic, 11
Navigator browser, 263
NCD, see Network Compute Devices
NCS, see NetWare Connect Services
NDS, see NetWare Directory Services
Nested NetWare interactive television, 61
Netizens, 319
Net surfers, 291
NetWare Connect Services (NCS), 60
NetWare Directory Services (NDS), 60, 63
NetWare Loadable Module (NLM), 61, 69
Network(s), 215
    access subscriptions, 223
    administration staff, 527
    automatic data transfer across, 64
    availability, 550
    commercial, 23
    Compute Devices (NCD), 489
    computer, 265, 373
    connections, 534
    data transfers, 92
    failure, 523
    fax applications, 62
    firewalls, 595
    interface card (NIC), 219
    links, external, 289
    local, 25
    management, 183, 251
    modifications, 575
    News Transport Protocol (NNTP), 40, 43, 90
    operating system (NOS), 584
    resources, 508
    security firewalls, 590
    shopping, 3
    speeds, 289
    strength, 289
Newbies, 306
Newsgroups, 315
    access to, 269
    specialized, 323
    Usenet, 294
News service, 290, 441
NIC, see Network interface card
NIST, see National Institute of Standards and Technology
NLM, see NetWare Loadable Module
NNTP, see Network News Transport Protocol
Nonreal-time audio, 101
Nonrepudiation, 361
Nonspecific subordinate reference, 208
North American Directory Forum (NADF), 25
NOS, see Network operating system
Notebook computer, 451, 455, 460
Notes
    business use of, 418
    limitations of, 421
Notification message, non-delivery type, 218
Novell messaging products, 59–82
    defining future of networking, 60–61
    GroupWise, 65–75
        administration, 72–73
        gateways, 71–72
        GroupWise 4.1 electronic messaging architecture, 68–69
        GroupWise 4.1 Message Server, 69–71
        GroupWise Telephony Access Server, 73–75
        integrated messaging services, 66–68
        positioning, 66
    GroupWise 5, 75–81
        business solutions, 77–78
        desktop, 78–81
        empowering people to act on information, 75–76
        solutions to problem, 76–77
    message handling services, 61–65
        global message handling services, 61–64

679

## Index

MHS services for NetWare 4.1, 64–65
NSA, *see* U.S. National Security Agency
NT server registry, 86

## O

Object
    classes (OCs), 120, 172
    identification (OID), 205
    store flexibility, 331
OCR, *see* Optical character recognition
OCs, *see* Object classes
Office
    equipment, 487
    furniture, 485
    setup, 462
    space
        design, 483
        hideaway, 484
        reduction, 471
        separate, 484
        shared, 484
Officing strategies, alternative, 463–464
Offline folder synchronization, 45
OID, *see* Object identification
OME, *see* Open messaging environment
One-call systems, 478
Online discussions, 79
Online magazines, 348
Online services, commercial, 2, 275–282
    communications personified, 275
    creating one whole online package, 275–279
        America Online, 277–278
        AT&T Interchange, 276
        CompuServe, 276–277
        Microsoft makes grand entrance, 278–279
        Prodigy, 277
    E-mail uses and abuses, 279–281
        advertising on Net, 280
        Telnetting to home, 280–281
    online for corporate users, 281–282
OpenMail
    directory services, 89
    server, 96
Open messaging environment (OME), 66
Open network model, 580
Open System Interconnection (OSI), 6, 191
Open Systems Environment (OSE), 516
OpenTime, 90
Operating system (OS), 584
    integration of messaging in, 39
    software, 220
    trusted, 88
Operations management, 146
Optical character recognition (OCR), 478
Order fulfillment, 405
Originator reference, 114, 117

OS, *see* Operating system
OSE, *see* Open Systems Environment
OSI, *see* Open System Interconnection
Outcomes measurement, 126
Outlook
    forms, 44
    Web Service, 49
Out-of-office message, 336
Overhead skylights, 487

## P

Pager services, 27
Paper
    evaluation, 543, 545
    mail, 348
Paperwork exchange, 350, 359
Partition management, 64
Pass-through messages, 72
Password, 73, 616
Payment processes, 355
PC
    client
        environment, 452
        model, 450
    interface, 428
    /LAN, *see* Personal computer/local area network
    use, growth in, 39
Peer-to-peer routing, 17
PEM, *see* Privacy Enhanced Mail
Performance measurement, 547–558
    messaging performance, 547–548
    messaging workload, 548–549
    performance reporting, 549–554
        addressibility, 550–551
        availability, 549–550
        consistency, 553–554
        delivery time, 552–553
        reliability, 551–552
    presentation, 556–558
    workload reporting, 554–556
        accessibility, 554–555
        accounts, 554
        messages, 555
        simultaneous connections, 554
        storage, 555–556
Peripheral equipment, 482
Personal computer/local area network (PC/LAN), 11
Personal directory, 432, 434
Personnel manager, 420
PGP, *see* Pretty Good Privacy
Phone equipment, 475
Pilot network, 157
PKCS, *see* Public Key Cryptosystem
PKI, *see* Public Key Infrastructure
PLAD, *see* Plain Language Address Designator

*Index*

Plain Language Address Designator (PLAD), 115
PO, *see* Purchase order
Point-to-point connections, 378
Point-to-point protocol (PPP), 53, 288, 293
Point-of-presence (POP), 279, 321
Point-of-sale transactions, 270, 271
Pollutant reduction trends, 471
Pollution problem, overcoming, 470
POP, *see* Point-of-presence
Postal system, 504
Postmark authority signature, 367
Post office, 437
  intelligent, 430, 436, 438
  servers, 98
Power Macintosh, 78
PPP, *see* Point-to-point protocol
PRDMDs, *see* Private Directory Management Domains
Presentation
  address, 206
  graphics, 56
Pretty Good Privacy (PGP), 605
Primary address designation, 138
Primary precedence, 113
Privacy
  conflicting views on, 617
  definition of, 625
  Enhanced Mail (PEM), 211, 248, 370, 591, 636
  parameters, 435
  policy, xiii, 533, 579
  in workplace, 502
Private Directory Management Domains (PRDMDs), 175
Private folders, 51
Private key, 600
Private management domain (PRMD), 106, 224
Private networks, 281, 383
PRMD, *see* Private management domain
Problem-solving model, 401
Process
  changes, 573
  control, 575
  improvement, 574
Prodigy, 277, 541
Product
  development decisions, 80
  evolution cycles, 379
  information libraries, 56
  management, 146
  selection process, 537
Production line capabilities, 385
Profiling user agent (PUA), 103, 105
Project
  management, 146, 147
  timeline, sample, 540
Protection plans, 579
Protocol conversion, 340

Proxy server, 287
PUA, *see* Profiling user agent
Publication infrastructure, coordination of, 137
Public electronic mail
  carrier, 167
  definition of, 626
Public folders, 43
Public information source, 51
Public Key Cryptosystem (PKCS), 140, 201
Public key encryption, 141, 358
Public Key Infrastructure (PKI), 371
Publishing traffic, heavy-volume, 322
Pull technology, 352
Purchase acknowledgments, 380
Purchase order (PO), 355, 376
Push technology, 352

**R**

Rapport Communications, 387
RAS, *see* Remote Access Service
RDBMS, *see* Relational database management system
RDN, *see* Relative Distinguished Name
Real-time meetings, 401
Receipt notifications, 116
Referrals, 192
Registration process, 363
Relational database management system (RDBMS), 5
Relative Distinguished Name (RDN), 188, 191, 208
Release authority, 122
Remote Access Service (RAS), 51, 53
Remote management, 235
Remote messaging applications, 9
Remote monitoring, 337
Remote procedure calls (RPCs), 51, 100, 284
Replication, 415
  agreements, 145
  engine, Microsoft Exchange Server, 46
  monitor, 339
  protocols, 169
  rules, 458
  technology, server-to-server, 45
Replying, 230
Repudiation, 388
Request for Comment (RFC), 31, 168, 244, 301
Request for information (RFI), 544
Request for Proposal (RFP), 111
Restricted delivery, 113
Retina scan, 362
Return receipts, 367
RFC, *see* Request for Comment
RFI, *see* Request for information
RFP, *see* Request for Proposal
Rich text format, conversion of Notes, 98
Risk-taking, approach to, 411

681

## Index

Root context, 207
Rotary phone, addressing from, 443
Routing
 -address mapping, 216
 paths, 16
RPCs, *see* Remote procedure calls
RSA, 585
 encryption, 203
 systems, 601, 606
Run of seven, 564
Rural physician, 126

## S

Sales tracking, 55, 80
Satellite offices, 474
Satellite television, 354
Scheduling, 342
 enterprisewide, 129
 information, 54
SCOS, *see* Single copy object store
Screening, 219
Screen resolution differences, 397
SDN, *see* Secure Data Network
SDNS, *see* Secure Data Network System
Search performance, 311
Secure Data Network (SDN), 102, 119
Secure Data Network System (SDNS), 120
Secure electronic transactions (SET), 349, 355
Secure Hash Algorithm (SHA), 119
Secure Hash Standard (SHS), 119
Secure socket layer (SSL), 381, 611
Security, 388
 classification, 105
 model, 336, 591
 services, xiii, 583
Selective replication, 335
Sensitivity indication, 113
Serial line Internet protocol (SLIP), 288, 293
Server
 -based folder, 42
 mailbox, 87
 maintenance, 338
 monitor, 339
 platform
  scalable, 329
  support, 333
 storage, 100
Service
 -level agreements, xiii, 137
 management, 146, 147
 provider(s), 137
  choosing, 284
  partnership between service user and, 518
SET, *see* Secure electronic transactions
Sewing machine industry, 375
SHA, *see* Secure Hash Algorithm

Shared code technology, 68
Shared folder synchronization, 96
Shareware options, 313
Shipment notices, electronic exchange of, 376
Shopping exchanges, 352
SHS, *see* Secure Hash Standard
Signature files, 29, 296
Simple Mail Transport Protocol (SMTP), xii, 4, 25, 244, 586, 608
 E-mail address, 123
 support, 37
Simple MAPI (sMAPI), 47
Simple Network Management Protocol (SNMP), 249, 338, 513, 514
Single address entry, 28
Single copy object store (SCOS), 332
SLIP, *see* Serial line Internet protocol
sMAPI, *see* Simple MAPI
Smileys, 303
SMP, *see* Symmetrical multiprocessor
SMTP, *see* Simple Mail Transport Protocol
Sneaker-net technology, 443
SNMP, *see* Simple Network Management Protocol
Society for Worldwide Interbank Financial Telecommunications (SWIFT), 602
SoftArc First Class, 11
Software
 affordability of, 160
 calendaring, 457
 client, 455
 companies, 75
 controls, 577
 development consultants, 59
 expertise, 409
 general-purpose, 414
 interpersonal messaging client, 537
 management, 407
 messaging, 553
 upgrading, 473
Spam, 280, 307
SPC, *see* Statistical process control, use of to manage message delivery
Speakerphones, 475
Speech-recognition capabilities, 443
Speed-dial codes, 434
Spell checking, 7
Spreadsheets, 18, 56, 81, 341, 468
Sprint, 6
SQL, *see* Structured query language
SSL, *see* Secure socket layer
Statistical process control (SPC), use of to manage message delivery, 559–575
 applying SPC to process improvement, 568–574
 control chart analysis, 562–566
 cross-platform messages, 566–567
 mathematical foundation, 560–562
 SPC as early warning system, 567–568

*Index*

Status tracking, 67
Storage
 requirements, reduction of, 86
 and retrieval services, 368
Stratification, 565
Streaming information, 436
Structured query language (SQL), 5
Subordinate reference, 208
Superior reference, 207
Supplier-buyer interaction, 384
Supply chain, 347
Support services, telecommuting and, 488
SWIFT, *see* Society for Worldwide Interbank Financial Telecommunications
Switch box, 425, 426
Symmetrical multiprocessor (SMP), 330
Synchronization, 385, 387
System
 administration, NetWare's, 82
 administrators, 77
 directories, local address books vs., 29
 management, 177
 pressure, 524
 variability, 564

# T

Tape-based machines, 476
TAS, *see* Telephony Access Server
Task
 automation, remote, 91
 management, 66, 67, 79
TCP/IP, *see* Transmission Control Protocol and Internet Protocol
TDCC, *see* Transportation Data Coordinating Committee
Team decision-making using groupware to enhance, 401–411
 critical success factors for using groupware, 406–409
  choosing groupware session facilitator, 407–409
  designing groupware session environment, 406–407
 group communication, 401–402
  information exchange tools, 402
  meeting management tools, 401–402
 group composition, 409–410
  boundaries for economy and manageability, 409–410
  diversity as asset, 409
  session objectives, 410
 groupware applications, 405–406
 groupware session, 402–403
 groupware tools and capabilities, 403–405
Technical staff, 289
Technical support, 79
Technology
 changes in, 3
 evolution, 324
 introduction, 177
Telecommunications Bill, 578
Telecommunication vendors, 243
Telecommuter(s)
 benefits, 465
 costs per, 491
 skill sets for, 472
 training, 472
Telecommuting, 326, 461–502
 acknowledgment form, sample, 494
 agreement form, sample, 496–497
 arrangement, 492
 assessment of, 499
 benefits and worries, 464–472, 500
  impact on company, 470–472
  impact on individual telecommuter, 465–469
  impact on manager/employer, 469–470
 candidate for, 498
 definition, 462
 demographics, 462–464
 establishing policy for telecommuting, 474
  company policy, 474
  telecommuter contract, 474
 formalizing telecommuting arrangement, 492–495
 getting started, 499
 guidelines for prospective telecommuter, 495–499
  assessment, 499
  getting started, 499
  good candidate for telecommuting, 498
  organizational approval, 498
  preparations for telecommuting, 498–499
 home office infrastructure, 474–483
  communications technology, 475–477
  data communications, 480–481
  electronic mail, 479–480
  fax technology, 477–478
  online connections, 482
  peripheral equipment, 482
  videoconferencing, 481–482
  words of advice, 482–483
 home office work environment, 483–489
  electrical power concerns, 486–487
  family issues, 488–489
  furniture, 485
  lighting, 487–488
  multifunctional equipment, 485–486
  office space, 484–485
  security, 488
  support services, 488
  zoning laws, 488
 institutionalizing, 490
 major U.S. financial services firm, technology division, 489–492
  first telecommuting pilot, 489–490
  overall lessons learned, 492

683

## Index

second telecommuting pilot, 490–492
   policy, 492
   preparations for, 498
   programs, 464
   rules, sample, 495
   skill sets, 472, 473
   training, 472–474
   Working Group, 489
Telemedicine, 10, 126
Telephone
   access server, administering, 74
   browsers, 383
   companies, 273
   directories, 181
   ordering systems, 579
Telephony Access Server (TAS), 73
Teleprocessing organizations, 179
Telex service provider, 32
Telnet, *see* Terminal emulation
Templates, 433
Terminal emulation (Telnet), 4, 259, 308
TGS, *see* Ticket granting service
Third-party
   directory services, 184
   products, 62, 329
   provider, 32
   software, 220
Three-way calling, 476
Ticket granting service (TGS), 585
Time-out values, 512
TP4, *see* Transport Protocol Class 4
Trade associations, 135
Training, package-based, 408
Transaction model, 71
Transient interest area, 439
Translation services, 162, 348, 428
Transmission
   Control Protocol and Internet Protocol (TCP/IP), 394
   processing, 219
   retry, 232
Transportation Data Coordinating Committee (TDCC), 346, 374
Transport Protocol Class 4 (TP4), 106
Trend analysis, 413
Two-way paging, 444, 456
Typewriter replacement, 131

### U

UA, *see* User agent
UAL, *see* Universal agent layer
UA/UI, *see* User agent/user interface
UMB, *see* Universal message box
UMS, *see* Universal message services
Unauthorized disclosure, 118
Uniform resource locator (URL), 49, 309
Universal agent layer (UAL), 85
Universal client, 8

Universal directory, 164
Universal In box, 41, 78, 424
Universal mailbox, 424, 480
Universal message box (UMB), 430
Universal message services (UMS), 423–444
   key components of, 429–442
      intelligent networks, 440–441
      intelligent post office, 436–440
      intelligent services, 441–442
      universal message box, 430–436
   levels of integration, 425–429
      existing services, switch box, 425–426
      single logical storage, 426–429
   nowhere to hide, 423–424
   overcoming complexity of, 442–443
   quick start definition, 424–425
Universal Out Box, 78
UNIX, 78
Urgent mail folder, 336
URL, *see* Uniform resource locator
U.S. Department of Defense (DOD), 101
Usenet
   filters, 307
   netiquette, 306
User
   agent (UA), 13, 102, 103, 246, 335
   agent/user interface (UA/UI), 13
   -friendly interface, 12
   lists, 7
U.S. National Security Agency (NSA), 118
U.S. Postal Service, 372

### V

Value-added networks (VANs), 213, 255–273
   business communication services, 257–262
      BCS selection criteria, 261–262
      current trends and future directions, 258–260
      market overview, 257–258
      use of, 260–261
   commercial online services, 269–271
      current trends and future directions, 270–271
      market overview, 269–270
      online service selection criteria, 271
      use of commercial online service, 271
   Internet service providers, 263–269
      comparison chart, 269
      current trends and future directions, 264–267
      market overview, 263–264
      selection criteria, 268
      use of, 267–268
   understanding need for services, 256–257
      point-to-point connectivity complexity, 256

# Index

value-added networks connectivity simplicity, 256–257
Value-added resellers (VARs), 278
Value-added service, public, 11
Values performance, 637
VANs, *see* Value-added networks
VARs, *see* Value-added resellers
VBA, *see* Visual Basic for Applications
Vendor Independent Messaging (VIM), 65, 233
VeriSign, 372
Video
  files, 81
  movies-on-demand, 352
  referrals, store-and-forward, 133
  rentals, 352
  teleconferencing, 101, 133
Videoconferencing, 469, 481
VIM, *see* Vendor Independent Messaging
Violation reconciliation procedures, 588
Virtual community, 319
Virtualization, 384
Virtual Machine (VM), 23
Virtual Memory System (VMS), 23
Virtual organization, 159
Virtual workgroup, 65
Virtual workplaces, 479
Virtual world
  business in, 273
  irony of, 271
Visual Basic, 9
  for Applications (VBA), 181
  Script, 46
VM, *see* Virtual Machine
VMS, *see* Virtual Memory System
Voice
  communications, 314, 475
  mail, 325, 468, 476, 500
    definition of, 626
    integration, 442
  messages, 423, 455, 482
  messaging, 209, 345
  record box, 427
Voting, 404

## W

WAIS, *see* Wide area information system
WAN
  connections, 273
  infrastructure, 20
  networking, 521
  traffic, long-distance, 44
Weather information, 429
Web
  -based technology, xiii
  browser, 83, 95, 263, 520
  pages, 281, 320
  pioneer developers, 321
  presence, establishment of, 320
  technology, 393
Web customers, netting, 317–326
  doing business on Internet, 317–318
  Internet potential, 323–324
  investing in commercial cyberspace, 318–319
  netting Web users, 319–320
  telecom reaches Net, 321
  using Web as marketing tool, 320
  Web pioneer developers, 321–323
    BBN Planet's Web advantage, 322
    Earthlink's Web solution, 321–322
    O'Reilly & Associates' WebSite, 322–323
WEDI, *see* WorkGroup for Electronic Data Interchange
WEMA, *see* World Electronic Messaging Association
Wide area information system (WAIS), 312, 313
Windows
  95, 78
  NT, 86
  platform, 35
Wireless communications, 27, 444
Wizard Mail, 129
Word processors, 18, 81, 232
Workaholic conditions, 473
Workday, typical, 76
Workflow, 7
  applications, 181
  complex, 414
  message-based, 95
  process, 438, 460
  routing, 66, 67
  technologies, database-centric, 90
Workgroup
  collaboration, 83
  computing, 65, 400
  scheduling, 8
WorkGroup for Electronic Data Interchange (WEDI), 127
Workload
  changes, 558
  reporting, 554
Worksheets, 569
Workstation hardware, 59
World Electronic Messaging Association (WEMA), 208
World Wide Web (WWW), 123, 310, 317
  browsing, 259
  globalization of, 266
  links to outside Internet via, 275
WWW, *see* World Wide Web

## X

X.400 applications programming interface (XAPI), 233

685

# Index

X.400 vs. SMTP, 243-253
  background of X.400 and SMTP, 243º244
  industrial-strength messaging requirements, 247-252
  SMTP, 244
  SMTP vs. X.400 debate, 245-247
    SMTP/MIME characteristics, 245-246
    X.400 characteristics, 246-247
  state of electronic mail, 244-245
X.500 directory services, business process enabler, 159-184
  directory applications, 179-184
    business tools, 181-182
    emerging network-related applications, 182-184
    messaging, 179-180
    phone books, 180-181
  driving forces, 160-161
  joint effort, 160
  organizational readiness, 162
  purpose of X.500 directory, 164-165
  technological readiness, 161-162
  uses and applications of X.500 directory, 176-179
    generic directory uses, 178-179
    initial planning, 177-178
  X.500 directory architecture, 165-168
    directory system agent, 167
    directory user agent, 165-167
    how X.500 works with X.400, 167-168
  X.500 directory information model, 170-174
    attributes, 172-173
    directory information base, 171
    directory information tree, 170-171
    information model updates, 174
    information objects, 172
    object classes, 172
    X.500 directory names, 173-174
  X.500 directory organizational model, 174-176
    administrative authority, 175
    naming authority, 175-176
  X.500 directory protocols, 168-170
    directory access protocol, 168
    directory system protocol, 168-169
    replication protocols, 169-170
  X.500 standard, 162-163
X.500 directory services, under covers, 185-212
  access control, 194-195
  authentication, 198-204
    data confidentiality, 203-204
    data integrity, 204
    nonrepudiation, 203
    simple authentication, 199-200
    strong authentication, 200-203
  choosing low-hanging fruit, 186
  directory entries, 186-188
  distributed directory knowledge references, 206-208
    cross-references, 208
    internal reference, 207
    nonspecific subordinate reference, 208
    root context, 207
    subordinate reference, 208
    superior reference, 207-208
  distributed X.500 directory operations services, 191-194
    chaining, 193
    multicasting, 193-194
    referrals, 192
    search, 194
  lightweight directory access protocol, 204-205
    LDAP uses and limitations, 204-205
    significance of LDAP, 204
  naming structure, 188-190
    alias names, 189
    distinguished name, 189
    name registration, 190
    relative distinguished name, 188-189
  understanding delivery mechanisms, 185-186
  WEMA 97 Directory Challenge, 208-211
    directory-enabled applications, 209
    operating parameters, 209-211
  X.500 directory security, 196-198
  X.500 directory service controls, 195-196
  X.500 directory services, 190-191
    directory interrogation operations, 190-191
    directory update operations, 191
  X.500 distributed directory system model, 205-206
XAPI, *see* X.400 applications programming interface
XFL, *see* Cross File

# Y

Yahoo guide, 319
Yellow Pages, 150, 178
Yellow-page search operations, 64

# Z

Zoning laws, 473, 488